Handbook of
Experimental Pharmacology

Volume 160

Editor-in-Chief

K. Starke, Freiburg i. Br.

Editorial Board

G.V.R. Born, London
M. Eichelbaum, Stuttgart
D. Ganten, Berlin
H. Herken[†], Berlin
F. Hofmann, München
L. Limbird, Nashville, TN
W. Rosenthal, Berlin
G. Rubanyi, Richmond, CA

Springer
Berlin
Heidelberg
New York
Hong Kong
London
Milan
Paris
Tokyo

Cardiovascular Pharmacogenetics

Contributors
E.D. Abel, A.R. Boobis, J.M.C. Connell, V.J. Dzau,
R.J. Edwards, D.B. Goldstein, D.O. Haskard, A.M. Henney,
M.M. Hoffmann, P.L. Huang, J.W. Knowles, D. Kong,
K. Lindpaintner, W. März, A.J. Marian, J.J.V. McMurray,
L.G. Melo, K.M. O'Shaughnessy, A.S. Pachori, R.M. Rao,
D.M. Roden, A.I. Russell, T. Shiga, T. Stanton, S.K. Tate,
E.G.D. Tuddenham, A. Vidal-Puig, T. Vyse, M.R. Wilkins,
B.R. Winkelmann

Editor
Martin R. Wilkins

Springer

Professor
Martin R. Wilkins
Experimental Medicine & Toxicology
Imperial College London
Hammersmith Campus
Du Cane Road
London W12 0NN
United Kingdom
e-mail: m.wilkins@ic.ac.uk

With 34 Figures and 28 Tables

ISSN 0171-2004

ISBN 3-540-40204-7 Springer-Verlag Berlin Heidelberg New York

Library of Congress Cataloging-in-Publication Data
Cardiovascular pharmacogenetics / contributors, E. D. Abel ... [et al.]. ; editor, Martin R. Wilkins. p. ; cm. – (Handbook of experimental pharmacology ; v. 160) Includes bibliographical references and index.
ISBN 3-540-40204-7 (alk. paper)
1. Heart–Diseases–Genetic aspects. 2. Heart–Diseases–Genetic aspects. 3. Heart–Effect of drugs on–Genetic aspects. 4. Pharmacogenetics. I. Abel, E. D. (E. Dale) II: Wilkins, Martin R., 1956- III. Series. [DNLM: 1. Cardiovascular Agents–therapeutic use. 2. Pharmacogenetics–methods. 3. Cardiovascular Diseases–drug therapy. 4. Cardiovascular Diseases–genetics. QV 150 C2745 2004]
QP905.H3 vol. 160   [RC682]   615′. s–dc21   [615′.71]   2003054275

This work is subject to copyright. All rights are reserved, whether the whole or part of the material is concerned, specifically the rights of translation, reprinting, re-use of illustrations, recitation, broadcasting, reproduction on microfilm or in any other way, and storage in data banks. Duplication of this publication or parts thereof is permitted only under the provisions of the German Copyright Law of September 9, 1965, in its current version, and permission for use must always be obtained from Springer-Verlag. Violations are liable for Prosecution under the German Copyright Law.

Springer-Verlag is a part of Springer Science+Business Media
springeronline.com

© Springer-Verlag Berlin Heidelberg 2004
Printed in Germany

The use of general descriptive names, registered names, etc. in this publication does not imply, even in the absence of a specific statement, that such names are exempt from the relevant protective laws and regulations and free for general use.

Product liability: The publishers cannot guarantee the accuracy of any information about dosage and application contained in this book. In every individual case the user must check such information by consulting the relevant literature.

Cover design: design & production GmbH, Heidelberg
Typesetting: Stürtz AG, 97080 Würzburg

Printed on acid-free paper   27/3150 hs – 5 4 3 2 1 0

Preface

The Human Genome Project was sold to the general public largely on the basis that a complete picture of the structure of human DNA would lead to new and better medicines. These medicines would be better because they would be tailored to individual patients, maximising the chances of a therapeutic response and minimising the risks of an adverse event. Taking the idea further, pundits have predicted that the time will come when we could carry our DNA on a card which could be read rapidly and enable the physician to choose the best drug. This is the future. This is pharmacogenetics.

When the draft human DNA sequence was announced and scientists were asked how this would help drug development, the example most frequently given was the debrisoquine model – where poor metabolisers of this hypotensive agent are exposed to higher plasma levels from a standard dose and at risk of collapse from excessive hypotension. This observation was made over 20 years ago and predated designs to sequence the human genome. Nonetheless, it raised awareness of variation in drug metabolism and was correctly assigned to genetic polymorphisms affecting CYP2D6. Together with the discovery of pseudocholinesterase deficiency, it marked the birth of pharmacogenetics.

The debrisoquine example is an interesting one and worthy of further analysis. Recognising the implications of impaired CYP2D6 activity, it has been argued that patients should be screened for their debrisoquine metaboliser status prior to the prescription of drugs metabolised predominantly by this enzyme. But this has never caught on – and one must ask why. One reason was that in the early days, assigning metaboliser status was dependent upon phenotyping individuals - calculating the ratio of the levels of metabolite and parent drug in a urine sample following a test dose. This is time consuming and is not widely available. It is now possible to assign status based on genotype. Several exons need to be screened to detect all poor metabolisers but with todays technology this is not a major challenge. And yet still it has not entered clinical practice.

Instead, pharmaceutical companies have used the observation to screen out drugs metabolised largely by CYP2D6 in an attempt to develop therapeutic agents with less inter individual variability in plasma levels. And herein may lie a lesson for the future. Genomic information has facilitated the pursuit of other genetic factors that influence the response to a drug, not only factors that affect drug kinetics but also those that impact on drug targets and pathways of disease. The initial expectation is that such information can be used to provide a more in-depth screen of patients prior to drug prescribing – to provide a more complete answer to the question, is this drug suitable for this patient? In this book, Klaus Lindpaintner suggests that the main value of genetic information will be in the molecular dissection of disease, the identification of new drug targets and the development of specific medicines for each disease subtype. In other words, physicians will not be screening patients with a view to excluding the use of drug(s) (i.e. finding the right medicine for the right patient), rather to select the most appropriate agent based upon a better understanding of the molecular pathogenesis of the condition (i.e. finding the right medicine for the disease-subtype).

This book considers the current status of pharmacogenetics in the management of cardiovascular disease. Lindpaintner and Winkelmann et al offer a definition of pharmacogenetics and define its relationship to pharmacogenomics. Tate and Goldstein discuss the concept of using haplotypes in association studies to examine drug response-genotype relationships. Boobis et al discuss genetic factors that influence cardiovascular drug kinetics. Thereafter follow a series of chapters dealing with specific manifestations of cardiovascular disease. Joshua Knowles discusses atherosclerosis, Wilkins and O'Shaughnessy explore hypertension, Marian dissects the genetic factors influencing cardiac mass, Stanton et al describe the use of genetics in the management of heart failure, Roden expands on insights from the ion channelopathies on the treatment of cardiac arrhythmias and Vidal-Puig and Abel survey new developments from genetic studies in insulin resistance. The potential for identifying novel drug targets emerges as a major theme from these chapters. Genetically determined perturbations of other significant players in cardiovascular disease are then considered in terms of their implications for drug therapy – Edward Tuddenham discusses thrombophilia, Haskard et al tour adhesions molecules and Henney enlarges on metalloproteinases. Paul Huang provides an example of how to explore the potential role of candidate genes in cardiovascular pathology. Finally, Melo et al speculate on the greater therapeutic potential of genetic information in forging new therapeutic approaches to cardiovascular disease.

It was my intention when compiling the list of topics for inclusion to be as wide-ranging as possible but, for practical reasons, the book could not be all inclusive. It is hoped, however, that this volume will provide a panorama of our current understanding of the role of genetic information in the management of cardiovascular disease. A recurring plea from the contributors is the need for more hypothesis-driven, rigorously conducted and analysed clinical research. Perhaps this compilation of knowledge will provide a springboard for the design of informative clinical studies. Whether the Human Genome Project will have the high impact in medicine that many have come to expect will depend on the results of such studies.

1st April 2003

Martin R. Wilkins
Experimental Medicine and Toxicology
Imperial College London
Hammersmith Hospital, London

List of Contributors

(Their addresses can be found at the beginning of their respective chapters.)

Abel, E.D.   243

Boobis, A.R.   39

Connell, J.M.C.   203

Dzau, V.J.   359

Edwards, R.J.   39

Goldstein, D.B.   25

Haskard, D.O.   323
Henney, A.M.   341
Hoffmann, M.M.   107
Huang, P.L.   281

Knowles, J.W.   79
Kong, D.   359

Lindpaintner, K.   1

März, W.   107
Marian, A.J.   177
McMurray, J.J.V.   203
Melo, L.G.   359

O'Shaughnessy, K.M.   149

Pachori, A.S.   359

Rao, R.M.   323
Roden, D.M.   223
Russell, A.I.   323

Shiga, T.   39
Stanton, T.   203

Tate, S.K.   25
Tuddenham, E.G.D.   311

Vidal-Puig, A.   243
Vyse, T.   323

Wilkins, M.R.   149
Winkelmann, B.R.   107

Contents

The Role of Genotyping in Pharmacological Therapy.................... 1
    K. Lindpaintner

Pharmacogenetics and the Treatment of Cardiovascular Disease........... 25
    S. K. Tate, D. B. Goldstein

Genetic Polymorphisms and Cardiovascular Drug Metabolism............ 39
    A. R. Boobis, T. Shiga, R. J. Edwards

Genes That Modify Susceptibility to Atherosclerosis:
Targets for Drug Action.. 79
    J. W. Knowles

Lipid-Lowering Responses Modified by Genetic Variation 107
    B. R. Winkelmann, M. M. Hoffmann, W. März

The Genetic Basis of Essential Hypertension and Its Implications
for Treatment 149
    K. M. O'Shaughnessy, M. R. Wilkins

Genetic Predisposition to Cardiac Hypertrophy 177
    A. J. Marian

Genetic Determinants of Susceptibility, Prognosis and Treatment
in Heart Failure 203
    T. Stanton, J. M. C. Connell, J. J. V. McMurray

The Genetics of Cardiac Channelopathies: Implications for Therapeutics... 223
    D. M. Roden

Insulin Resistance and Cardiovascular Disease: New Insights from Genetics 243
    A. Vidal-Puig, E. D. Abel

Genetic Disruption of Nitric Oxide Synthases and Cardiovascular Disease:
Lessons from a Candidate Gene 281
    P. L. Huang

Association of Thrombotic Disease with Genetic Polymorphism
of Haemostatic Genes: Relevance to Pharmacogenetics 311
  E.G.D. Tuddenham

The Influence of Genetic Factors on Leukocyte
and Endothelial Cell Adhesion Molecules............................ 323
  R.M. Rao, A.I. Russell, T. Vyse, D.O. Haskard

Genetic Regulation of Metalloproteinase Activity:
Pathogenic and Therapeutic Implications 341
  A.M. Henney

Current Perspectives on Gene and Cell-Based Therapies
for Myocardial Protection, Rescue and Repair........................ 359
  L.G. Melo, A.S. Pachori, D. Kong, V.J. Dzau

Subject Index 405

The Role of Genotyping in Pharmacological Therapy

K. Lindpaintner

VP Research, Roche Genetics, F. Hoffmann-La Roche, Bldg 93/532,
4070 Basel, Switzerland
e-mail: klaus.lindpaintner@roche.com

1	Introduction	2
2	Definition of Terms	3
2.1	Pharmacogenetics	3
2.2	Pharmacogenomics	4
3	Pharmacogenomics: Finding New Medicines Quicker and More Efficiently	5
4	Pharmacogenetics: More Targeted, More Effective Medicines for Our Patients	6
4.1	Genes and Environment	6
4.2	An Attempt at a Systematic Classification of Pharmacogenetics	7
4.2.1	Classical Pharmacogenetics	8
4.2.2	Pharmacogenetics as a Consequence of Molecular Differential Diagnosis	12
4.2.3	Different Classes of Markers	14
4.2.4	Complexity Is to Be Expected	14
5	Incorporating Pharmacogenetics into Drug Development Strategy	14
6	Regulatory Aspects	17
7	Pharmacogenetic Testing for Drug Efficacy Versus Safety	18
8	Ethical and Societal Aspects of Pharmacogenetics	19
9	Conclusion	21
	References	22

Abstract This communication is intended to provide a view of what the disciplines of genetics and genomics stand to contribute (and how they have actually contributed for many years) to drug discovery and development and, more broadly, to the practice of health care. Particular emphasis will be placed on examining the role of genetics, that is, acquired or inherited variations at the level of DNA-encoded information, with regard to common complex diseases. A realistic understanding of this role is essential for a balanced assessment of the impact of genetics on health care in the future. Definitions for some of the terms that are in wide and often unreflected use today will be provided. A more systematic classification of pharmacogenetics will be attempted. It is important to be aware that what will be discussed is to a large extent still uncharted territory. So by necessity, many of the positions taken on today's understanding and knowledge must be viewed as somewhat speculative in nature. Where appropriate and possible, select examples will be provided, although it should be pointed out that much of the literature in the area of genetic epidemiology and pharmacogenetics lacks the stringent standards normally applied to peer-reviewed research, and replicate data are generally absent.

Keywords Pharmacogenetics · Pharmacogenomics · Toxicogenetics · Drug discovery · Drug regulation

1
Introduction

The advances made over the last 30 years in molecular biology, molecular genetics and genomics and the development and refinement of associated methods and technologies have had a major impact on our understanding of biology, including the action of drugs and other biologically active xenobiotics. The tools that have been developed to allow these advances, and the knowledge of fundamental principles underlying cellular function thus derived, have become indispensable to almost any field of biological research, including future progress in biomedicine and health care.

It is important to realize that with regard to pharmacology and drug discovery, these accomplishments, starting sometime in the last third or quarter of the 20th century, have led gradually to a fundamental shift from the chemical paradigm to a biological paradigm. Whereas previously medicinal chemistry drove new developments in drug discovery, with biology almost an ancillary service that examined new molecules for biological function, biology has now taken the lead, based on a new-found understanding of physiological effects of biomolecules and pathways, requesting from the chemist compounds that modulate the function of these biomolecules or pathways, with—at least theoretically—a predictable functional impact in the setting of integrated physiology.

One particular aspect has uniquely captured the imagination of both scientists and the public, namely our understanding of genetics, especially our cataloguing of genome sequences. While understandable—given the austere beauty of Mendel's laws, the compelling esthetics of the double helix structure, and the awe-inspiring accomplishment (coupled with an unprecedented public relations campaign) of the human genome project—the public excitement about genetics and genomics and the high expectations regarding the impact they will have on the practice of health care are almost certainly unrealistic. Thus, at the interface between genetics/genomics and pharmacology, pharmacogenetics and pharmacogenomics (usually in the most loosely defined terms) are commonly touted as heralding a revolution in medicine. Yet, as soon as one begins to probe more carefully, little substance is yet to be found to support these enthusiastic claims.

Table 1 Terminology

Pharmacogenetics
Differential effects of a drug, in vivo, in different patients, dependent on the presence of inherited gene variants
Assessed primarily genetic (SNP) and genomic (expression) approaches
A concept to provide more patient/disease-specific health care
One drug, many genomes (i.e., different patients)
Focus: patient variability
Pharmacogenomics
Differential effects of compounds, in vivo or in vitro, on gene expression, among the entirety of expressed genes
Assessed by expression profiling
A tool for compound selection/drug discovery
Many drugs (i.e., early-stage compounds), one genome [i.e., normative genome (database, technology platform)]
Focus: compound variability

Indeed, as pointed out above, the major change in how we discover drugs, from the chemical to the biological paradigm, already occurred some time ago; what the current advances promise to allow us to do in due time is to move from a physiology-based to a (molecular) pathology-based approach towards drug discovery, promising the advancement from a largely palliative to a more cause/contribution-targeting pharmacopoeia.

2
Definition of Terms

There is widespread indiscriminate use of the terms "pharmacogenetics" and "pharmacogenomics", causing some confusion. While no universally accepted definition exists, there is an emerging consensus on their differential meaning and use (Table 1).

2.1
Pharmacogenetics

The term "genetics" relates etymologically to the presence of individual properties, and inter-individual differences in these properties, due to inheritance. The term "pharmacogenetics" describes the interactions between a drug and an individual's (or perhaps more accurately, groups of individuals) response to it as it relates to differences in DNA-based information. It is concerned with the assessment of clinical efficacy and/or the safety and tolerability profile; in other words, the pharmacological response phenotype of a drug in groups of individuals that differ with regard to certain DNA-encoded characteristics. It tests the hypothesis that these differences may allow prediction of individual drug response. Assessment of DNA-encoded characteristics is based most commonly on the presence or absence of polymorphisms at the level of nuclear DNA. However, this assessment may occur also at different levels where such DNA variation translates into different characteristics, such as differential mRNA expression or splicing, protein levels or functional characteristics, or even physiological phenotypes, all of which may be seen as surrogate or more highly integrated markers of the underlying genetic variant. It should be noted, however, that some authors continue to subsume all applications of expression profiling under the term "pharmacogenomics", in a definition of the terms that is more driven by the technology used rather than by functional context.

2.2
Pharmacogenomics

In contrast, the terms "pharmacogenomics", and its close relative, "toxicogenomics", are etymologically linked to "genomics", the study of the genome and of the entirety of expressed and non-expressed genes in any given physiological state. These two fields of study are concerned with a comprehensive, genome-wide assessment of the effects of pharmacological agents, including toxins/toxicants, on gene expression patterns. Pharmacogenomic studies are thus used to evaluate the differential effects of a number of chemical compounds (in the process of drug discovery commonly applied to lead selection) with regard to inducing or suppressing gene transcription in an experimental setting. Except for situations in which pharmacogenetic considerations are front-loaded into the discovery process, inter-individual variations in gene sequence are not usually taken into account in this process. Therefore, unlike pharmacogenetics, pharmacogenomics does not focus on differences among individuals with regard to the drug's effects, but rather examines differences among several (prospective) drugs or compounds with regard to their biological effects across the entire genome or some significant part thereof. The basis of comparison is quantitative measures of expression, using a number of more or less comprehensive gene-expression-profiling methods, commonly based on microarray formats. By extrapolation from the experimental results to theoretically desirable patterns of activation or inactivation of gene expression in the setting of integrative pathophysiology, this approach is expected to provide a faster, more comprehensive, and perhaps even more reliable way to assess the likelihood of finding an ultimately successful drug than previously available schemes, involving mostly in vivo animal experimentation.

Thus, although both pharmacogenetics and pharmacogenomics refer to the evaluation of drug effects using (primarily) nucleic acid markers and technology, the directionalities of their approaches are distinctly different: pharmacogenetics represents the study of differences among a number of individuals with regard to clinical response to a particular drug ("one drug, many genomes"), whereas pharmacogenomics represents the study of differences among a number of compounds with regard to gene expression response in a single (normative) genome/expressome ("many drugs, one genome"). Accordingly, the fields of intended use are distinct: the former will help, in the clinical setting, to find the medicine most likely to be optimal for a patient (or to find the patients most likely to respond to a drug), the latter will aid in the setting of pharmaceutical research to find the most suitable drug candidate from a given series of compounds under evaluation.

3
Pharmacogenomics: Finding New Medicines Quicker and More Efficiently

Once a screen (assay) has been set up in a drug discovery project and lead compounds are identified, the major task becomes the identification of an optimized clinical candidate molecule among the many compounds synthesized by medicinal chemists. Conventionally, such compounds are screened in a number of animal or cell models for efficacy and toxicity, experiments that, while having the advantage of being conducted in the in vivo setting, commonly take significant amounts of time and depend entirely on the similarity between the experimental animal condition/setting and its human counterpart, i.e., the validity of the model.

Although such experiments will never be entirely replaced by expression profiling at either the nucleic acid (genomics) or the protein (proteomics) level, the latter technique offers powerful advantages and complimentary information. First, the efficacy and profile of induced changes can be assessed in a comprehensive fashion (within the limitations, primarily sensitivity and completeness of transcript representation, of the technology platform used). Second, these assessments of differential efficacy can be carried out much more expeditiously than in conventionally used, (patho)physiology-based animal models. Third, the complex pattern of expression changes revealed by such experiments may provide new insights into possible biological interactions between the actual drug target and other biomolecules, and thus reveal new elements or branch-points of a biological pathway that may be useful as surrogate markers, novel diagnostic analytes, or as additional drug targets. Fourth, and increasingly important, these tools serve to determine specificity of action among members of gene families that may be highly important for both the efficacy and safety of a new drug. It must be borne in mind that any and all such experiments are limited by the coefficient of correlation with which the expression patterns determined are linked to the desired in vivo physiological action of the compound.

A word of caution regarding micro-array-based expression profiling would appear to be in order: It is important to remain aware of the fact that all microarray expression data are of only associative character, i.e., they do not infer causation, and must be interpreted mindful of this limitation.

As a subcategory of this approach, toxicogenomics is evolving as a powerful adjuvant to classic toxicological testing. As pertinent databases are being created from experiments with known toxicants, revealing expression patterns that may be predictive of the longer-term toxic liabilities of compounds, future drug discovery efforts should benefit from insights allowing earlier rejection of compounds likely to cause such complications.

When using these approaches in drug discovery, even if implemented with proper biostatistics and analytical rigor, it is imperative to understand the probabilistic nature of such experiments: a promising profile on pharmacogenomic and toxicogenomic screens will enhance the likelihood of having selected an ultimately successful compound, and will achieve this goal quicker than conventional animal experimentation, but will do so only with a certain likelihood of success. The less reductionist approach of the animal experiment will still be needed to evaluate the chosen compound. It is to be anticipated, however, that such approaches will constitute an important time- and resource-saving first evaluation or screening step that will help to focus and reduce the number of animal experiments that will ultimately need to be conducted.

4
Pharmacogenetics: More Targeted, More Effective Medicines for Our Patients

4.1
Genes and Environment

It is common knowledge that today's pharmacopoeia, although representing enormous progress compared with what our physicians had only 15 or 20 years ago, is far from perfect. Many patients respond only partially, or fail to respond altogether to the drugs they are given, and others suffer adverse events that range form unpleasant to serious and life-threatening.

There is an emerging consensus that all common complex diseases are multifactorial in nature, i.e., that they are brought upon by the coincidence of certain intrinsic (inborn or acquired) predispositions and susceptibilities on the one hand, and extrinsic, environment-derived influences on the other. The relative importance of these two influences varies across a broad spectrum. In some diseases external factors appear to be more important, while in others intrinsic predispositions prevail. In almost all cases, a number of both intrinsic (genetic) as well as extrinsic factors appear to contribute, although it is not clear from the currently available literature how much this reflects the requirement of several intrinsic and extrinsic factors to coincide in any one individual, or how much this reflects the causative heterogeneity of each of today's conventional clinical diagnoses. In either case, the disease-causing (or better, -contributing) role that intrinsic, genetically encoded properties play with regard to the occurrence of the disease is fundamentally different in these common, complex diseases as compared to the classic monogenic mendelian diseases. While in the latter the impact of the genetic variant is typically categorical in nature, i.e., deterministic, in the former case, the presence of a disease-associated genetic variant is merely of probabilistic influence, raising (or lowering) the likelihood of disease occurrence to some extent but never predicting it in a black-and-white fashion.

If we regard a pharmacological agent as an extrinsic, environmental factor with a potential to affect the health-status of the individual to whom it is administered, then individual differences in response to such an agent would be expected, under the paradigm just elaborated upon, to be based on differences regarding the intrinsic characteristics of these patients, as long as we can exclude variation in the exposure to the drug (this is important, as in clinical practice non-adherence to prescribed regimens of administration, or drug–drug interactions interfering with bioavailability of the drug, are perhaps the most likely culprits when such differences in response phenotype are observed). The influence of such intrinsic variation on drug response may be more easily recognizable and more relevant in drugs with a steep dose–response curve. The argument for the greater likelihood of observing environmental factor/gene interactions with drugs as compared to, say, food-stuffs, goes along the same lines.

Clearly a better fundamental and mechanistic understanding of the molecular pathology of disease and of the role of intrinsic, biological properties predisposing to such diseases, as well as of drug action at the molecular level, will be essential for future progress in health care. Current progress in molecular biology and genetics has provided us with some of the prerequisite tools that should help us reach the goal of a more refined understanding.

4.2
An Attempt at a Systematic Classification of Pharmacogenetics

Two conceptually quite different categories of inter-individually differential drug response may be distinguished on the basis of the underlying biological variance (Table 2):

1. In the first case, the underlying biological variation is *in itself not disease-causing* or *-contributing*, and becomes clinically relevant *only* in response to the exposure to the drug in question (classical pharmacogenetics).

Table 2 Pharmacogenetics systematic classification

Classic pharmacogenetics
Pharmacokinetics
Absorption
Metabolism
Activation of prodrugs
De-activation
Generation of biologically active metabolites
Distribution
Elimination
Pharmacodynamics
Palliative drug action (modulation of disease-symptoms or disease signs by targeting physiologically relevant systems, without addressing those mechanisms that cause or causally contribute to the disease)
Molecular differential-diagnosis-related pharmacogenetics
Causative drug action (modulation of actual causative of contributory mechanisms

2. In the second case, the biological variation is *directly disease-related*, is of pathological importance per se, and represents a subgroup of the overall clinical disease/diagnostic entity. The differential response to a drug is thus related to how well this drug addresses or is matched to the presence or relative importance of the pathological mechanism it targets in different patients, i.e., the molecular differential diagnosis of the patient (disease-mechanism-related pharmacogenetics).

Although these two scenarios are conceptually rather different, they result in similar practical consequences with regard to the administration of a drug, namely stratification of patients based on a particular, DNA-encoded marker. It seems therefore legitimate to subsume both under the umbrella of "pharmacogenetics".

4.2.1
Classical Pharmacogenetics

This category includes differential pharmacokinetics and pharmacodynamics.

Pharmacokinetics. Drug response may vary due to inter-individual differences in absorption, distribution, metabolism (with regard to both activation of pro-drugs, inactivation of the active molecule, and generation of derivative molecules with biological activity) or excretion of the drug. In any of these cases, the differential effects observed are due to the presence—at the intended site of action—either of inappropriate concentrations of the pharmaceutical agent, or of inappropriate metabolites, or of both, resulting either in lack of efficacy or in toxic effects. Pharmacogenetics, as it relates to pharmacokinetics, has been recognized as an entity for more than 100 years, going back to the observation, commonly credited to Archibald Garrod, that a subset of psychiatric patients treated with the hypnotic, sulphonal, developed porphyria. We have since then come to understand the underlying genetic causes for many of the previously known differences in enzymatic activity, most prominently with regard to the P450 enzyme family (Tables 3 and 4), and these have been the subject of recent reviews (Dickins and Tucker 2001; Evans and Relling 1999). However, such pharmacokinetic effects are also seen with membrane transporters, such as in the case of differential activity of genetic variants of MDR-1 that affects the effective intracellular concentration of anti-retrovirals (Fellay et al. 2002), or of the purine-analogue-metabolizing enzyme, thiomethyl-purine-transferase (Dubinsky et al. 2000).

Despite the widespread recognition of isoenzymes with differential metabolizing potential since the middle of the 20th century, the practical application and implementation of this knowledge has been minimal so far. This may be the consequence, on one hand, of the irrelevance of such differences in the presence of relatively flat dose-effect-curves (i.e., a sufficiently wide therapeutic window), as well as, on the other hand, the fact that many drugs are subject to complex, parallel metabolizing pathways, where in the case of underperformance of one

Table 3 Pharmacogenetics: chronology

Pharmacogenetic phenotype	Described	Underlying gene/mutation	Identified
Sulphonal porphyria	ca. 1890	Porphobilinogen deaminase?	1985
Suxamethonium hypersensitivity	1957–1960	Pseudocholinesterase	1990–9192
Primaquine hypersensitivity; favism	1958	G-6-PD	1988
long QT syndrome	1957–1960	*Herg*, etc.	1991–1997
Isoniazid slow/fast acetylation	1959–1960	*N*-acetyltransferase	1989–1993
Malignant hyperthermia	1960–1962	Ryanodine receptor	1991–1997
Fructose intolerance	1963	Aldolase B	1988–1995
Vasopressin insensitivity	1969	Vasopressin receptor2	1992
Alcohol susceptibility	1969	Aldehyde dehydrogenase	1988
Debrisoquine hypersensitivity	1977	CYP2D6	1988–1993
Retinoic acid resistance	1970	PML-RARA fusion-gene	1991–1993
6-Mercaptopurin-toxicity	1980	Thiopurine methyltransferase	1995
Mephenytoin resistance	1984	CYP2C19	1993–1994
Insulin insensitivity	1988	Insulin receptor	1988–1993

Table 4 Pharmacogenetics: pharmacological phenotyping

Phase I enzyme	Testing substance
Aldehyde dehydrogenase	Acetaldehyde
Alcohol dehydrogenase	Ethanol
CYP1A2	Caffeine
CYP2A6	Nicotine, coumarin
CYP2C9	Warfarin
CYP2C19	Mephenytoin, omeprazole
CYP2D6	Dextromethorphan, debrisoquine, sparteine
CYP2E1	Chlorzoxazone, caffeine
CYP3A4	Erythromycin
CYP3A5	Midazolam
Serum cholinesterase	Benzoylcholine, butyrylcholine
Paraoxonase/arylesterase	Paraoxon
Phase II enzyme	Testing substance
Acetyltransferase (NAT1)	*Para*-aminosalizylsäure
Acetyltransferase (NAT2)	Isoniazid, sulfamethazine, caffeine
Dihydropyrimidine dehydrogenase	5-fluorouracil
Glutathione transferase (GST-M1)	*Trans*-stilbene-Oxid
Thiomethyltransferase	2-mercaptoethanol, d-penicillamine, captopril
Thiopurine methyltransferase	6-mercaptopurine, 6-thioguanine, 8-azathioprine
UDP-glucuronosyl transferase (UGT1A)	Bilirubin
UDP-glucuronosyl transferase (UGT2B7)	Oxazepam, ketoprofen, oestradiol, morphine

enzyme, another one may compensate. Such compensatory pathways may well have somewhat different substrate affinities, but allow plasma levels to remain within therapeutic concentrations. Thus, the number of such polymorphisms that have found practical applicability is rather limited and, by and large, so far restricted to determinations of the presence of functionally deficient variants of the enzyme, thiopurine-methyl-transferase, in patients prior to treatment with purine-analogue chemotherapeutics.

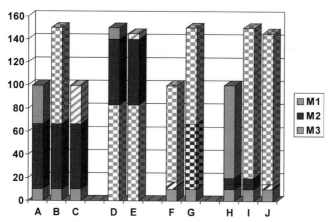

Fig. 1 *A*, Normal physiology: three molecular mechanisms (*M1, M2, M3*) contribute to a trait. *B*, Diseased physiology D1: derailment (cause/contribution) of molecular mechanism 1 (*M1*). *C*, Diseased physiology D1: causal treatment T1 (aimed at *M1*). *D*, Diseased physiology D3: derailment (cause/contribution) of molecular mechanism 3 (*M3*). *E*, Diseased physiology D3, treatment T1: treatment does not address cause. *F*, Diseased physiology D1, palliative treatment T2 (aimed at *M2*). *G*, Diseased physiology D1, palliative treatment T2; T2-refractroy gene variant in *M2*. *H*, Normal physiology variant: differential contribution of *M1* and *M2* to normal trait. *I*, Diseased physiology D1-variant: derailment of mechanism *M1*. *J*, Diseased physiology D1-variant: treatment with T2. *Solid colors* indicate normal function, *stippling* indicates pathologic dysfunction, *hatching* indicates therapeutic modulation

Pharmacodynamics. Pharmacodynamic effects, in contrast, may lead to inter-individual differences in a drug's effects despite the presence of appropriate concentrations of the intended active (or activated) drug compound at the intended site of action. Here, DNA-based variation in how the target molecule, or another (downstream) member of the target molecule's mechanistic pathway, can respond to the medicine modulates the effects of the drug. This will apply primarily to palliatively working medicines that improve a condition symptomatically by modulating disease-phenotype-relevant (but not disease-cause-relevant) pathways that are not dysfunctional but can be used to counterbalance the effect of a dysfunctional, disease-causing pathway and therefore allow mitigation of symptoms. A classic example of such an approach is the acute treatment of thyrotoxicity with beta-adrenergic blocking agents: even though the sympathetic nervous system does not in this case contribute causally to tachycardia and hypertension, dampening even its baseline tonus through this class of drugs relieves the cardiovascular symptoms and signs of this condition, before the causal treatment (in this case available through partial chemical ablation of the hyperactive thyroid gland) can take effect. Notably, the majority of today's pharmacopoeia actually belongs to this class of palliatively acting medicines.

A schematic (Fig. 1) is provided to help clarify these somewhat complex concepts. A hypothetical case of a complex trait/disease is depicted where excessive, dysregulated function of one of the trait-controlling/-contributing pathways (Fig. 1, A, B) causes symptomatic disease; the example used refers to blood pressure as the trait, and hypertension as the disease in question, respectively (for the case of a defective or diminished function of a pathway, an analogous schematic could be constructed and again for a deviant function). A palliative treatment would be one that addresses one of the pathways that, while not dysregulated, contributes to the overall deviant physiology (Fig. 1, F), while the respective pharmacogenetic-pharmacodynamic scenario would occur if this particular pathway was, due to a genetic variant, not responsive to the drug chosen (Fig. 1, G). A palliative treatment may also be ineffective if the particular mechanism targeted by the palliative drug (due to the presence of a molecular variant) provides less than the physiologically expected baseline contribution to the relevant phenotype (Fig. 1, H). In such a case, modulating an a-priori unimportant pathway in the disease scenario will not yield successful palliative treatment results (Fig. 1, I, J).

Some of the most persuasive examples we have to date of such a palliative drug-related pharmacogenetic effect are in the field of asthma. The treatment of asthma relies on an array of drugs aimed at modulating different generic pathways, thus mediating bronchodilation or anti-inflammatory effects, often without regard to the possible causative contribution of the targeted mechanism to the disease. One of the mainstays of the treatment of asthma is activation of the beta-2-adrenoceptor by specific agonists, which leads to relaxation of bronchial smooth muscles and, consequently, bronchodilation. Recently, several molecular variants of the beta-2-adrenoceptor have been shown to be associated with differential treatment response to beta-2-agonists (Martinez et al. 1997; Tan et al. 1997). Individuals carrying one or two copies of a variant allele that contains a glycine in place of arginine in position 16 were found to have a three- and five-fold reduced response to the agonist, respectively. This was shown in both in vitro (Green et al. 1994, 1995) and in vivo (Green et al. 1995) studies to correlate with an enhanced rate of agonist-induced receptor down-regulation, but not with any difference in transcriptional or translational activity of the gene, or with agonist binding. In contrast, a second polymorphism affecting position 19 of the beta upstream peptide was shown to affect translation (but not transcription) of the receptor itself, with a 50% decrease in receptor numbers associated with the variant allele, which happens to be in strong linkage disequilibrium with the variant allele at position 16 in the receptor. The simultaneous presence of both mutations would be predicted to result in low expression and enhanced down-regulation of an otherwise functionally normal receptor, depriving patients carrying such alleles of the benefits of effective bronchodilation as a palliative (i.e., non-causal) counter-measure to their pathological airway hyper-reactivity. Importantly, there is no evidence that any of the allelic variants encountered are associated with the prevalence or incidence, and thus potentially the etiology of the underlying disease (Reihsaus et al.1993; Dewar et al. 1998). This would reflect the scenario depicted in Fig. 1, H.

Inhibition of leukotriene synthesis, another palliative approach towards the treatment of asthma, proved clinically ineffective in a small fraction of patients who carried only non-wild-type alleles of the 5-lipoxygenase promoter region

(Drazen et al. 1999). These allelic variants had previously been shown to be associated with decreased transcriptional activity of the gene (In et al. 1997). It stands to reason, and consistent with clinical observations, that in the presence of already reduced 5-lipoxygenase activity, pharmacological inhibition may be less effective (Fig. 1, H–J). Of note again, there is no evidence for a primary, disease-causing or -contributing role of any 5-lipoxygenase variants; all of them were observed at equal frequencies in disease-affected and non-affected individuals (In et al. 1997).

Pharmacogenetic effects may not only account for differential efficacy, but also contribute to the differential occurrence of adverse effects. An example of this scenario is provided by the well-documented pharmacogenetic association between molecular sequence variants of the 12S rRNA, a mitochondrion-encoded gene, and aminoglycoside-induced ototoxicity (Fischel-Ghodsian et al. 1999). Intriguingly, the mutation that is associated with susceptibility to ototoxicity renders the sequence of the human 12S rRNA similar to that of the bacterial 12S rRNA gene, and thus effectively turns the human 12S rRNA into the (bacterial) target for aminoglycoside drug action, presumably mimicking the structure of the bacterial binding site of the drug (Hutchin and Cortopassi et al. 1994). As in the other examples, presence of the 12S rRNA mutation per se has no primary, drug-treatment-independent pathological effect per se.

By analogy, one may speculate that such molecular mimicry may occur within one species: adverse events may arise if the selectivity of a drug is lost because a gene that belongs to the same gene family as the primary target, loses its identity vis-à-vis the drug and attains, based on its structural similarity with the principal target, similar or at least increased affinity for the drug. Depending on the biological role of the imposter molecule, adverse events may occur, even though the variant molecule may be quite silent with regard to any contribution to disease causation. Although we currently have no obvious examples for this scenario, it is certainly plausible for various classes of receptors and enzymes.

4.2.2
Pharmacogenetics as a Consequence of Molecular Differential Diagnosis

As alluded to earlier, there is general agreement today that any of the major clinical diagnoses in the field of common complex disease, such as diabetes, hypertension or cancer, are comprised of a number of etiologically (i.e., at the molecular level) more or less distinct subcategories. In the case of a causally acting drug, this may imply that the agent will only be appropriate, or will work best, in that fraction of all the patients who carry the (all-inclusive and imprecise) clinical diagnosis in whom the dominant molecular etiology, or at least one of the contributing etiological factors, matches the mechanism of action of the drug in question (Fig. 1, C). If the mechanism of action of the drug addresses a pathway that is not disease relevant, perhaps because it is already down-regulated as an appropriate physiological response to the disease, then logically, the drug would be expected not to show efficacy (Fig. 1, D, E). Thus, unrecognized and undiagnosed disease heterogeneity, disclosed indirectly by the presence or absence of response to a drug targeting a mechanism that contributes to only one of several molecular subgroups of the disease, provides an important explanation for differential drug response and likely represents a substantial fraction of what we today somewhat indiscriminately subsume under the term "pharmacogenetics".

Currently, the most frequently cited example for this category of pharmacogenetics is trastuzumab (Herceptin), a humanized monoclonal antibody directed against the *her-2* oncogene. This breast cancer treatment is prescribed based on the level of *her-2*-oncogene expression in the patient's tumor tissue. Differential diagnosis at the molecular level not only provides an added level of diagnostic sophistication, but also actually represents the prerequisite for choosing the appropriate therapy. Because trastuzumab specifically inhibits a gain-of-function variant of the oncogene, it is ineffective in the two-thirds of patients who do not over-express the drug's target, whereas it significantly improves survival in the one-third of patients who constitute the subentity of the broader diagnosis of breast cancer in whom the gene is expressed (Baselga et al. 1996). Some have argued against this being an example of pharmacogenetics, because the parameter for patient stratification (i.e., for differential diagnosis) is the somatic gene expression level rather than particular genotype data (Haseltine 1998). This is a difficult argument to follow, since in the case of a treatment-effect-modifying germ-line mutation it would obviously not be the nuclear gene variant per se, but also its specific impact on either structure/function or on expression of the respective gene/gene product that would represent the actual physiological corollary underlying the differential drug action. Conversely, an a-priori observed expression difference is highly likely to reflect a potentially, as yet undiscovered, sequence variant. Indeed, as pointed out earlier, there are a number of examples in the field of pharmacogenomics where the connection between genotypic variant and altered expression has already been demonstrated (In et al. 1997; McGraw et al. 1998).

Another example, although still hypothetical, of how proper molecular diagnosis of relevant pathological mechanisms will significantly influence drug efficacy is in the evolving class of anti-AIDS/HIV drugs that target the CCR5 cell-surface receptor (Huang et al. 1996; Dean et al. 1996; Samson et al. 1996). These drugs would be predicted to be ineffective in those rare patients who carry the delta-32 variant, but who nevertheless have contracted AIDS or test HIV-positive (most likely due to infection with an SI-virus phenotype that utilizes CXCR4) (O'Brien et al. 1997; Theodorou et al. 1997).

It should be noted that the pharmacogenetically relevant molecular variant need not affect the primary drug target, but may equally well be located in another molecule belonging to the system or pathway in question, both upstream and downstream in the biological cascade with respect to the primary drug target.

4.2.3
Different Classes of Markers

Pharmacogenetic phenomena, as pointed out previously, need not be restricted to the observation of a direct association between allelic sequence variation and phenotype, but may extend to a broad variety of indirect manifestations of underlying, but often (as yet) unrecognized sequence variation. Thus, differential methylation of the promoter region of O6-methylguanine-DNA-methylase has recently been reported to be associated with differential efficacy of chemotherapy with alkylating agents. If methylation is present, expression of the enzyme that rapidly reverses alkylation and induces drug-resistance is inhibited, and therapeutic efficacy is greatly enhanced (Esteller et al. 2000).

4.2.4
Complexity Is to Be Expected

In the real world, it is likely that a combination of the scenarios depicted affect how well a patient responds to a given treatment, or how likely it is that he or she will suffer an adverse event. Thus, a fast-metabolizing patient with poor-responder pharmacodynamics may be particularly unlikely to gain any benefit from taking the drug in question, while a slow-metabolizing status may counterbalance in another patient the same inopportune pharmacodynamics, and a third patient, who is a slow metabolizer and displaying normal pharmacodynamics, may be more likely to suffer adverse events. In all of them, both the pharmacokinetic and pharmacodynamic properties may result from the interaction of several of the mechanisms described above. In addition, we know of course that co-administration of other drugs, or even the consumption of certain foods, may affect and further complicate the picture for any given treatment.

5
Incorporating Pharmacogenetics into Drug Development Strategy

It is important to note that despite the public hyperbole and the high expectations surrounding the use of pharmacogenetics to provide personalized care, these approaches are likely to be applicable only to a fraction of medicines that are being developed. Further, if and when such approaches are used, they will represent no radical new direction or concept in drug development but simply a stratification strategy akin to others which we have been using it all along.

The opportunity to subdivide today's clinical diagnosis into molecular subtypes, based on a deeper, more differentiated understanding of pathology at the molecular level, will permit a more sophisticated and precise diagnosis of disease and foster medical advances which will appear as pharmacogenetic phenomena. However, the sequence of events that is today often presented as characteristic for a pharmacogenetic scenario—namely, exposing patients to a drug, recognizing a differential [i.e. (quasi-)bimodal-] response pattern, discovering a marker that predicts this response, and creating a diagnostic product to be co-marketed with the drug henceforth—is likely to be reversed. Rather, the search for new drugs will be based specifically, and a priori, on a new mechanistic understanding of disease causation or contribution (i.e., a newly found ability to diagnose a molecular subentity of a previously more encompassing, broader, and less precise clinical disease definition). Thus, pharmacogenetics will not be so much about finding the "right medicine for the right patient", but about finding the correct medicine for a given disease (subtype), as we have aspired to do all along throughout the history of medical progress. This is, in fact, good news: the conventional pharmacogenetic scenario would invariably present major challenges from both a regulatory and a business development and marketing standpoint, as it will confront development teams with a critical change in the drug's profile at a very late point during the development process. In addition, the timely development of an approvable diagnostic in this situation is difficult at best, and its marketing as an add-on to the drug is a less than attractive proposition to diagnostics business. Thus, the practice of pharmacogenetics will, in many instances, be marked by progress along the very same path that has been one of the main avenues of medical progress all along: differential diagnosis first, followed by the development of appropriate, more specific treatment modalities.

Thus, the first step in the sequence of events in this case is likely to involve the development of an in vitro diagnostic test as a stand-alone product that may be marketed on its own merits, allowing the physician to establish an accurate, state-of-the-art diagnosis of the molecular subtype of the patient's disease. Sometimes such a diagnostic may prove helpful, even in the absence of specific therapy, by guiding the choice of existing medicines and/or of non-drug treatment modalities such as specific changes in diet or lifestyle. The availability of such a diagnostic, as part of the more sophisticated understanding of disease, will undoubtedly foster and stimulate the search for new, more specific drugs; and once such drugs are found, the availability of the specific diagnostic test will be important for carrying out the appropriate clinical trials. This will allow a prospectively planned, much more systematic approach towards clinical and business development, with a commensurate greater chance of actual realization and success.

In practice, some degree of guesswork will remain, due to the nature of common complex disease. First, all diagnostic approaches, including those based on DNA analysis in common complex disease, as stressed above, will provide only a measure of probability. Although the variances of drug response among patients who do (or do not) carry the drug-specific subdiagnosis will be smaller, there will still be a distribution of differential responses: although by and large the drug will work better in the responder group, there will be some patients in this subgroup who will respond less or not at all, and conversely, not everyone belonging to the non-responder group will fail completely to respond, depending perhaps on the relative magnitude with which the particular mechanism

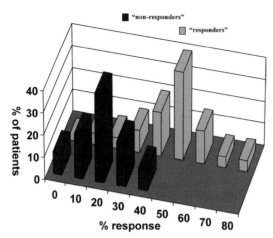

Fig. 2 Hypothetical example of bimodal distribution according to marker that indicates non-responder or responder status. Note that in both cases a distribution is present, with overlaps; thus the categorization into responders or non-responders based on the marker must be understood to convey only the probability of belonging to one or the other group

contributes to the disease. It is important to bear in mind, therefore, that even in the case of fairly obvious bi-modality, patient responses will still show distribution patterns and that all predictions as to responder or non-responder status will only have a certain probability of being accurate (Fig. 2). The terms "responder" and "non-responder" as applied to groups of patients stratified based on a DNA marker represent mendelian-thinking-inspired misnomers that should be replaced by more appropriate terms that reflect the probabilistic nature of any such classification, e.g., likely (non-) responder.

In addition, based on our current understanding of the polygenic and heterogeneous nature of complex disorders, we will only be able to exclude in any one patient those genetic variants that do not appear to contribute to the disease, and therefore deselect certain treatments, even in an ideal world where we would know about all possible susceptibility gene variants for a given disease and have treatments for them. We will, however, most likely find ourselves left with a small number, perhaps two to four, potential disease-contributing gene variants whose relative contribution to the disease will be very difficult, if not impossible, to rank in an individual patient. It is likely then that trial and error, and this great intangible quantity, physician experience, will still play an important role, albeit on a more limited and subselective basis.

Where differential drug response and/or safety occurs as a consequence of a pathologically irrelevant, purely drug-response-related pharmacogenetics scenario, there will be greater difficulty in planning and executing a clinical development program because it will be more difficult to anticipate or predict differential responses a priori. In this situation, it may also be more difficult to find the relevant marker(s), unless it happens to be among the obvious candidate genes implicated in the disease physiopathology or the treatment's mode of action. Although screening for molecular variants of these genes, and testing for their possible associations with differential drug response, is a logical first step, if this is unsuccessful, it may be necessary to embark on an unbiased genome-wide screen for such a marker or markers. Despite recent progress in high-throughput genotyping, the obstacles that will have to be overcome on the technical, data-analysis and cost levels are formidable. They will limit the deployment of such programs, at least for the foreseeable future, to select cases in which there are very solid indications for doing so, based on clinical data showing a near-categorical (e.g., bi-modal) distribution of treatment outcomes. Even then we may expect to encounter for every success, due a favorably strong linkage disequilibrium across considerable genomic distance in the relevant chromosomal region, as many or more failures, where the culpable gene variant cannot be found due to the higher recombination rate or other characteristics of the stretch of genome on which it is located.

6
Regulatory Aspects

At the time of writing, regulatory agencies in both Europe and the United States are beginning to show keen interest in the potential role that pharmacogenetic approaches may play in the development and clinical use of new drugs and in the potential challenges that such approaches may present to the regulatory approval process. While no formal guidelines have been issued, the pharmaceutical industry has already been reproached, albeit in a rather non-specific manner, for not being more proactive in the use of pharmacogenetic markers. It will be of key importance for all concerned to engage in an intensive dialogue at the end of which, it is hoped, will emerge a joint understanding that stratification according to DNA-based markers is fundamentally nothing new, and not different from stratification according to any other clinical or demographic parameter, as has been used all along.

Still, based on the perception that DNA-based markers represent a different class of stratification parameters, a number of important questions will need to be addressed and answered, hopefully always in analogy to conventional stratification parameters, including those referring to ethical aspects. Among the most important ones are questions concerning:

- The need and/or ethical justification (or lack thereof) to include likely non-responders in a trial for the sake of meeting safety criteria, which, given the restricted indication of the drug, may indeed be excessively broad
- The need to use active controls if the patient/disease stratum is different from that in which the active control was originally tested
- The strategies to develop and gain approval for the applicable first-generation diagnostic, as well as for the regulatory approval of subsequent generations of tests to be used to determine eligibility for prescription of the drug,

as well as a number of ethical and legal questions relating to the unique requirements regarding privacy and confidentiality for genetic testing that may raise novel problems with regard to regulatory audits of patient data (see below).

A concerted effort to avoid what has been termed genetic exceptionalism—the differential treatment of DNA-based markers as compared with other personal medical data—should be made so as not to further complicate the already very difficult process of obtaining regulatory approval. This seems justified based on the recognized fact that in the field of common complex disease, DNA-based markers are not at all different from conventional medical data in all relevant aspects, namely specificity, sensitivity, and predictive value.

7
Pharmacogenetic Testing for Drug Efficacy Versus Safety

In principle, pharmacogenetic approaches may be useful both to raise efficacy and to avoid adverse events, by stratifying patient eligibility for a drug according to appropriate markers. In both cases, clinical decisions and recommendations must be supported by data that have undergone rigorous biostatistical scrutiny. Based on the substantially different prerequisites and opportunities for acquiring such data, and applying them to clinical decision-making, we expect the use of pharmacogenetics for enhanced efficacy to be considerably more common than for the avoidance of adverse events.

The chances of generating adequate data on efficacy in a subgroup is reasonably high, given the fact that unless the drug is viable in a reasonably sizeable number of patients, it will probably not be developed for lack of a viable business case, or at least only under the protected environment of orphan drug guidelines. Implementation of pharmacogenetic testing to stratify for efficacy, provided that safety in the non-responder group is not an issue, will primarily be a matter of physician preference and sophistication, and potentially of third-party payer directives, but would appear less likely to become a matter of regulatory mandate, unless a drug has been developed selectively in a particular stratum of the overall indication (in which case the indication label will be restricted to this stratum). Indeed, an argument can be made against depriving those who carry the likely non-responder genotype regarding eligibility for the drug, but who individually, of course, may respond to the drug with a certain, albeit lower probability. From a regulatory aspect, the use of pharmacogenetics for efficacy, if adequate safety data exist, appears largely unproblematic; the worst-case scenario (a genotypically inappropriate patient receiving the drug) would result in treatment without expected beneficial effect, but with no increased odds to suffer adverse consequences, i.e., much of what one would expect under conventional paradigms.

The usefulness and clinical application of pharmacogenetic strategies for improving safety, particularly with regard to serious adverse events, will meet with considerably greater hurdles and is less likely to become practical. A number of reasons are cited for this. First, in the event of serious adverse events associated with the use of a widely-prescribed medicine, withdrawal of the drug from the market is usually based largely on anecdotal evidence from a rather small number of cases, in accordance with the Hippocratic mandate *primum non nocere*. If the sample size is insufficient to demonstrate a statistically significant association between drug exposure and event, as is typically the case, it will most certainly be insufficient to allow meaningful testing for genotype–phenotype correlations; the biostatistical hurdles become progressively more difficult as many markers are tested and the number of degrees of freedom applicable to the analysis for association continues to rise. Therefore, the fraction of attributable risk shown to be associated with a given at-risk (combination of) genotype(s) would have to be very substantial for regulators to accept such data. Indeed, the low prior probability of the adverse event, by definition, can be expected to yield an equally low positive (or negative) predictive value.

Second, the very nature of safety issues raises the hurdles substantially because in this situation the worst-case scenario, administration of the drug to the wrong patient, will result in a higher probability of harm to the patient. Therefore, it is likely that the practical application of pharmacogenetics for the purpose of limiting adverse events will be restricted to diseases with a dire prognosis, where a high medical need exists, where the drug in question offers unique potential advantages (usually bearing the characteristics of a life-saving drug), and where, therefore, the tolerance even for relatively severe side effects is much greater than for other drugs. This applies primarily to areas such as oncology or HIV/AIDS. In most other indications, the sobering biostatistical and regulatory considerations discussed represent barriers that are unlikely to be overcome easily; and the proposed, conceptually highly attractive, routine deployment of pharmacogenetics as a generalized drug surveillance or pharmaco-vigilance practice following the introduction of a new pharmaceutical agent (Roses 2000) faces these scientific as well as formidable economic hurdles.

8
Ethical and Societal Aspects of Pharmacogenetics

No discussion about the use of genetic/genomic approaches to health care can be complete without considering their impact on ethics, society and the law.

Much of the discussion about ethical and legal issues relating to pharmacogenetics is centered on the issue of genetic testing, a topic that has recently been the focus of a number of guidelines, advisories, white papers, etc., issued by a number of committees in both Europe and the United States. It is interesting to note that the one characteristic that almost all these documents share is a studious avoidance of defining exactly what a genetic test is. Where definitions are given, they tend to be very broad, including not only the analysis of DNA but also of transcription and translation products affected by inherited variation. In as much as the most sensible solution to this dilemma would be a consensus to treat all personal medical data in a similar fashion regardless of the degree to which DNA-encoded information affects it (noting that there really is not any medical data that are not to some extent affected by intrinsic patient properties), it may, for the time being, be helpful to let the definition of what constitutes genetic data be guided by the public perception of genetic data, in as much as the whole discussion of this topic is prompted by these public perceptions.

In the public eye, a genetic test is usually understood either (1) as any kind of test that establishes the diagnosis (or predisposition) of a classic monogenic, heritable disease, or (2) as any kind of test based on nucleic acid analysis. This includes the (non-DNA-based) Guthrie test for phenylketonuria as well as forensic and paternity testing and the DNA-based test for Lp(a), but not the plasma-protein-based test for the same marker (even though the information derived is identical). Since monogenic disease is, in effect, excluded from this discussion, it stands to reason to restrict the definition of genetic testing to the analysis of (human) DNA sequence.

Based on the perceived particular sensitivity of genetic data, institutional review boards commonly apply a specific set of rules for granting permission to test for DNA-based markers in the course of drug trials or other clinical research, including (variably) separate informed consent forms, the anonymization of samples and data, specific stipulations about availability of genetic counseling, provision to be able to withdraw samples at any time in the future, etc.

Arguments have been advanced (Roses 2000) that genotype determinations for pharmacogenetic characterization, in contrast to genetic testing for primary disease risk assessment, are less likely to raise potentially sensitive issues with regard to patient confidentiality, the misuse of genotyping data or other nucleic-acid-derived information, and the possibility of stigmatization. While this is certainly true when pharmacogenetic testing is compared to predictive genotyping for highly penetrant mendelian disorders, it is not apparent why in common complex disorders, issues surrounding predictors of primary disease risk would be any more or less sensitive than those pertaining to predictors of likely treatment success or failure. Indeed, two lines of reasoning may actually indicate an increased potential for ethical issues and complex confrontations among the various stakeholders to arise from pharmacogenetic data.

First, while access to genotyping and other nucleic acid-derived data related to disease susceptibility can be strictly limited, the very nature of pharmacogenetic data calls for a rather more liberal position regarding use: if this information is to serve its intended purpose, i.e., improving the patient's chance for successful treatment, then it is essential that it is shared among at least a somewhat wider circle of participants in the health care process. Thus, the prescription for a drug that is limited to a group of patients with a particular genotype will inevitably disclose those patients' genotype to anyone of a large number of individuals involved in the care of those patients at the medical and administrative level. The only way to limit this quasi-public disclosure of this type of patient genotype data would be if he or she were to sacrifice the benefits of the indicated treatment for the sake of data confidentiality.

Second, patients profiled to carry a high disease probability along with a high likelihood for treatment response may be viewed, from the standpoint of insurance risk, for example, as comparable to patients displaying the opposite profile, i.e., a low risk to develop the disease, but having a high likelihood not to respond to medical treatment, if the disease indeed occurs. For any given disease risk, then patients less likely to respond to treatment would be seen as a more unfavorable insurance risk, particularly if non-responder status is associated with chronic, costly illness rather than with early mortality, the first case having much more far-reaching economic consequences. The pharmacogenetic profile may thus, under certain circumstances, become a more important (financial) risk-assessment parameter than primary disease susceptibility, and would be expected, in as much as it represents but one stone in the complex disease mosaic, to be treated with similar weight, or lack thereof, as other genetic and environmental risk factors.

Practically speaking, the critical issue is not only, and perhaps not even predominantly, the sensitive nature of the information and how it is disseminated and disclosed, but how and to what end it is used. Obviously, the generation and acquisition of personal medical information must always be contingent on the individual's free choice and consent, as must be all the application of such data for specific purposes. Beyond this, however, there is today an urgent need for the requisite dialogue and discourse among all stakeholders within society to develop and endorse a set of criteria by which the use of genetic, and indeed of all personal medical information, should occur. It will be critically important that society as a whole endorses, in an act of solidarity with those destined to develop a certain disease, guidelines that support the beneficial and legitimate use of the data in the patient's interest while at the same time prohibiting their use in ways that may harm the individual, personally, financially, or otherwise. As long as we trust our political decision processes to reflect the consensus of society, and as long as such consensus reflects the principles of justice and equality, the resulting set of principles should assert such proper use of medical information. Indeed, both aspects, data protection and patient/subject protection, are seminal components of the mandates included in the WHO's "Proposed International Guidelines on Ethical Issues in Medical Genetics and Genetic Services" (http://www.who.int/ncd/hgn/hgnethic.htm) which mandate autonomy, beneficence, no maleficence, and justice.

9
Conclusion

Pharmacogenetics, in the different scenarios included in this term, will represent an important new avenue towards understanding disease pathology and drug action, and will offer new opportunities of stratifying patients to achieve optimal treatment success. As such, it represents a logical, consequent step in the history of medicine—but an evolutionary, rather than a revolutionary one. Its implementation will take time and will not apply to all diseases and all treatments equally. If society finds ways to sanction the proper use of this information, thus allowing and protecting its unencumbered use for the patient's benefit, important progress in health care will be made.

References

Baselga J, Tripathy D, Mendelsohn J et al (1996) Phase II study of weekly intravenous recombinant humanized anti-p185(HER2) monoclonal antibody in patients with HER2/neu-overexpressing metastatic breast cancer. J Clin Oncol 14:737–744

Dean M, Carrington M, Winkler C et al (1996) Genetic restriction of HIV-1 infection and progression to AIDS by a deletion of the CKR5 structural gene. Science 273:1856–1862

Dewar JC, Wheatley AP, Venn A et al (1998) β2 adrenoceptor polymorphisms are in linkage disequilibrium, but are not associated with asthma in an adult population. Clin Exp All 28:442–448

Dickins M, Tucker G (2001) Drug disposition: To phenotype or genotype. Int J Pharm Med15:70–73; also see: http://www.imm.ki.se/CYPalleles/

Drazen JM, Yandava CN, Dube L et al (1999) Pharmacogenetic association between ALOX5 promoter genotype and the response to anti-asthma treatment. Nat-Genet 22:168–170

Dubinsky M, Lamothe S, Yang HY et al (2000) Pharmacogenomics and Metabolite Measurement for 6-Mercaptopurine Therapy in Inflammatory Bowel Disease. Gastroenterology 118:705–713

Esteller M, Garcia-Foncillas J, Andion E et al (2000) Inactivation of the DNA-repair gene mgmt and the clinical response of gliomas to alkylating agents. N Engl J Med 343:1350–1354

Evans WE, Relling MV (1999) Pharmacogenomics: Translating functional genomics into rational therapies. Science 206:487–491; also see: http://www.sciencemag.org/feature/data/1044449.shl/

Fellay J, Marzolini C, Meaden ER et al (2002) Response to antiretroviral treatment in HIV-1-infected individuals with allelic variants of the multidrug resistance transporter 1: a pharmacogenetics study. Lancet 359:30–36

Fischel-Ghodsian N (1999) Genetic factors in aminoglycoside toxicity. Ann NY Acad Sci 884:99–109

Green SA, Turki J, Innis M et al (1994) Amino-terminal polymorphisms of the human beta 2-adrenergic receptor impart distinct agonist-promoted regulatory properties. Biochemistry 33:9414–9419

Green SA, Turki J, Bejarano P et al (1995) Influence of beta 2-adrenergic receptor genotypes on signal transduction in human airway smooth muscle cells. Am J Respir Cell Mol Biol 13:25–33

Haseltine WA (1998) Not quite pharmacogenomics (letter; comment). Nat Biotechnol 16:1295

Huang Y, Paxton WA, Wolinsky SM et al (1996) The role of a mutant CCR5 allele in HIV-1 transmission and disease progression. Nat Med 2:1240–1243

Hutchin T, Cortopassi G (1994) Proposed molecular and cellular mechanism for aminoglycoside ototoxicity. Antimicrob Agents Chemother 38:2517–2520.

In KH, Asano K, Beier D et al (1997) Naturally occurring mutations in the human 5-lipoxygenase gene promoter that modify transcription factor binding and reporter gene transcription. J Clin Invest 99(5):1130–1137

Martinez FD, Graves PE, Baldini M et al (1997) Association between genetic polymorphisms of the beta2-adrenoceptor and response to albuterol in children with and without a history of wheezing. J Clin Invest 100:3184–3188

McGraw DW, Forbes SL, Kramer LA et al (1998) Polymorphisms of the 5' leader cistron of the human beta2-adrenergic receptor regulate receptor expression. J Clin Invest 102:1927–32

O'Brien TR, Winkler C, Dean M et al (1997) HIV-1 infection in a man homozygous for CCR5 32. Lancet 349:1219

Reihsaus E, Innis M, MacIntyre N et al (1993) Mutations in the gene encoding for the beta 2-adrenergic receptor in normal and asthmatic subjects. Am J Respir Cell Mol Biol 8:334–349

Roses A (2000) Pharmacogenetics and future drug development and delivery. Lancet 355:1358–1361

Samson M, Libert F, Doranz BJ et al (1996) Resistance to HIV-1 infection in Caucasian individuals bearing mutant alleles of the CCR-5 chemokine receptor gene. Nature 382:722–725

Tan S, Hall IP, Dewar J et al (1997) Association between beta 2-adrenoceptor polymorphism and susceptibility to bronchodilator desensitisation in moderately severe stable asthmatics. Lancet 350:995–999

Theodorou I, Meyer L, Magierowska M et al (1997) HIV-1 infection in an individual homozygous for CCR5 32. Lancet 349:1219–1220

WHO guidelines on ethical issues in medical genetics and genetic services http://www.who.int/ncd/hgn/hgnethic.htm

Pharmacogenetics and the Treatment of Cardiovascular Disease

S. K. Tate · D. B. Goldstein

Department of Biology (Galton Lab), University College London,
The Darwin Building, Gower Street, London, WC1E 6BT, UK
e-mail: d.goldstein@ucl.ac.uk

1	Introduction	26
2	Haplotype Mapping	27
2.1	Linkage Disequilibrium	27
2.2	The Genomic Structure of Linkage Disequilibrium	27
2.3	The Frequency of Causal Variants	28
3	Haplotype Mapping Using Tags	29
3.1	Determining Haplotype Structure and Selection of Tagging SNPs	30
3.2	Checks for Population Stratification	30
3.3	Assessing Population Structure	31
3.4	Candidate Genes and Pathways	31
3.5	Candidate Gene Approach vs Whole Genome Scan	32
3.6	Framework for Healthy Volunteers	33
4	Interpreting Associations: What Comes Next?	34
4.1	Determine Associated Interval	34
4.2	Itemize and Study Possible Causal Variants	34
5	Conclusion	35
	References	36

Abstract In this chapter we describe the emerging framework for genetic association studies in pharmacogenetics, including a framework for working with healthy volunteers. The basic approach for case–control studies is to compare the genetic makeup of populations of patients with different response profiles. We describe how recent studies of the pattern of linkage disequilibrium in the human genome have led to the idea of using a subset of tagging single nucleotide polymorphisms (SNPs) to represent the common haplotypes within a population. This approach greatly reduces the economic costs of association studies and allows consideration of multiple candidate genes. Many drugs target multigenic pathways, e.g. the renin–angiotensin system (RAS), and it is desirable to include all the components of the pathway in pharmacogenetic studies. We also compare the candidate gene approach and a whole genome scanning approach and argue that even in the future, when whole genome scans become feasible, in most cases a candidate gene approach will still be the preferred method. Finally, we emphasize that finding an association is only the beginning and there are many steps before information from an association study can be used diagnostically in the clinic.

Keywords Linkage disequilibrium · Association studies · Haplotype · Single nucleotide polymorphism · Population stratification

1
Introduction

There is considerable inter-individual variation in the effectiveness of cardiovascular drugs. For example, diversity in the response to various classes of antihypertensive agent, including angiotensin converting enzyme (ACE) inhibitors and β-adrenergic blockers, is well documented. Less than 50% of hypertensive patients achieve adequate blood pressure control with ACE inhibitor monotherapy (Materson et al. 1995). Drug response (encompassing efficacy and adverse reactions) shows complex dependence on environmental factors including drug–drug interactions. However, genetic differences among individuals, for example in drug metabolism and disposition and genetic polymorphisms in drug targets, also play a role in determining drug response. The genetic basis of drug response can be studied using association studies. In case–control association studies, genetic differences are evaluated between, for example, responders and non-responders; other association studies may use a quantitative representation of drug response.

Genetic variation plays a significant role in inter-individual variation in drug response. Associations have been found between genetic polymorphisms and response to many classes of cardiovascular drugs including β-agonists, β-blockers, ACE inhibitors, anti-arrhythmics, anti-thrombotics, AT_1-receptor blockers, diuretics, statins and lipid-lowering drugs (reviewed in Humma and Terra 2002). The associations, however, are often inconsistent, and in most cases the underlying mechanisms remain poorly known. For the majority of studies, only a few polymorphisms at most have been characterized in the relevant candidate genes, making it unclear which polymorphisms are responsible for the detected association.

It is now clear that incorporating detailed information on patterns of linkage disequilibrium, or the haplotype structure, of candidate genes can greatly increase the power and reliability of association studies. Here we describe the emerging framework for genetic association studies in pharmacogenetics, including a framework for working with healthy volunteers. We will also discuss the steps that should be taken following positive associations to facilitate the eventual clinical application of these reported associations.

2
Haplotype Mapping

2.1
Linkage Disequilibrium

Linkage disequilibrium (LD) is the non-random association between alleles at different loci. For example, there is linkage disequilibrium between two single nucleotide polymorphisms (SNPs) (e.g. with alleles *Aa* and *Bb* at the two loci) if the frequency of the *AB* haplotype in a population deviates from its expected frequency (i.e. the product of the frequencies for *A* and *B* alleles in the population). Average LD declines with chromosomal distance, though very unevenly, and genomic regions with high levels of LD have a reduced number of haplotypes. There is considerable variation in the pattern of LD across the genome and among populations. In association studies it is neither currently possible nor desirable to exhaustively test every SNP. As many SNPs will carry redundant information because they are in strong LD with other SNPs, it is possible to use haplotypes that contain a subset of non-redundant loci in association studies.

There are several methods of measuring LD. One of the most useful is r^2, which is a measure of the association between pairs of alleles. This is most relevant to association studies as it is related to power. If the sample size required for a specific level of statistical power in typing the causal variant directly is N, then the sample size required for the same level of power is N/r^2, where r^2 is the LD between the causal variant and the typed associated marker (see Pritchard and Przeworski 2001).

2.2
The Genomic Structure of Linkage Disequilibrium

Recent studies indicate that parts of the human genome can be partitioned into blocks of sequence showing high LD among common SNPs punctuated by regions where LD has broken down, sometimes corresponding to recombination hotspots (Daly et al. 2001). As SNPs in the same block are in strong LD, many will be redundant for mapping purposes. LD blocks also exhibit low haplotype diversity. For example, in one of Daly's two blocks of LD, two haplotypes account for 96% of chromosomes observed (Daly et al. 2001). These observations led to the idea of using tagging SNPs in association studies, to represent the common haplotypes within a block. An initial mapping study could locate a causal variant to within a block of LD with a small subset of SNPs being sufficient to represent most common haplotypes. This makes the tagging approach economical and allows analyses of large data sets.

Selecting the essential tagging SNPs ensures that the majority of common variation within a particular region of LD is accounted for. The tagging method is consequently more powerful than previous haplotype mapping methods

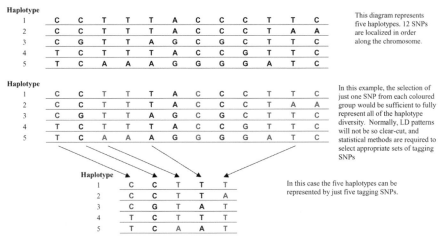

Fig. 1 Illustration of haplotype tagging SNPs

based on SNP spacing alone, which had no guarantee of capturing the common variation.

Further studies on LD pattern in the human genome are likely to reveal more variation in structure. Simulations have predicted the existence of punctuated LD without hotspots of recombination (Zhang et al. 2002), and even with hotspots, LD may not be functionally block-like in some populations (Stumpf and Goldstein 2003). Despite this apparent complexity, tags can still be used to efficiently represent common haplotypes, even in the absence of LD blocks (see Fig. 1).

2.3
The Frequency of Causal Variants

Haplotype mapping will not be effective in identifying rare variants with only modest effects on phenotype. The common disease common variant (CDCV) hypothesis proposes that much of the genetic component of variation in disease predisposition is due to variants with moderate to high frequency in the population. There is, however, considerable disagreement in the community about the validity of the CDCV hypothesis (e.g. Pritchard and Cox 2002). Whatever the reality in the case of predisposition to common disease, it is clear that common variants play a significant role in drug response. Table 1 lists examples of common variants known to influence cardiovascular drug response. Variable drug response therefore appears more genetically tractable in terms of association studies than the study of common disease predisposition.

Table 1 Examples of common variants associated with response to cardiovascular drugs or associated with altered cardiovascular function, and their frequencies (frequency of allele in bold type)

Gene/ Polymorphism	Frequency (%)		Reference(s)	Associated phenotype
	Caucasians	African-Americans		
Angiotensin-converting enzyme I/D	44	40	Wang and Staessen 2000	Genotype associated with ACE inhibitor response. Most data inconsistent (e.g. Stavroulakis et al. 2000; Ohmichi et al. 1997).
Angiotensinogen M235**T**	42	77	Wang and Staessen 2000	*235T* allele associated with higher plasma AGT levels (Schunkert et al. 1997) and greater reduction of blood pressure with ACE inhibitors (Hingorani et al. 1995)
Angiotensin II type 1-receptor A1166**C**	29	5	Wang and Staessen 2000; Gained et al 1997	*C* allele associated with decreased aortic stiffness in hypertensives receiving ACE inhibitor therapy (Benetos et al. 1996)
β1 adrenergic receptor Arg389**Gly**	27	42	Liggett 2000	*Arg/Arg* genotype associated with resting heart rate, systolic and diastolic blood pressure and double product (Humma et al. 2001)
β-2 adrenergic receptor			Xie et al. 2000	
Arg16Gly	43	49		Genotype associated with agonist response. Data inconsistent (e.g. Gratze et al. 1999; Cockcroft et al. 2000)
Gln27**Glu**	35	18		*Glu* allele associated with increased agonist response (Dishy et al. 2001)
Apolipoprotein E**4**	13.4	19.6	Lee et al. 2001	*E4* allele associated with lesser response to statins (Carmena et al. 1993)
CYP2C9			Lee et al. 2002	
***2**	15	1		Carriers may have reduced warfarin clearance and daily dose requirement (Reviewed in Lee et al. 2002)
***3**	10	0.5		

3
Haplotype Mapping Using Tags

One general framework for a haplotype mapping study using tags is described in this section. Briefly, it involves (1) determining candidate gene(s) haplotype structure in the population of interest, (2) the selection of tags and (3) genotyping tags in clinical material. This approach has recently been used by Weale and colleagues (Weale et al., in preparation) for the neuronal sodium channel gene

SCN1A where it was found that just five tags were sufficient to represent the common variation in the gene.

3.1
Determining Haplotype Structure and Selection of Tagging SNPs

The first step in an association study using tags is to determine the underlying haplotype structure of the candidate gene(s) in controls. This is done by resequencing and needs to be done for each population of interest. Although Weale and colleagues (Weale et al., in preparation) found the same five tags were sufficient in two study populations, if the major haplotypes differ between populations then the appropriate tags may also differ. Resequencing can be carried out in either singleton material or trios (the latter having the advantage of an error check). Ideally introns and extended regulatory regions will be considered in addition to coding regions.

Once haplotype structure has been determined, appropriate tagging SNPs must be selected. The power of the study will depend on the choice of tags. In some circumstances (e.g. for very small genes) it will be possible to identify tags by eye. For the majority of genes though, particularly if they are large and/or regulatory regions are also considered, it is necessary to use formal statistical criteria. We have recently introduced a package called the TagIT program (Weale and Goldstein 2003, see http://popgen.biol.ucl.ac.uk/), which runs in the MATLAB programming environment and implements a number of statistical criteria for selecting and evaluating tags based on genotype data from trios or based on resolved haplotype frequencies. Other programs exist also (Johnson et al. 2001, see David Clayton's website http://www-gene.cimr.cam.ac.uk/clayton/software/). These programs allow the statistical evaluation of tags to ensure they sufficiently cover those that will not be typed. Additionally, it is important for there to be sufficient tags for a resampling approach to evaluate how well unselected SNPs are represented by the tags (Weale and Goldstein 2003). Finally, tags must be genotyped in clinical material. To capture variants of small effect, a large study population must be used.

3.2
Checks for Population Stratification

Ideally all associations should be checked for possible stratification effects. Although the general importance of stratification remains unclear (Ardlie et al. 2002), it can lead to spurious associations and is thought to be responsible for some of the inconsistent or unreplicated case–control studies. Surprisingly, very few reported associations check for stratification. Stratification can occur if the study population is structured and the subgroups have different drug responses. This structure can lead to significant associations at loci that are unlinked to any causal sites, but that have frequency differences between the subgroups. One approach to control for stratification is called Genome Control (Reich and

Goldstein 2001). This approach is based on the assumption that if stratification exists, it should raise association statistics at unlinked markers in addition to at candidate gene markers. Detection of stratification using Genome Control involves genotyping unlinked markers in the set of cases and controls in which the association is found. It is then possible to quantify the level of association that is solely due to population stratification and to correct for this stratification. Only a moderate number of markers are needed for this method, making it practical for most studies. It would be helpful to develop a standard panel of markers to use in stratification checks. An alternative approach is called structured association which involves statistical identification of the subgroups, and effective evaluation of the evidence for association within each identification subgroup (Pritchard and Rosenberg 1999; Pritchard et al. 2000).

3.3
Assessing Population Structure

A genetically structured population comprises two or more subpopulations with average differences in the distribution of certain variants. Many genetic variants, including many drug metabolizing enzyme polymorphisms (Evans and Relling 1999), and many drug target polymorphisms (see Table 1) vary in frequency among populations. This potentially affects average drug efficacy, toxicity and optimal dosage within different populations. For example, the ACE inhibitor enalapril was shown to have lower efficacy in Caucasian patients when compared to those of African ancestry in two randomized large-scale trials (Exner et al. 2001). Because of these observations, it will often be important to describe population structure when evaluating drugs. There are currently two main strategies for doing this: racial labelling and explicit genetic inference.

Ethnic or racial labelling uses racial labels to describe the structure of human genetic variation. Risch and colleagues (Risch et al. 2002) propose five major racial groups based on continental ancestry. Whilst this method is easy to implement, it assumes a rather simplistic view of human genetic history. Explicit genetic inference ignores geographic, racial or ethnic labels and instead groups individuals using genetic data. For example, Wilson and colleagues (Wilson et al. 2001) used explicit genetic inference to separate 354 individuals from eight global populations into four genetically distinct clusters based on 39 unlinked microsatellites (using the program STRUCTURE, Pritchard et al. 2000). It is currently unclear whether the scheme proposed by Risch and colleagues will be generally sufficient to represent human genetic structure in the context of drug trials.

3.4
Candidate Genes and Pathways

Haplotype mapping using tags is sufficiently economical to allow consideration of many genes. A conventional starting point for a candidate gene study would be the direct protein target of a drug or other elements that interact directly with the drug (e.g. drug metabolizing enzymes). This can now be extended to include the genes encoding all components of a particular pathway. Additionally the haplotype framework allows testing for epistatic interactions between combinations of different genes with different haplotypes.

The renin-angiotensin system (RAS) is a multigenic pathway targeted by cardiovascular drugs including ACE inhibitors, AT_1 blockers and certain β-blockers. An estimated 20–30 proteins are required to generate the eventual physiological response to ACE inhibitors, yet most RAS pharmacogenetic studies have focused on SNPs in single genes (and often a single SNP). It is not surprising there have been inconsistencies across studies.

A multigenic approach to the RAS system is supported by data from a study (only published in abstract form) where combinations of 45 polymorphisms in the ACE, AT_1-receptor and angiotensinogen genes were studied for their association with ACE inhibitor response. Certain combinations of ten polymorphisms were found to be better predictors of drug response than any single SNP (Lithell et al. 1999). The model could be further improved by including additional genes of the pathway and by using tags.

3.5
Candidate Gene Approach vs Whole Genome Scan

Candidate gene studies, using haplotype tagging SNPs, provide an economical method of capturing common variants influencing drug response, especially given that haplotype structures only need to be determined once in the population of interest. The biological actions of many cardiovascular drugs are well enough understood that a set of appropriate candidate genes can be chosen. Unfortunately, in most or all cases, the candidate gene list will not be complete, motivating genome-wide studies.

Statistically powerful genome-wide association studies, however, are not currently feasible, both for technical and economic reasons. There are an estimated 10 million SNPs (with minor allele frequency >1%) in the human genome; fewer than 4 million have been catalogued to date. It is clearly not possible to type all SNPs in an association study. Instead, the current approach is to focus on a subset that is more or less uniformly spaced, or to focus on a set that represents common haplotypes identified in a preliminary screen. This approach is very costly and currently not exhaustive, in the sense that candidate gene screens as described above can be.

To screen the genome exhaustively, it will be necessary to fully describe genome-wide patterns of linkage disequilibrium. The International HapMap Project is an international research effort to construct a genome-wide haplotype map for this purpose. The HapMap project proposes to identify a set of genome-wide tags based on as-yet undefined criteria. Gabriel and colleagues (Gabriel et al. 2002) estimate that a fully powered genome-wide association study may need up to 1,000,000 SNPs for an African population or up to 300,000 for a non-African population. The actual number will depend of course on the criteria used. It should be noted that even when the tags become available for the whole genome, the cost of genome-wide scans will still be prohibitive until genotyping costs fall dramatically. The necessary adjustments for multiple comparisons in screening the whole genome will also sharply reduce power. For these reasons, in many cases, it will remain preferable to focus on biologically motivated candidate genes. This will also have the advantage of not requiring as severe a correction for multiple comparisons as would be required in genome wide analyses. Thus, when appropriate candidate genes are known, these should be analysed first.

3.6
Framework for Healthy Volunteers

Association studies in pharmacogenetics have usually focused on patients. In many cases, association studies of drug response in healthy volunteers will provide a valuable complement to clinical studies. For example, studies that consider many candidate genes, and possible interactions between them, require an increased sample size to include representatives of each multilocus genotype, but often it is not possible to collect clinical phenotypic data from sufficient patients.

The major advantage of using healthy volunteers is access to a large cohort of individuals. It is possible to screen large groups of volunteers and probe for a large number of haplotype combinations. This will also facilitate fine mapping of causal variants. Fine localization utilizes rare recombinant haplotypes to find associations with a causal variant and requires a large study population. Using healthy individuals also minimizes environmental factors that complicate patient studies, e.g. drug–drug interactions, impaired organ function, and concomitant illnesses, and allows individuals to be matched to other environmental factors, e.g. age or lifestyle.

We illustrate the approach with metabolism. For example, the cytochrome P450 enzyme CYP3A4 is known to metabolize approximately 50% of all prescribed drugs, including certain calcium channel blockers, anti-arrhythmics and cholesterol-lowering agents used in the treatment of cardiovascular disease. Many drug metabolizing enzymes exhibit considerable inter-individual variation in rate of metabolism. It would be possible, by administering a suitable probe drug, to relate drug metabolizing phenotype to haplotype in an association study using healthy volunteers. Clearly not all drug targets will be amenable to studies using healthy volunteers but certain cardiovascular drugs, e.g. β-blockers, can be given in single doses to healthy volunteers with (relative) safety.

4
Interpreting Associations: What Comes Next?

4.1
Determine Associated Interval

For a number of reasons it will be much easier to use causal variants diagnostically as opposed to using markers that are in association with causal variants. For example, ambiguous results from studies of the ACE insertion/deletion (I/D) polymorphism may be due to altered patterns of association in the gene across populations. Most studies find the *DD* genotype to be associated with the greatest response to ACE inhibitors (e.g. Stavroulakis et al. 2000). However, some show no association between response and *ACE I/D* genotype, whereas others show the *II* genotype to be associated with greater drug response (e.g. Ohmichi et al. 1997). A possible explanation for this is that the *I/D* locus is in LD with a causal variant in some populations but not in others.

It is therefore critical that pharmacogenetics studies track down the variants that are responsible for correlations between gene variants and drug responses. One important step in this effort is assessment of the associated interval surrounding polymorphisms that have been associated with phenotypes (Goldstein 2003). This involves looking at the structure of LD in the region. The associated interval can be delimited by examining the pattern of r^2 in the region surrounding the polymorphism that has been associated with phenotype and determining the points upstream and downstream where LD with the associated variant drops below a threshold.

The length of the associated interval would be expected to vary from region to region and population to population, since the length of LD blocks has been found to range from just a few kilobases to 100 kb. Long stretches of LD are advantageous for the coarse mapping of a causal variant to within a particular associated interval but can create problems in fine mapping.

It may be possible to exploit differences in LD among human populations to map causal variants more finely. As a result of LD differences, haplotype diversity and frequency may vary between populations. Coarse mapping using a haplotype tagging approach can be undertaken in populations with more extensive LD to localize a causal variant to within a block of LD; then fine mapping can be undertaken in a population with less extensive LD to possibly map it to within a shorter block of LD. This approach assumes that genetic causation is similar across the populations.

4.2
Itemize and Study Possible Causal Variants

However carefully done, genetic association studies will often identify a set of candidate causal variants that must be carefully sifted through. Even when there is strong evidence implicating a given site in influencing drug response, clinical use of this information will be greatly facilitated by an understanding of the biological effect of the variant(s). Where possible, functional assays can be used to formulate a model to describe how the variants exert their effects. In many cases, experimental functional assays are possible. These can be carried out broadly in two ways, either using site-directed mutagenesis to look at the effects of single variants, or by looking at the effects of variants in the haplotypic background by transfecting cells with a region of DNA with a particular haplotype.

Studies using mRNA allow variants to be studied in their haplotypic context. It is possible that there may be interactions between polymorphisms which would not be detected outside of the haplotype. This method was used by Drysdale and colleagues (Drysdale et al. 2000) to study certain β_2-adrenergic receptor haplotypes found to be associated with response to the β-agonist albuterol. Expression vectors representing two haplotypes associated with divergent in vivo responsiveness were used to study the effects of the mRNA in vivo. It was found that β_2-adrenergic receptor mRNA levels and receptor density were approximately 50% greater in cells transfected with the haplotype associated with the greater physiological response. Furthermore, these results were different from results previously obtained (by the same authors) where individual SNPs had been studied out of context of a verified haplotype. This emphasizes the importance of studying SNPs in vitro in the context of their haplotype.

Site directed mutagenesis is used to study the effects of a single variant. By assessing the effects of polymorphisms separately, it may be possible to determine the effects, if any, of the polymorphisms on physiological function. This approach could be valuable in efforts to distinguish those variants that are causal from those which are merely correlated with a particular phenotype. If the variant is simply in LD with the causal variant, then studying it in isolation would not reveal any plausible functional effects. For example, the β_2-adrenergic receptor (B_2AR) is the target of β-agonists (frequently used for inotropic support in critical cardiovascular disease). Polymorphisms in the B_2AR gene have been associated with varying response to β-agonists. Experimental assessment of coding variants has shown them to cause changes in coupling (Ile164) or agonist-promoted down-regulation (Gly 16 and Glu27) (Ligget 2000).

It is important to be cautious when the causal variant(s) is not known. For example, because LD varies across populations then an association that is diagnostic in one population may not be in another. Population structure will need to be considered, especially if there are inconsistencies across different studies. Results need to be replicated and any putative causal variants need follow-up functional studies.

5
Conclusion

We have described a framework for genetic association studies in cardiovascular disease which uses tags to represent the common variation within a population. With appropriately chosen tags, this approach is a powerful way of finding common variants that influence drug response, and there is clear evidence that common variants play a significant role in cardiovascular drug response. Haplotype mapping using tags is also sufficiently economical to allow consideration of multigene pathways such as the RAS. It should therefore be a priority for future pharmacogenetic studies to take a multigenic approach.

The future for pharmacogenetic studies therefore looks very promising, but it should be emphasized that there is much work to be done before an association has clinical utility.

References

Ardlie KG, Lunetta KL, Seielstad M (2002) Testing for population subdivision and association in four case-control studies. Am J Hum Genet 71:304–311

Benetos A, Cambien F, Ricard S et al (1996) Influence of the angiotensin II receptor gene polymorphism on the effects of perindopril and nitrendipine on arterial stiffness in hypertensive individuals. Hypertension 28:1081–1084

Carmena R, Roederer G, Mailloux H et al (1993) The response to lovastatin treatment in patients with heterozygous familial hypercholesterolemia is modulated by apolipoprotein E polymorphism. Metabolism 42:895–901

Cockcroft JR, Gazis AG, Cross DJ et al (2000) Beta(2)-adrenoceptor polymorphism determines vascular reactivity in humans. Hypertension 36:371–375

Daly MJ, Rioux JD, Schaffner SF et al (2001) High-resolution haplotype structure in the human genome. Nat Genet 29:229–232

Dishy V, Sofowora GG, Xie HG et al (2001) The effect of common polymorphisms of the beta2-adrenergic receptor on agonist-mediated vascular desensitization. N Engl J Med 345:1030–1035

Drysdale CM, McGraw DW, Stack CB et al (2000) Complex promoter and coding region beta 2-adrenergic receptor haplotypes alter receptor expression and predict in vivo responsiveness. Proc Natl Acad Sci USA 97:10483–10488

Exner DV, Dries DL, Domanski MJ et al (2001) Lesser response to angiotensin-converting enzyme inhibitor therapy in black as compared with white patients with left ventricular dysfunction. N Engl J Med 344:1351–1357

Evans WE, Relling MV (1999) Pharmacogenomics: translating functional genomics into rational therapeutics. Science 286:487–491

Gabriel SB, Schaffner SF, Nguyen H et al (2002) The structure of haplotype blocks in the human genome. Science 298:2225–2229

Gainer JV, Hunley TE, Kon V et al (2001) Angiotensin II type I receptor polymorphism in African Americans lower frequency of the C1166 variant. Biochem Mol Biol Int 43:227–231

Goldstein DB (2003) Pharmacogenetics in the laboratory and the clinic. N Engl J Med 348:553–556

Gratze G, Fortin J, Labugger R et al (1999) Beta-2 Adrenergic receptor variants affect resting blood pressure and agonist-induced vasodilation in young adult Caucasians. Hypertension 33:1425–1430

Hingorani AD, Stevens PA, Hopper R et al (1995) Renin-angiotensin system gene polymorphisms influence blood pressure and the response to angiotensin converting enzyme inhibition. J Hypertens 13:1602–1609

Humma LM, Terra SG (2002) Pharmacogenetics and cardiovascular disease: Impact on drug response and applications to disease management. Am J Health-Syst Pharm 59:1241–1252

Humma LM, Puckett BJ, Richardson HE et al (2001) Effects of beta1-adrenoceptor genetic polymorphisms on resting hemodynamics in patients undergoing diagnostic testing for ischemia. Am J Cardiol 88:1034–1037

Johnson GC, Esposito L, Barratt BJ et al (2001) Haplotype tagging for the identification of common disease genes. Nat Genet 29:233–237

Johnson JA (2001) Drug Target Pharmacogenomics. Am J Pharmacogenomics 1:271–281

Lee CR, Goldstein JA, Pieper JA (2002) Cytochrome P450 2C9 polymorphisms: a comphrehensive review of in-vitro and human data. Pharmacogenetics 12:251–263

Lee JH, Tang MX, Schupf N et al (2001) Mortality and apolipoprotein E in Hispanic, African-American, and Caucasian elders. Am J Med Genet 103:121–127

Liggett SB (2000) Pharmacogenetics of Beta-1- and Beta-2-Adrenergic Receptors. Pharmacology 61:167–173

Lithell HO, Berglund L, Jonsson J et al (1999) Prediction of blood pressure-response to ACE inhibitors by using patterns of genetic variation in the RAAS. Circulation 100:I-755. Abstract

Materson BJ, Reda DJ, Cushman WC (1995) Department of veterans Affairs single-drug therapy of hypertension study. Revised figures and new data. Department of Veterans Affairs Cooperative Study Group on Antihypertensive Agents. Am J Hypertens 8:189–192

Ohmichi N, Iwai N, Uchida Y et al (1997) Relationship between the response to the angiotensin-converting enzyme inhibitor imidapril and the angiotensin-converting enzyme genotype. Am J Hypertens 10:951–955

Pritchard JK, Cox NJ (2002) The allelic architecture of human disease genes: common disease-common variant...or not? Human Molecular Genetics 11:2417–2423

Pritchard JK, Stephens M, Rosenberg NA et al (2000) Association mapping in structured populations. Am J Hum Genet 67:170–181

Pritchard JK, Rosenberg NA (1999) Use of unlinked genetic markers to detect population stratification in association studies. Am J Hum Genet 65:220–228

Pritchard JK, Przeworski M (2001) Linkage Disequilibrium in Humans: Models and Data. Am J Hum Genet 69:1–14

Reich DE, Goldstein DB (2001) Detecting association in a case-control study while correcting for population stratification. Genetic Epidemiology 20:4–16

Risch N, Burchard E, Ziv E et al (2002) Categorization of humans in biomedical research: genes, race and disease. Genome Biology 3:1–12

Schunkert H, Hense HW, Gimenez-Roqueplo AP et al (1997) The angiotensinogen T235 variant and the use of antihypertensive drugs in a population-based cohort. Hypertension 29:628–633

Stavroulakis GA, Makris TK, Krespi PG et al (2000) Predicting response to chronic antihypertensive treatment with fosinopril: the role of angiotensin-converting enzyme gene polymorphism. Cardiovasc Drugs Ther 14:427–432

Stumpf MPH, Goldstein DB (2003) Demography, recombination hotspot intensity, and the block-like structure of linkage disequilibrium. Curr Biol 13:1–8

Wang JG, Staessen JA (2000) Genetic polymorphisms in the renin-angiotensin system: relevance for susceptibility to cardiovascular disease. Eur J Pharmacol 410:289–302

Weale ME, Depondt C, Macdonald S et al (in preparation) Selection and evaluation of haplotype tagging SNPs in the neuronal sodium channel gene SCN1A: Implications for linkage disequilibrium gene mapping

Weale ME, Goldstein DB (2003) TagIT Version 1.14 http://popgen.biol.ucl.ac.uk/software

Wilson JF, Weale ME, Smith AC et al (2001) Population genetic structure of variable drug response. Nat Genet 29:265–269

Xie HG, Stein CM, Kim RB et al (2000) Human beta2-adrenergic receptor polymorphisms: no association with essential hypertension in black or white Americans. Clin Pharmacol Ther 67:670–675

Zhang K, Calabrese P, Nordburg N et al (2002) Haplotype block structure and its applications to association studies: power and study designs. Am J Hum Genet 71:1386–1394

Genetic Polymorphisms and Cardiovascular Drug Metabolism

A. R. Boobis · T. Shiga · R. J. Edwards

Section on Clinical Pharmacology, Division of Medicine, Imperial College London, Hammersmith Campus, Du Cane Road, London, W12 0NN, UK
e-mail: a.boobis@imperial.ac.uk
e-mail: r.edwards@imperial.ac.uk

1	Introduction	40
2	Drug Metabolism	43
3	Enzymes of Drug Metabolism	45
3.1	The P450 Superfamily	45
3.1.1	CYP1 Family	46
3.1.2	CYP2 Family	46
3.1.3	CYP3 Family	48
3.1.4	CYP4 Family	49
3.2	P450 and Eicosanoid Metabolism	49
3.3	Flavin-Containing Monooxygenase	49
3.4	Monoamine Oxidase	50
3.5	Alcohol Dehydrogenase and Aldehyde Dehydrogenase	50
3.6	Hydrolases	51
3.6.1	Epoxide Hydrolase	51
3.7	Phase II Metabolism	52
3.7.1	UDP-Glucuronosyltransferase	52
3.7.2	Sulphotransferase	52
3.7.3	*N*-Acetyltransferase	53
3.7.4	Glutathione-*S*-transferase	53
3.8	Thiopurine Methyltransferase	54
3.9	Catechol-*O*-Methyltransferase	54
3.10	Histamine *N*-Methyltransferase	55
4	Drug Transporters	55
5	Polymorphisms of Enzymes of Drug Metabolism	55
5.1	Genotype and Phenotype	55
5.2	CYP2D6	59
5.3	CYP2C19	63
5.4	CYP2C9	64
5.5	Polymorphisms of Other CYPs	65
5.6	Other Genetic Variants of CYPs	65
5.7	Polymorphisms of Other Phase I Enzymes	66
5.8	Polymorphisms of Phase II Enzymes	67
6	Polymorphisms in P-Glycoprotein	68

7	Impact of Genetic Polymorphisms of Drug Metabolism on Overall Variability in Drug Effect	68
8	Conclusions	70
	References	72

Abstract Both the therapeutic and the adverse effects of drugs can be profoundly influenced by the activity of enzymes of drug metabolism. Biotransformation is effected by a number of families of structurally and functionally related enzymes, which inactivate and ultimately enable the excretion of therapeutic agents. The enzymes of drug metabolism exhibit broad and overlapping substrate specificities. Many of these enzymes are now known to exhibit polymorphic expression, due often to point mutations in the structural gene. The result may be a large difference in metabolic activity, and consequently in the pharmacokinetics, between phenotypes. Examples of polymorphic enzymes include CYP2D6, CYP2C9, CYP2C19 and NAT2. Some drug transporters such as ABCB1 (P-glycoprotein) are also polymorphic. The consequences of such polymorphism will depend upon a number of factors, including the steepness of the concentration–effect curve, the contribution of the polymorphic pathway to overall elimination and the magnitude of other sources of variability in the effect of the polymorphic enzyme. In this last case, whilst it might be possible to demonstrate clear differences in effect between phenotypes in healthy volunteers, other sources of variability, including the disease itself, may dominate differences in response in patients. There are some clear examples of the importance of polymorphic drug metabolism in the effects of drugs, such as CYP2C9 and warfarin and NAT2 and hydralazine. However, it is likely that the extent to which genotyping might benefit the individual patient will be established only in adequate clinical studies.

Keywords Pharmacokinetics · Metabolism · Enzymes · CYP450 · Drug transporters

1
Introduction

Generally, the effect of a drug is determined by the concentration present at its site of action (see Fig. 1). This may be a receptor, ion channel, enzyme or other effector site within a cell or tissue. Interaction with such targets usually involves the free (unbound) drug, and hence reversible non-specific binding to macromolecules and other non-effector sites will reduce the effective concentration at the site of action. Drug effect is therefore proportional to the free concentration present. As bound drug is unable to diffuse across lipid membranes, the distribution of drug from plasma also depends on the unbound concentration. At equilibrium, the free concentration in the tissues is the same as in the plasma,

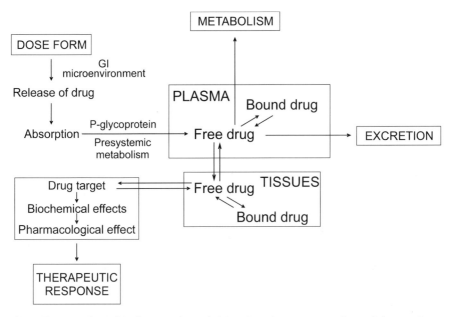

Fig. 1 The inter-relationships between drug administration, plasma concentration and therapeutic response. Following oral administration, the microenvironment of the gastrointestinal tract (e.g. pH, luminal fluid composition) will determine the rate and extent of release of the drug from the formulation in which it is administered. The extent of intestinal absorption will depend upon factors such as the presence and composition of food, motility, intestinal pathology. Bioavailability (systemically available fraction) is determined by the extent of absorption, the degree to which the drug is transported by P-glycoprotein and other transporters, and by any pre-systemic metabolism in the gut wall or in the liver. In plasma, the drug may bind reversibly to proteins, notably albumen and alph1-acid-glycoprotein. Only free drug can diffuse across lipid membranes and hence penetrate tissues, where there may be reversible binding to non-target sites such as proteins or lipids. Usually it is the free drug that interacts with the pharmacological target, for example an enzyme, receptor or ion channel. This interaction is usually, but not always, reversible. Hence, for most compounds there is an equilibrium between levels of the drug in plasma and target occupancy, and thus response. The biochemical events triggered by interaction with the target result in a pharmacological effect, for example interaction within an ion channel may result in altered myocardial signalling and ultimately in cardiac slowing. This then leads to the desired therapeutic response, in the previous example correction of cardiac rhythm. The duration of action of drugs is limited by their elimination, sometimes by direct excretion but more often by metabolism. Again, it is the free compounds that interact with the processes of elimination. At every stage in this process there will be variability, due to intrinsic (genetic, physiological, pathological) and extrinsic (environment, diet, concurrent therapy) factors, together with patient behaviour

and there is a dynamic relationship between the free and bound forms in tissues and in plasma and between the free concentration in the tissues and plasma. As a consequence, the free concentration in plasma reflects the active concentration in tissues, and hence the effect of the drug. Thus, there is a direct relationship between the plasma free concentration and therapeutic effect (concentration–effect relationship). As the bound form in plasma is in equilibrium with the free form, there is also a relationship (albeit less direct) between total plasma concentration and effect.

In treating a patient, the physician is unable to influence the concentration of the drug directly. Rather, the dose and/or frequency of administration can be adjusted. Hence, there is a need to be able to relate these three variables to each other in a quantitative manner, and this is the role of pharmacokinetics. Pharmacokinetics thus describes the dosimetric and temporal relationships between drug administration and concentration at the effector site. Plasma concentration, particularly of the free, unbound form of the drug, serves as a surrogate for the active site concentration. The relationship between the concentration of a drug and its therapeutic effect(s) is described by the pharmacodynamics of the compound.

The pharmacokinetics of a drug are dictated by the processes of distribution from the plasma compartment into the tissues, and elimination, removal of the active moiety from the effector site (Fig. 1). Since free drug is generally in equilibrium throughout the body, in effect elimination means removal of the active form from the body. This can occur through the process of excretion, via the bile (and hence into the faeces) or, more often, the urine. However, for the majority of drugs, elimination occurs via the process of metabolism, or biotransformation, into an inactive (or less active) product. The reason that metabolism is so important in the disposition of drugs is largely a consequence of the physicochemical properties selected for in the design of an efficacious molecule, i.e. those favouring lipid solubility enabling good tissue penetration. The corollary of this is that whilst filterable by the kidney, most drugs are extensively reabsorbed in the proximal tubule. Hence, without metabolism to inactivate them, and to increase their water solubility, such compounds would persist in the body for a very prolonged period and thereby make adjustment to the therapeutic regimen very difficult.

For obvious reasons, many drugs are designed for administration by the oral route. This adds an additional process to their pharmacokinetics, that of absorption (Fig. 1). Again, properties increasing lipid solubility favour good absorption. However, in addition to absorption itself, oral administration leads to the possibility of metabolism by enzymes in the intestinal mucosa or in the liver[1] prior to availability of the drug to the systemic circulation. Such pre-systemic metabolism can influence the fraction of the dose administered that reaches the circulation and hence the concentration at the effector site. Thus, for orally administered drugs, metabolism plays a key role in determining both the bioavailable dose and the persistence of the drug in the body, with consequential effects on therapeutic effect.

Few drugs are without unwanted effects at higher concentrations. This may be a consequence of either exaggerated primary pharmacology or secondary pharmacology. For some compounds, there is a large margin between concen-

---

[1] All drug absorbed from the small intestine first enters the portal vein, which passes through the liver before it reaches the systemic circulation.

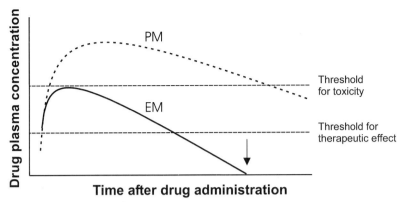

Fig. 2 Possible influence of phenotype on the pharmacokinetics of a drug and the importance of the therapeutic window. Data shown are for the plasma concentration, on a logarithmic scale, plotted against time after oral administration of a drug. The *horizontal lines* illustrate, respectively, the thresholds for therapeutic effect and for toxicity, due either to primary pharmacology or to secondary effects. The difference in metabolic activity between phenotypes is such that there is an appreciable difference in the pharmacokinetics of the compound between the extensive metabolizer (*EM*) phenotype and the poor metabolizer (*PM*) phenotype. Kinetic parameters affected in PM subjects include the AUC and C_{max}, due to decreased presystemic metabolism, with resultant increase in bioavailability and prolonged half-life, due to reduced systemic clearance (there is no change in volume of distribution). These changes will result in increased plasma concentrations, greater persistence, an increase in the time to reach steady state and greater steady state concentrations on repeated administration. It is apparent that the dosing regimen (time of next administration indicated by *vertical arrow*) results in therapeutic concentrations being achieved in EM subjects, whist the therapeutic window is such that the concentration never exceeds the threshold for toxicity. In contrast, in PM subjects the concentration is such that it exceeds the threshold for toxicity for an appreciable interval. In addition, whilst the interval between doses is such that there will be no accumulation in EM subjects, in PM subjects there will be significant accumulation

trations producing the desired effect and those producing unwanted effects, whilst for others the margin is much narrower. This margin between the minimum effective concentration and the minimum concentration associated with unwanted effects is termed the therapeutic index (or window) (see Fig. 2). Hence, a key objective of the therapeutic regimen is to achieve a concentration within this therapeutic window (and to maintain it for an appropriate period of time). Given the importance of metabolism as a determinant of the plasma concentration of drugs, it should be evident that understanding this process can be very important in both drug development and therapeutic use.

2
Drug Metabolism

Drug metabolism, also known as biotransformation, is the process whereby a compound is altered chemically via an enzyme-catalysed reaction. The product is often inactive, although it may have some residual activity. In some instances, the product is more active, has altered activity or is toxic. Indeed, sometimes the activity of a metabolite of a drug can serve as a lead for new drug development; for example, the acid metabolite of terfenadine, fexofenadine. Biotransformation reactions can be classified chemically into two phases, phase I and phase II. In the first phase the drug undergoes functionalization, by the introduction or revelation of a functional group in the molecule. Examples of phase I reactions include oxidations, reductions and hydrolyses. The second phase involves synthetic or conjugating reactions, in which an endogenous molecule (or part of one) is combined with the drug (or more often with a phase I metabolite). Examples of phase II reactions include glucuronidation, sulphation and *N*-acetylation. Phase II metabolism usually represents the major change in lipid solubility of the drug, most conjugated metabolites being very water soluble and readily excreted from the body.

Biotransformation reactions are catalysed by a number of families of drug metabolizing enzymes. These are usually characterized by broad substrate specificity, so that one enzyme can often metabolize many different drugs. Also, one drug can often be metabolized by more than one enzyme, either at the same or different sites on the molecule. This can lead to very complex metabolic pathways, with reactions occurring both in series and in parallel, some metabolites being intermediates and others being metabolically stable end products.

The members of each family (or superfamily) of drug metabolizing enzymes are related to each other both functionally and structurally. They are encoded by the members of a gene family (or superfamily), sharing sequence similarity, and often gene structure, with one another. In almost all cases, each drug metabolizing enzyme is subject to independent regulation. Some of the enzymes exhibit marked interindividual variability in activity, levels of expression or both. This variability is due to genetic, environmental and pathophysiological factors, each acting to a greater or lesser extent depending on the enzyme in question.

Several environmental factors contribute substantially to the variability of some drug-metabolizing enzymes. These include the constituents of the diet, both micronutrients and macronutrients, environmental inducers[2] such as tobacco smoke, alcohol and concurrent drug ingestion and inhibition, which again could arise from compounds present in the diet, the environment or from concurrently ingested therapy. In addition, many pathophysiological factors are now recognized to influence some drug-metabolizing enzymes relatively selectively and hence contribute to interindividual variation in biotransformation capacity. Such factors include: the extremes of age, particularly early age in that the various enzymes mature at different rates; pregnancy where hormonal and other influences can alter the activity of some enzymes; other hormonal factors

---

[2] Induction is a regulatory process resulting in increased levels of the enzyme protein. It usually involves de novo transcription via the interaction of induction receptors (transcription factors) with specific upstream response elements (enhancers). In a few instances, transcription is not increased but rather there is mRNA or protein stabilization.

such as thyroid hormone, growth hormone and insulin, all of which have been shown to differentially affect some drug-metabolizing enzymes; organ dysfunction, particularly diseases of the liver, but also of the kidney and perhaps the CNS and other organs, where again the effects are not universal but confined to only some biotransformation enzymes; and finally infection and inflammation where cytokines and other mediators have been shown to alter the expression or activity of some drug-metabolizing enzymes more than others. Finally, it should be borne in mind that several of the environmental and pathophysiological factors that contribute to variability in these enzymes act through altering the levels of transcription and hence there may be indirect genetic influences on this variability; for example, in the nature of the upstream regulatory sequences or in the structure or amount of transcriptional factors mediating the effect.

A major source of variability in the expression of some drug-metabolizing enzymes is polymorphism in expression due to relatively subtle mutations in the structural gene. These result in an almost complete absence of enzyme and hence associated activity in affected individuals. An additional genetic factor contributing to variability in enzyme activity is the allele frequency found in different ethnic populations. It is becoming increasingly apparent that the dominant allele in one population may differ from that in another and hence produce differences in activity of the respective enzyme.

3
Enzymes of Drug Metabolism

3.1
The P450 Superfamily

Whilst numerous families of enzymes are involved in the metabolism of drugs, by far the most important of these in determining the effect of drugs are the members of the P450 superfamily. All P450 enzymes are products of a single superfamily of genes derived from the same ancestral gene estimated to be at least 3 billion years old. P450 enzymes are haem-thiolate proteins containing protoporphyrin IX at their active site. They are organized into families and sub-families based on primary sequence similarity. In general, members of one family are less than or equal to 36% similar to members of other families whilst members of one sub-family are 40%–60% similar to members of other sub-families. Members of the same sub-family share more than 60% identity (Nebert and Russell 2002).

Humans have 18 families of P450, within which a total of approximately 60 genes have been identified, and with the availability of the complete sequence of the human genome it is now apparent that at most, only one or two new members will be added to these. Each P450 enzyme is the product of a discrete genetic locus and is subject to independent genetic regulation. Hence, the various forms of P450 are differentially affected by genetic, environmental, physiological and pathological factors. Only the products of the first three P450 families

(CYP1, CYP2 and CYP3) are involved, to a significant extent, in the metabolism of drugs (and other foreign compounds). Products of the remaining families are involved in the biosynthesis and degradation of key endogenous compounds, such as steroids, fatty acids and eicosanoids (Nebert and Russell 2002).

P450 enzymes are found in every tissue of the body but those responsible for the metabolism of drugs are located primarily in the liver and to a lesser extent in the so-called portals of entry, in particular the small intestine. P450 enzymes involved in the synthesis or degradation of endogenous compounds are located in those tissues or cells most appropriate to their function. For example, certain forms of P450 are found at particularly high levels in the heart, the platelet, the adrenal gland and vascular endothelial cells.

3.1.1
CYP1 Family

The CYP1 family comprises CYP1A1, CYP1A2 and CYP1B1. Amongst these, only CYP1A2 is expressed constitutively to any extent (Hines and McCarver 2002). Expression is restricted mainly to the liver (Pelkonen and Raunio 1997). CYP1A2 contributes to the metabolism of a range of environmental and dietary chemicals. In addition, this enzyme contributes to the oxidative metabolism of the antiarrhythmic drugs mexiletine and propafenone (depropylation), the beta-blocker propranolol (hydroxylation and deisopropylation), the inotrope pimobendan, the antiplatelet drug cilostazol and the anticoagulant *R*-warfarin. The calcium channel blocker verapamil is an inhibitor of CYP1A2, but is not metabolized by this enzyme to an extent that is clinically significant (Fuhr et al. 1992). Expression of all members of the CYP1 family is inducible via the aryl hydrocarbon (Ah) receptor.

3.1.2
CYP2 Family

CYP2 is the largest of the CYP families in humans (Nebert and Russell 2002). A number of members of this family are expressed both in the liver and in extra-hepatic tissues, e.g. CYP2E1, although there is tissue-specific expression of several forms. CYP2A6 plays a role in the metabolism of a number of environmental chemicals, particularly when they are inhaled (Hines and McCarver 2002; Liu et al. 1996). This enzyme is also known to metabolize the anticoagulant coumarin (hydroxylation). As yet, the function of CYP2B6 has not been well studied (Hollenberg 2002), although it is known to play a significant role in the metabolism of some anti-cancer drugs such as cyclophosphamide.

The CYP2C sub-family comprises four members in humans, CYP2C8, CYP2C9, CYP2C18 and CYP2C19, amongst which CYP2C9 is the most highly expressed in liver. CYP2C18 is either expressed at very low levels or not at all. CYP2C9 metabolizes a wide range of drugs of clinical significance, including the diabetic drug tolbutamide, the anticonvulsant phenytoin and numerous anti-inflammatory drugs such as ibuprofen and diclofenac (Goldstein and De Morais 1994). It also plays a role in the metabolism of several cardiovascular drugs such as the HMG-CoA reductase inhibitor fluvastatin and the diuretic torsemide. CYP2C9 oxidizes the angiotensin-II receptor blocker losartan to an active metabolite, and contributes to the metabolism of irbesartan and candesartan although not to a clinically significant extent (Kazierad et al. 1997; Taavitsainen et al. 2000). The S-enantiomer of warfarin, which is a more potent anticoagulant than the R-enantiomer, is metabolized mainly by CYP2C9. CYP2C9 is inhibited by several drugs, including sulphaphenazole and desethylamiodarone, the active metabolite of the antiarrhythmic amiodarone (Ohyama et al. 2000). CYP2C19 metabolizes proton pump inhibitors, such as omeprazole, the benzodiazepine diazepam and the anti-malarial proguanil to its active form cycloguanil. Some cardiovascular drugs such as propranolol (side-chain oxidation to naphthoxylactic acid) and R-warfarin are metabolized to a minor extent by CYP2C19. The anti-platelet drug, ticlopidine, is an inhibitor of this enzyme. CYP2C8 has not been as well studied as CYP2C9 or CYP2C19. It is known to metabolize both enantiomers of the calcium channel blocker, verapamil, and its metabolite norverapamil. However, this does not represent a very significant clearance mechanism in vivo, as other enzymes such as CYP3A4, although catalytically less active, play a much more important role because of their much higher levels of expression (Tracy et al. 1999; Shimada et al. 1994).

CYP2D6 is involved in the metabolism of a large number of drugs, often playing a dominant role. With respect to cardiovascular drugs, CYP2D6 is important because of its contribution to the elimination of many antiarrhythmic drugs and beta-blockers (Eichelbaum et al. 1997). It also plays a role in the oxidation of some calcium channel blockers and HMG-CoA reductase inhibitors. A number of drugs are known to inhibit CYP2D6, either competitively or non-competitively, such as quinidine. CYP2D6 does not appear to be inducible.

CYP2E1, together with alcohol dehydrogenase and aldehyde dehydrogenase, plays a role in the oxidation of ethanol (Lieber 1997). Substrates tend to be low-molecular-weight solvents, such as benzene and halothane. Although some cardiovascular drugs are metabolized by CYP2E1, this is a minor reaction, unlikely to be of any clinical significance. CYP2E1 is inhibited by disulfiram (Antabuse), which also inhibits aldehyde dehydrogenase (Frye and Branch 2002). CYP2E1 is inducible by solvents such as ethanol.

The mechanism of induction of members of the CYP2 family is less well understood than that of the CYP1 and CYP3 families, although significant progress has been made on some of the members such as CYP2E1. Some members such as CYP2B6 and some of the CYP2C sub-family can be induced via the constitutive androstane receptor (CAR), although there is significant cross-talk with the PXR receptor (see Sect. 3.4.1.3).

3.1.3
CYP3 Family

The CYP3A subfamily, the only sub-family in family 3 in humans, comprises four members: CYP3A4, CYP3A5, CYP3A7 and CYP3A43. CYP3A4 is expressed at high, but variable, levels in liver and small intestine. CYP3A5 is expressed at lower levels in liver, but exhibits a wider tissue distribution than CYP3A4. CYP3A7 is a fetal form that is not expressed at appreciable levels in the adult. CYP3A43 is expressed only at extremely low levels in liver and other tissues.

CYP3A4 is the most abundantly expressed P450 enzyme in human liver and gastrointestinal tract and is known to metabolize more than 120 frequently prescribed drugs (Nebert and Russell 2002), as well as many endogenous compounds, such as steroids and bile acids. CYP3A5 is present at lower levels than CYP3A4 and appears to share considerable specificity with CYP3A5, although it is often catalytically less active. Hence, for the majority of compounds metabolized by CYP3A, it is CYP3A4 that plays the dominant role. CYP3A4 and CYP3A5 have a very broad substrate specificity and can metabolize a diverse range of structures, often of a multi-ring nature. Indeed, there appears to be some overlap in those physicochemical properties favouring desirable absorption characteristics and those supporting CYP3A metabolism. However, the same characteristics favour ligand occupancy of P-glycoprotein, which tends to reduce intestinal absorption.

Classic substrates for CYP3A4 include erythromycin, midazolam and cyclosporine. Amongst cardiovascular drugs, CYP3A4 is responsible for the metabolism of numerous calcium channel blockers, particularly those possessing a dihydropyridine ring, some antiarrhythmic drugs such as amiodarone, and lipid-lowering drugs including HMG Co-A reductase inhibitors. The specificity of CYP3A5 has not been as thoroughly characterized as that of CYP3A4, but it has been shown to metabolize the calcium channel blocker verapamil, the HMG CoA reductase inhibitor lovastatin and the benzodiazepines diazepam and midazolam.

The location of CYP3A4 in both the liver and the small intestine means that the enzyme contributes to both presystemic and systemic elimination of drugs, with potential consequences for both bioavailability (e.g. terfenadine) and half-life (e.g. midazolam). CYP3A4 is potently inhibited by a number of azole antifungal drugs such as ketoconazole and itraconazole. Other inhibitors include cimetidine and diltiazem.

CYP3A4 is highly inducible by ligand-activated transcription via a steroid family receptor, the pregnane X receptor (PXR). The PXR response element, located in the upstream region of the CYP3A4 gene, is activated by dexamethasone and rifampicin-type of agents. CYP3A4 is also responsive to phenobarbital-like compounds, which act through the CAR receptor (Sueyoshi and Negishi 2001; Rushmore and Kong 2002). CYP3A5 is much less inducible than CYP3A4.

3.1.4
CYP4 Family

Members of the CYP4 family, CYP4A11, CYP4B1, CYP4F2 and CYP4F3, are involved primarily in the metabolism of fatty acids and arachidonic acids, although they can metabolize some drugs (Nebert and Russell 2002). Some CYP4A and CYP4B forms are expressed in the distal convoluted tubules of the kidney, and defects in some CYP4A cause alterations in salt metabolism and water balance (Simpson 1997).

3.2
P450 and Eicosanoid Metabolism

Members of all four families, CYP1, CYP2, CYP3 and CYP4, can metabolize arachidonic acid to epoxyeicosatrienoic acids (EETs), hydroxyeicosatetraenoic acids (HPETEs) and hydroperoxyeicosatetraenoic acids (HPETTs), and can also catalyse the oxidation of prostaglandin H2 to prostaglandins D2, F2α and E2. These eicosanoids play an important role in the cardiovascular system, the respiratory system, platelet aggregation and several other key endogenous processes. However, the relevance of most of these forms to the overall regulation of end-product levels is not yet known, but in many cases is likely to be trivial. In contrast, specific forms of P450 are uniquely responsible for both thromboxane A2 synthase activity, i.e. CYP5A1 and for prostacyclin synthase activity, i.e. CYP8A1. The respective products of these enzymes have opposing roles in platelet aggregation and on the cardiovascular system (Funk 2001).

3.3
Flavin-Containing Monooxygenase

Flavin-containing monooxygenses (FMOs) are a family of FAD, NADPH- and O_2-dependent enzymes that play a role in the oxygenation of nucleophilic compounds, include nitrogen-, sulphur-, phosphorous- and selenium-heteroatom-containing drugs, xenobiotics and endogenous substrates. To date, six different FMO genes have been identified. FMO3 is prominent in adult liver and is associated with the majority of FMO-mediated metabolism (Cashman and Zhang 2002). FMO1 is expressed in the kidney and intestine (Yeung et al. 2000) and FMO2 is dominantly expressed in lung (Krueger 2002). The roles of these forms in the metabolism of drugs, if any, are not yet known. Substrates for FMO3 include (S)-nicotine, trimethylamine, clozapine, cimetidine and ranitidine. FMO3 is not inducible (Cashman and Zhang 2002).

3.4
Monoamine Oxidase

Monoamine oxidase (MAO), which oxidizes amines such as tyramine and catecholamines from endogenous and exogenous sources, is present as two forms, MAO A and MAO B. These are located in the outer mitochondrial membrane of neuronal and other cells in the brain and the periphery, although there is tissue variability in the expression of the two forms. MAO enzymes are encoded by separate genes on the X chromosome. MAO has been a therapeutic target for many years, e.g. for the treatment of Parkinson's disease and depression (Cesura and Pletscher 1992), and hence a number of form-specific inhibitors are available, acting either reversibly or irreversibly. These include clorgyline (MAO A) and L-deprenyl (MAO B). There is considerable overlap in substrate specificities. This extends to the biogenic amines serving as neurotransmitters, dopamine, norepinephrine and serotonin (Binda et al. 2002). However, MAO A preferentially deaminates adrenaline, noradrenaline and serotonin, whilst MAO B preferentially deaminates phenylethylamine and benzylamine. Dopamine and tyramine serve as substrates for both forms of MAO. Apart from neuronal cells, in which MAO A is the predominant form, and circulating platelets, in which MAO B predominates, the two forms show a similar widespread tissue and cellular distribution, with particular expression in placenta, heart, liver, kidney, lung and small intestine. MAO B in the peripheral tissues may play a role in the oxidation of exogenous catecholamines such as intravenous dopamine (Yan et al. 2002).

3.5
Alcohol Dehydrogenase and Aldehyde Dehydrogenase

Alcohol dehydrogenases (ADHs) are NAD*H*-dependent enzymes involved in the elimination of endogenous and exogenous ethanol and other alcohols, aldehydes, including acetaldehyde, products of lipid peroxidation, other xenobiotics and products of their metabolism (Ashmarin et al. 2000). The conversion of alcohols to aldehydes is a reversible reaction, and ADH can catalyse the reaction in either direction, depending on reactant concentrations. There are five classes of ADH, each with a characteristic tissue-specific expression. Class I enzymes (ADH1A, B, C) are abundant in liver. These all possess substantial ethanol-oxidizing activity and a wide substrate specificity. Substrates include bile acids, testosterone, neurotransmitters, retinol, peroxidatic aldehydes and mevalonate (Holmes 1994). Class IV (ADH4) has high activity in stomach and cornea, and is involved in the metabolism of alcohols and retinoids. ADH4 is the only form other than ADH1 that plays an important role in the metabolism of ingested ethanol (Jörnvall et al. 2000). Class II (ADH2) is involved in the metabolism of peroxidatic aldehydes, norepinephrine, mevalonate and congeners. Class III (ADH3) functions in formaldehyde and omega-hydroxyfatty acid metabolism. Class V (ADH5) has been less well studied.

Aldehyde dehydrogenases (ALDHs) comprise a large family of NADH(P)-dependent enzymes and catalyse the oxidation of a wide range of endogenous and exogenous aliphatic and aromatic aldehydes. ALDH2 plays a role in acetaldehyde oxidation (Vasiliou and Pappa 2000). Several forms of ADH (e.g. ADH1, ADH2, ADH3) and of ALDH are inducible by alcohol and other low-molecular-weight solvents. ADH is capacity limited at the range of doses of ethanol that are ingested, which gives rise to the potential for competitive inhibition by alternate substrates.

3.6
Hydrolases

Hydrolases catalyse the addition of water to a wide range of compounds and most exhibit a very broad substrate specificity, restricted often more by functional group (e.g. epoxide, amide, ester) than overall chemical structure. Hydrolytic enzymes are widely distributed throughout the body, including plasma, although expression does vary with cell and tissue type. Such enzymes, particularly in the liver, small intestine and sometimes plasma, play an important role in the hydrolysis of several cardiovascular drugs including acetylsalicylic acid, clofibrate and the angiotensin-converting enzyme inhibitor ester-prodrugs, enalapril, benazepril, delapril and temocapril. There are at least three different carboxylesterases that participate in the hydrolysis of esters and/or amides. In addition, there are several cholinesterases and at least two epoxide hydrolases (EPHXs) that can participate in the metabolism of drugs.

3.6.1
Epoxide Hydrolase

Oxidation by one or more enzymes often results in the formation of reactive, xenobiotic epoxides. Epoxide hydrolases (EPHXs) are important for the ultimate detoxication of these intermediates and are a subcategory of a broad group of hydrolytic enzymes that include esterases, proteases, dehalogenases, and lipases. Substrates of microsomal EPHX (EPHX1) include toxic and procarcinogenic compounds, as well as commonly used anticonvulsant drugs, phenytoin and carbamazepine. Cytosolic EPHX (EPHX2) is a xenobiotic metabolizing enzyme that also participates in the metabolism of endogenously derived fatty acid epoxides (Fretland and Omiecinski 2000).

3.7
Phase II Metabolism

3.7.1
UDP-Glucuronosyltransferase

Glucuronidation is a major pre-excretory metabolic pathway leading to conjugates that are more water soluble than the unconjugated substrates, resulting in their more ready excretion from the body. Glucuronidation is catalysed by UDP-GA-dependent, endoplasmic reticular UDP-glucuronosyltransferases (UGTs). Although exhibiting wide tissue distribution, they are most active in the liver. UGTs comprise two multi-member families, and vary in their substrate specificity and regulation. UGTs play an important role in the deactivation and elimination of a large number of endogenous compounds such as bilirubin, steroids and bile acids (McGurk et al. 1998). They also metabolize a wide variety of drugs, and more particularly the products of their phase I metabolism. Drugs conjugated directly by UGTs include nonsteroidal anti-inflammatory agents, morphine and tocainide. Many drugs, for example mexiletine, have to be oxidized before conjugation with glucuronide and final elimination from the body. The drug most commonly involved in inhibitory interactions is probenecid, although furosemide, salicylic acid and oxazepam have also been reported to inhibit glucuronidation (Grancharov et al. 2001; Irshaid et al. 1990). Some of the UGTs are inducible, but generally not to the same extent as P450 enzymes.

3.7.2
Sulphotransferase

Sulphotransferases (SULTs) catalyse the transfer of the sulphate moiety from the cofactor 3′-phosphoadenosine-5′-phosphosulphate (PAPS) to nucleophilic groups of suitable substrates, which include drugs and endogenous small molecules such as hormones and neurotransmitters. Nine SULTs in three families (SULT1, 2, 4) have been described in humans. There is very broad and overlapping substrate specificity, together with wide tissue distribution, although there are important differences amongst the forms. There is considerable overlap between SULTs and UGTs in their substrate specificity, many compounds acting as substrates equally well for the two enzyme systems. Like UGTs, the SULTs conjugate a range of both endogenous and exogenous compounds, often products of phase I metabolism. The sulphate conjugates are much more water soluble and are readily excreted. SULT1A1 expression is very high in liver, although there is some expression in most other tissues including brain. Substrates include 4-nitrophenol, minoxidil, dopamine and epinephrine. SULT1A3 is highly expressed in the jejunum and colon, but again there is some expression in other tissues including brain and platelets. However, expression of this form in adult liver is negligible. Substrates include catecholamines, such as dopamine, norepinephrine and isoprenaline. SULT1B1 is expressed particularly in colon, liver and intestine. It is involved in the conjugation of iodothyronines. SULT2A1 is expressed in liver and adrenal gland. Substrates include dehydroepiandrosterone and minoxidil (Glatt et al. 2001). In general, SULTs do not appear to be inducible, although there are reports that expression of some forms can be increased by glucocorticoids.

3.7.3
N-Acetyltransferase

Two forms of arylamine N-acetyltransferase (NAT) have been well characterized, NAT1 and NAT2, which are localized to chromosome 8. Recently, another two NAT genes have been reported, *NAT5* and *NAT8* but nothing is known about the function of these two genes. NAT1 and NAT2 catalyse the acetylation of amino-, hyrazino- and N-hydroxy moieties present in hydrazine and arylamine drugs. NAT1 is widely distributed in the body, whereas NAT2 expression is confined largely to the liver. NAT1 acetylates simple arylamines, such as 4-aminobenzoic acid. It also conjugates a catabolite of folic acid (Grant et al. 2000). Substrates for NAT2 include isoniazid, hydralazine, procainamide and a metabolite of caffeine. The acetylation product of procainamide, N-acetylprocainamide is a biologically active metabolite and has antiarrhythmic properties. Both NAT1 and NAT2 play an important role in the metabolism of environmental chemicals, including aromatic and heterocyclic amines. For example, the NAT2 slow acetylator phenotype has been associated with higher risk of bladder cancer, whilst the NAT2 fast acetylator phenotype has been reported to be at higher risk of colon cancer in many studies (Grant et al. 2000; Hein 2002).

3.7.4
Glutathione-S-transferase

Glutathione is the major non-protein thiol in cells, comprising the tri-peptide gamma-glutamyl-cysteinyl-glycine, and has many roles in cellular defence and metabolism, including protecting the cell from oxidative stress, detoxication of electrophilic species, and maintenance of thioldisulphide status of cellular proteins (Meister 1995). Glutathione-S-transferases (GSTs) are a superfamily of at least eight different enzyme families. The active enzyme is a dimer, often but not always, of identical sub-units. Gene families are designated by a letter A, K, M, O, P, S, T and Z, to give *GSTA*, etc. Individual members are indicated by an Arabic numeral, e.g. *GSTA1*. The gene products are referred to as alpha, kappa, mu, omega, pi, sigma, theta and zeta, respectively. These enzymes catalyse the conjugation of electrophilic, often cytotoxic agents, to glutathione, producing less reactive chemical species. Resistance to cancer chemotherapeutic drugs, such as the alkylating agents, has been directly correlated with the overexpression of GSTs. In particular, overexpression of GST π has been linked to a number of different human cancers, and this enzyme plays a non-catalytic role in cellular pathways of proliferation, stress response and apoptosis (Tew and Ronai

1999). GST α and GST σ synthesize PGD2 and PGE2, and microsomal GSTs metabolize arachidonic acids (Hayes and Strange 2000).

Ethacrynic acid is both an inhibitor and a substrate for GSTs, and its glutathione conjugate is also an enzyme inhibitor (Pleomen et al. 1990). Clofibrate also inhibits GSTs (Foliot et al. 1984). GST inhibitors would be expected to alter the efficacy of alkylating agents by interfering with GST-mediated conjugation.

3.8
Thiopurine Methyltransferase

Thiopurine methyltransferase (TPMT) is a cytoplasmic enzyme that catalyses the S-methylation of a variety of toxic thiopurine drugs such as 6-mercaptopurine, 6-thioguanine and azathioprine, mainly in liver. There have been reports that low TPMT activity is associated with an increased risk of toxicity of these drugs used to treat acute lymphocytic leukaemia, autoimmune disorders, inflammatory bowel disease and transplantation rejection (Weinshilboum 2001). It was also reported that a TPMT-deficient heart transplant recipient died of sepsis with leukopenia after treatment with a conventional dose of azathioprine (Schütz et al. 1993). Benzoic acid derivatives, such as salicylic acid, are potent inhibitors of TPMT. After a therapeutic dose of aspirin, plasma concentrations of salicylic acid are within the range for TPMT inhibition. Sulfasalazine and its metabolite 5-aminosalicylic acid inhibit TPMT, and concurrent furosemide therapy could influence the S-methylation of thiopurines (Lennard 1998).

3.9
Catechol-O-Methyltransferase

Catechol-O-methyltransferase (COMT) is an intracellular enzyme widely distributed in the body and probably best known for its activity in methylating catecholamines (Evans 1993). The substrates of COMT include L-DOPA, norepinephrine, epinephrine, dopamine, dobutamine, isoprenaline, and $α$-methyldopa. The function of COMT is generally the elimination of active or toxic catechols and some other hydroxylated metabolites. COMT also acts as an enzymatic detoxication barrier between the blood and other tissues such as the intestinal mucosa and the kidney. COMT may also modulate the dopaminergic tone in brain, kidney and intestine (Mannisto and Kaakkola 1999). The role of COMT in the brain has been well studied. The relative importance of the enzymes involved in the metabolism and the uptake of endogenous catecholamine are clear, but their role in extraneuronal transport is less clear (Mannisto and Kaakkola 1999). Inhibition of COMT failed to show a significant alteration of the elimination of infused catecholamines (Friedgen et al. 1996).

3.10
Histamine N-Methyltransferase

This enzyme catalyses the *N*-methylation of histamine, and is the major pathway for histamine metabolism. It may constitute an important modulating factor in some histamine-related diseases such as allergy and asthma (Evans 1993).

4
Drug Transporters

In addition to the enzymes of xenobiotic metabolism, the disposition of drugs can be affected by the activity of drug transporters. This superfamily comprises almost 50 members, organized into seven families. All are membrane-spanning ATP-binding cassettes (ABC transporters). Some members, such as ABCB1 (MDR1 gene product, P-glycoprotein), are involved in the translocation of drugs and other foreign compounds across cell membranes, with appreciable expression in the small intestine, placenta, blood–brain and blood–testis barrier. Substrates include anticancer agents, cardiac drugs (e.g. digoxin, quinidine), HIV protease inhibitors, immunosuppressants (e.g. cyclosporine), and β-adrenoceptor antagonists (Borst et al. 2000). The presence of P-glycoprotein in the small intestine limits the oral bioavailability of substrates, whilst that in the blood–brain barrier reduces penetration to the CNS. There is substantial overlap in substrate specificity between P-glycoprotein and CYP3A4. Like CYP3A4, P-glycoprotein is inducible by compounds such as rifampicin, with similar consequences for the oral bioavailability of substrates. Some ligands act as inhibitors of P-glycoprotein, for example quinidine, leading to increased bioavailability of substrates such as digoxin.

5
Polymorphisms of Enzymes of Drug Metabolism

5.1
Genotype and Phenotype

The possibility that enzymes of drug metabolism might be subject to genetic variation was first suggested by Garrod almost a century ago (Garrod 1914). However, it was not until the late 1950s that specific examples of genetic variation in enzymes of drug metabolism were identified, with the discovery of anomalous succinylcholine hydrolysis and the polymorphic acetylation of isoniazid. Despite the interest raised by such observations, and increasing interest in the genetic basis of disease and drug response, it was not until the mid 1970s that the modern era of pharmacogenetics was born, with the simultaneous and independent discovery of the debrisoquine/sparteine oxidation polymorphism by groups in the UK and Germany, respectively. It was subsequently demonstrated that this is due to genetic heterogeneity of a specific form of P450, later identified as CYP2D6. In the years subsequent to the first published report of the CYP2D6 polymorphism in 1977, there has been an ever increasing number of reports of polymorphisms of enzymes of drug metabolism, with CYP2C19 in 1984, glutathione S-transferase is 1986, FMO3 in 1990 and CYP2C9 in 1995, although evidence for genetic variation in this enzyme was reported as early as 1964 (Kutt et al. 1964).

In each of these cases, the polymorphism was identified phenotypically, i.e. from a functional change in the gene product such that it was reflected in either a response to the drug (e.g. succinylcholine-induced paralysis, debrisoquine-induced orthostatic hypotension) or in its metabolic fate (e.g. acetylation of isoniazid, hydrolysis of paraoxon). Enzyme activity and drug response may vary for any one of a number of reasons, not always genetic, but once a genetic basis for such variation has been established, for example by family and population studies, it is axiomatic that the genotype has a functional consequence.

With the output of the human genome project, there is considerable interest in identifying the nature and extent of genetic variation between individuals and populations. This increasingly relies on intensive sequencing efforts, to identify differences in single nucleotides in the genomes from different individuals. Where these occur with a frequency greater than 1% within the population, they are termed polymorphisms (single nucleotide polymorphisms, SNPs[3]). The total number of such SNPs within the population is estimated to be several million, with a density of up to approximately 1 per 300 nucleotide pairs. Both broad range SNP mapping exercises and specific candidate gene analysis are being performed, resulting in the publication of large numbers of polymorphisms for almost any gene of interest. One of the challenges facing biomedical science in the coming years is to determine the functional significance, if any, of polymorphisms in such genes. Whilst identification of SNPs is now relatively rapid, establishing their consequences is often much more time-consuming (see Fig. 3).

SNPs can occur anywhere in the genome. With respect to specific genes, SNPs may occur in upstream regulatory regions (affecting transcription), introns and downstream regions (affecting mRNA processing) or in exons (affecting amino acid sequence). Hence, the consequences of a SNP may be altered expression or function of the gene product. Where a gene itself codes for a DNA or RNA regulatory protein, functional changes can lead indirectly to altered expression of drug metabolizing enzymes. Whilst changes in amino acid coding or in transcription binding sites raise suspicion that they will give rise to a phenotype, current knowledge is such that this is not yet routinely predictable. In addition, not all DNA regulatory sequences have yet been identified, so that some SNPs in

---

[3] The information on SNPs described here has been obtained from the following databases http://www.ncbi.nlm.nih.gov/SNP/, http://www.genome.utah.edu/genesnps/, and http://snp.cshl.org/ and is correct at the time of writing. However, as these databases are updated regularly, it is highly likely that there will be changes in these SNP databases over time.

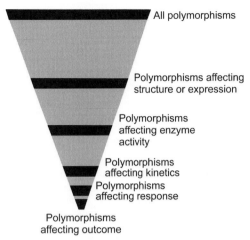

Fig. 3 The importance of genetic polymorphisms in the outcome of therapeutic intervention. The human genome is very polymorphic. Throughout the entire genome, there are several million different polymorphisms within the population. Only some of these will occur in either regulatory regions or in the coding region of a protein. Of these, only a sub-set will affect the activity of the enzyme, either because expression is not significantly affected, the mutation is such that the new codon codes for the same amino acid, the substituted amino acid is in a non-functional domain of the protein or the substitution is for an amino acid sufficiently similar to the wild-type that it has no effect on protein function. Of those polymorphisms affecting enzyme activity, only some will alter the kinetics of a drug, often because of competing pathways in the elimination of the compound. Only some pharmacokinetic polymorphisms will affect the pharmacological response to the compound, perhaps because of the shape of the concentration–effect curve, or because there is no direct relationship between plasma concentration and effect. Even when the pharmacological response exhibits a difference between phenotypes, this does not always translate into a demonstrable difference in therapeutic outcome, because of the importance of other factors as determinants of outcome. Hence, the impact of any given polymorphism at the molecular level can be determined only by consideration of the functional and clinical consequences of that polymorphism

inter-genic regions, whilst less likely to lead to a phenotype, may still have functional consequences.

It is important to distinguish between population-based genetic analysis of a disease or drug response and that of the individual patient. Whilst linkage analysis and SNP mapping may be extremely useful in identifying subtle and/or multi-genic factors in modulating response, individualization of patient therapy requires more careful consideration of the likely impact of a specific genotype (or cluster of genotypes, i.e. haplotype) on therapeutic outcome, in terms of patient benefit. This may only become apparent after appropriate clinical studies (Fig. 3).

As indicated in the introduction, drug effect depends on target tissue concentration, and this often parallels plasma concentration. Hence, for drugs relying on metabolism for a significant component of their elimination, changes in the activity of the enzyme(s) involved will lead to a corresponding change in effect. Where that enzyme is subject to genetic polymorphism, this may be reflected in

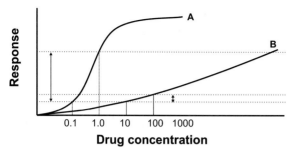

Fig. 4 The importance of the shape of the concentration–effect curve in determining the consequences of a pharmacogenetic polymorphism. Two drugs (A and B) both exhibit a good relationship between drug concentration in plasma and pharmacological response. However, the nature of this relationship differs for the two compounds, with B showing a much flatter response. A polymorphism resulting in a tenfold increase in plasma concentration would result in an increase in response shown by the vertical arrows. Whereas this is appreciable for A, for B the increase in response in trivial, and as such often would not translate into a discernible difference between the phenotypes

a genetically determined difference in effect. The magnitude of this effect, if any, will depend on the steepness of the concentration-effect curve (Fig. 4) and on the width of the therapeutic window (Fig. 2). Hence, polymorphisms of drug-metabolizing enzymes may lead to under- or over-dosing, with decreased or increased primary pharmacology, or to increased adverse effects due to secondary pharmacology or toxicology.

Until the start of the last decade (early 1990s), genetic analysis of enzymes of drug metabolism was largely on the basis of phenotype, using suitable probe substrates for the purpose. As the molecular basis of many polymorphisms was determined, and molecular biological techniques became more widely disseminated, phenotyping gave way to genotyping. For most genes, a relatively small number of mutations are responsible for the majority of affected alleles. Usually, genotyping strategies analyse for all of these. In large-scale studies, such as population screening, a range of high-throughput methods is becoming available, often based on mass spectrometry. There are advantages and disadvantages to both phenotyping and genotyping. Phenotype assesses function, which may be regarded as a major advantage, but this can be affected by factors other than genetics, and is sometimes not stable over time, even within an individual. Genotype overcomes this difficulty, but does depend on assessment of relevant alleles. Whilst studies to date have focused on alleles implicated in phenotypic expression, increasingly SNPs are being investigated where the phenotypic consequence, if any, is not known. As indicated earlier, SNPs can affect either expression or function. However, this is not necessarily an absolute effect, so that whilst for some alleles the result is the complete absence of functional protein, due to non-expression, instability of product or to critical changes in protein structure, for other alleles the result is altered (increased or decreased) expression or function (increased or decreased). Changes can be quite subtle, with a change in specificity or differential changes in activity, depending on substrate, as exemplified in some of the polymorphisms of the esterase PON1 (Fig. 3).

When assessing phenotype with a probe drug (a substrate specific for the polymorphic enzyme or one that produces a metabolite specific to that enzyme), the drug is often administered a few hours before collecting a plasma and/or urinary sample. The exact details of the protocol should be determined in optimization studies to establish the most robust conditions for the test. The parameter used to define phenotype should be established in population studies designed for that purpose, preferably including subjects of defined genotype. The most suitable parameter will vary with the probe. Parameters that have been used include clearance of parent drug, plasma or metabolite/parent or parent/metabolite ratio, urinary metabolite levels, urinary metabolite/parent or parent/metabolite ratio and even the enantiomeric ratio of racemic compounds subject to stereoselective metabolism by the polymorphic enzyme. There are usually sound pharmacokinetic and/or logistical reasons for the choice of a specific parameter, and these have been discussed in detail elsewhere (e.g. see Streetman et al. 2000). One of the most widely used parameters is the metabolic ratio (MR), which is the ratio of parent drug to metabolite in urine over a defined period of time. This ratio varies inversely with metabolic activity, as the metabolite is in the denominator. Whilst the ratio is widely used as an inverse measure of metabolic activity, it is important to recognize that the numerator reflects the renal clearance of parent compound. Although this generally shows much less variation than metabolism, there are of course factors that can selectively alter this parameter, and hence the MR, independent of genetics.

5.2
CYP2D6

CYP2D6 is probably the polymorphic enzyme with the greatest impact on the metabolism of drugs. It is responsible for the metabolism of more than 50 frequently prescribed drugs, including many used to treat cardiovascular conditions (Eichelbaum et al. 1997). Often CYP2D6 is one of several CYP enzymes responsible for the metabolism of a drug, such as is the case for mexiletine and propranolol. However, there are also many drugs that are metabolized almost exclusively by CYP2D6, including a number of those used as beta-adrenoceptor antagonists, antiarrhythmics, and sodium channel blockers (Table 1). In these cases, the effect of the CYP2D6 polymorphism may have important clinical consequences.

The activity of CYP2D6 can be determined in vivo using debrisoquine or dextromethorphan (or other suitable compounds) as a probe drug. A number of studies using this method have shown a distinct bimodal distribution of metabolic ratios in Caucasians, with 5%–10% being poor metabolizers (PMs) of CYP2D6 substrates (Eichelbaum and Gross 1990; Meyer 1994). In contrast, the PM phenotype in Orientals is rare (0%–1%) (Sohn et al. 1991). The phenotype in Africans has been reported to vary widely (0%–19%) between different coun-

Table 1 Examples of cardiovascular drugs subject to polymorphic metabolism

Drug	Pharmacological action	Metabolism	Enzyme(s)
Alprenolol	Beta-blocker	Aromatic hydroxylation	2D6
Amiodarone	Antiarrhythmic	De-ethylation	2C9, 3A4, 1A2
Amlodipine	Ca channel blocker	Dihydropyridine ring oxidation	3A4
Aprindine	Antiarrhythmic	Aromatic hydroxylation	2D6
Dihydropyridine	Ca channel blocker		3A4, 2D6, 2C9
Carvedilol	Beta-blocker	Hydroxylation	2D6
Debrisoquine (withdrawn)	Antihypertensive	4-Hydroxylation	2D6
Diltiazem	Ca channel blocker	Demethylation	3A4
Encainide	Antiarrhythmic	O-demethylation to active metabolite	2D6
Felodipine	Ca channel blocker	Dihydropyridine ring oxidation	3A4
Flecainide	Na channel blocker	O-dealkylation	2D6
Isradipine	Ca channel blocker	Dihydropyridine ring oxidation	3A4
losartan	Angiotensin II inhibitor	Oxidation to active metabolite	2C9 (activation)
Metoprolol	Beta-blocker	Aliphatic hydroxylation and O-dealkylation	2D6
Mexiletine	Na channel blocker	Hydroxylation	2D6, 1A2
Mibefradil (now withdrawn)	Selective T-type Ca blocker	Hydroxylation, demethylation, depropylation	3A4, P-glycoprotein
Nicardipine	Ca channel blocker	Dihydropyridine ring oxidation	3A4
Nifedipine	Ca channel blocker	Dehydrogenation	3A4
Nisoldipine	Beta-blocker	Dihydropyridine ring oxidation	3A4
Propafenone		Aromatic hydroxylation, N-dealkylation, glucuronidation	2D6, 3A4, UGT
N-propylajmaline	Antiarrhythmic	Benzylic hydroxylation	2D6
Procainamide	Na channel blocker	N-hydroxylation, N-de-ethylation	2D6
Propafenone	Antiarrhythmic	Depropylation	1A2
Propranolol	Beta-blocker	4-Hydroxylation, N-dealkylation, side chain oxidation, gluronidation	2D6, 2D6, 1A2, 2C19, UGT
Quinidine	Antiarrhythmic	3-Hydroxylation, N-oxidation	3A4, 3A5
Sparteine	Antiarrhythmic	Oxidation	2D6
Ticlopidine	Platelet aggregation inhibitor	Hydroxylation	2C19
Timolol	Beta-blocker	Dealkylation	2D6
Verapamil	Ca channel blocker	N-demethylation, O-demethylation	3A4, 1A2, 2C9
(S)-warfarin	Anticoagulant	Hydroxylation	2C9
(R)-warfarin	Anticoagulant	Hydroxylation	1A2, 2C19

Note that under "Enzyme(s)" the number/letter combinations refer to P450 enzymes and should be preceded by CYP.

tries (Eichelbaum and Gross 1990), although most recent studies suggest that in the majority of black Africans the occurrence is relatively low (0%–2%) (Gaedigk et al. 1991). Inter-ethnic differences have also been observed for extensive metabolizers (EMs). Caucasians have a mean MR of 0.6, whereas Orientals have a mean value of 1.0, reflecting a slower rate of metabolism of CYP2D6 substrates (Fig. 5). Overall, Africans also metabolize CYP2D6 substrates more slowly than Caucasians, with a mean MR of 1.0. However, studies performed in dif-

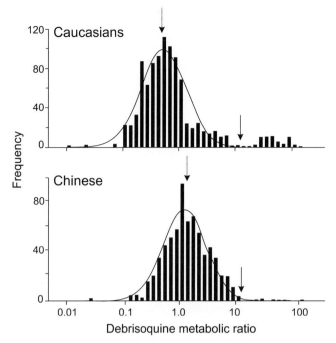

Fig. 5 Interethnic differences in the population distribution curve for CYP2D6-dependent metabolism. Data have been taken from Bertilsson et al. (1992). Several hundred Caucasian and Chinese subjects were administered debrisoquine to determine their CYP2D6 phenotype. Data have been expressed as the urinary debrisoquine metabolic ratio (*MR*), the ratio of parent compound to 4-hydroxydebrisoquine. This parameter is inversely related to the metabolic activity of CYP2D6, so that subjects towards the right of the distribution curves have low activity. The distribution in Caucasians is clearly bimodal, with an antimode at an MR of 12 (*black vertical arrow*). Subjects with an MR above 12 are classified as PMs whilst those with MRs below 12 are classified as EMs. This phenotypic classification has been verified in other studies by genotyping. Subjects with MRs around 12 cannot be classified by phenotyping. Note the presence of a small number of ultra-rapid metabolizers with MRs less than 0.1. In the Chinese population, there are far fewer subjects with MRs above 12, and the relative absence of PM subjects in oriental populations has been confirmed by genotyping in other studies. The *grey vertical arrows* indicate the median values for the two populations. It is apparent that whilst there are very few PM subjects in the Chinese population, the median MR in EM subjects is greater than that in Caucasian EM subjects. It is now apparent that this is due to the high prevalence of an allele in Chinese populations (*CYP2D6*10*) with reduced activity compared with the wild-type protein (encoded by *CYP2D6*1*), which is prevalent in EMs in the Caucasian population (see Table 2)

ferent countries showed marked variations (0.6–1.5) underlining the extent of ethnic heterogeneity in African populations (Masimirembwa and Hasler 1997). In addition to this overall trend, there is also a subgroup of subjects that have a very low MR (<0.2) known as ultrarapid metabolizers (Dahl et al. 1995). This phenotype is found in Caucasians, but not in Orientals or most black Africans, although it does occur in Ethiopians and Egyptians (Mahgoub et al. 1979; Aklillu et al. 1996).

Table 2 Inter-ethnic differences in frequencies of some common alleles for P450 genes

Allele	Caucasians	Orientals	Africans	Ethiopians/Saudis
	Allele frequency (%)			
CYP2C9*2	8–13	0	2–3	4
CYP2C9*3	4–9	2–3	1	2
CYP2C19*2	13	23–32	13–18	14–15
CYP2C19*3	0	6–10	0–1	0–2
CYP2D6*2xn	1–5	0–2	2	10–16
CYP2D6*4	12–21	1	2	1–4
CYP2D6*5	2–7	6	4	1–3
CYP2D6*10	1–2	51	6	3–9
CYP2D6*17	0	ND	34	3–9

Data in this table have been compiled from Aklilliu et al. 2002; Dickmann et al. 2001; Goldstein et al. 1997; Ingelman-Sundberg 2001; Masimirembwa et al. 1996; Persson et al. 1996; Scordo et al. 2001; Xie et al. 1999.

The low frequency of inactivating alleles of CYP2D6 (*4 and *5) in Chinese and African populations is reflected in the small number of PM subjects in these populations. Similarly, there are fewer CYP2C9 alleles coding for enzyme with reduced activity in Chinese and Black populations, with correspondingly fewer PMs for this for of P450. The high frequency of alleles coding for CYP2D6 enzyme with reduced activity (*10 and *17, respectively) in Chinese and African populations results in a shift in the median activity of CYP2D6 in EM subjects in these populations to lower values than in Caucasians (higher metabolic ratios). The frequency of inactivating alleles of CYP2C19 in Chinese subjects is greater than in Caucasians, resulting in more PM subjects for this enzyme in this population.

The molecular genetic basis of the CYP2D6 polymorphism has been studied extensively. At present, well over 50 allelic variants of the CYP2D6 gene have been identified, although many of these occur in only a very small number of individuals. SNP analysis of the CYP2D6 gene has revealed at least 135 polymorphic sites, the consequences of many of which are unknown. From the alleles identified in PM subjects, it has been shown that the PM phenotype arises mainly because of inheritance of the *CYP2D6*4* allele (Table 2). This allele contains a splice-site mutation and as a consequence does not produce a functional enzyme (Kagimoto et al. 1990). In addition, the *CYP2D6*3* allele, which contains a frame shift mutation, also produces a PM phenotype (Kagimoto et al. 1990). Both of these alleles occur in Caucasians, but not in Orientals or black Africans (Table 2). However, a third allele, *CYP2D6*5*, which is a gene deletion mutant (Gaedigk et al. 1991; Steen et al. 1995), occurs at a similar frequency in Caucasians, Orientals and black Africans (Table 2). Hence, the low occurrence of the PM phenotype in Orientals and black Africans is a result of inheritance of only the *CYP2D6*5* allele amongst these ethnic populations and the higher rate amongst Caucasians is due to inheritance of *CYP2D6*3*, *CYP2D6*4* and *CYP2D6*5* alleles.

The higher MR in Orientals compared with Caucasians is due to the *CYP2D6*10* allele (Table 2). This allele contains a mutation at codon 34 resulting in a single amino acid change *Pro→Ser* and as a consequence produces an unstable enzyme with a lower activity (Yokota et al. 1993). This allele is present at a frequency of 50% in Oriental populations, but only 5% in Caucasians and Africans (Armstrong et al. 1994). The slower rate of metabolism in Africans is due to the *CYP2D6*17* allele (Table 2). This allele contains three mutations causing amino acid changes, i.e. *107Ser→Thr, 206Arg→Cys* and *486Ser→Thr* and produces an enzyme with a lower affinity for CYP2D6 substrates (Oscarson et al. 1997). This allele is present in 15%–34% of Africans, but is essentially absent in both Caucasians and Orientals (Masimirembwa et al 1996; Droll et al. 1998).

Ultrarapid metabolizers occur as a result of duplication of the *CYP2D6*2* allele. The enzyme produced by the *CYP2D6*2* allele has a similar activity to that produced by the wild-type *CYP2D6*1* allele. However, certain individual may inherit 2, 3, 4, 5 or as many as 13 gene copies arranged in tandem and thus produce proportionately higher amounts of enzyme (Johansson et al. 1993).

There is a 2- to 5-fold difference between PM and EM individuals in their capacity to metabolize drugs that are substrates for CYP2D6. Therefore, for drugs such as debrisoquine and sparteine that are oxidized by CYP2D6 and not to any great extent by other CYP enzymes, the drug serum concentration is highly dependent on this activity. So, individuals that have a PM phenotype will require 50%–80% less drug to achieve the same serum levels as individuals with an EM phenotype. By the same token, if the dose administered to PM subjects is not reduced then they are more likely to risk severe side-effects or drug toxicity. Several such drugs, including debrisoquine, have been withdrawn from the market or had their use restricted as a result of such complications. At the other extreme, ultrarapid metabolizers are likely to metabolize drugs that are CYP2D6 substrates so rapidly that therapeutic serum concentrations can be reached only with difficulty. Conversely, for compounds such as encainide where pharmacological activity resides mainly in the metabolite (in this case *O*-desmethylencainide), it is EM subjects that might have an increased risk of proarrhythmic events and this may be particularly so for ultrarapid metabolizers. Even amongst EMs, a range of activities is found, with Orientals and Africans having lower activities than Caucasians. This suggests that the doses given to individuals from different ethnic backgrounds might need to be adjusted appropriately.

5.3
CYP2C19

CYP2C19 catalyses the 4-hydroxylation of *S*-mephenytoin, but not R-mephenytoin (Goldstein and de Morais 1994). The activity of CYP2C19 in vivo can be determined following the administration of a dose of the racemic drug and measurement of the urinary ratio of the *S*- and R-enantiomers or measurement of the level of 4-hydroxy-*S*-mephenytoin compared to *S*-mephenytoin (Wedlund et al. 1984). Such determinations have shown that a polymorphism in CYP2C19 activity exists, with 2%–6% of Caucasians (Kupfer and Preisig 1984; Wedlund et al. 1984), 14%–22% of Orientals (Wilkinson et al. 1989) and 4%–8% of black Africans (Herrlin et al. 1998; Bathum et al. 1999) having a PM phenotype.

Molecular genetic analysis has shown that the PM phenotype is due mostly to two mutated alleles, although at least fifteen alleles for this gene are now known (approx. 80 SNPs). The *CYP2C19*2* allele contains an aberrant splice site mutation resulting in a defective CYP2C19 protein with no enzymatic activity. This allele accounts for the PM phenotype in 83% of Caucasians, 75% of Orientals and 75% of black Africans (de Morais et al. 1994). The *CYP2C19*3* allele contains a premature stop codon and consequently no functional CYP2C19 protein. This allele is present in most Japanese PM subjects who do not carry the *CYP2C19*2* allele (Table 2). Therefore, the PM phenotype in Orientals is due to inheritance of either of these two alleles. However, *CYP2C19*3* is not found in Caucasians or black Africans and other *CYP2C19* alleles identified so far are quite rare (Table 2). Therefore, it appears that other alleles of functional significance remain to be discovered.

The clinical significance of polymorphisms in CYP2C19 for cardiovascular drugs is limited as few are substrates for this enzyme. Although propranolol is metabolized by CYP2C19, other CYP enzymes, including CYP2D6, are also involved. Consequently, propranolol metabolism is likely to be reduced only in those individuals who are PMs of both CYP2C19 and CYP2D6.

5.4
CYP2C9

Two major allelic variants (from a total of 12 described to date; 60 SNPs) of the CYP2C9 gene that affect enzyme activity have been identified in the Caucasian population. *CYP2C19*2* contains a point mutation resulting in *144Arg→Cys* and *CYP2C19*3* contains a different point mutation resulting in *359Ile→Leu* (Rettie et al. 1994; Sullivan-Klose et al. 1996). The enzymes produced by these allelic variants have approximately 12% and 5%, respectively, of the activity of the wild-type enzyme produced by the *CYP2C9*1* allele (Rettie et al. 1994; Haining et al. 1996). The *CYP2C9*2* allele is present in 8%–12% of Caucasians but is rare amongst Orientals, and the *CYP2C9*3* allele occurs in 6%–10% of Caucasians and approximately 3% of Orientals (Table 2).

The most notable significance of this polymorphism is the reduced ability of subjects that have the *CYP2C9*2* and/or *CYP2C9*3* alleles to metabolize the anticoagulant warfarin. Warfarin is a racemic mixture of the *S*- and *R*- enantiomers, but *S*-warfarin is three to five times more potent than R-warfarin as an anticoagulant. *S*-warfarin is metabolized to 7-hydroxywarfarin by CYP2C9, whereas R-warfarin is metabolized by several other CYP enzymes, including CYP2C19. It has been shown that patients that are homozygous for the *CYP2C9*3* genotype have extreme sensitivity to *S*-warfarin, but also a reduced rate of metabolism is found in heterozygotes. Patients carrying the *CYP2C9*2* genotype exhibit a milder reduction in activity (Scordo et al. 2002).

Losartan, which is used as an angiotensin II receptor antagonist, has to be oxidized to its carboxylic acid metabolite, known as E-3174, before it is active. This reaction is catalysed by CYP2C9. It has been shown that in patients homozygous for the *CYP2C9*3* genotype the rate of conversation of losartan to E-3174 is much reduced and is also somewhat lower in patients with *CYP2C9*1/*3* and *CYP2C9*2/*3* genotypes (Yasar et al. 2002). As a consequence, the levels of the bioactive metabolite are reduced in such patients.

5.5
Polymorphisms of Other CYPs

CYP2A6 characteristically catalyses the oxidation of coumarin to 7-hydroxycoumarin (Yun et al. 1991) and is also subject to polymorphism. Two main mutated alleles (of 16 alleles reported to date; 80 SNPs) that result in loss of CYP2A6 activity have been found. These are *CYP2A6*2*, which contains a point mutation leading to $160Leu \rightarrow His$ in the enzyme. This form of CYP2A6 is inactive (Yamano et al. 1990). The other is *CYP2A6*3*, which contains several alterations as a result of gene conversion between CYP2A6 and CYP2A7, and the enzyme is also catalytically inactive (Fernandez-Salguero et al. 1995). However, few drugs and none that are commonly used in cardiovascular medicine are known substrates for CYP2A6.

The CYP2B6 enzyme is expressed at detectable levels in the liver of about 20% of individuals, suggesting that expression may be due to polymorphism. A number of alleles of CYP2B6 have been described (nine to date), and those (*CYP2B6*3* and *CYP2B6*5*) carrying a $487Arg \rightarrow Cys$ mutation appear to result in reduced expression levels of the enzyme. Other alleles have been reported to code for enzyme with reduced activity. As yet there is insufficient information to draw any definite conclusions concerning the impact of the various polymorphisms reported for this enzyme. Furthermore, CYP2B6 has not been implicated as a key enzyme in the metabolism of any pharmaceutical drug.

CYP3A5 is expressed at high levels in a minority (approximately 10%–30%) of Caucasians. Such individuals carry at least one *CYP3A5*1* allele. Single-nucleotide polymorphisms in the *CYP3A5*3* and *CYP3A5*6* alleles have been shown to cause alternative splicing and protein truncation and loss of CYP3A5 expression (Kuehl et al. 2001; Hustert et al. 2001). CYP3A5 catalyses the metabolism of the same drugs as CYP3A4; CYP3A4 is often the predominant form in liver and its expression does not appear to be subject to any major functional polymorphisms. This suggests that CYP3A5 polymorphisms are of little relevance.

5.6
Other Genetic Variants of CYPs

There are a large number of other known genetic variants of CYP genes, including the genes mentioned above and others that have not been mentioned such as *CYP1A1*, *CYP1A2* and *CYP2E1*. Most of the variants, although polymorphic, have no obvious effect on the level of expression or activity of the enzymes that they encode. Hence these changes are considered as non-functional polymorphisms.

5.7
Polymorphisms of Other Phase I Enzymes

Alcohol dehydrogenase (ADH) is a dimeric enzyme that catalyses the conversion of alcohol into acetaldehyde. Seven ADH genes have been reported, three of which (class I ADH) encode alpha, beta and gamma subunits, respectively, which dimerize in various combinations. These genes (*ADH1A*, *ADH1B* and *ADH1C*) were previously designated *ADH1*, *ADH2* and *ADH3*, respectively. Three alleles of the *ADH1B* gene (formerly *ADH2*) have been well characterized. These were previously designated $\beta1$, $\beta2$, and $\beta3$ and are now known as *ADH1B*1*, *ADH1B*2* and *ADH1B*3*, respectively (Osier et al. 2002). The *2 allele encodes for a high activity subunit that is the product of a single mutation (*47Arg→His*). Individuals inheriting at least one *2 allele have a high enzyme activity (Osier et al. 2002). Whereas only 5%–20% of Caucasians carry the *2 allele, it is present in approximately 85% of Asians, who therefore metabolize alcohol more rapidly (Goedde et al. 1992). Acetaldehyde is further metabolized to acetic acid by another polymorphic enzyme, acetaldehyde dehydrogenase 2 (ALDH2), a member of a family of at least three similar enzymes. A point mutation in a mutant allele for ALDH2 (*ALDH2*2*) leading to *487Glu→Lys* produces an enzyme with reduced activity (Yoshida et al. 1984). Approximately 50% of Asians are homozygous for the mutant allele (Goedde et al. 1992). If such individuals (especially those with an *ADH1B*2* genotype) consume ethanol this will be converted into acetaldehyde, which, not being cleared rapidly, will cause flushing due to dilatation of facial blood vessels and in extreme cases may cause acetaldehyde toxicity.

A polymorphism in flavin monooxygenase was first identified in chickens that produced bad smelling eggs. This defect was found to be due to a deficiency in the *FMO3* gene (one of at least six genes in this family) and causes trimethylaminuria or fish odour syndrome (Treacy et al. 1998; Krueger et al. 2002). However, in humans this condition is very rare. Nevertheless, a number of other alleles resulting in a less dramatic phenotype have been described (Furnes et al. 2002). Substrates for which FMO3 plays a significant role in their metabolism include nicotine, cimetidine, ranitidine and clozapine.

Catchol-*O*-methyltransferase (COMT) is responsible for the biotransformation of neurotransmitters such as norepinephrine and drugs such as L-DOPA. A mutant allele, *COMT*2*, which is due to a point mutation (*58Val→Met*), results in an enzyme with lower activity and reduced stability. The allele frequency is about 50% and the resultant polymorphism is thought to be related to a number of neuropsychiatric conditions (Lachman et al. 1996).

Thiopurine methyl transferase (TPMT) catalyses the conjugation of the methyl group of *S*-adenosyl methionine to aromatic and heterocyclic sulphydryl substrates, for example, to the antihypertensive drug captopril. It has been reported that 11% of Caucasians are heterozygous and 0.3% are homozygous with respect to TPMT deficiency (McLeod and Siva 2002). This is due to at least two mutant alleles. The *TPMT*2* allele encodes for a point mutation (*80Ala→Pro*) that results in an enzyme with lower catalytic activity (Krynetski et al. 1996). The *TPMT*3* allele encodes for a variant enzyme with *154Ala→Thr* and *240Tyr→Cys*, which is unstable and rapidly degraded resulting in levels 100-fold lower than those found for the wild type enzyme (Tai et al. 1996).

5.8
Polymorphisms of Phase II Enzymes

UDP-Glucuronosyltransferases (UGTs) catalyse the glucuronidation of a wide variety of substrates. Genetic polymorphisms have been described for six of the 16 UGT genes, i.e. *UGT1A1, 1A6, 1A7, 2B4, 2B7* and *2B15*. However, of these only that affecting *UGT1A1* has been demonstrated to be of functional significance (Miners et al. 2002).

Members of the sulphotransferase (SULT) superfamily catalyse the sulphation of a multitude of xenobiotics, hormones and neurotransmitters. Humans have at least ten functional *SULT* genes, and genetic polymorphisms have been described for three of them. Several allelic variants have been described that differ in their functional properties, but at present little is known about *SULT* allele frequencies in different population groups or their relevance for the clinical use of drugs subject to metabolism by sulphation (Coughtrie and Johnston 2001).

Glutathione S-transferases (GSTs) catalyse the addition of a glutathione moiety onto a variety of drugs. Population studies have shown that both the *GSTM1* and *GSTT1* genes are frequently deleted to produce null genotypes. The allele frequencies for *GSTM1*-null and *GSTT1*-null are 74% and 38%, respectively. The *GSTP1* gene also has several allelic variants containing single amino acid changes that affect both protein stability and substrate specificity (Daly 1995; Hayes and Pulford 1995). Whilst of considerable importance for environmental chemicals, conjugation with glutathione is much less important in the disposition of therapeutic drugs and there are no good examples of a clinically relevant consequence of such a polymorphism.

N-Acetyltransferase (NAT) enzymes are responsible for the acetylation of a large number of drugs. A polymorphism in acetylation activity was first discovered from use of the anti-tuberculosis antibiotic isoniazid when it was found that 60% of patients excreted the drug unchanged in the urine and 25% excreted inactive metabolites (Vatsis et al. 1995). It is now known that there are at least two *NAT* genes, *NAT1* and *NAT2*, and it is multiple alleles at the *NAT2* gene locus that are responsible for the acetylation phenotype. The wild-type gene is *NAT2*4* and mutant *NAT2* alleles are named as *NAT2*5–19* (Hein 2002). Examples of cardiovascular drugs subject to polymorphic acetylation include procainamide and hydralazine. Slow acetylators are at increased risk of developing anti-nuclear antibodies on treatment with hydralazine and procainamide, and some subjects may develop drug-induced lupus. The evidence for a link between lupus and acetylator phenotype is stronger with hydralazine than with procainamide.

6
Polymorphisms in P-Glycoprotein

Amongst the ABC transporters, P-glycoprotein appears to be that most involved in the transport of xenobiotic compounds. It has been the most extensively studied for polymorphisms, and to date almost 30 SNPs have been described. Several of the mutations are thought to be silent but some are predicted to have functional consequences by coding for a different amino acid in an important domain of the protein. One of the silent mutations, *C3435T* results in reduced expression of P-glycoprotein, possibly through linkage disequilibrium with a functional polymorphism. There is a prevalence of 25%–30% for subjects homozygous for the *C3435T* polymorphism in Caucasian populations. Prevalence in Asians may be slightly lower, at 20%–25%, whereas in black Africans, in both Africa and America, it is much reduced, at 0%–5% (Ameyaw et al. 2002).

Digoxin plasma concentration is increased by approximately 40% and clearance is reduced by approximately 25% in subjects homozygous for the *C3435T* allele (Hoffmeyer et al. 2000). However, not all substrates for P-glycoprotein were affected by this polymorphism, e.g. cyclosporine, talinolol. Hence, the overall clinical significance of this polymorphism of P-glycoprotein has yet to be determined.

7
Impact of Genetic Polymorphisms of Drug Metabolism on Overall Variability in Drug Effect

Genetic polymorphism is just one of several sources of variability in the metabolism and kinetics of drugs used to treat cardiovascular diseases (see Fig. 6). As such, care should be taken in interpreting its implications for patient therapy. As indicated in Sect. 3.6.1, the magnitude of the impact will depend on a number of different factors.

1. The penetrance of the polymorphism: i.e. the magnitude of the difference in metabolism between phenotypes. Whilst some polymorphisms result in a complete absence of activity in the PM phenotype, others result in a more modest reduction in activity or even a change in specificity. The impact that this will have on drug effect will depend on the factors discussed below.
2. The frequency of the poor metabolizer phenotype: the fewer the number of PMs the less the overall impact of phenotypic variation on population response. The issue of whether identifying such individuals would be advantageous would then depend on the other factors discussed here such as the magnitude of any benefit and the risk (if any) of treating blindly.
3. The steepness of the concentration-effect curve: where this is shallow, even an appreciable change in concentration, for example such as might occur between genotypes, would be associated with only a very modest change in effect.

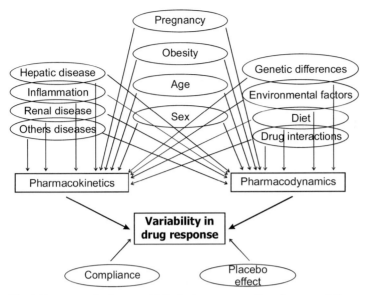

Fig. 6 Factors contributing to variability in drug response. Many factors can influence variability in drug response through an effect on the pharmacokinetics, pharmacodynamics, drug use or subtle psychological factors. Sometimes, one factor can dominate, but often, random variability in other factors will tend to reduce the clear separation between sub-groups classified on the basis of a single variable

4. The width of the therapeutic window: this is related to the preceding item, the shape of the concentration–effect curve. When the therapeutic window is wide, even relatively large changes in drug concentration can be tolerated without any toxicity.
5. The contribution of the polymorphic pathway to the elimination of the drug: it is almost axiomatic that for a polymorphism to play a significant role in the effect of a drug, that pathway must make a significant contribution to the fate of the drug. Where it represents a more modest fate of the compound, say not more than 50%, even the complete absence of the pathway would result in, at most, a twofold decrease in clearance. Where a drug is subject to extensive presystemic elimination, this might result in a more than doubling of bioavailability. However, in most cases, there will be little impact on the effects of the compound. Exceptions would be when elimination is normally close to saturation, or when there is a very narrow margin between acceptable and toxic levels.
6. The pharmacodynamics of the drug: even a dramatic difference in metabolism between phenotypes may not be of clinical importance when the affected pathway is responsible for the generation of a pharmacologically active metabolite. The extent to which this is the case will depend upon the efficacy of the metabolite and its kinetics. If it is much more short-lived than the parent, the polymorphism could well have an effect on outcome.

7. Other sources of variability in either the kinetics or dynamics of the compound: polymorphic drug metabolism is only one possible source of variability that might influence effect. However, there are numerous other possible sources, including the influence of the disease process itself. When trying to individualize patient therapy, an obvious consideration is the extent to which accounting for genotype will influence therapeutic outcome. If substantial variability remains after allowing for genotype, then the benefit to the patient will be minimal, and some other strategy needs to be considered such as plasma level monitoring.
8. The extent to which the effect of the drug can be monitored: where the end point of concern can be readily assessed and the kinetics of the drug are such that dose adjustments are rapidly effective, it may be argued that to some extent genetic heterogeneity in drug metabolism is not relevant, as the pharmacological effect is more proximal to clinical outcome than plasma concentration or indeed the alleles coding for even the major pathway of elimination.

Whilst there are certainly examples where a genetic polymorphism in drug metabolism can be shown to result in an appreciable difference in response between phenotypes, this is often in healthy volunteers (see Figs. 1 and 6). The extent to which this translates into a similar difference in patients needs to be established in adequate investigations before any therapeutic strategy incorporates dose adjustment on the basis of genotype. In a number of instances, other sources of variability may be such that there will be no benefit to the patient by stratifying on the basis of genotype for a drug-metabolizing enzyme.

It is important to distinguish between population and individual impact of polymorphisms in drug metabolism. For example, it might be possible to demonstrate that a significantly greater number of PM subjects exhibit side effects of a drug in a population study of perhaps 100 patients in each phenotype. Perhaps 20% of EM subjects and 30% of PM subjects experience side effects, with a *p*-value of 0.001, i.e. the difference is certainly real. However, these data clearly demonstrate a) not all (or even the majority) of PM subjects experience side effects and b) even an appreciable number of EM subjects experience side effects. The question then arises as to whether there is any benefit to the individual patient in genotyping, and this can only really be answered in adequate clinical trials.

8
Conclusions

It is now apparent that the genes for essentially all of the enzymes of drug metabolism exhibit genetic heterogeneity. To date, most of the polymorphisms identified have been in the coding region of the gene, but increasingly polymorphisms in regulatory regions are being identified, and this will increase as the various SNP projects mature. However, the presence of a single nucleotide polymorphism does not necessarily mean that there will be a functional consequence. Even when a polymorphism is functional, it does not necessarily result in a significant change in the kinetics of a drug, much less in the response to it. This will depend on several factors, including the extent to which enzyme activity is affected, the proportion of the clearance of the drug for which the polymorphic enzyme is responsible, the steepness of the concentration–effect curve and the width of the therapeutic window. Careful consideration of each of these issues often allows rational prediction of clinical relevance. Ultimately, this can be evaluated in focused clinical trials. It is important for the physician not to labour under the misconception that all drugs subject to metabolism by a polymorphic enzyme will necessarily exhibit marked variability between genotypes. Indeed, sometimes intra-genotype variability can be considerable and pose significant therapeutic challenges.

When the metabolism of a drug relies to such an extent on a single enzyme that genetic polymorphism of that enzyme results in marked inter-genotypic differences, this indicates the potential for drug–drug interactions in EM subjects. This is because clearance in these subjects is heavily reliant on a single enzyme and hence modulating the activity of that enzyme by a co-administered drug can markedly affect elimination, either by induction or inhibition. In PM subjects, no such interaction is anticipated, as the induced/inhibited enzyme is not present.

It is often argued that the availability of extensive information on genetic control of drug-metabolizing enzymes will lead to individualization of drug therapy. However, there are reasons to question this view as a widespread consequence of such knowledge. The pharmaceutical industry is striving to develop drugs with better margins of safety, so that some variation in plasma concentration could be more readily tolerated. The need to match each patient individually for genotype, drug and dose schedule will have substantial implications for the costs of health care. That is not to say that there would not be some benefit, but this needs to be assessed in adequate clinical trials, not just assumed as an article of faith.

Pharmacogenetics has undergone rapid expansion over that last few years. A major consequence of this has been an attempt by most pharmaceutical companies to eliminate any candidate compound that is subject to metabolism by polymorphic enzymes as early as possible in drug discovery. The wisdom of this is perhaps open to some question. Of course, the information needed to assess accurately the likely clinical impact is not available until relatively well into the R & D process. However, by eliminating a role for polymorphic enzymes, there may a tendency to select for molecules more likely to be substrates for other enzymes such as CYP3A4 or indeed for P-glycoprotein transport. Given the interindividual variability that exists in the activity of CYP3A4 and its liability to drug–drug interactions, this may still result in attrition, and indeed has claimed several recent casualties. Substrates for polymorphic enzymes should probably be eliminated only when this is the only enzyme responsible for the major part of their elimination (75% or more).

References

Aklillu E, Persson I, Bertilsson L et al (1996) Frequent distribution of ultrarapid metabolizers of debrisoquine in an Ethiopian population carrying duplicated and multiduplicated functional CYP2D6 alleles. J Pharmacol Exp Ther 278:441–446

Aklillu E, Herrlin K, Gustafsson LL et al (2002) Evidence for environmental influence on CYP2D6-catalysed debrisoquine hydroxylation as demonstrated by phenotyping and genotyping of Ethiopians living in Ethiopia or in Sweden. Pharmacogenetics 12:375–383

Ameyaw MM, Regateiro F, Li T et al (2001) MDR1 pharmacogenetics: frequency of the C3435T mutation in exon 26 is significantly influenced by ethnicity. Pharmacogenetics 11:217–221

Armstrong M, Fairbrother K, Idle JR et al (1994) The cytochrome P450 CYP2D6 allelic variant CYP2D6J and related polymorphisms in a European population. Pharmacogenetics 4:73–81

Ashmarin IP, Danilova RA, Obukhova MF et al (2000) Main ethanol metabolizing alcohol dehydrogenases (ADH I and ADH IV): biochemical functions and the physiological manifestation. FEBS Lett 486:49–51

Bathum L, Skjelbo E, Mutabingwa TK et al (1999) Phenotypes and genotypes for CYP2D6 and CYP2C19 in a black Tanzanian population. Br J Clin Pharmacol 48:395–401

Bertilsson L, Lou YQ, Du YL et al (1992) Pronounced differences between native Chinese and Swedish populations in the polymorphic hydroxylations of debrisoquin and S-mephenytoin. Clin Pharmacol Ther 51:388–397

Binda C, Mattevi A and Edmondson DE (2002) Structure-function relationships in flavoenzyme-dependent amine oxidations: a comparison of polyamine oxidase and monoamine oxidase. J Biol Chem 277:23973–23976

Borst P, Evers R, Kool M et al (2000) A family of drug transporters: the multidrug resistance-associated proteins. J Natl Cancer Inst 92:1295–1302

Cashman JR and Zhang J (2002) Interindividual differences of human flavin-containing monooxygenase 3: genetic polymorphisms and functional variation. Drug Metab Dispos 30:1043–1052

Cesura AM and Pletscher A (1992) The new generation of monoamine oxidase inhibitors. Prog Drug Res 38:171–297

Coughtrie MW and Johnston LE (2001) Interactions between dietary chemicals and human sulfotransferases-molecular mechanisms and clinical significance. Drug Metab Dispos 29:522–528

Dahl ML, Johansson I, Bertilsson L et al (1995) Ultrarapid hydroxylation of debrisoquine in a Swedish population. Analysis of the molecular genetic basis. J Pharmacol Exp Ther 274:516–520

Daly AK (1995) Molecular basis of polymorphic drug metabolism. J Mol Med 73:539–553

De Morais SM, Wilkinson GR, Blaisdell J et al (1994) The major genetic defect responsible for the polymorphism of S-mephenytoin metabolism in humans. J Biol Chem 269:15419–15422

Dickmann LJ, Rettie AE, Kneller MB et al (2001) Identification and functional characterization of a new CYP2C9 variant (CYP2C9*5) expressed among African Americans. Mol Pharmacol 60:382–387

Droll K, Bruce-Mensah K, Otton SV et al (1998) Comparison of three CYP2D6 probe substrates and genotype in Ghanaians, Chinese and Caucasians. Pharmacogenetics 8:325–333

Eichelbaum M and Gross AS (1990) The genetic polymorphism of debrisoquine/sparteine metabolism-clinical aspects. Pharmacol Ther 46:377–394

Eichelbaum M, Kroemer HK and Fromm MF (1997) Impact of P450 genetic polymorphism on the first-pass extraction of cardiovascular and neuroactive drugs. Adv Drug Deliv Rev 27:171–199

Evans DA (1993) Methylation reactions. In 'Genetic factors in drug therapy: Clinical and molecular pharmacogenetics'. Cambridge University Press, New York. pp. 330–345

Fernandez-Salguero P, Hoffman SM, Cholerton S et al (1995) A genetic polymorphism in coumarin 7-hydroxylation: sequence of the human CYP2A genes and identification of variant CYP2A6 alleles. Am J Hum Genet 57:651–660

Foliot A, Touchard D and Celier C (1984) Impairment of hepatic glutathione S-transferase activity as a cause of reduced biliary sulfobromophthalein excretion in clofibrate-treated rats. Biochem Pharmacol 33:2829–2834

Fretland AJ and Omiecinski CJ (2000) Epoxide hydrolases: biochemistry and molecular biology. Chem Biol Interact 129:41–59

Friedgen B, Wolfel R, Russ H et al (1996) The role of extraneuronal amine transport systems for the removal of extracellular catecholamines in the rabbit. Naunyn Schmiedebergs Arch Pharmacol 354:275–286

Frye RF and Branch RA (2002) Effect of chronic disulfiram administration on the activities of CYP1A2, CYP2C19, CYP2D6, CYP2E1, and N-acetyltransferase in healthy human subjects. Br J Clin Pharmacol 53:155–162

Fuhr U, Woodcock BG and Siewert M (1992) Verapamil and drug metabolism by the cytochrome P450 isoform CYP1A2. Eur J Clin Pharmacol 42:463–464

Funk CD (2001) Prostaglandins and leukotrienes: advances in eicosanoid biology. Science 294:1871–1875

Furnes B, Feng J, Sommer SS et al (2003) Identification of novel variants of the flavin-containing monooxygenase gene family in African Americans. Drug Metab Dispos 31:187–193

Gaedigk A, Blum M, Gaedigk R et al (1991) Deletion of the entire cytochrome P450 CYP2D6 gene as a cause of impaired drug metabolism in poor metabolizers of the debrisoquine/sparteine polymorphism. Am J Hum Genet 48:943–950

Garrod AE (1914) Medicine from the Chemical Standpoint. Lancet ii:281–289

Glatt H, Boeing H, Engelke CE et al (2001) Human cytosolic sulphotransferases: genetics, characteristics, toxicological aspects. Mutat Res 482:27–40

Goedde HW, Agarwal DP, Fritze G et al (1992) Distribution of ADH2 and ALDH2 genotypes in different populations. Hum Genet 88:344–346

Goldstein JA and De Morais SM (1994) Biochemistry and molecular biology of the human CYP2C subfamily. Pharmacogenetics 4:285–299

Goldstein JA, Ishizaki T, Chiba K et al (1997) Frequencies of the defective CYP2C19 alleles responsible for the mephenytoin poor metabolizer phenotype in various Oriental, Caucasian, Saudi Arabian and American black populations. Pharmacogenetics 7:59–64

Grancharov K, Naydenova Z, Lozeva S et al (2001) Natural and synthetic inhibitors of UDP-glucuronosyltransferase. Pharmacol Ther 89:171–186

Grant DM, Goodfellow GH, Sugamori K et al (2000) Pharmacogenetics of the human arylamine N-acetyltransferases. Pharmacology 61:204–211

Haining RL, Hunter AP, Veronese ME et al (1996) Allelic variants of human cytochrome P450 2C9: baculovirus-mediated expression, purification, structural characterization, substrate stereoselectivity, and prochiral selectivity of the wild-type and I359L mutant forms. Arch Biochem Biophys 333:447–458

Hayes JD and Pulford DJ (1995) The glutathione S-transferase supergene family: regulation of GST and the contribution of the isoenzymes to cancer chemoprotection and drug resistance. Crit Rev Biochem Mol Biol 30:445–600

Hayes JD and Strange RC (2000) Glutathione S-transferase polymorphisms and their biological consequences. Pharmacology 61:154–166

Hein DW (2002) Molecular genetics and function of NAT1 and NAT2: role in aromatic amine metabolism and carcinogenesis. Mutat Res 506–507:65–77

Herrlin K, Massele AY, Jande M et al (1998) Bantu Tanzanians have a decreased capacity to metabolize omeprazole and mephenytoin in relation to their CYP2C19 genotype. Clin Pharmacol Ther 64:391–401

Hines RN and McCarver DG (2002) The ontogeny of human drug-metabolizing enzymes: phase I oxidative enzymes. J Pharmacol Exp Ther 300:355–360

Hoffmeyer S, Burk O, von Richter O et al (2000) Functional polymorphisms of the human multidrug-resistance gene: multiple sequence variations and correlation of one allele with P-glycoprotein expression and activity in vivo. Proc Natl Acad Sci USA 97:3473–3478

Hollenberg PF (2002) Characteristics and common properties of inhibitors, inducers, and activators of CYP enzymes. Drug Metab Rev 34:17–35

Holmes RS (1994) Alcohol dehydrogenases: a family of isozymes with differential functions. Alcohol Alcohol Suppl 2:127–130

Hustert E, Haberl M, Burk O et al (2001) The genetic determinants of the CYP3A5 polymorphism. Pharmacogenetics 11:773–779

Ingelman-Sundberg M (2001) Genetic variability in susceptibility and response to toxicants. Toxicol Lett 120:259–268

Irshaid YM, Gharaybeh KI, Ammari FF et al (1990) Glucuronidation of 7-hydroxy-4-methylcoumarin by human liver microsomes. Inhibition by certain drugs. Eur J Drug Metab Pharmacokinet 15:295–301

Johansson I, Lundqvist E, Bertilsson L et al (1993) Inherited amplification of an active gene in the cytochrome P450 CYP2D locus as a cause of ultrarapid metabolism of debrisoquine. Proc Natl Acad Sci U S A 90:11825–11829

Jornvall H, Hoog JO, Persson B et al (2000) Pharmacogenetics of the alcohol dehydrogenase system. Pharmacology 61:184–191

Kagimoto M, Heim M, Kagimoto K et al (1990) Multiple mutations of the human cytochrome P450IID6 gene (CYP2D6) in poor metabolizers of debrisoquine. Study of the functional significance of individual mutations by expression of chimeric genes. J Biol Chem 265:17209–17214

Kazierad DJ, Martin DE, Blum RA et al (1997) Effect of fluconazole on the pharmacokinetics of eprosartan and losartan in healthy male volunteers. Clin Pharmacol Ther 62:417–425

Krueger SK, Martin SR, Yueh MF et al (2002) Identification of active flavin-containing monooxygenase isoform 2 in human lung and characterization of expressed protein. Drug Metab Dispos 30:34–41

Krynetski EY, Tai HL, Yates CR et al (1996) Genetic polymorphism of thiopurine S-methyltransferase: clinical importance and molecular mechanisms. Pharmacogenetics 6:279–290

Kuehl P, Zhang J, Lin Y et al (2001) Sequence diversity in CYP3A promoters and characterization of the genetic basis of polymorphic CYP3A5 expression. Nat Genet 27:383–391

Kupfer A and Preisig R (1984) Pharmacogenetics of mephenytoin: a new drug hydroxylation polymorphism in man. Eur J Clin Pharmacol 26:753–759

Kutt H, Wolk M, Scherman R and McDowell F (1964) Insufficient parahydroxylation as a cause of diphenylhydantoin toxicity. Neurology 14:542–548

Lachman HM, Papolos DF, Saito T et al (1996) Human catechol-O-methyltransferase pharmacogenetics: description of a functional polymorphism and its potential application to neuropsychiatric disorders. Pharmacogenetics 6:243–250

Lennard L (1998) Clinical implications of thiopurine methyltransferase–optimization of drug dosage and potential drug interactions. Ther Drug Monit 20:527–531

Lieber CS (1997) Cytochrome P-4502E1: its physiological and pathological role. Physiol Rev 77:517–544

Liu C, Zhuo X, Gonzalez FJ et al (1996) Baculovirus-mediated expression and characterization of rat CYP2A3 and human CYP2A6: role in metabolic activation of nasal toxicants. Mol Pharmacol 50:781–788

Mahgoub A, Idle JR and Smith RL (1979) A population and familial study of the defective alicyclic hydroxylation of debrisoquine among Egyptians. Xenobiotica 9:51–56

Mannisto PT and Kaakkola S (1999) Catechol-O-methyltransferase (COMT): biochemistry, molecular biology, pharmacology, and clinical efficacy of the new selective COMT inhibitors. Pharmacol Rev 51:593–628

Masimirembwa C, Hasler J, Bertilssons L et al (1996) Phenotype and genotype analysis of debrisoquine hydroxylase (CYP2D6) in a black Zimbabwean population. Reduced enzyme activity and evaluation of metabolic correlation of CYP2D6 probe drugs. Eur J Clin Pharmacol 51:117–122

Masimirembwa CM and Hasler JA (1997) Genetic polymorphism of drug metabolising enzymes in African populations: implications for the use of neuroleptics and antidepressants. Brain Res Bull 44:561–571

McGurk KA, Brierley CH and Burchell B (1998) Drug glucuronidation by human renal UDP-glucuronosyltransferases. Biochem Pharmacol 55:1005–1012

McLeod HL and Siva C (2002) The thiopurine S-methyltransferase gene locus—implications for clinical pharmacogenomics. Pharmacogenomics 3:89–98

Meister A (1995) Glutathione metabolism. Methods Enzymol 251:3–7

Meyer UA (1994) The molecular basis of genetic polymorphisms of drug metabolism. J Pharm Pharmacol 46 Suppl 1:409–415

Miners JO, McKinnon RA and Mackenzie PI (2002) Genetic polymorphisms of UDP-glucuronosyltransferases and their functional significance. Toxicology 181–182:453–456

Nebert DW and Russell DW (2002) Clinical importance of the cytochromes P450. Lancet 360:1155–1162

Ohyama K, Nakajima M, Suzuki M et al (2000) Inhibitory effects of amiodarone and its N-deethylated metabolite on human cytochrome P450 activities: prediction of in vivo drug interactions. Br J Clin Pharmacol 49:244–253

Oscarson M, Hidestrand M, Johansson I et al (1997) A combination of mutations in the CYP2D6*17 (CYP2D6Z) allele causes alterations in enzyme function. Mol Pharmacol 52:1034–1040

Osier MV, Pakstis AJ, Soodyall H et al (2002) A global perspective on genetic variation at the ADH genes reveals unusual patterns of linkage disequilibrium and diversity. Am J Hum Genet 71:84–99

Pelkonen O and Raunio H (1997) Metabolic activation of toxins: tissue-specific expression and metabolism in target organs. Environ Health Perspect 105 Suppl 4:767–774

Persson I, Aklillu E, Rodrigues F et al (1996) S-mephenytoin hydroxylation phenotype and CYP2C19 genotype among Ethiopians. Pharmacogenetics 6:521–526

Ploemen JH, van Ommen B and van Bladeren PJ (1990) Inhibition of rat and human glutathione S-transferase isoenzymes by ethacrynic acid and its glutathione conjugate. Biochem Pharmacol 40:1631–1635

Rettie AE, Wienkers LC, Gonzalez FJ et al (1994) Impaired (S)-warfarin metabolism catalysed by the R144C allelic variant of CYP2C9. Pharmacogenetics 4:39–42

Rushmore TH and Kong AN (2002) Pharmacogenomics, regulation and signaling pathways of phase I and II drug metabolizing enzymes. Curr Drug Metab 3:481–490

Schutz E, Gummert J, Mohr F et al (1993) Azathioprine-induced myelosuppression in thiopurine methyltransferase deficient heart transplant recipient. Lancet 341:436-

Scordo MG, Aklillu E, Yasar U et al (2001) Genetic polymorphism of cytochrome P450 2C9 in a Caucasian and a black African population. Br J Clin Pharmacol 52:447–450

Scordo MG, Pengo V, Spina E et al (2002) Influence of CYP2C9 and CYP2C19 genetic polymorphisms on warfarin maintenance dose and metabolic clearance. Clin Pharmacol Ther 72:702–710

Shimada T, Yamazaki H, Mimura M et al (1994) Interindividual variations in human liver cytochrome P-450 enzymes involved in the oxidation of drugs, carcinogens and toxic chemicals: studies with liver microsomes of 30 Japanese and 30 Caucasians. J Pharmacol Exp Ther 270:414–423

Simpson AE (1997) The cytochrome P450 4 (CYP4) family. Gen Pharmacol 28:351–359

Sohn DR, Shin SG, Park CW et al (1991) Metoprolol oxidation polymorphism in a Korean population: comparison with native Japanese and Chinese populations. Br J Clin Pharmacol 32:504–507

Steen VM, Molven A, Aarskog NK et al (1995) Homologous unequal cross-over involving a 2.8 kb direct repeat as a mechanism for the generation of allelic variants of human cytochrome P450 CYP2D6 gene. Hum Mol Genet 4:2251–2257

Streetman DS, Bertino JS Jr, Nafziger AN (2000) Phenotyping of drug-metabolizing enzymes in adults: a review of in-vivo cytochrome P450 phenotyping probes. Pharmacogenetics 10:187–216

Sueyoshi T and Negishi M (2001) Phenobarbital response elements of cytochrome P450 genes and nuclear receptors. Annu Rev Pharmacol Toxicol 41:123–143

Sullivan-Klose TH, Ghanayem BI, Bell DA et al (1996) The role of the CYP2C9-Leu359 allelic variant in the tolbutamide polymorphism. Pharmacogenetics 6:341–349

Taavitsainen P, Kiukaanniemi K and Pelkonen O (2000) In vitro inhibition screening of human hepatic P450 enzymes by five angiotensin-II receptor antagonists. Eur J Clin Pharmacol 56:135–140

Tai HL, Krynetski EY, Yates CR et al (1996) Thiopurine S-methyltransferase deficiency: two nucleotide transitions define the most prevalent mutant allele associated with loss of catalytic activity in Caucasians. Am J Hum Genet 58:694–702

Tew KD and Ronai Z (1999) GST function in drug and stress response. Drug Resist Updat 2:143–147

Tracy TS, Korzekwa KR, Gonzalez FJ et al (1999) Cytochrome P450 isoforms involved in metabolism of the enantiomers of verapamil and norverapamil. Br J Clin Pharmacol 47:545–552

Treacy EP, Akerman BR, Chow LM et al (1998) Mutations of the flavin-containing monooxygenase gene (FMO3) cause trimethylaminuria, a defect in detoxication. Hum Mol Genet 7:839–845

Vasiliou V and Pappa A (2000) Polymorphisms of human aldehyde dehydrogenases. Consequences for drug metabolism and disease. Pharmacology 61:192–198

Vatsis KP, Weber WW, Bell DA et al (1995) Nomenclature for N-acetyltransferases. Pharmacogenetics 5:1–17

Wedlund PJ, Aslanian WS, McAllister CB et al (1984) Mephenytoin hydroxylation deficiency in Caucasians: frequency of a new oxidative drug metabolism polymorphism. Clin Pharmacol Ther 36:773–780

Weinshilboum R (2001) Thiopurine pharmacogenetics: clinical and molecular studies of thiopurine methyltransferase. Drug Metab Dispos 29:601–605

Wilkinson GR, Guengerich FP and Branch RA (1989) Genetic polymorphism of S-mephenytoin hydroxylation. Pharmacol Ther 43:53–76

Xie HG, Kim RB, Stein CM et al (1999) Genetic polymorphism of (S)-mephenytoin 4'-hydroxylation in populations of African descent. Br J Clin Pharmacol 48:402–408

Yamano S, Tatsuno J and Gonzalez FJ (1990) The CYP2A3 gene product catalyzes coumarin 7-hydroxylation in human liver microsomes. Biochemistry 29:1322–1329

Yan M, Webster LT, Jr. and Blumer JL (2002) Kinetic interactions of dopamine and dobutamine with human catechol-O-methyltransferase and monoamine oxidase in vitro. J Pharmacol Exp Ther 301:315–321

Yasar U, Forslund-Bergengren C, Tybring G et al (2002) Pharmacokinetics of losartan and its metabolite E-3174 in relation to the CYP2C9 genotype. Clin Pharmacol Ther 71:89–98

Yeung CK, Lang DH, Thummel KE et al (2000) Immunoquantitation of FMO1 in human liver, kidney, and intestine. Drug Metab Dispos 28:1107–1111

Yokota H, Tamura S, Furuya H et al (1993) Evidence for a new variant CYP2D6 allele CYP2D6J in a Japanese population associated with lower in vivo rates of sparteine metabolism. Pharmacogenetics 3:256–263

Yoshida A, Huang IY and Ikawa M (1984) Molecular abnormality of an inactive aldehyde dehydrogenase variant commonly found in Orientals. Proc Natl Acad Sci USA 81:258–261

Yun CH, Shimada T and Guengerich FP (1991) Purification and characterization of human liver microsomal cytochrome P-450 2A6. Mol Pharmacol 40:679–685

Genes That Modify Susceptibility to Atherosclerosis: Targets for Drug Action

J. W. Knowles

Curriculum in Genetics and Molecular Biology,
University of North Carolina at Chapel Hill, 701 Brinkhous-Bullitt Bldg,
CB#7525, Chapel Hill, NC 27599, USA
e-mail: knowlej@stanford.edu

1	Introduction	80
2	Genetic Analysis of Atherosclerosis: Risk Stratification and Pharmacogenetics	82
2.1	Rationale	82
2.2	Lipid Pathways: Apolipoprotein E, Lipoprotein Lipase, Cholesterol Ester Transfer Protein	84
2.3	Blood Pressure Pathways: Angiotensinogen, Angiotensin-Converting Enzyme, G-Protein $\beta 3$ Subunit, α-Adducin	85
3	The Interplay Between Genetic Analyses of Atherosclerosis in Humans and Animal Models: Pathogenic Insights and New Targets for Therapy	86
3.1	Rationale	86
3.2	Lipid Pathways: Adenosine Triphosphate-Binding Cassette Transporter A1, Cholesterol 7α-hydroxylase, Sterol Regulatory Element-Binding Proteins, Apolipoprotein AV, Fatty-Acid-Binding Protein aP2, Plasma Phospholipid-Transfer Protein	88
3.3	Blood Pressure Pathways: Endothelial Nitric Oxide Synthase, Neuropeptide Y, $\beta 2$-Adrenergic Receptor	90
3.4	Inflammatory Pathways: Macrophage Colony-Stimulating Factor, Monocyte Chemoattractant Protein-1, A2a Adenosine Receptor, Myeloperoxidase, 12/15 Lipoxygenase, p47phox	92
4	The Future of Genetic Analyses in Human and Animal Models of Atherosclerosis	94
5	Conclusions	96
References		96

Abstract Atherosclerotic cardiovascular disease is the leading cause of morbidity and mortality in the developed world and is a rapidly growing epidemic in less developed countries. Atherosclerosis is a complex disease with both environmental and genetic determinants. Genetic susceptibility is governed by an unfavorable combination of variants in genes involving multiple pathways, each with small quantitative effects on gene function. However, our understanding of the genes and mutations involved is limited. This chapter will use specific examples from both humans and murine models of atherosclerosis to illustrate the utility of genetic analysis to (1) identify polymorphisms in atherosclerotic susceptibility genes that will enhance our ability to both stratify patients according to risk and to develop targeted therapeutics based on an individual genotype and (2) identify new genes/pathways that contribute to atherosclerosis that may be new targets for drug action.

Keywords Arteriosclerosis · Pharmacogenetics · Mice · Polymorphism · Heart · Cardiac

1
Introduction

Atherosclerotic cardiovascular disease (CVD) is the leading cause of morbidity and mortality in the developed world, with 50% of deaths directly attributable to atherosclerosis through coronary artery disease (CAD), myocardial infarction (MI), or cerebrovascular accident (CVA). Worldwide, CVD causes 17 million deaths/year and by the early part of the twenty-first century CVD is projected to be the leading cause of death, accounting for an estimated 40% of all deaths (Anonymous 1999, 2002a, b). Atherosclerosis also plays a major role in the pathogenesis of congestive heart failure, renal failure and severe peripheral vascular disease as well as contributing to some forms of dementia and impotence (Ross 1993, 1999). The consequences of atherosclerosis place an enormous burden on health care systems worldwide; in the US alone costs are estimated at an annual rate of $250 billion (Anonymous 2002a).

Atherosclerosis and arteriosclerosis refer to the same process in different vessels (the former refers to large conduit arteries whereas the latter refers to smaller resistance vessels within end organs (Turner and Boerwinkle 2000), but only the term "atherosclerosis" will be used in this chapter. The pathological process of atherosclerosis is continuously progressive, yet the natural history is so orderly that discrete stages of lesion development can be defined by the presence of certain cell types and by other histopathological characteristics (Stary et al. 1995). Early atherosclerotic lesions, called fatty streaks, consist of lipid-laden macrophages within the intima of the vessel wall and are usually present by early adulthood. Intermediate atherosclerotic lesions contain multiple layers of foamy macrophages and vascular smooth muscle cells. These lesions progress to advanced plaques characterized by necrotic cores, cholesterol clefts and fibrous caps (Davies 2001; Glass and Witztum 2001; Lusis 2000; Ross 1993).

The progression of atherosclerotic lesions over time is highly variable, but our current understanding of atherosclerosis, articulated in a landmark review in 1993 (Ross 1993), is that atherosclerosis is a self-perpetuating inflammatory disease driven by the response to vascular injury. By extension, any stimulus that causes vascular injury (such as hypertension or diabetes mellitus) is a po-

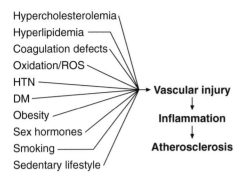

Fig. 1 Factors that influence atherosclerosis. Atherosclerosis results from a complex amalgam of genetic and environmental determinants that initiate or modify the response to vascular injury

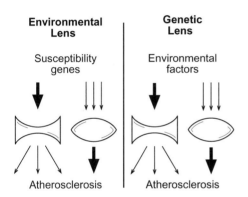

Fig. 2 Atherosclerosis through the lens. *Left panel*: Genetic susceptibility to atherosclerosis, whether favorable *small arrows*) or unfavorable (*heavy dark arrow*), which can be altered by environmental conditions. These environmental conditions can either magnify or dissipate the effect of the genetic background. *Right panel*: Environmental susceptibility to atherosclerosis can be magnified or dissipated by the genetic makeup of the individual

tential risk factor and any gene that can initiate or alter the response to vascular injury is a potential modifier of atherosclerosis (Fig. 1). Injurious stimuli can be both extrinsic (as in the case of dietary fat intake or smoking) and intrinsic (as in the case of genetically determined hypercholesterolemia). Thus, as illustrated in Fig. 2, the susceptibility to atherosclerosis is a complex amalgam of gene/environment interactions (Boerwinkle et al. 1996; Ellsworth et al. 1999; Glass and Witztum 2001; Lusis 2000; Peyser 1997; Ross 1999; Rubin and Tall 2000; Winkelmann and Hager 2000; Winkelmann et al. 2000). Through large epidemiological studies, we understand some of the risk factors for the progression of atherosclerosis such as hypercholesterolemia, hyperlipidemia, smoking, age, obesity, male sex, and family history. Although we use these risk factors (as in the Framingham scheme) to estimate the lifetime relative risk of MI, it is thought that only about 50% of the possible risk factors have been identified (Castelli 1996; Valdes et al. 2001). Some degree of hypercholesterolemia is required for the development of atherosclerosis as shown by studies demonstrating that CVD rarely develops in those with total cholesterol in the range of 125–140 mg/dl (Castelli 2001). Equally clear is that extreme elevations of cholesterol are universally associated with early, advanced atherosclerotic disease. However, most MIs and CVAs happen in people with normal or moderately elevated cholesterol and

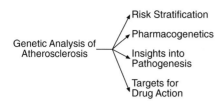

Fig. 3 Genetic analysis of atherosclerosis is useful for multiple purposes

at present we have a very limited ability to identify the patients at greatest risk (Castelli 2001).

The purpose of this chapter is to illustrate the utility of genetic analysis to provide insight into the pathogenesis and treatment of atherosclerosis. To that end, the first major theme of this chapter will be an overview of known genetic polymorphisms that may prove useful both for risk stratification and for the development of targeted therapeutics based on individual genotype. Emphasis will be on the most well-validated polymorphisms that have been evaluated by pharmacological studies. The second major theme is the utility of parallel experiments in humans and animal models to increase our understanding of the basic pathogenesis of atherosclerosis and identify new targets for drug action (Fig. 3). The final theme is a discussion of both the future of genetic analyses in humans and mice, as well as some of the limitations of these studies.

2
Genetic Analysis of Atherosclerosis: Risk Stratification and Pharmacogenetics

2.1
Rationale

The importance of genetic susceptibility to atherosclerosis is illustrated by familial aggregation of coronary artery disease (CAD), even after controlling for established risk factors, and by studies in monozygotic twins which show an 8- to 15-fold increased risk of death from CAD in the remaining twin when the other twin died prematurely from this disease (Colditz et al. 1991; Marenberg et al. 1994; Nora et al. 1980; Williams et al. 1993). The heritability of atherosclerosis (the percentage of the disease that can be accounted for by genetics) is estimated at more than 50% (Lusis 2000; Sing and Moll 1989) and is thought to be due to an unfavorable combination of variations in multiple genes. However, the genes that control this variability in susceptibility are not well defined (Lusis 2000; Peyser 1997; Winkelmann and Hager 2000; Winkelmann et al. 2000). Therefore, the identification of the genetic mutations that modify the susceptibility to atherosclerosis is of paramount importance for the purposes of risk assessment and pharmacogenetics.

Traditionally, pharmacogenetics refers to differences in drug metabolism caused by genetic variation (see the chapter by Boobis et al., this volume, and

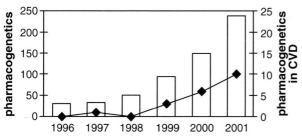

Fig. 4 Increasing importance of pharmacogenetics. In the last 6 years, the number of Medline citations including the word "pharmacogenetics" (*columns*). The number of citations in the pharmacogenetics of heart disease is also increasing (*line*). The latter was determined by a Medline search using the word "pharmacogenetics" plus any of the following: "atherosclerosis," "arteriosclerosis," "CVD," "CAD," "MI," "heart"

Ellsworth et al. 1999; Turner et al. 2001b), but in the postgenomic era the term has come to include tailoring drug therapy based on genetic variants in atherosclerotic modifiers (Wilkins et al. 2000). Although pharmacogenetics in the latter sense is a field in its infancy, particularly for CVD (Fig. 4), it holds considerable interest for both the CVD scientific community and for the lay public (Gorman 2001; Kolata 1999). With the cost of sequencing an individual human decreasing rapidly, it is reasonable to assume better individualized drug therapy for the prevention of disease will become increasingly important. To fulfill this promise, however, the identification of specific genetic modifiers of atherosclerosis is essential.

The study of genetic modifiers of atherosclerosis is conducted using association and linkage analyses, which are reviewed extensively elsewhere (see the chapters by Tate and Goldstein and Winkelmann et al., this volume, and Boerwinkle et al. 1996, 2000; Broeckel et al. 2002; Collins et al. 1997; Hauser and Pericak-Vance 2000; Klos et al. 2001; Kraus 2000; Peacock et al. 2001; Peltonen and McKusick 2001; Peyser 1997; Risch and Merikangas 1996; Winkelmann and Hager 2000; Winkelmann et al. 2000). Through these types of studies, numerous genes in humans have been tested for their relationship to atherosclerotic susceptibility. While dozens of gene polymorphisms have been associated with atherosclerosis, few have been validated with large-scale studies or tested for pharmacogenetic effects. Thus, although we are beginning to see a tidal wave of genetic information emerging from the genome project and large-scale single nucleotide polymorphism (SNP) mapping projects (Cargill et al. 1999; Halushka et al. 1999; Lander et al. 2001; Venter et al. 2001), this data has not yet been translated to the clinical setting. Nonetheless, there are several good examples of polymorphisms that are well-substantiated by both human and animal research. This section of the chapter will focus on a few prototypical examples from pathways involved in lipid metabolism or blood pressure (BP) homeostasis (also see the chapters by Winkelmann et al. and O'Shaughnessy and Wilkins, this volume) to illustrate the potential clinical use of genetic analysis for risk stratification and pharmacogenetics.

2.2
Lipid Pathways: Apolipoprotein E, Lipoprotein Lipase, Cholesterol Ester Transfer Protein

Apolipoprotein E (apoE) is an amphipathic protein that plays a pivotal role in lipoprotein trafficking. ApoE is a constituent of chylomicrons, VLDL, LDL, and HDL, and acts as a ligand for the receptor mediated clearance of these particles (Mahley 1988; and the chapter by Winkelmann et al., this volume). The human *APOE* gene is polymorphic at two nucleotides, resulting in three alleles, $\varepsilon2$, $\varepsilon3$, $\varepsilon4$, which code for three isoforms E2, E3, E4. Among other things, these isoforms differ with respect to their affinity for hepatic lipoprotein receptors (E4>E3>E2), and to their effect on plasma cholesterol levels (E2 tends to lower and E4 tends to raise cholesterol compared to E3). Experiments in mice that have been humanized, in a sense, to express human E2, E3 or E4 instead of the native apoE have shown that the small differences in protein structure coded by these alleles are functional and they differ markedly in plasma retention of lipoproteins and atherosclerosis risk (Knouff et al. 1999; Sullivan et al. 1998). In humans, the $\varepsilon4$ allele is one of the few polymorphisms that has been shown repeatedly to be a predictor of CVD/MI (Boerwinkle et al. 1996; Eichner et al. 1993; Stengard et al. 1995; Wilson et al. 1994; Winkelmann et al. 2000) and thus is a potentially important gene for risk stratification. The *APOE* gene variants also differ with respect to the effect of statin therapy: individuals with an $\varepsilon4$ allele have a lesser response and those with an $\varepsilon2$ allele have a greater response to statins versus $\varepsilon3/\varepsilon3$ individuals (Carmena et al. 1993; Ojala et al. 1991; Ordovas et al. 1995).

Another gene with variants that may be important for assessing risk and tailoring therapy is the lipoprotein lipase (*LPL*) gene. LPL is involved in the production of HDL through the hydrolysis of triglycerides in chylomicrons and in VLDL. The most common polymorphism in the *LPL* gene (Ser447Ter) has a beneficial effect on HDL cholesterol and has been associated with a decreased risk of CAD based on both a population study and a meta-analysis of 20,903 subjects (Galtonet al. 1996; Winkelmann et al. 2000; Wittrup et al. 1999). Thus, this relatively common variant (carrier frequency about 20%) may become important for risk stratification.

A final example from lipid pathways is the gene for cholesteryl ester transfer protein (CETP), which is a key player in the metabolism of high-density lipoprotein. In one study of 807 patients, those that were homozygous for a TaqI polymorphism in intron 1 of this gene not only had more severe coronary disease, but also had a greater response to statin therapy (Kuivenhoven et al. 1998). While these results are somewhat controversial (Winkelmann et al. 2000), further studies are needed to demonstrate whether the *CETP* gene will prove useful for risk assessment and pharmacogenetics.

2.3
Blood Pressure Pathways: Angiotensinogen, Angiotensin-Converting Enzyme, G-Protein $\beta 3$ Subunit, α-Adducin

The renin/angiotensin/aldosterone system (RAAS) plays a crucial role in BP and blood volume homeostasis and is one of the most well-characterized pathways in human physiology. Angiotensin (AGT) sits at the head of this pathway, and a polymorphism in the *AGT* gene (Met235Thr) has been shown to increase circulating levels of AGT and, through linkage analysis, to be associated with essential hypertension (HTN) (Jeunemaitre et al. 1992). The initial linkage of *AGT* to HTN was an important milestone, yet the association of this locus with HTN did not prove that quantitative changes in AGT levels could cause changes in BP. As will be discussed in more detail later in the chapter, genetic studies in the mouse are very effective at testing causation. By using an ingenious gene titration approach in mice, Kim and Smithies (Kim et al. 1995) were able to demonstrate that varying the copy number of the *Agt* gene (from one to four copies of the gene) resulted in linear changes in both AGT plasma levels and BP. The relationship of the *AGT* polymorphism to human CVD/MI is less certain, with several studies reporting a positive association (Katsuya et al. 1995; Winkelmann et al. 1999) but others reporting no association (Ludwig et al. 1997; Tiret et al. 1995). Nevertheless, this variant may have effects on the response to anti-hypertensive therapy, as the *Thr235* allele is associated with a greater response to angiotensin converting enzyme (ACE) inhibitors (Hingorani et al. 1995).

Another gene in this pathway that has been extensively investigated is the *ACE* gene, partly because of the clinical efficacy of drugs that interfere with this pathway and decrease BP and CVD mortality (Yusuf et al. 2000b). One insertion/deletion (*I/D*) mutation in the *ACE* gene (defined by either the presence or absence of a 250-bp fragment in the coding sequence) has received particular attention because of its influence on plasma ACE activity (Rigat et al. 1990). While controversy exists as to the contribution of this *ACE* polymorphism to MI/CVD, most large-scale studies have been negative (Agerholm-Larsen et al. 1997; Cambien et al. 1992; Katsuyaet al. 1995; Keavney et al. 2000; Lindpaintner et al. 1995; Winkelmann and Hager 2000; Winkelmann et al. 1996), suggesting that this polymorphism is not important in risk stratification. In mice, varying the copy number of the *Ace* gene (from one to three copies) results in a change in plasma ACE activity but does not affect either BP or atherosclerosis risk (Krege et al. 1997a, b; Smithies et al. 2000), suggesting that moderate changes in ACE activity, analogous to the situation in humans homozygous for the *I* allele, may not result in an overt clinical phenotype. Nevertheless, the human I/D polymorphism may play a role in pharmacogenetics. Although some studies have not demonstrated a difference in blood pressure response to ACE inhibitors, calcium channel blockers or β-adrenergic blockade (β-blockers) (Dudley et al. 1996; Hingorani et al. 1995; Sasaki et al. 1996; Turner et al. 2001b), a recent study of 328 patients with heart failure demonstrated that patients with a *D* allele of the *ACE* gene had a worse prognosis, which could be improved by β-adrenergic blockade (McNamara et al. 2001).

Although a link to atherosclerosis has not been established, two further examples of genetic polymorphisms that may affect the response to drug therapy emphasize the expanding role of pharmacogenetics. Recently, a polymorphism (Cys825Thr) in the $β_3$-subunit of G proteins encoded by the gene *GNB3* was found to influence the response of BP to monotherapy with hydrochlorothiazide (HCTZ) in a study of almost 400 subjects: *TT* homozygotes had a greater response (by 6 mmHg) compared to *CC* homozygotes, with heterozygotes displaying an intermediate phenotype (Turner et al. 2001a). Thus, this polymorphism may be important in tailoring antihypertensive therapy. Guided by findings in Milan hypertensive rats, similar studies involving adducin (a heterodimeric cytoskeletal protein involved in renal tubular reabsorption of sodium) have demonstrated that patients that are heterozygotes for a Trp460 variant of the α-adducin gene have a greater response to diuretic therapy than Gly460 homozygotes (Cusi et al. 1997; Glorioso et al. 1999), although its relationship to essential HTN and atherosclerosis remain in question (Bray et al. 2000b).

3
The Interplay Between Genetic Analyses of Atherosclerosis in Humans and Animal Models: Pathogenic Insights and New Targets for Therapy

3.1
Rationale

Since pioneering epidemiological studies in the 1940s–1960s, major strides have been made in the treatment of CVD (Braunwald 1997). In the last 50 years, morbidity and mortality have decreased in the developed world (Tyroler 2000). Most of the early gains were due to the peri-MI rescue of patients with advanced cardiac disease and paralleled the rise in coronary care units (Braunwald 1997). Later gains (especially over the last 20 years) can be partly credited to better control of modifiable risk factors and partly to advances in pharmacological and interventional therapy (Rosamond et al. 1998). As in all areas, drug therapy for CVD has been guided by a better understanding of the pathogenesis of atherosclerosis, and our understanding is increasingly governed by our knowledge of the genetic basis of the disease.

Despite great progress in the development of certain highly effective medications for the prevention of CVD, our pharmacological armamentarium is still limited to medications in a handful of classes that work though a relatively small number of pathways. For instance, treatment of hyperlipidemia is mostly limited to the use of statins, niacin (or niacin derivatives), bile acid sequestrants, and fibrates (such as fenofibrate and gemfibrozil). Similarly, chronic HTN is managed with drugs acting principally through four major mechanisms: blockade of the RAAS with ACE-I or angiotensin receptor blockers (ARBs); calcium channel blockade; induction of diuresis (with thiazide or loop diuretics); and

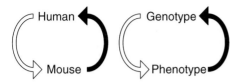

Fig. 5 The dynamic relationship between human and mouse studies. Genetic analysis of atherosclerosis benefits from going back and forth between humans and mice. This approach is useful both for testing hypotheses, finding new targets, and reaching a better understanding of pathogenesis

blockade of adrenergic signaling. The use of genetic analysis to identify new players and pathways in the pathogenesis of atherosclerosis is of crucial importance in developing new therapeutic options (Fig. 3).

The pathogenesis of atherosclerosis is complex, polygenic, and multifactorial. These factors, along with the difficulties inherent with human investigation, have made the study of atherosclerosis in humans costly and laborious. Even with these obstacles, human studies have yielded remarkable results. The most famous example is the elucidation of cholesterol homeostasis pathways based on studies in patients with familial hypercholesterolemia and the subsequent development of HMG-CoA reductase inhibitors (statins) as therapy for hypercholesterolemia (Brown and Goldstein 1986).

Nevertheless, most advances in the identification of the genetic and environmental determinants of atherosclerosis have been greatly facilitated by the interplay between studies in humans and in various animal models including monkeys, rabbits, and rats. Even the studies of familial hypercholesterolemia were helped by concomitant studies in Watanabe hyperlipidemic rabbits (Brown and Goldstein 1986). In terms of genetic studies, recent efforts at sequencing the rat genome along with the comprehensive use of phenotyping with linkage analysis have yielded promising data (Jacob and Kwitek 2002; Stoll et al. 2001). However, the unique ability to modify genes through the use of gene targeting has made the mouse the most important model organism for the study of the genetics of atherosclerosis. In particular, mice deficient for either apolipoprotein E (*Apoe*-deficient) or the LDL receptor (*Ldlr*-deficient) have been used extensively in the study of atherosclerosis (Piedrahita et al. 1992; Plump et al. 1992; Zhang et al. 1992). These mutated mice have proven especially valuable in two areas: (1) identification and characterization of genes and pathways involved in the disease process and (2) evaluating genes for causation rather than just association (Fig. 5).

For considerations of space, it is not possible to review all the genetic polymorphisms that have been associated with atherosclerosis, which have been reviewed extensively in other places (see the chapters by Winkelmann et al., O'Shaughnessy and Wilkins, Vidal-Puig and Abel, Huang, Tuddenham, Haskard, and Henney, this volume; Glass and Witztum 2001; Kadowaki 2000; Knowles and Maeda 2000; Lusis 2000; Moller 2001; Plump 1997; Rader and Pure 2000; Reardon and Getz 2001; Saltiel and Kahn 2001; Smith and Breslow 1997;

Winkelmann et al. 2000). Very few polymorphisms initially associated with atherosclerosis in humans have been definitively validated by later studies with large-scale databases. Moreover, most of the polymorphisms associated with atherosclerosis have not been proven to play a causative role in the disease (Winkelmann and Hager 2000; Winkelmann et al. 2000). However, parallel and iterative work in humans and animal models, particularly mice, has begun to shed light on the molecular mechanisms of atherosclerosis. Several prototypical examples from the areas of lipid metabolism, HTN, and inflammation serve to illustrate the utility of human and murine studies in evaluating potentially important genes and the importance of the dynamic relationship between human and animal studies for providing novel pathogenic insights as well as new targets for therapy.

3.2
Lipid Pathways: Adenosine Triphosphate-Binding Cassette Transporter A1, Cholesterol 7α-hydroxylase, Sterol Regulatory Element-Binding Proteins, Apolipoprotein AV, Fatty-Acid-Binding Protein aP2, Plasma Phospholipid-Transfer Protein

Perhaps the most spectacular story in the genetic analysis of atherosclerosis in the last several years is the discovery of the gene responsible for Tangier disease, because of the implications both for understanding pathogenesis and for therapeutic options. Tangier disease is a rare condition of defective reverse cholesterol transport, and characterized in the homozygous state by the virtual absence of HDL, decreased LDL levels, and moderate elevations in triglycerides (TG). The abnormal lipid metabolism results in cholesterol deposition in multiple organs, peripheral neuropathy and early CVD (Hobbs and Rader 1999; Serfaty-Lacrosniere et al. 1994). In 1999, three independent groups identified the underlying cause as a defect in the adenosine triphosphate-binding cassette transporter A1 (*ABCA1*, also called *ABC-1*) gene (Bodzioch et al. 1999; Brooks-Wilson et al. 1999; Rust et al. 1999). While homozygous Tangier disease is rare, heterozygote defects in *ABCA1* have also been linked to a more common disorder, familial HDL deficiency (FHA) (Marcil et al. 1999; Owen 1999). Importantly, heterozygotes for *ABCA1*-null defects also have both a biochemical (decreased HDL, increased TG) and clinical phenotype (increased CAD) (Clee et al. 2000). Subsequent studies in vitro and in mice (Lawn et al. 1999; Singaraja et al. 2002) have demonstrated the key role that ABCA1 plays in reverse cholesterol efflux and atherosclerosis. Pharmacological interventions in this pathway to increase HDL levels may provide a new strategy for the treatment of CVD (Schmitz and Langmann 2001).

Mice have been integral in the study of *ABCA1*, since the cloning of the gene by Luciani and colleagues in 1994 (Luciani et al. 1994). The discovery that mutations in the *ABCA1* gene were responsible for Tangier disease opened up a new direction in lipid research that has benefited from parallel work in mice, which focused on the mechanisms of action of this protein. Mice deficient in

ABCA1 have a phenotype that mimics the Tangier disease phenotype in humans, including absence of HDL, decreased total cholesterol, decreased cholesterol efflux from cells and lipid deposition in several tissues (Christiansen-Weber et al. 2000; Orso et al. 2000). *Apoe*-deficient mice crossed with mice that overexpress human ABCA1 develop much smaller lesions than *Apoe*-deficient control mice and have greater cholesterol efflux from isolated macrophages (Singaraja et al. 2002).

Pullinger and colleagues used a candidate gene approach in a cohort of patients with known hyperlipidemia and discovered that mutations in *CYP7A1*, a gene critical in bile acid synthesis, is associated with hypercholesterolemia, markedly decreased bile acid excretion and premature gallbladder disease. Heterozygotes for the mutation are also hyperlipidemic. This gene also has potential implications for pharmacogenetics, as the nature of this defect makes these patients relatively insensitive to statin therapy (Pullinger et al. 2002). The phenotype of mice deficient in *Cyp7a1* does not exactly mirror that of humans; the mice are not universally hyperlipidemic and *Cyp7a1* appears to be more critical in the postnatal period for mice than for humans (Beigneux et al. 2002). Nevertheless, it will be interesting to test this mutant as a potential modifier of atherosclerosis.

Pathways that surround the sterol regulatory element-binding proteins (SREBPs) show great promise for the development of anti-atherosclerotic therapeutics. As reviewed by Horton and colleagues, SREBPs are transcription factors that activate the expression of multiple genes involved in lipid metabolism (Horton et al. 2002). To exert their effects in the nucleus, SREBPs must be released from the endoplasmic reticulum membrane, and the proteins involved in this processing include two proteases designated Site-1 and Site-2 proteases (or S1P and S2P), as well as SREBP cleavage-activating protein (SCAP). SCAP also serves as a sensor of sterol levels, and SCAP ligands have recently been shown to be potent lipid-lowering drugs (Grand-Perret et al. 2001; Rader 2001). Knockout and transgenic mouse studies have been crucial in the elucidation of the mechanisms of this intricate pathway (ten different mouse models of this pathway have been generated and characterized (Horton et al. 2002). Although there is no data in humans that directly links mutations in these genes to atherosclerosis, further work using these models will be important in the development of new therapeutic options.

Comparative sequencing of the human and mouse genomes is another useful approach in identifying new candidate genes and pathways. This type of analysis by Pennacchio and colleagues serendipitously revealed a new apolipoprotein gene, *APOAV*, that may play a role in atherosclerosis (Pennacchio et al. 2001). Subsequent studies in transgenic mice expressing human *APOAV* demonstrated that these mice have a decrease in TG levels compared to controls, while mice deficient in *Apoav* have TG levels four times that of control mice. Follow-up studies in humans demonstrated that SNPs across this locus are significantly associated with TG levels in two independent studies. The authors suggest that modulation of APOAV is a potential strategy for reducing TG levels.

Finally, studies in mice have also revealed other potential modifiers of atherosclerosis. Adipocyte fatty-acid-binding protein, aP2 (encoded by *Ap2*), is primarily expressed in adipose tissue and has a crucial role in glucose and lipid metabolism. Mice deficient in *Ap2* have protection against hyperinsulinemia and insulin resistance in the setting of obesity (Hotamisligil et al. 1996). *Apoe*-deficient mice also deficient for *Ap2* are protected from atherosclerosis, and macrophages from these mice have a reduced ability to accumulate cholesterol esters. Thus, this gene is a potential therapeutic target for the prevention of metabolic syndrome and atherosclerosis (Makowski et al. 2001). Similar studies with mice deficient for plasma phospholipid-transfer protein (PLTP), a protein involved in the transfer of excess phospholipids from chylomicrons and VLDL to HDL, have demonstrated that this protein is another potential target for drug development. *Apoe*-deficient mice that lack PLTP had a marked decrease in both apoB-containing lipoproteins, as well as in atherosclerosis (Jiang et al. 2001).

3.3
Blood Pressure Pathways: Endothelial Nitric Oxide Synthase, Neuropeptide Y, β2-Adrenergic Receptor

A considerable body of evidence suggests that HTN (defined at a BP >140/90 mmHg) contributes to the development and progression of atherosclerosis. Atherosclerosis occurs in high-pressure arteries, but not lower pressure veins, with lesions localized at areas of high wall stress. Patients with HTN are three times more likely to develop atherosclerosis, and treatment with antihypertensive medications decreases the risk of death from atherosclerotically mediated cardiovascular events (Chobanian and Alexander 1996; Doyle 1990; Gimbrone 1999; Group 1991; Yusuf et al. 2000b). Rare forms of HTN are inherited in a mendelian fashion (Lifton et al. 2001), but the genetic basis of essential HTN (which affects >90% of hypertensive patients) is largely undefined although, like atherosclerosis, is likely to be a complex disease caused by the cumulative effects of small quantitative changes in the function of numerous susceptibility genes (Smithies et al. 2000). It has proven difficult to isolate genes that contribute to the burden of essential HTN (see the chapter by O'Shaughnessy et al., this volume). Nevertheless, besides *AGT* and *ACE*, which are discussed above, there are several genes that have recently been implicated from human and murine studies. A more comprehensive list of mouse models of HTN has recently been published (Smithies et al. 2000).

One gene that has received considerable attention as a candidate for human HTN and atherosclerosis is the gene for endothelial nitric oxide synthase (eNOS or NOS III) (see the chapter by Huang, this volume). eNOS serves important basal regulatory functions in the vasculature. In response to stimuli such as sheer stress or acetylcholine, eNOS catalyzes the production of NO, which diffuses across the endothelial cell membrane to smooth muscle cells to induce vasodilation. It also acts locally to prevent platelet and leukocyte adhesion

(Moncada and Higgs 1993). Direct evidence that lack of eNOS could cause HTN was provided by Huang et al. and Shesely et al. (Huang et al. 1995; Shesely et al. 1996). However, evidence that a deficit of eNOS could contribute to enhanced atherosclerosis was missing until mice were developed that lacked both apoE and eNOS. Mice deficient in both of these genes develop much larger lesions than *Apoe*-deficient control mice (including coronary vessel atherosclerosis), have aortic aneurysms, and have evidence of renal damage, both of which can be prevented by treatment with an ACE-I and to a lesser extent by a diuretic (Knowles and Maeda 2001; Knowles et al. 2000). This work has been extended to show that these double knockout mice develop evidence of myocardial ischemia if fed an atherosclerotic diet (Kuhlencordt et al. 2001). Recently, a common variant of eNOS (Glu298Asp) has been shown to be a major risk factor for CAD (Hingorani et al. 1999). Thus, parallel experiments in humans and mice have reinforced the notion that eNOS may play an important role in human HTN and atherosclerosis.

In a population-based sample of 966 men in Finland, a polymorphism (Leu7-Pro) in the neuropeptide Y (*NPY*) gene was significantly associated with slightly increased BPs and an increase in atherosclerosis. NPY is a neurotransmitter with multiple actions including regulating appetite, insulin release, vasoconstriction, and vascular smooth muscle cell mitogenesis (Karvonen et al. 2001). Although NPY-null mice are remarkably normal, neither this gene, nor the genes for NPY receptors, have yet been tested for effects on atherosclerosis in mice (Naveilhan et al. 1999; Palmiter et al. 1998). The pleiotropic nature of this gene makes it an attractive candidate for moderating parameters associated with the metabolic syndrome including HTN, thereby potentially affecting atherosclerosis.

Using genome-wide linkage analysis followed by sequencing of positional candidate genes, Bray et al. found that two variations, Arg16Gly and Gln27Glu, in the β_2-adrenergic receptor gene (*ADRB2*) were significantly associated with HTN (Bray et al. 2000a). These polymorphisms are estimated to only account for about 2% of the variation in BP in this sample population, once again demonstrating the polygenic nature of essential HTN and highlighting the difficulties of identifying genetic effectors. Mice deficient in the β_2-adrenergic receptor have normal resting BP but an enhanced hypertensive response to exercise or epinephrine challenge and a blunted hypotensive response to isoproterenol challenge (Chruscinski et al. 1999), suggesting that this gene may moderate BP under stressful circumstances. It will be important to ascertain whether these polymorphisms contribute to atherosclerosis.

3.4
Inflammatory Pathways: Macrophage Colony-Stimulating Factor, Monocyte Chemoattractant Protein-1, A2a Adenosine Receptor, Myeloperoxidase, 12/15 Lipoxygenase, p47phox

The final common pathway in the development and progression of atherosclerosis is the inflammatory response to vascular damage. Inflammatory cells are found in every stage of lesion development and inflammatory markers in plasma such as C-reactive protein (CRP), fibrinogen, and leukocyte count are also elevated in patients with atherosclerosis (Blake and Ridker 2001; Danesh et al. 1998; Rodker et al. 2000a; Tracy et al. 1997). It is therefore not surprising that there has been considerable interest in the search for polymorphisms in inflammatory genes in atherosclerotic populations. While levels of some of these inflammatory markers are highly heritable (Pankow et al. 2001), it has proven difficult to isolate the genes responsible. Data from association studies are preliminary and partly conflicting (reviewed in Andreotti et al. 2002). Nevertheless, some intriguing results such as polymorphisms in toll-like receptor 4 (TLR4), E selectin, and high levels of CRP and interleukin-6 (IL-6) being associated with atherosclerosis (Blake and Ridker 2001; Kiechl et al. 2002; Ridker et al. 2000a,b; Taubes 2002; Zee and Ridker 2002) should be the basis of further exploration.

Finding a specific cell type or cytokine in an atherosclerotic lesion does not reveal whether this cell/molecule plays a role in the disease or is just an innocent bystander. By using genetic manipulations in *Apoe-* and *Ldlr-*deficient mice, it has been possible to prove the causal role of certain inflammatory genes in the modification of atherosclerosis. Many of these experiments have been reviewed (Knowles and Maeda 2000), and others are extensively covered in the chapters by Winkelmann et al. and Haskard in the present volume. However, some past and recent examples of this approach may provide insight.

The critical role of macrophages in the development of atherosclerosis has been buttressed by experiments in mice. Osteopetrotic mice (*op*) have a defect in macrophage colony stimulating factor (MCSF) that causes a severe deficiency in macrophages and monocytes. *Apoe-*deficient mice crossed with *op* mice have much smaller atherosclerotic lesions (one-fifth as large) than control animals (Smith et al. 1995). The role of the monocyte chemoattractant protein-1 (MCP-1) signaling pathway in atherosclerosis has been repeatedly shown in murine systems. MCP-1 is a cytokine that acts through its receptor (CC chemokine receptor 2 or CCR2) on monocytes, macrophages and T lymphocytes. Irradiated *Apoe-*deficient mice in which the marrow has been replaced with cells that overexpress MCP-1 have an increase in lesion size (Aiello et al. 1999). In the opposite direction, the absence of MCP-1 decreases the lesion size five times in LDLR-deficient mice (GU et al. 1998), and the absence of CCR2 also acts to decrease lesion size in *Apoe-*deficient mice (Boring et al. 1998; Dawson et al. 1999).

Murine experiments have also identified a novel class of anti-inflammatory molecules that may function as modifiers of atherosclerosis. Agonists for cell surface purinergic receptors act by increasing levels of cAMP in immune cells.

It was thought that this mechanism could decrease inflammation in vivo, but it was unproven until mice deficient in A2a adenosine receptors gene (*Adora2a*) were generated and characterized. *Adora2a*-deficient mice had a runaway immune response to several inflammatory challenges, including extensive tissue damage and death (Ohta and Sitkovsky 2001). This type of research has already stimulated interest into drugs that interact with this system.

Experiments in mice can also yield unexpected results that can change our way of thinking about the disease process. Myeloperoxidase (MPO) generates oxidants critical to host defense and local tissue damage. MPO is present in atherosclerotic plaques and is an attractive candidate for modification of plaques through oxidant damage. However, the causal role of MPO had not been investigated until Brennan et al. irradiated *Ldlr*-deficient mice to remove their native bone marrow and repopulated the marrow with MPO-deficient or wild-type cells. Unexpectedly, *Ldlr*-deficient mice with MPO-deficient marrow developed lesions 50% larger than control mice (Brennan et al. 2001). The authors suggest a protective role for MPO-generated intermediates in atherosclerosis. Further investigation of this paradoxical result will be intriguing, especially in light of recent work from Eiserich and colleagues demonstrating that MPO can directly modulate vascular inflammatory responses by regulating NO bioavailability (Eiserich et al. 2002).

Since it was discovered that oxidization of LDL is atherogenic, oxidative stress has been hypothesized to be an important contributor to the pathophysiology of atherosclerosis, perhaps as a common effector of many pro-atherosclerotic pathways (Heinecke 1998). Inflammatory and hypertensive pathways as well as hyperglycemia have been shown to increase oxidative stress through the generation of reactive oxygen species (ROS) (Berliner et al. 1995; Navab et al. 1996). However, proving a causal role for oxidation in atherosclerosis has been difficult. Studies of antioxidant drugs in animal models have been conflicting (Heinecke 1998; Steinberg 1997; Witting et al. 1999; Zhang et al. 1997), as have studies in humans (Rimm et al. 1993; Stephens et al. 1996; Yusuf et al. 2000a). The fact that oxidation has been so closely linked to atherogenesis yet large scale trials have not shown a benefit to antioxidant therapy has been referred to as the oxidative paradox (Patterson et al. 2000).

Recent work in animal models has begun to help decipher this paradox, with several genes involved in inflammation/oxidation tested for atherosclerotic causality in mice. For instance, 12/15-lipoxygensase is involved in the formation of the inflammatory mediators 12-HETE and 15-HETE from arachidonic acid. It can also act to oxidize cholesterol esters and phospholipids. Because LDL oxidation plays a role in atherosclerosis and because 12/15-lipoxygenase colocalizes in atherosclerotic plaques, this gene was an attractive (but unproven) candidate for modifying atherosclerosis. *Apoe*-deficient mice also deficient in 12/15 lipoxygenase had much smaller lesions than control animals. The authors suggest that inhibition of this enzyme may decrease atherosclerosis, although the mechanism of action remains uncertain (Cyrus et al. 1999).

Along this same line are experiments in mice lacking p47*phox*, a subunit of NADPH oxidase which is responsible for generation of reactive oxidation species (ROS) in vascular cells. Smooth muscle cells from mice deficient in p47*phox* produce less ROS in response to growth factor stimulation and mice doubly deficient in p47*phox* and *Apoe* have smaller atherosclerotic lesions than *Apoe*-deficient mice (Barry-Lane et al. 2001). Thus, in this area work in mice is opening new avenues of exploration for study in humans.

4
The Future of Genetic Analyses in Human and Animal Models of Atherosclerosis

Genetic analysis of complex diseases such as atherosclerosis is entering a new era. Large scale sequencing efforts in humans and mice are expected to lead to the identification of new genes and pathways that modify atherosclerosis, and by extension, to the identification of unique targets for drug action (Charron and Komajda 2001; Komajda and Charron 2001a,b; Rubin and Tall 2000; Turner and Boerwinkle 2000; Turner et al. 2001b). In addition, with the cost of sequencing the entire genome of an individual expected to be in the thousands of dollars in the next 10 years, it will be possible to pursue a different approach to genetic epidemiology. Dense SNP analysis on a large scale, augmented by haplotype analysis, will identify patterns of SNPs within populations that are associated with atherosclerosis (see the chapter by Tate and Goldstein, this volume). This kind of pattern recognition may be especially fruitful for several reasons (Daly et al. 2001; Goldstein 2001; Johnson et al. 2001; Nickerson et al. 2000; Olivier et al. 2001; Patil et al. 2001). First, these studies may yield important clues about combinations of alleles that increase susceptibility to atherosclerosis. Second, these studies may reveal elements of noncoding DNA that have effects on atherosclerosis; the relevance of these noncoding regions is emphasized by recent studies showing a high degree of conservation between humans and mice in some chromosomal areas (Copeland et al. 2002; Mural et al. 2002; Pennisi 2002). Third, if these kinds of large scale SNP analyses are undertaken in populations that have already been well-characterized in terms of environmental exposure, we will learn a great deal about gene/environment interactions (Boerwinkle et al. 2000; Hauser and Pericak-Vance 2000; Kraus 2000; Turner et al. 2001b). For example, association between the null polymorphism of glutathione *S*-transferase and cigarette smoking in relation to CAD has been recently described (Li et al. 2000).

Efforts could yield great progress if the problem is attacked with rigor. However, failure to design adequate studies will be costly in terms of time, effort, and money. The pitfalls of this type of analysis as by Winkelmann et al. (Winkelmann and Hager 2000; Winkelmann et al. 2000) include underpowered studies and poorly controlled studies (Dahlman et al. 2002). In addition, genes associated with atherosclerosis must be validated for causation in humans, in vitro, or more commonly in animal models.

Genetic mouse models of atherosclerosis have been incredibly powerful in helping to dissect the nature of the disease, and the pathogenesis of the disease is remarkably similar across species (Rubin and Smith 1994). Nevertheless, like human studies, murine studies must be pursued with rigor. Although humans and mice share many similarities, it is important to recognize that there are some differences between murine and human physiology that may have important consequences for certain types of atherosclerotic studies. For instance, while human males are more affected by atherosclerosis, neither *Apoe-* nor *Ldlr-*deficient mice have this same gender predilection. There are also differences in lipid metabolism between mice and humans, resulting in somewhat different lipid profiles. For example, mice lack cholesterol ester transfer protein (CETP) and apolipoprotein(a), and the plasma cholesterol levels in both *Apoe-* and *Ldlr-*deficient mice are greater than those of a typical human population (especially if atherogenic diets are used), with marked elevations in VLDL and remnant cholesterol (Rader and Fitzgerald 1998).

There are also well-characterized differences between various strains of mice which can have profound effects on atherosclerotic susceptibility, thus these experiments must be rigorously controlled (Dansky et al. 1999; Knowles and Maeda 2000; Krege 1996; Shi et al. 2000; Smithies and Maeda 1995). Finally, while the general progression of atherosclerosis in mice and humans is very similar, plaque rupture and subsequent thrombosis in mice is very rare and only seen in extreme circumstances (Caligiuri et al. 1999), making it difficult to analyze this stage of atherosclerosis in mice. However, a recently reported double knockout mouse of *Apoe* and the SR-BI receptor does develop coronary artery occlusion and MI (Braun et al. 2002) and may be a major advance in this arena.

It is also important to emphasize that mouse knockout experiments are the genetic equivalent of recessively inherited mutations in humans caused by a loss of gene function. This type of mutation makes only a small contribution to atherosclerosis in humans (which is more likely due to the total burden of a combination of small quantitative changes in gene function). The polygenic nature of atherosclerosis in humans has several important implications for studies in mice. First, it is important to study the effects of quantitative changes in gene function in mice, analogous to the effect of a heterozygous knockout as well as overproduction at physiological levels. Common allelic variants that cause quantitative alterations in gene expression are more likely to play a larger role in complex human disease because they affect a large number of people. Combining numerous mutations, each with small quantitative effects, will remain a major challenge (Knowles and Maeda 2000; Smithies et al. 2000). Finally, the identification of polymorphisms that have small quantitative effects in mouse genes that may play a role in CVD also remains a challenge, and ethyl-nitrosourea (ENU) mutagenesis screens with proper phenotyping are likely to be helpful (Chen et al. 2000; Hrabe De Angelis et al. 2000; Nolan et al. 2000).

5
Conclusions

The next few years are potentially a period of great advancement in our understanding and treatment of CVD. Getting a handle on the massive amounts of genetic information and putting this information to use is going to be the great scientific challenge of the early twenty-first century. In the meantime, it is absolutely imperative to use all currently available means to treat CVD. It is a terrible tragedy that 50% of individuals with CVD, HTN, and DM are not being treated adequately with currently available regimens (Braunwald 1997; Waters 2001). It is also critically important to note that simply understanding the genetics of atherosclerosis will never fully enable us to stop the epidemic of disease. Environmental risk factors, such as low socioeconomic status and smoking, also play an important role and must be addressed in order to have the greatest impact on CVD. A balanced approach of addressing life-style changes along with genetically based therapies will be the most effective in disease prevention (Willett 2002). By extension, less direct pharmacological intervention, such as treatments that decrease smoking, could have considerable influence on CVD. Finally, the new era of genetic discovery also has its own unique problems. The adverse ethical implications of being able to identify susceptibility genes for atherosclerosis cannot be overestimated, especially in a time when discrimination based on genetics has not been outlawed (Reilly 2000). Nevertheless, our rapidly expanding knowledge of the genetic determinants of atherosclerosis is likely to lead to an enhanced ability to identify disease-prone individuals and design specific preventative measures and tailored therapies.

Acknowledgements. I would like to thank Dr. Nobuyo Maeda for her guidance, including helpful discussions and comments. Drs. Kari North and Cam Patterson also provided useful comments.

References

Agerholm-Larsen B, Nordestgaard BG, Steffensen R et al (1997) ACE gene polymorphism: ischemic heart disease and longevity in 10,150 individuals. A case-referent and retrospective cohort study based on the Copenhagen City Heart Study. Circulation 95:2358–2367

Aiello RJ, Bourassa PA, Lindsey S et al (1999) Monocyte chemoattractant protein-1 accelerates atherosclerosis in apolipoprotein E-deficient mice. Arterioscler Thromb Vasc Biol 19:1518–1525

Andreotti F, Porto I, Crea F et al (2002) Inflammatory gene polymorphisms and ischaemic heart disease: review of population association studies. Heart 87:107–112

Anonymous (1999) The World Health Report. World Health Organization, Geneva

Anonymous (2002a) 2002 heart and stroke statistical update. American Heart Association, Dallas

Anonymous (2002b) 2002 international cardiovascular disease statistics. American Heart Association, Dallas

Barry-Lane PA, Patterson C, van der Merwe M et al (2001) p47phox is required for atherosclerotic lesion progression in ApoE(-/-) mice. J Clin Invest 108:1513–1522

Beigneux A, Hofmann AF, Young SG (2002) Human CYP7A1 deficiency: progress and enigmas. J Clin Invest 110:29–31

Berliner JA, Navab M, Fogelman AM et al (1995) Atherosclerosis: basic mechanisms. Oxidation, inflammation, and genetics. Circulation 91:2488–2496

Blake GJ, Ridker PM (2001) Novel clinical markers of vascular wall inflammation. Circ Res 89:763–771

Bodzioch M, Orso E, Klucken J et al (1999) The gene encoding ATP-binding cassette transporter 1 is mutated in Tangier disease. Nat Genet 22:347–351

Boerwinkle E, Ellsworth DL, Hallman DM et al (1996) Genetic analysis of atherosclerosis: a research paradigm for the common chronic diseases. Hum Mol Genet 5:1405–1410

Boerwinkle E, Hixson JE, Hanis CL (2000) Peeking under the peaks: following up genome-wide linkage analyses. Circulation 102:1877–1878

Boring L, Gosling J, Cleary M et al (1998) Decreased lesion formation in CCR2-/- mice reveals a role for chemokines in the initiation of atherosclerosis. Nature 394:894–897

Braun A, Trigatti BL, Post MJ et al (2002) Loss of SR-BI expression leads to the early onset of occlusive atherosclerotic coronary artery disease, spontaneous myocardial infarctions, severe cardiac dysfunction, and premature death in apolipoprotein E-deficient mice. Circ Res 90:270–276

Braunwald E (1997) Shattuck lecture–cardiovascular medicine at the turn of the millennium: triumphs, concerns, and opportunities [see comments]. N Engl J Med 337:1360–1369

Bray MS, Krushkal J, Li L, Ferrell R et al (2000a) Positional genomic analysis identifies the beta(2)-adrenergic receptor gene as a susceptibility locus for human hypertension. Circulation 101:2877–2882

Bray MS, Li L, Turner ST et al (2000b) Association and linkage analysis of the alpha-adducin gene and blood pressure. Am J Hypertens 13:699–703

Brennan ML, Anderson MM, Shih DM et al (2001) Increased atherosclerosis in myeloperoxidase-deficient mice. J Clin Invest 107:419–430

Broeckel U, Hengstenberg C, Mayer B et al (2002) A comprehensive linkage analysis for myocardial infarction and its related risk factors. Nat Genet 30:210–214

Brooks-Wilson A, Marcil M, Clee SM et al (1999) Mutations in ABC1 in Tangier disease and familial high-density lipoprotein deficiency. Nat Genet 22:336–345

Brown MS, Goldstein JL (1986) A receptor-mediated pathway for cholesterol homeostasis. Science 232:34–47

Caligiuri G, Levy B, Pernow J et al (1999) Myocardial infarction mediated by endothelin receptor signaling in hypercholesterolemic mice. Proc Natl Acad Sci U S A 96:6920–6924

Cambien F, Poirier O, Lecerf L et al (1992) Deletion polymorphism in the gene for angiotensin-converting enzyme is a potent risk factor for myocardial infarction [see comments]. Nature 359:641–644

Cargill M, Altshuler D, Ireland J et al (1999) Characterization of single-nucleotide polymorphisms in coding regions of human genes. Nat Genet 22:231–238

Carmena R, Roederer G, Mailloux H et al (1993) The response to lovastatin treatment in patients with heterozygous familial hypercholesterolemia is modulated by apolipoprotein E polymorphism. Metabolism 42:895–901

Castelli WP (1996) Lipids, risk factors and ischaemic heart disease. Atherosclerosis 124 Suppl: S1–9

Castelli WP (2001) Making practical sense of clinical trial data in decreasing cardiovascular risk. Am J Cardiol 88:16F-20F

Charron P, Komajda M (2001) Are we ready for pharmacogenomics in heart failure? Eur J Pharmacol 417:1-9

Chen Y, Yee D, Dains K et al (2000) Genotype-based screen for ENU-induced mutations in mouse embryonic stem cells. Nat Genet 24:314–317

Chobanian AV, Alexander RW (1996) Exacerbation of atherosclerosis by hypertension. Potential mechanisms and clinical implications [see comments]. Arch Intern Med 156:1952–1956

Christiansen-Weber TA, Voland JR, Wu Y et al (2000) Functional loss of ABCA1 in mice causes severe placental malformation, aberrant lipid distribution, and kidney glomerulonephritis as well as high-density lipoprotein cholesterol deficiency. Am J Pathol 157:1017–1029

Chruscinski AJ, Rohrer DK, Schauble E et al (1999) Targeted disruption of the beta2 adrenergic receptor gene. J Biol Chem 274:16694–16700

Clee SM, Kastelein JJ, van Dam M et al (2000) Age and residual cholesterol efflux affect HDL cholesterol levels and coronary artery disease in ABCA1 heterozygotes. J Clin Invest 106:1263–1270

Colditz GA, Rimm EB, Giovannucci E et al (1991) A prospective study of parental history of myocardial infarction and coronary artery disease in men. Am J Cardiol 67:933–938

Collins FS, Guyer MS, Charkravarti A (1997) Variations on a theme: cataloging human DNA sequence variation. Science 278:1580–1581

Copeland NG, Jenkins NA, O'Brien SJ (2002) Genomics. Mmu 16–comparative genomic highlights. Science 296:1617–1618

Cusi D, Barlassina C, Azzani T et al (1997) Polymorphisms of alpha-adducin and salt sensitivity in patients with essential hypertension. Lancet 349:1353–1357

Cyrus T, Witztum JL, Rader DJ et al (1999) Disruption of the 12/15-lipoxygenase gene diminishes atherosclerosis in apo E-deficient mice [see comments]. J Clin Invest 103:1597–1604

Dahlman I, Eaves IA, Kosoy R et al (2002) Parameters for reliable results in genetic association studies in common disease. Nat Genet 30:149–150

Daly MJ, Rioux JD, Schaffner SF et al (2001) High-resolution haplotype structure in the human genome. Nat Genet 29:229–232

Danesh J, Collins R, Appleby P et al (1998) Association of fibrinogen, C-reactive protein, albumin, or leukocyte count with coronary heart disease: meta-analyses of prospective studies. Jama 279:1477–1482

Dansky HM, Charlton SA, Sikes JL et al (1999) Genetic background determines the extent of atherosclerosis in ApoE- deficient mice. Arterioscler Thromb Vasc Biol 19:1960–1968

Davies MJ (2001) Going from immutable to mutable atherosclerotic plaques. Am J Cardiol 88:2F-9F

Dawson TC, Kuziel WA, Osahar TA, et al (1999) Absence of CC chemokine receptor-2 reduces atherosclerosis in apolipoprotein E-deficient mice. Atherosclerosis 143:205–211

Doyle AE (1990) Does Hypertension Predispose to Coronary Disease? Conflicting Epidemiological and Experimental Evidence. In: Brenner JHLaBM (ed) Hypertension: Pathophysiology, Diagnosis and Management, vol 1. Raven Press, Ltd, New York, pp 119–125

Dudley C, Keavney B, Casadei B et al (1996) Prediction of patient responses to antihypertensive drugs using genetic polymorphisms: investigation of renin-angiotensin system genes. J Hypertens 14:259–262

Eichner JE, Kuller LH, Orchard TJ et al (1993) Relation of apolipoprotein E phenotype to myocardial infarction and mortality from coronary artery disease. Am J Cardiol 71:160–165

Eiserich JP, Baldus S, Brennan ML et al (2002) Myeloperoxidase, a leukocyte-derived vascular NO oxidase. Science 296:2391–2394

Ellsworth DL, Sholinsky P, Jaquish C et al (1999) Coronary heart disease. At the interface of molecular genetics and preventive medicine. Am J Prev Med 16:122–133

Galton DJ, Mattu R, Needham EW et al (1996) Identification of putative beneficial mutations for lipid transport. Z Gastroenterol 34 Suppl 3:56–58

Gimbrone MA, Jr. (1999) Vascular endothelium, hemodynamic forces, and atherogenesis [comment]. Am J Pathol 155:1-5

Glass CK, Witztum JL (2001) Atherosclerosis. The Road Ahead. Cell 104:503–516

Glorioso N, Manunta P, Filigheddu F et al (1999) The role of alpha-adducin polymorphism in blood pressure and sodium handling regulation may not be excluded by a negative association study. Hypertension 34:649–654

Goldstein DB (2001) Islands of linkage disequilibrium. Nat Genet 29:109–111

Gorman C (2001) Heart disease. Rethinking treatments for the heart. Time 157:85–86

Grand-Perret T, Bouillot A, Perrot A et al (2001) SCAP ligands are potent new lipid-lowering drugs. Nat Med 7:1332–1338

Group SCR (1991) Prevention of stroke by antihypertensive drug treatment in older persons with isolated systolic hypertension. Final results of the Systolic Hypertension in the Elderly Program (SHEP). [see comments]. Jama 265:3255–3264

Gu L, Okada Y, Clinton SK et al (1998) Absence of monocyte chemoattractant protein-1 reduces atherosclerosis in low density lipoprotein receptor-deficient mice. Mol Cell 2:275–281

Halushka MK, Fan JB, Bentley K et al (1999) Patterns of single-nucleotide polymorphisms in candidate genes for blood-pressure homeostasis. Nat Genet 22:239–247

Hauser ER, Pericak-Vance MA (2000) Genetic analysis for common complex disease. Am Heart J 140: S36–44

Heinecke JW (1998) Oxidants and antioxidants in the pathogenesis of atherosclerosis: implications for the oxidized low density lipoprotein hypothesis. Atherosclerosis 141: 1-15

Hingorani AD, Jia H, Stevens PA et al (1995) Renin-angiotensin system gene polymorphisms influence blood pressure and the response to angiotensin converting enzyme inhibition. J Hypertens 13:1602–1609

Hingorani AD, Liang CF, Fatibene J et al (1999) A common variant of the endothelial nitric oxide synthase (Glu298- >Asp) is a major risk factor for coronary artery disease in the UK. Circulation 100:1515–1520

Hobbs HH, Rader DJ (1999) ABC1: connecting yellow tonsils, neuropathy, and very low HDL. J Clin Invest 104:1015–1017

Horton JD, Goldstein JL, Brown MS (2002) SREBPs: activators of the complete program of cholesterol and fatty acid synthesis in the liver. J Clin Invest 109:1125–1131

Hotamisligil GS, Johnson RS, Distel RJ et al (1996) Uncoupling of obesity from insulin resistance through a targeted mutation in aP2, the adipocyte fatty acid binding protein. Science 274:1377–1379

Hrabe de Angelis MH, Flaswinkel H, Fuchs H et al (2000) Genome-wide, large-scale production of mutant mice by ENU mutagenesis. Nat Genet 25:444–447

Huang PL, Huang Z, Mashimo H et al (1995) Hypertension in mice lacking the gene for endothelial nitric oxide synthase [see comments]. Nature 377:239–242

Jacob HJ, Kwitek AE (2002) Rat genetics: attaching physiology and pharmacology to the genome. Nat Rev Genet 3:33–42

Jeunemaitre X, Soubrier F, Kotelevtsev YV et al (1992) Molecular basis of human hypertension: role of angiotensinogen. Cell 71:169–180

Jiang XC, Qin S, Qiao C et al (2001) Apolipoprotein B secretion and atherosclerosis are decreased in mice with phospholipid-transfer protein deficiency. Nat Med 7:847–852

Johnson GC, Esposito L, Barratt BJ et al (2001) Haplotype tagging for the identification of common disease genes. Nat Genet 29:233–237

Kadowaki T (2000) Insights into insulin resistance and type 2 diabetes from knockout mouse models. J Clin Invest 106:459–465

Karvonen MK, Valkonen VP, Lakka TA et al (2001) Leucine7 to proline7 polymorphism in the preproneuropeptide Y is associated with the progression of carotid atherosclerosis, blood pressure and serum lipids in Finnish men. Atherosclerosis 159:145–151

Katsuya T, Koike G, Yee TW et al (1995) Association of angiotensinogen gene T235 variant with increased risk of coronary heart disease. Lancet 345:1600–1603

Keavney B, McKenzie C, Parish S et al (2000) Large-scale test of hypothesised associations between the angiotensin- converting-enzyme insertion/deletion polymorphism and myocardial infarction in about 5000 cases and 6000 controls. International Studies of Infarct Survival (ISIS) Collaborators. Lancet 355:434–442

Kiechl S, Lorenz E, Reindl M et al (2002) Toll-like receptor 4 polymorphisms and atherogenesis. N Engl J Med 347:185–192

Kim HS, Krege JH, Kluckman KD et al (1995) Genetic control of blood pressure and the angiotensinogen locus. Proc Natl Acad Sci U S A 92:2735–2739

Klos KL, Kardia SL, Ferrell RE et al (2001) Genome-wide linkage analysis reveals evidence of multiple regions that influence variation in plasma lipid and apolipoprotein levels associated with risk of coronary heart disease. Arterioscler Thromb Vasc Biol 21:971–978

Knouff C, Hinsdale ME, Mezdour H et al (1999) Apo E structure determines VLDL clearance and atherosclerosis risk in mice. J Clin Invest 103:1579–1586

Knowles JW, Maeda N (2000) Genetic modifiers of atherosclerosis in mice. Arterioscler Thromb Vasc Biol 20:2336–2345

Knowles JW, Maeda N (2001) Dyslipidemia and hypertension: twin killers in renal vascular disease. Am J Kidney Dis 37:1322–1323

Knowles JW, Reddick RL, Jennette JC et al (2000) Enhanced atherosclerosis and kidney dysfunction in eNOS(-/-)Apoe(-/-) mice are ameliorated by enalapril treatment. J Clin Invest 105:451–458

Kolata G (1999) Using gene tests to customize medical treatment. NY Times (Print): A1, A34

Komajda M, Charron P (2001a) The heart of genomics. Nat Med 7:287–288

Komajda M, Charron P (2001b) How will the human genome project change cardiovascular medicine? Heart 86:123–124

Kraus WE (2000) Genetic approaches for the investigation of genes associated with coronary heart disease. Am Heart J 140: S27–35

Krege J (1996) Mouse systems for studying the genetics of hypertension and related disorders. Trends in Cardiovascular Medicine 6:232–238

Krege JH, Kim HS, Moyer JS et al (1997a) Angiotensin-converting enzyme gene mutations, blood pressures, and cardiovascular homeostasis. Hypertension 29:150–157

Krege JH, Moyer JS, Langenbach LL et al (1997b) Angiotensin-converting enzyme gene and atherosclerosis. Arterioscler Thromb Vasc Biol 17:1245–1250

Kuhlencordt PJ, Gyurko R, Han F et al (2001) Accelerated atherosclerosis, aortic aneurysm formation, and ischemic heart disease in apolipoprotein E/endothelial nitric oxide synthase double-knockout mice. Circulation 104:448–454

Kuivenhoven JA, Jukema JW, Zwinderman AH et al (1998) The role of a common variant of the cholesteryl ester transfer protein gene in the progression of coronary atherosclerosis. The Regression Growth Evaluation Statin Study Group. N Engl J Med 338:86–93

Lander ES, Linton LM, Birren B et al (2001) Initial sequencing and analysis of the human genome. Nature 409:860–921

Lawn RM, Wade DP, Garvin MR et al (1999) The Tangier disease gene product ABC1 controls the cellular apolipoprotein-mediated lipid removal pathway. J Clin Invest 104: R25–31

Li R, Boerwinkle E, Olshan AF et al (2000) Glutathione S-transferase genotype as a susceptibility factor in smoking-related coronary heart disease. Atherosclerosis 149:451–462

Lifton RP, Gharavi AG, Geller DS (2001) Molecular Mechanisms of Human Hypertension. Cell 104:545–556

Lindpaintner K, Pfeffer MA, Kreutz R et al (1995) A prospective evaluation of an angiotensin-converting-enzyme gene polymorphism and the risk of ischemic heart disease [see comments]. N Engl J Med 332:706–711

Luciani MF, Denizot F, Savary S et al (1994) Cloning of two novel ABC transporters mapping on human chromosome 9. Genomics 21:150–159

Ludwig EH, Borecki IB, Ellison RC et al (1997) Associations between candidate loci angiotensin-converting enzyme and angiotensinogen with coronary heart disease and myocardial infarction: the NHLBI Family Heart Study. Ann Epidemiol 7:3–12

Lusis AJ (2000) Atherosclerosis. Nature 407:233–241

Mahley RW (1988) Apolipoprotein E: cholesterol transport protein with expanding role in cell biology. Science 240:622–630

Makowski L, Boord JB, Maeda K et al (2001) Lack of macrophage fatty-acid-binding protein aP2 protects mice deficient in apolipoprotein E against atherosclerosis. Nat Med 7:699–705

Marcil M, Brooks-Wilson A, Clee SM et al (1999) Mutations in the ABC1 gene in familial HDL deficiency with defective cholesterol efflux. Lancet 354:1341–1346

Marenberg ME, Risch N, Berkman LF et al (1994) Genetic susceptibility to death from coronary heart disease in a study of twins. N Engl J Med 330:1041–1046

McNamara DM, Holubkov R, Janosko K et al (2001) Pharmacogenetic interactions between beta-blocker therapy and the angiotensin-converting enzyme deletion polymorphism in patients with congestive heart failure. Circulation 103:1644–1648

Moller DE (2001) New drug targets for type 2 diabetes and the metabolic syndrome. Nature 414:821–827

Moncada S, Higgs A (1993) The L-arginine-nitric oxide pathway. N Engl J Med 329:2002–2012

Mural RJ, Adams MD, Myers EW et al (2002) A comparison of whole-genome shotgun-derived mouse chromosome 16 and the human genome. Science 296:1661–1671

Navab M, Berliner JA, Watson AD et al (1996) The Yin and Yang of oxidation in the development of the fatty streak. A review based on the 1994 George Lyman Duff Memorial Lecture. Arterioscler Thromb Vasc Biol 16:831–842

Naveilhan P, Hassani H, Canals JM et al (1999) Normal feeding behavior, body weight and leptin response require the neuropeptide Y Y2 receptor. Nat Med 5:1188–1193

Nickerson DA, Taylor SL, Fullerton SM et al (2000) Sequence diversity and large-scale typing of SNPs in the human apolipoprotein E gene. Genome Res 10:1532–1545

Nolan PM, Peters J, Strivens M et al (2000) A systematic, genome-wide, phenotype-driven mutagenesis programme for gene function studies in the mouse. Nat Genet 25:440–443

Nora JJ, Lortscher RH, Spangler RD et al (1980) Genetic–epidemiologic study of early-onset ischemic heart disease. Circulation 61:503–508

Ohta A, Sitkovsky M (2001) Role of G-protein-coupled adenosine receptors in downregulation of inflammation and protection from tissue damage. Nature 414:916–920

Ojala JP, Helve E, Ehnholm C et al (1991) Effect of apolipoprotein E polymorphism and XbaI polymorphism of apolipoprotein B on response to lovastatin treatment in familial and non-familial hypercholesterolaemia. J Intern Med 230:397–405

Olivier M, Bustos VI, Levy MR et al (2001) Complex high-resolution linkage disequilibrium and haplotype patterns of single-nucleotide polymorphisms in 2.5 Mb of sequence on human chromosome 21. Genomics 78:64–72

Ordovas JM, Lopez-Miranda J, Perez-Jimenez F et al (1995) Effect of apolipoprotein E and A-IV phenotypes on the low density lipoprotein response to HMG CoA reductase inhibitor therapy. Atherosclerosis 113:157–166

Orso E, Broccardo C, Kaminski WE et al (2000) Transport of lipids from golgi to plasma membrane is defective in tangier disease patients and Abc1-deficient mice. Nat Genet 24:192–196

Owen JS (1999) Role of ABC1 gene in cholesterol efflux and atheroprotection. Lancet 354:1402–1403

Palmiter RD, Erickson JC, Hollopeter G et al (1998) Life without neuropeptide Y. Recent Prog Horm Res 53:163–199

Pankow JS, Folsom AR, Cushman M et al (2001) Familial and genetic determinants of systemic markers of inflammation: the NHLBI family heart study. Atherosclerosis 154:681–689

Patil N, Berno AJ, Hinds DA et al (2001) Blocks of limited haplotype diversity revealed by high-resolution scanning of human chromosome 21. Science 294:1719–1723

Patterson C, Madamanchi NR, Runge MS (2000) The oxidative paradox : another piece in the puzzle [In Process Citation]. Circ Res 87:1074–1076

Peacock JM, Arnett DK, Atwood LD et al (2001) Genome scan for quantitative trait loci linked to high-density lipoprotein cholesterol: The NHLBI Family Heart Study. Arterioscler Thromb Vasc Biol 21:1823–1828

Peltonen L, McKusick VA (2001) Genomics and medicine. Dissecting human disease in the postgenomic era. Science 291:1224–1229

Pennacchio LA, Olivier M, Hubacek JA et al (2001) An apolipoprotein influencing triglycerides in humans and mice revealed by comparative sequencing. Science 294:169–173

Pennisi E (2002) Genomics. Charting a genome's hills and valleys. Science 296:1601–1603

Peyser PA (1997) Genetic epidemiology of coronary artery disease. Epidemiol Rev 19:80–90

Piedrahita JA, Zhang SH, Hagaman JR et al (1992) Generation of mice carrying a mutant apolipoprotein E gene inactivated by gene targeting in embryonic stem cells. Proc Natl Acad Sci U S A 89:4471–4475

Plump A (1997) Atherosclerosis and the mouse: a decade of experience. Ann Med 29:193–198

Plump AS, Smith JD, Hayek T et al (1992) Severe hypercholesterolemia and atherosclerosis in apolipoprotein E- deficient mice created by homologous recombination in ES cells. Cell 71:343–353

Pullinger CR, Eng C, Salen G et al (2002) Human cholesterol 7alpha-hydroxylase (CYP7A1) deficiency has a hypercholesterolemic phenotype. J Clin Invest 110:109–117

Rader DJ (2001) A new feature on the cholesterol-lowering landscape. Nat Med 7:1282–1284

Rader DJ, FitzGerald GA (1998) State of the art: atherosclerosis in a limited edition [news; comment]. Nat Med 4:899–900

Rader DJ, Pure E (2000) Genetic susceptibility to atherosclerosis: insights from mice. Circ Res 86:1013–1015

Reardon CA, Getz GS (2001) Mouse models of atherosclerosis. Curr Opin Lipidol 12:167–173

Reilly PR (2000) Ethical and legal issues in genetic testing to predict risk of heart disease. Am Heart J 140: S6–10

Ridker PM, Hennekens CH, Buring JE et al (2000a) C-reactive protein and other markers of inflammation in the prediction of cardiovascular disease in women. N Engl J Med 342:836–843

Ridker PM, Rifai N, Stampfer MJ et al (2000b) Plasma concentration of interleukin-6 and the risk of future myocardial infarction among apparently healthy men. Circulation 101:1767–1772

Rigat B, Hubert C, Alhenc-Gelas F et al (1990) An insertion/deletion polymorphism in the angiotensin I-converting enzyme gene accounting for half the variance of serum enzyme levels. J Clin Invest 86:1343–1346

Rimm EB, Stampfer MJ, Ascherio A et al (1993) Vitamin E consumption and the risk of coronary heart disease in men [comment] [see comments]. N Engl J Med 328: 1450–1456

Risch N, Merikangas K (1996) The future of genetic studies of complex human diseases. Science 273:1516–1517

Rosamond WD, Chambless LE, Folsom AR et al (1998) Trends in the incidence of myocardial infarction and in mortality due to coronary heart disease, 1987 to 1994. N Engl J Med 339:861–867

Ross R (1993) The pathogenesis of atherosclerosis: a perspective for the 1990 s. Nature 362:801–809

Ross R (1999) Atherosclerosis–an inflammatory disease [see comments]. N Engl J Med 340:115–126

Rubin EM, Smith DJ (1994) Atherosclerosis in mice: getting to the heart of a polygenic disorder. Trends Genet 10:199–203

Rubin EM, Tall A (2000) Perspectives for vascular genomics. Nature 407:265–269

Rust S, Rosier M, Funke H et al (1999) Tangier disease is caused by mutations in the gene encoding ATP-binding cassette transporter 1. Nat Genet 22:352–355

Saltiel AR, Kahn CR (2001) Insulin signalling and the regulation of glucose and lipid metabolism. Nature 414:799–806

Sasaki M, Oki T, Iuchi A et al (1996) Relationship between the angiotensin converting enzyme gene polymorphism and the effects of enalapril on left ventricular hypertrophy and impaired diastolic filling in essential hypertension: M- mode and pulsed Doppler echocardiographic studies. J Hypertens 14:1403–1408

Schmitz G, Langmann T (2001) Structure, function and regulation of the ABC1 gene product. Curr Opin Lipidol 12:129–140

Serfaty-Lacrosniere C, Civeira F, Lanzberg A et al (1994) Homozygous Tangier disease and cardiovascular disease. Atherosclerosis 107:85–98

Shesely EG, Maeda N, Kim HS et al (1996) Elevated blood pressures in mice lacking endothelial nitric oxide synthase. Proc Natl Acad Sci U S A 93:13176–13181

Shi W, Wang NJ, Shih DM et al (2000) Determinants of Atherosclerosis Susceptibility in the C3H and C57BL/6 Mouse Model : Evidence for Involvement of Endothelial Cells but Not Blood Cells or Cholesterol Metabolism. Circ Res 86:1078–1084

Sing CF, Moll PP (1989) Genetics of variability of CHD risk. Int J Epidemiol 18: S183–195

Singaraja RR, Fievet C, Castro G et al (2002) Increased ABCA1 activity protects against atherosclerosis. J Clin Invest 110:35–42

Smith JD, Breslow JL (1997) The emergence of mouse models of atherosclerosis and their relevance to clinical research. J Intern Med 242:99–109

Smith JD, Trogan E, Ginsberg M et al (1995) Decreased atherosclerosis in mice deficient in both macrophage colony- stimulating factor (op) and apolipoprotein E. Proc Natl Acad Sci U S A 92:8264–8268

Smithies O, Kim HS, Takahashi N et al (2000) Importance of quantitative genetic variations in the etiology of hypertension. Kidney Int 58:2265–2280

Smithies O, Maeda N (1995) Gene targeting approaches to complex genetic diseases: atherosclerosis and essential hypertension. Proc Natl Acad Sci U S A 92:5266–5272

Stary HC, Chandler AB, Dinsmore RE et al (1995) A definition of advanced types of atherosclerotic lesions and a histological classification of atherosclerosis. A report from the Committee on Vascular Lesions of the Council on Arteriosclerosis, American Heart Association. Arterioscler Thromb Vasc Biol 15:1512–1531

Steinberg D (1997) Lewis A. Conner Memorial Lecture. Oxidative modification of LDL and atherogenesis. Circulation 95:1062–1071

Stengard JH, Zerba KE, Pekkanen J et al (1995) Apolipoprotein E polymorphism predicts death from coronary heart disease in a longitudinal study of elderly Finnish men. Circulation 91:265–269

Stephens NG, Parsons A, Schofield PM et al (1996) Randomised controlled trial of vitamin E in patients with coronary disease: Cambridge Heart Antioxidant Study (CHAOS) [see comments]. Lancet 347:781–786

Stoll M, Cowley AW, Jr., Tonellato PJ et al (2001) A genomic-systems biology map for cardiovascular function. Science 294:1723–1726

Sullivan PM, Mezdour H, Quarfordt SH et al (1998) Type III hyperlipoproteinemia and spontaneous atherosclerosis in mice resulting from gene replacement of mouse Apoe with human Apoe*2. J Clin Invest 102:130–135

Taubes G (2002) Cardiovascular disease. Does inflammation cut to the heart of the matter? Science 296:242–245

Tiret L, Ricard S, Poirier O et al (1995) Genetic variation at the angiotensinogen locus in relation to high blood pressure and myocardial infarction: the ECTIM Study. J Hypertens 13:311–317

Tracy RP, Lemaitre RN, Psaty BM et al (1997) Relationship of C-reactive protein to risk of cardiovascular disease in the elderly. Results from the Cardiovascular Health Study and the Rural Health Promotion Project. Arterioscler Thromb Vasc Biol 17:1121–1127

Turner ST, Boerwinkle E (2000) Genetics of hypertension, target-organ complications, and response to therapy. Circulation 102: IV40–45

Turner ST, Schwartz GL, Chapman AB et al (2001a) C825T polymorphism of the G protein beta(3)-subunit and antihypertensive response to a thiazide diuretic. Hypertension 37:739–743

Turner ST, Schwartz GL, Chapman AB et al (2001b) Antihypertensive pharmacogenetics: getting the right drug into the right patient. J Hypertens 19:1–11

Tyroler HA (2000) Coronary heart disease epidemiology in the 21st century. Epidemiol Rev 22:7–13

Valdes AM, Wolfe ML, Tate HC et al (2001) Association of traditional risk factors with coronary calcification in persons with a family history of premature coronary heart disease: the study of the inherited risk of coronary atherosclerosis. J Investig Med 49:353–361

Venter JC, Adams MD, Myers EW et al (2001) The sequence of the human genome. Science 291:1304–1351

Waters DD (2001) Are we aggressive enough in lowering cholesterol? Am J Cardiol 88:10F-15F

Wilkins MR, Roses AD, Clifford CP (2000) Pharmacogenetics and the treatment of cardiovascular disease. Heart 84:353–354

Willett WC (2002) Balancing life-style and genomics research for disease prevention. Science 296:695–698

Williams RR, Hunt SC, Hopkins PN et al (1993) Genetic basis of familial dyslipidemia and hypertension: 15-year results from Utah. Am J Hypertens 6:319S-327S

Wilson PW, Myers RH, Larson MG et al (1994) Apolipoprotein E alleles, dyslipidemia, and coronary heart disease. The Framingham Offspring Study. Jama 272:1666–1671

Winkelmann BR, Hager J (2000) Genetic variation in coronary heart disease and myocardial infarction: methodological overview and clinical evidence. Pharmacogenomics 1:73–94

Winkelmann BR, Hager J, Kraus WE et al (2000) Genetics of coronary heart disease: current knowledge and research principles. Am Heart J 140: S11–26

Winkelmann BR, Nauck M, Klein B et al (1996) Deletion polymorphism of the angiotensin I-converting enzyme gene is associated with increased plasma angiotensin-converting enzyme activity but not with increased risk for myocardial infarction and coronary artery disease. Ann Intern Med 125:19–25

Winkelmann BR, Russ AP, Nauck M et al (1999) Angiotensinogen M235T polymorphism is associated with plasma angiotensinogen and cardiovascular disease. Am Heart J 137:698–705

Witting P, Pettersson K, Ostlund-Lindqvist AM et al (1999) Dissociation of atherogenesis from aortic accumulation of lipid hydro(pero)xides in Watanabe heritable hyperlipidemic rabbits [see comments]. J Clin Invest 104:213–220

Wittrup HH, Tybjaerg-Hansen A, Nordestgaard BG (1999) Lipoprotein lipase mutations, plasma lipids and lipoproteins, and risk of ischemic heart disease. A meta-analysis. Circulation 99:2901–2907

Yusuf S, Dagenais G, Pogue J et al (2000a) Vitamin E supplementation and cardiovascular events in high-risk patients. The Heart Outcomes Prevention Evaluation Study Investigators. N Engl J Med 342:154–160

Yusuf S, Sleight P, Pogue J et al (2000b) Effects of an angiotensin-converting-enzyme inhibitor, ramipril, on cardiovascular events in high-risk patients. The Heart Outcomes Prevention Evaluation Study Investigators [see comments]. N Engl J Med 342:145–153

Zee RY, Ridker PM (2002) Polymorphism in the human C-reactive protein (CRP) gene, plasma concentrations of CRP, and the risk of future arterial thrombosis. Atherosclerosis 162:217–219

Zhang SH, Reddick RL, Avdievich E et al (1997) Paradoxical enhancement of atherosclerosis by probucol treatment in apolipoprotein E-deficient mice. J Clin Invest 99:2858–2866

Zhang SH, Reddick RL, Piedrahita JA et al (1992) Spontaneous hypercholesterolemia and arterial lesions in mice lacking apolipoprotein E. Science 258:468–471

Lipid-Lowering Responses Modified by Genetic Variation

B. R. Winkelmann[1] · M. M. Hoffmann[2] · W. März[3]

[1] Kooperationseinheit Pharmakogenomik/Angewandte Genomforschung, Universitätsklinikum Heidelberg, Im Neuenheimer Feld 221, 69120 Heidelberg, Germany
e-mail: bernhard_winkelmann@med.uni-heidelberg.de

[2] Universitätsklinikum, Zentrale Klinische Forschung, Breisacher Str. 66, 79106 Freiburg, Germany

[3] Clinical Institute of Medical and Chemical Laboratory Diagnostics, University Hospital, Auenbrugger Platz 15, 8036 Graz, Austria

1	Introduction	108
2	**Pharmacogenetics, Pharmacogenomics and Methodological Issues**	109
2.1	Definition of Pharmacogenetics and Pharmacogenomics	109
2.2	Methodological Issues	110
2.2.1	Family and Association Studies	110
2.2.2	Candidate Genes, SNPs and Complexity of Drug Response	111
2.2.3	Sample Size Issues	112
3	**Lipid Metabolism Pathways and Lipid-Lowering Drugs**	112
3.1	Brief Overview of Lipid Metabolism Pathways	112
3.2	The Exogenous Pathway	114
3.3	The Endogenous Pathway	115
3.4	Statins: The Most Potent Class of Lipid-Lowering Agents	116
3.5	Fibrates: Cellular Mode of Action and Candidate Genes for Pharmacodynamic Effect	116
4	**Candidate Genes and Pharmacokinetic Pathways**	118
5	**Candidate Genes and Lipid Pathways**	119
5.1	HMG-CoA Reductase	119
5.2	LDL Receptor	119
5.3	SREBP: Regulator of LDL Receptor Activity	120
5.4	Apolipoprotein A-I	123
5.5	Apolipoprotein A-IV	123
5.6	Apolipoprotein B	124
5.7	Apolipoprotein E	125
5.8	Cholesteryl Ester Transfer Protein	130
5.9	Hepatic Triglyceride Lipase	131
5.10	Lipoprotein Lipase	132
5.11	Lipoprotein (a)	133
5.12	Cholesterol-7α-Hydroxylase	134
5.13	Peroxisome Proliferator-Activated Receptors	134
6	**Non-lipid Pathway Genes**	135
6.1	Stromelysin-1	135
7	**Limitations and Conclusions**	136
	References	137

Abstract Elevated cholesterol and other dyslipidaemias are major risk factors for atherosclerotic cardiovascular disease—the major cause of death in North America and Europe. Correction of dyslipidaemia with diet or lipid-lowering agents has been shown to reduce the risk of future coronary events. However, the response to diet or lipid-lowering drugs is not uniform within any population. Even among carefully selected patients in clinical trials, individual responses to a lipid-modifying intervention vary greatly. On the one hand, factors such as gender, age, concomitant disease and concomitant medication may modify the pharmacokinetics or pharmacodynamics of lipid-lowering therapy. On the other hand, genetic factors are also important. Polymorphisms in genes regulating the metabolism of lipoproteins (e.g. apolipoprotein E, lipoprotein lipase, cholesteryl ester transfer protein) are associated with differences in plasma lipoprotein concentrations and can explain a substantial fraction of their variance in the general population, as demonstrated in measurements of low-density lipoprotein (LDL) or high-density lipoprotein (HDL). With the widespread availability of molecular genetic testing, the focus has shifted to the study of genetic determinants of drug response and their role in optimizing the choice of agent with regard to efficacy and tolerability. At present, despite several positive, but in general isolated examples, the overall impact of such gene variants in predicting individual response to a lipid-lowering intervention still needs clarification in well-designed confirmatory studies. Advances in pharmacogenomics will help to deepen our understanding of lipid and lipoprotein metabolism and the consideration that needs to be give to genetic factors in prescribing lipid-lowering therapy.

Keywords Lipids · Statins · Fibrates · Polymorphisms · Candidate genes · Pharmacogenetics

1
Introduction

Changes in circulating plasma lipoproteins, in particular increases in low-density lipoproteins (LDL) and triglyceride-rich lipoproteins and decreases in high-density lipoproteins (HDL), are among the most important causes of atherosclerosis. Although the benefits of lipid-lowering therapy have been shown in many patient populations, the individual variation in response is large. In the case of LDL lowering by statins, responses in individuals may vary from decreases by 10%–70% (Aguilar-Salinas et al. 1998). It is reasonable to assume that these differences, at least in part, relate to the genetic diversity among individuals.

The genes that modify the response to lipid-lowering response to drugs or dietary components may be grouped into two classes:

a. Genes involved in the pharmacokinetics of a drug (or dietary component), that is, in the ADME (absorption, distribution, metabolism and excretion) process, that will modulate the availability of the active compound (the mother drug or a metabolite or a dietary component) in the body.

b. Genes involved in the pharmacodynamics of drug (diet) action, i.e. genes regulating the biosynthesis, transport, processing and catabolism of lipoproteins. Variants of these genes have been scrutinized extensively during the last 2 decades, mainly in order to identify genetic factors affecting atherogenesis. This has unravelled the molecular mechanisms of monogenetic disorders that severely affect lipoprotein metabolism, for example, familial hypercholesterolaemia (due to mutations in the LDL receptor gene) or Tangier disease (due to mutations in the *ABCA1* gene). These inborn errors of metabolism are too rare (well below 1% in the population) to make a significant contribution to the variance of LDL or HDL cholesterol concentrations observed in the general population. In contrast, gene polymorphisms (i.e. the prevalence of the mutation is above 1% in the population) involved in lipoprotein metabolism may determine a substantial fraction of the variance in the concentration of circulating lipoproteins and may thus provide good candidates for the investigation of the impact of gene variants on the pharmacodynamics of lipid-lowering intervention.

2
Pharmacogenetics, Pharmacogenomics and Methodological Issues

2.1
Definition of Pharmacogenetics and Pharmacogenomics

Pharmacogenetics is the study of the variability in drug response among humans due to inheritable factors. The term "pharmacogenetics" was coined in the 1950s (Vogel 1959), although research in this area commenced before that time (for further historical details, see Vesell 2000). To begin with, the research effort focused on drug metabolizing enzymes, especially cytochrome *P450* isoenzymes. The study of variation in genes involved in the pharmacodynamic effect of a drug has emerged as a major focus in academia and in the pharmaceutical industry relatively recently (Anonymous 2000).

Pharmacogenomics is a relatively new field that has expanded from pharmacogenetics (Vesell 2000; Wieczorek et al. 2001). Pharmacogenomics studies the effects of small molecules, in general using genomic techniques such as microarrays, for gene expression in humans, animals or model organisms. Thus, pharmacogenomics represents a more global analysis of drug effects in biological systems with the study of entire pathways or even of all expressed genes in cells or tissue (Altman and Klein 2002), seeking to predict drug efficacy and safety (Murphy 2000). Pharmacogenetics, in contrast, copes with the study of drug response in relation to genetic diversity, and historically typically dealt with one or a few candidate genes per investigation. In this sense, pharmacogenetics may be considered a subspecialty within pharmacogenomics (Bailey et al. 1998; Rusnak et al. 2001; for a detailed discussion of both terms see the chapter by Lindpaintner, this volume).

Traditionally in drug development and registration, the emphasis has been placed on mean values of quantitative measures or responder rates. However, optimum dose requirements for many drugs are known to vary among individuals; for example, the daily required dose varies 40-fold for propranolol and 20-fold for warfarin (Lu 1998). Similarly, genetic predisposition for adverse drug reactions is a major concern. Thus, pharmacogenomics/pharmacogenetics represents the research effort directed at elucidating the genomic basis of drug action and identifying individual genetic patterns in order to understand and predict differences in responses to drugs between individuals. Ultimately, such information could be used to tailor drug prescriptions to individual genotypes (McLeod et al. 2001).

Lately, Kalow, a key researcher instrumental in the development of the field of pharmacogenetics, suggested yet another term, "pharmacobiology", to describe the concept that drug effects are the result of an interaction of genetic predisposition with the environment (Kalow 2001). The inherent beauty of the current research focus on DNA markers (individual genotypes) is, of course, the stability of such markers over an individual's life time, when compared to environmental factors that may change and are often difficult, if not impossible, to quantify. Nevertheless, one should always keep in mind that although knowledge of the particular genotype with respect to candidate genes may help to optimize drug dose and improve the chances of a better response and lessen the risk of side effects, the response to drugs involves a complex sequence of events regulated by many genes and that in many instances environmental factors may be more decisive than the genotypic influence.

2.2
Methodological Issues

2.2.1
Family and Association Studies

Pharmacogenetic studies emerged from the investigation of abnormal drug responses that tracked in families, such as an adverse response to suxamethonium due to an inherited deficiency of plasma cholinesterase (Kalow 1956) or haemolysis after antimalarial therapy due to inherited differences in erythrocyte glucose-6-phosphate activity (Carson et al. 1956). But while family studies have been instrumental in the beginning of pharmacogenetic research, their impact in today's pharmacogenetic and pharmacogenomic research has declined and been largely replaced by association studies. There are two major reasons for this. First, it is rather cumbersome, if not impossible, to study drug response in families since virtually all clinical trials studying the pharmacokinetics and pharmacodynamics of drugs are carried out in unrelated individuals. Secondly, with the annotation of almost all genes in the human genome, an unsurpassed pool of candidate genes is accessible in databases and these genes provide new candidates for any association with drug response. With the availability of single nucleotide polymorphism (SNP) maps, genome-wide association studies may even be possible in the future to identify chromosomal loci without any prior knowledge of the genetic pathway involved. The latter has always been a major advantage of family studies in which genome-wide linkage studies based on differences in microsatellite repeat markers in affected/non-affected (responders/non-responders) are possible. SNPs have the added advantage that they occur more frequently throughout the human genome than microsatellites, roughly every 500–1,000 base pairs (bp) or 3–6 million SNPs in total, and are less prone to germline mutations, and so their inheritance is more stable (Destenaves et al. 2000).

2.2.2
Candidate Genes, SNPs and Complexity of Drug Response

The selection of candidate genes for study, based on a priori knowledge, is still hampered by a limited knowledge of drug pharmacokinetics and mechanisms of action. Once a candidate gene has been selected, genetic association studies involve the analysis of either a continuous trait (i.e. change in cholesterol level) or of groups (responder, non-responder derived from the application of a pre-specified cutoff for drug efficacy).

The individual response to a drug is a complex trait. Thus, the concept of studying one allelic variant in one candidate gene at time will ultimately be of limited value. Nevertheless, previous research has shown that rare mutations explaining a major proportion of the variability of drug response exist. With refinements in phenotyping study subjects and in searching relevant genetic variants (i.e. SNPs or other allelic variants that have an impact on gene function, be it by changing the amino acid sequence, or by having an impact on gene expression) more common mutations in the population (i.e. polymorphisms with a frequency $\geq 10\%$ of the rare variant) will be identified that contribute to the interindividual variability of drug response.

The concept is emerging that ordered arrays of SNPs along the gene as inherited from father and mother, that is haplotypes, will be better markers of drug response than an individual SNP alone (see the chapter by Tate and Goldstein, this volume, for an in-depth discussion). Currently, the availability of clinical studies on haplotypes for drug response is low. However, it is well known that although allelic variation may be an essential determinant of gene function, it is not the only one. Regulation of gene expression and post-translational modification modulates or supersedes the functional impact of a genetic variation. In order to disentangle the impact of variation in the genome (i.e. individual genotype), the armamentarium of molecular genomic tools in pharmacogenomics needs to include gene expression and proteomics besides genotyping (Kafatos 2001) plus the knowledge of major other environmental factors (concomitant drugs, non-drug factors).

2.2.3
Sample Size Issues

Clinical trials investigating endpoints of drug efficacy or adverse drug reactions, both intermediate (biochemical or physiological traits) and clinical endpoints (clinical event rates), typically base sample size estimates on adequate power (i.e. generally 80%–95%) for a given prespecified error rate to miss an effect (i.e. generally 1% or 5%). In contrast, many of the early trials investigating the association of certain genotypes with drug response parameters were retrospective analyses from phase II or III clinical trials, where in general only a subset of the entire study population has given consent and participated in the respective genetics substudy. Most such studies are underpowered to detect such effects and the chance of false-positive findings is high.

In order to provide a sound basis for the interpretation of the results of such genotype–drug response association studies, a sample size estimate of the power of the study for a given allelic frequency of the gene variant to be studied is mandatory. Cardon et al. (2000) calculated that for a fixed twofold response rate of the gene variant compared to the wild type (common) genotype, and a given alpha error of 5% and a power of 90%, the group size will require a sample in excess of 1,000 individuals for relatively rare alleles (<10%), and about 200–400 individuals if the frequency of the susceptibility allele is 30% or greater (Fig. 1). Thus, confirmatory candidate gene studies are feasible if the investigator has prior knowledge about allelic frequency and the magnitude of genotypic effects. However, studies involving a large number of loci of unknown frequency may suffer decisively from a lack of statistical power to detect any effects once results are available. Interestingly, if more than one genetic marker is tested, as would be the case in genome-wide SNP association screening studies, the increase in sample size due to adjustment for multiple testing with the number of SNPs tested is not nearly as dramatic as one might expect. Typically testing 10–100 SNPs requires samples 1.5–2 times larger than those estimated for a single SNP association study, and testing 100,000 markers would increase the sample size about threefold, based on assumptions of an allele frequency of 50% and a genotype relative risk of sixfold (Fig. 2) (Cardon et al. 2000).

3
Lipid Metabolism Pathways and Lipid-Lowering Drugs

3.1
Brief Overview of Lipid Metabolism Pathways

There are three major pathways of lipid transport in humans that are tightly interrelated, the exogenous, the endogenous and the reverse cholesterol transport pathway: (a) transport of dietary (exogenous fat) with uptake in the intestine and transport in chylomicrons via the lymph to the liver, (b) transport of hepatic (endogenous) fat via triglyceride-rich very-low-density lipoproteins (VLDL),

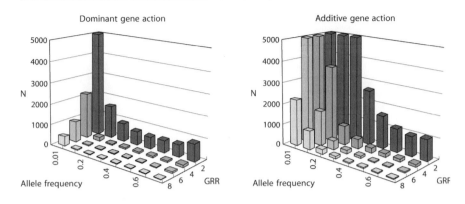

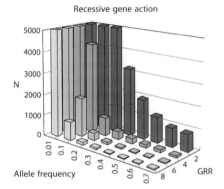

Fig. 1 Sample sizes required in a clinical trial under genetic stratification. Sampling requirements under dominant, additive (co-dominant), and recessive gene action are shown. Sample sizes have been truncated at a maximum of 5,000 to show the variability in sampling at feasible levels of ascertainment. All sample size calculations were performed using a significance level of 0.05 with power 0.90. (From Cardon et al. 2000, with permission)

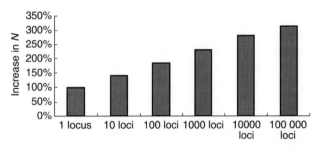

Fig. 2 The effect of testing multiple loci in a pharmacogenomic study. Bars represent the expected relative sample sizes required when 1, 10, 1,000, 10,000 and 100,000 loci are tested. The significance level desired after approximate Bonferroni correction for multiple tests is α 0.025 corresponding to a single di-allelic locus. The relative increases are shown using an allele frequency of 0.5 and a genotype relative risk (GRR) of 6. Other models of gene action, allele frequency and GRR yield similar results (Witte et al. 1999). (From Cardon et al. 2000, with permission)

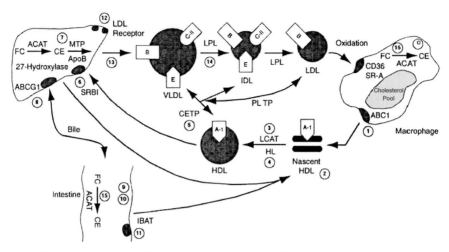

Fig. 3 Lipoprotein metabolism. *ABC*, adenosine triphosphate binding cassette; *ABCG*, adenosinetriphosphate binding cassette protein G1 transport; *Apo A1*, apolipoprotein A1; *CE*, cholesterol ester; *CETP*, cholesterol ester transfer protein; *FC*, free cholesterol; *HDL*, high-density lipoprotein; *HL*, hepatic lipase; *HMG-CoA*, 3-hydroxy-3-methylglutaryl coenzyme A; *IBAT*, ileal bile acid transport; *LCAT*, lecithin cholesterol acyl transferase; *LDL*, low-density lipoprotein; *MTP*, microsomal transfer protein; *SR-A*, scavenger receptor A; *SR-B1*, scavenger receptor B1; *VLDL*, very-low-density lipoprotein. (From Davidson 2001, with permission)

intermediate-density lipoproteins (IDL) and LDL, and (c) reverse cholesterol transport pathway, the transfer and uptake of free cholesterol from the peripheral tissues such as the arterial wall and its subsequent delivery to the liver (Brown, 2001; Shah et al. 2001). The two endogenous pathways are depicted in Fig. 3.

3.2
The Exogenous Pathway

This pathway transports dietary lipids absorbed from the gut and incorporated into chylomicrons into the intestinal lymph. Chylomicrons enter the bloodstream via the thoracic duct and bypass the liver. The triglyceride moiety of the chylomicrons is hydrolysed by lipoprotein lipase (LPL), which resides on the luminal surface of the capillary endothelium. The liberated free fatty acids are taken up by muscle cells for oxidation (energy delivery) and by adipose tissue for energy storage. The rate of intravascular lipolysis is regulated by peroxisome proliferator-activated receptors (PPAR), which belong to the family of nuclear hormone receptors. Both PPAR alpha (predominantly expressed in the liver, muscle, kidney, and heart) and PPAR gamma (mainly found in adipose tissues) stimulate the expression of LPL. PPAR alpha down-regulates the synthesis of apolipoprotein C-III (apoC3) which is an inhibitor of LPL.

During the hydrolysis of chylomicrons, smaller remnant particles evolve and the excess surface components (phospholipids and apolipoproteins) are transferred to HDL particles. Subsequently, the remnant particles become enriched in cholesterol and acquire apolipoprotein E (apoE) from HDL. ApoE is needed as ligand for lipoprotein receptors in the liver, because apolipoprotein B-48, the major apolipoprotein of chylomicrons, lacks the receptor binding domain of apolipoprotein B-100. The three major steps in the exogenous pathway are thus fat absorption from the intestine, hydrolysis of triglycerides in the circulation and receptor-mediated catabolism of cholesterol in the liver.

3.3
The Endogenous Pathway

In the endogenous pathway, the liver secretes triglycerides and cholesterol via VLDL particles into the circulation. Like chylomicrons, VLDL particles undergo lipolysis and turn into intermediate-density lipoproteins (IDL). There is a sizeable reuptake of IDL particles in the liver, with the remainder undergoing further lipolysis by LPL and hepatic lipase (HL), also known as hepatic triglyceride lipase (HTGL), to form LDL. The majority of circulating cholesterol is transported in the LDL fraction, which has apolipoprotein B (apoB) as the only surface protein for receptor docking. Cholesterol is delivered to peripheral cells for the synthesis of cell membranes and steroid hormones via the LDL receptor. However, it should be noted that reuptake of cholesterol by the liver catabolizes roughly two-thirds of the circulating LDL (not shown in Fig. 3).

The reverse cholesterol transport pathway is mediated by HDL-particles, which are formed from apolipoprotein A-I (apoA1) containing precursor particles originating from the intestine and the liver, and from surface components derived from the catabolism of chylomicrons. Nascent HDL particles mobilize free cholesterol from peripheral cells. A major step forward in the understanding of the reverse cholesterol transport pathway was the discovery of mutations in the *ABCA1* gene (ATP binding cassette transporter A1), leading to low HDL states (Rust et al. 1999). Along with other transporters, such as the scavenger receptor (SR) class B type 1 (SR-B1/ClA-1), *ABCA1* is involved in the transfer of unesterified (free) cholesterol from peripheral cells to nascent HDL. Cholesteryl esters are formed by HDL-associated enzyme lecithin:cholesterol acyltransferase (LCAT) and apoA1 serving as co-factor (Fig. 3). Cholesteryl ester transfer protein (CETP) is responsible for the transfer of esterified cholesterol to apolipoprotein B-100 containing lipoproteins (LDL, IDL, VLDL), while phospholipids are moved from apolipoprotein B-100 containing lipoproteins to HDL by phospholipid transfer protein (PLTP). Cholesteryl esters associated with HDL can also be delivered into hepatocytes by selective transfer. This process is mediated by the HDL scavenger–receptor class B type 1 (SRB1) and its human homologue (CLA-1). Finally, cholesterol is excreted from the body via the bile, either directly or after conversion into bile acids. Many genes involved in reverse cholesterol transport are directly or indirectly regulated by PPARs: apoA1 and apolipoprotein A-II (apoA2), PLTP, LCAT and cholesterol 7α-hydroxylase (CYP7A1) by PPAR alpha, SRBI/CLA-1, and ABCA1 by both PPAR alpha and gamma (Tall 1990; Staels et al. 1998a, b; Barbier et al. 2002).

3.4
Statins: The Most Potent Class of Lipid-Lowering Agents

Hydroxymethylglutaryl coenzyme A (HMGCoA) reductase inhibitors, better known as statins, are the most potent lipid-lowering agents currently available and have emerged as first-line therapy for hypercholesterolaemia. Statins inhibit HMGCoA reductase, the rate-limiting enzyme in cholesterol synthesis in the liver. By reducing cellular cholesterol synthesis, statins lead to an enhanced cellular uptake of LDL via the LDL receptor, thus ultimately lowering circulating LDL cholesterol in plasma. However, statins appear to affect other pathways as well. Although the exact mechanism needs to be clarified, statins may reduce VLDL production (and thus total triglycerides) by reducing apoB and apoC3 synthesis, which are both required for the assembly of VLDL particles (Ginsberg 1998). Further, statins modestly increase HDL cholesterol, possibly by indirectly stimulating PPAR alpha (Fruchart 2001) or by down-regulating hepatic lipase activity (Hoogerbrugge and Jansen 1999).

A wealth of prospective clinical trials consistently demonstrate that statins prevent or reduce cardiovascular events in primary and secondary prevention (LaRosa et al. 1999; Heart Protection Study Collaborative Group 2002; Serruys et al. 2002; Shepherd et al. 2002). It was only recently that the British Heart Protection study in 20,000 adults demonstrated that prevention of morbidity and mortality from ischaemic events can be achieved at a broad range of baseline LDL cholesterol levels and across all patient risk-subgroups (Heart Protection Study Collaborative Group 2002). Treatment with statins reduces lifetime risk (Ulrich et al. 2000), though at a cost of about 18,000 euros per year of life gained (van Hout et al. 2000).

The therapeutic potential of this drug class is probably far greater than previously anticipated (Ichihara and Satoh 2002). Many of the so-called pleiotropic (non-lipid-lowering) effects of statins could be of major relevance to a variety of disease processes. For example, statins enhance nitric oxide production and improve endothelial function, display anti-inflammatory potency, inhibit integrins and lower circulating adhesion molecules (Frenette 2001; März and Winkelmann 2002).

3.5
Fibrates: Cellular Mode of Action and Candidate Genes for Pharmacodynamic Effect

Although generally not as potent as statins in lowering LDL cholesterol, fibrates are more effective in lowering triglycerides and increasing HDL cholesterol. The HDL-raising effect is particularly pronounced in cases of very low pre-treatment levels (Després 2001). In mixed type IIB hyperlipidaemia, fenofibrate reduced the number of circulating triglyceride-rich VLDL-1 and VLDL-2 particles, which subsequently results in a marked decrease in cellular lipid loading (Milosavjevic et al. 2001). Even though LDL cholesterol is only moderately affected, fibrates profoundly decrease atherogenic dense LDL. Hence, fibrates are particularly suited for the treatment of the atherogenic lipid triad (high triglycerides, low HDL and dense LDL), the hallmark of patients with the metabolic syndrome or type 2 diabetes.

The molecular mode of action of fibrates is complex. Fibrates act by binding to PPARα (Staels et al. 1998a, b; Fruchart et al. 1998, 1999; Corton et al. 2000). PPARα is predominantly expressed in tissues that metabolize high amounts of fatty acids and mediates the action of fibrates on genes involved in lipid metabolism. Lipoprotein lipase is up-regulated (Schoonjans et al. 1996; Desager et al. 1996), whereas apolipoprotein CIII, an inhibitor of LPL, is down-regulated (Staels et al. 1995). Fibrates thus promote the hydrolysis of triglycerides in the plasma. Moreover, fibrates decrease the hepatic synthesis of triglycerides, apo B and VLDL (Lamb et al. 1993). In humans, the expression of apo A-I, apo AII, ABC-A1 (Desager et al. 1996; Chinetti et al. 2001) and SRBI/CLA-1 is stimulated, explaining the rise in HDL cholesterol and the activation of reverse cholesterol transport. However, the HDL effects of fibrates may also be due, partially, to a drug-induced reduction of CETP activity (Guerin et al. 1996) and an increase of PLPT activity (Tu et al. 1999; Bouly et al. 2001). In addition, fibrates increase receptor-mediated clearance of LDL, most likely due to changes in the composition of LDL towards more receptor-active particles rather than by up-regulation of the LDL receptor itself (Caslake et al. 1993). In a study of mRNA expression in animals, fenofibrate down-regulated hepatic lipase gene expression (Staels et al. 1992). However, in normolipidaemic healthy volunteers, hepatic lipase activity was only marginally affected (Desager et al. 1996).

Hypertriglyceridaemia plays an important role in atherogenesis (Hodis 1999). It is therefore likely, but not entirely clear, that the effects of fibrates on circulating lipoproteins translate into clinical benefit, as defined by a reduction in cardiovascular events. Twelve placebo-controlled trials of fibrate therapy published between 1966 and 1996 indicated no benefit in terms of reduction in risk of coronary deaths (Bucher et al. 1999). The period since 1996, however, has seen the publication of four additional trials (LOCAT, VA-HIT and BIP) in which 6,144 patients were treated with fibrates or placebo: bezafibrate in BECAIT (Ericsson et al. 1996), gemfibrozil in LOCAT (Frick et al. 1997), gemfibrozil in VA-HIT (Rubins et al. 1999), and bezafibrate in BIP (The BIP Study Group, 2000). Two of them were major trials. In VA-HIT, a secondary prevention trial in individuals with low HDL cholesterol and only mildly elevated LDL cholesterol, gemfibrozil produced a significant reduction in the incidence of fatal or nonfatal coronary events and in stroke (Rubins et al. 1999). BIP, a secondary prevention trial as well, demonstrated no significant overall effect of treatment with bezafibrate. Intriguingly, however, a significant reduction by almost 40% in the primary endpoint (fatal or nonfatal myocardial infarction or sudden death) was seen in a subgroup of individuals presenting with triglycerides ≥200 mg/dl at baseline (The BIP Study Group, 2000). In the Diabetes Atherosclerosis Intervention Study (DAIS), progression of atherosclerosis was significantly delayed by fenofibrate in patients with type 2 diabetes (DAIS 2001). The impact of micronized fenofibrate is being evaluated further in two large ongoing trials in patients with type 2 diabetes. The Fenofibrate Intervention and Event Lowering in Diabetes (FIELD) study has enrolled 9,795 patients to receive either placebo or fenofibrate 200 mg per day for at least 5 years. The Action to Control Cardiovascular Risk in Diabetes (ACCORD) will evaluate fenofibrate on top of statin treatment in approximately 10,000 patients with type 2 diabetes.

As with statins, fibrates have other potentially beneficial non-lipid effects with respect to atherosclerosis prevention, including antithrombotic effects (decrease in fibrinogen and PAI-1), anti-inflammatory activity (inhibition of TNFα-induced endothelial expression of VCAM-1 and interleukin-6, decrease in plasma uric acid) (De la Serna and Cadarso 1999).

4
Candidate Genes and Pharmacokinetic Pathways

Current knowledge about the impact of genetic variants of cytochrome P450s on the pharmacokinetics of statins is rather limited. Atorvastatin, cerivastatin, lovastatin, and simvastatin are all substrates of cytochrome P450 (CYP) 3A4 (Hermann 1999). Given the low bioavailability of statins, ranging from less than 5% for lovastatin and simvastatin to about 10%–20% for atorvastatin, CYP3A4 inhibition in the gastrointestinal tract may be of major importance (Sica and Gehr 2001; Ucar et al. 2000). No major functionally relevant mutations have been described for CYP3A4 (Evans and Johnson 2001), but it is an important site of interaction for environmental factors, including other pharmacological substrates for CYP3A4. Cerivastatin, withdrawn from the market in 2001 due to an increased rate of rhabdomyolysis (Staffa JA et al. 2002), is in addition metabolized by CYP2C8, while pravastatin is not significantly metabolized by any of the CYPs (Knopp 1999). Fluvastatin is metabolized by CYP2C9, which to a minor degree also contributes to the metabolism of lovastatin and simvastatin (Hermann 1999). CYP2D6, a monooxygenase displaying several genetic variants with important functional consequences on its metabolizing activity plays a minor role in the metabolism of statins (Linder et al. 1997; Tanaka 1999).

In the clinic, information about the effect of genetic factors on the pharmacokinetics of statins may be helpful in the identification of patients susceptible to rare but potentially life-threatening adverse events of statin therapy, such as myopathy and rhabdomyolysis. However, although these adverse events are related to higher doses and generally associated with very high plasma concentrations (i.e. 5–10 times the normal range), there is no evidence at the moment of an association with any genetic variation of candidate genes involved in the pharmacokinetic pathway of the statins. More important in this regard are interactions with drugs that inhibit or induce CYPs. Antibiotics, such as erythromycin and cyclosporine inhibit CYP3A4 and thus retard statin metabolism, while inducers of CYP3A4 such as barbiturates or carbamazepine will reduce the plasma concentration of the mother compound (Beaird 2000). The overall effect of inhibiting or inducing metabolizing enzymes will depend on whether the statin is a prodrug and/or in the case of most statins (i.e. lovastatin, simvastatin, atorvastatin and cerivastatin) whether these agents have active metabolites which may or may not (i.e. fluvastatin) be metabolized (Bottorff et al. 2000; Ucar et al. 2000).

5
Candidate Genes and Lipid Pathways

5.1
HMG-CoA Reductase

To our knowledge there have been no reports of variants of the HMG-CoA reductase gene itself that influence the efficacy of treatment with statins.

5.2
LDL Receptor

Up-regulation of the hepatic LDL receptor is the major therapeutic consequence of inhibiting HMG-CoA reductase in liver cells. It leads to an increased hepatic uptake of cholesterol and a decrease in circulating LDL cholesterol levels. Familial hypercholesterolaemia (FH) is autosomal dominant and caused by mutations in the gene for the LDL receptor. Until now, more than 680 distinct mutations distributed over the entire gene have been described (Heath et al. 2001). Heterozygous FH individuals express only half the number of functional LDL receptors, have a markedly raised plasma cholesterol and usually present with premature coronary artery disease. Homozygous FH individuals are more severely affected and may succumb before the age of maturity. The prevalence of homozygous FH is rare (~1:1,000,000), whereas heterozygous FH is relatively common (1:500 in Caucasians). Heterozygous FH subjects have been treated successfully with statins (Karayan et al. 1994; Vuorio et al. 1995; Heath et al. 1999), and cholesterol lowering has also been observed in *LDL-receptor*-negative, homozygous carriers (Feher et al. 1993).

The type of mutation has been shown to influence the cholesterol-lowering effect of statins in some studies (Leitersdorf et al. 1993; Couture P et al. 1998; Vohl et al. 2002; Heath et al. 1999; Chaves et al. 2001), though not consistently (Sijbrands et al. 1998; Brorholt-Petersen et al. 2001). *LDL receptor* mutations leading to the complete absence or truncation of the protein (null mutations) have been associated with higher baseline cholesterol levels, a poor response to statins and a higher risk of CHD. For example, carriers of *LDL receptor* mutations predicted to be severe (such as null mutations or mutations that affect exon 4 repeat 5) will poorly respond to statins compared to carriers of mild mutations. As much as 13% and 20% of the LDL cholesterol response to a statin was explained by the variation at the mutant *LDL receptor* locus in the study of Couture et al. (1998) and Leitersdorf and co-workers (1993), respectively. However, the observed differences in the LDL cholesterol response to statin therapy are not completely understood (Vergopoulus et al. 2002).

The sample size of such studies was small and only a few were randomized double-blind placebo-controlled trials (Table 1). Another problem is the multitude of mutations present in the *LDL-receptor* gene in an outbred population, so that only a few FH patients included in a study share identical mutations. In conclusion, whether the characterization of the molecular defect in FH is relevant to the immediate clinical management and whether particular mutations may need more aggressive lipid-lowering treatment in order to reach achieve a level of LDL-cholesterol lowering adequate for a sufficient reduction in coronary heart disease mortality remains unanswered. Furthermore, the mortality risk of FH subjects seems to be heavily modulated by environmental factors (Sijbrands et al. 2001).

5.3
SREBP: Regulator of LDL Receptor Activity

The expression of the *LDL receptor* gene is regulated by the intracellular cholesterol pool through sterol responsive element binding proteins (SREBPs) 1 and 2 (Brown and Goldstein 1997; Osborne 2000). Precursors of SREBPs are anchored in the membrane of the endoplasmic reticulum. When the sterol content of a cell decreases, SREBP processing proteins including SREBP cleavage activating protein (SCAP), site 1 protease (S1P) and site 2 protease (S2P) act synergistically to release the amino-terminal domain of SREBP by proteolysis. These active domains are transferred into the nucleus to activate the transcription of the genes of the *LDL receptor* and of enzymes involved in the biosynthesis of cholesterol. Furthermore, SREBPs up-regulate genes involved in the production of free fatty acids, including acetyl-CoA carboxylase and fatty acid synthase (Bennett et al. 1995; Lopez et al. 1996).

Mutations within the *SREBPs* and the SREBP processing proteins (*SCAP, S1P, S2P*) have been sought intensively, especially in patients with familial hypercholesterolaemia. Four polymorphic sites within *SCAP* (Nakajima et al. 1999; Iwaki et al. 1999), one within the promoter of *SREBP-1a* (Vedie et al. 2001) and five mutations in *SREBP-2* (Muller et al. 2001; Miserez et al. 2002) have been published. The *SCAP* gene has an exonic polymorphism (A→G transition) leading to isoleucine (A) to valine (G) substitution at codon 796 (Iwaki et al. 1999). The lipid-lowering response to 40 mg pravastatin was independent of the Ile796-Val polymorphism in a placebo-controlled 6-month study of 51 hypercholesterolaemic men (Fan et al. 2001). In the 372 participants of the Lipoprotein Coronary Atherosclerosis Study (LCAS), Salek and co-workers (Salek et al. 2002) detected no difference in lipid-lowering response relating to *SCAP Ile796Val* (*A2386G*) genotype, but found a strongly graded interaction between *SREBP-1α*

Table 1 Studies in patients with heterozygous familial hypercholesterolaemia (FH) investigating the LDL-lowering response to statin treatment depending on the underlying LDL receptor defect

Gene	Variant	Clinical phenotype	Intervention	Design	Reported effect	Number of subjects	Significance	First author, year
LDL-R	Trp23stop (receptor-negative)	Hetero-zygous FH	Fluvastatin 40 mg/placebo	Double-blind cross-over	Trend for less effective LDL lowering if receptor-binding defective	28	NS	Brorholt-Petersen 2001
	Trp66gly (receptor-binding-defective)					30	NS	
LDL-R	Del 15 kb exon1 (receptor-negative)	Hetero-zygous FH	Simvastatin 20 mg	Double-blind	Less effective LDL lowering in carriers of Trp66Gly mutation	31 (23)	0.05 (Three-group comparison)	Couture 1998
	Cys646tyr (receptor-negative)		Placebo (3:1 rando-mization)	Parallel group		13 (10)	0.01 (Two-group comparison)	Vohl 2002
	Trp66gly (receptor-defective)					19 (14) (=number that finished the study)		
LDL-R	Severe mutation (exon 4 repeat 5)	Hetero-zygous FH	Simvastatin 10 mg	Open label	Less effective LDL lowering in carriers of a severe mutation	14 (8/11/6)[a]	0.02	Heath 1999
	Mild mutation (other)		20 mg 40 mg	Observational		16 (7/13/11)[a]		

Table 1 (continued)

Gene	Variant	Clinical phenotype	Intervention	Design	Reported effect	Number of subjects	Significance	First author, year
LDL-R	Cys660stop ("Lebanese")	Hetero-zygous FH	Fluvastatin 40 mg	Single-blind	Less effective LDL lowering in carriers of Sephardic or Lithuanian mutations	21	0.005 (Comparison among five groups)	Leitersdorf 1993
	Asp147his ("Sephardic")					15[b]		
	652delGGT ("Lithuanian")					5		
	Other haplotypes					9		
	Unclassified					14		
LDL-R	mRNA-positive	Hetero-zygous FH	Simvastatin 20 mg	Open label	Similar LDL lowering response	13	NS	Sijbrands 1998
	mRNA-negative***			Observational		14		
LDL-R	Defective mutation	Hetero-zygous FH	Simvastatin 20 mg	Open label	Less effective LDL lowering in carriers of null mutations	20	0.04	Chaves 2001
	Null mutations[c]			Observational		22		

[a] Number of subjects actually treated with 10, 20 or 40 mg.
[b] Subjects with an *asp147his* mutation had the highest baseline LDL and the least LDL response to fluvastatin.
[c] Null mutation, no mRNA/no protein (e.g. early stop codon) or truncated protein, as opposed to mRNA-positive mutations resulting in a defective LDL receptor.

−36del/G genotype and response of apoA1 and apoC3 to fluvastatin and a modest interaction with HDL cholesterol. However, the authors rightly emphasized the possibility of a false positive finding, since the α error of $p<0.05$ needs to be adjusted in this study for multiple testing due to the variety of lipid parameters and endpoints assessed. Thus, their findings will need replication in other studies.

5.4
Apolipoprotein A-I

Apolipoprotein A-I (apoA1) is the major apolipoprotein of HDL and plays an important role in the formation of mature HDL and in reverse cholesterol transport. HDL concentrations are largely determined by the rate of synthesis of apo AI in the liver. As a consequence, deficiency of apo AI results in an almost complete absence of HDL and in accelerated atherosclerosis. A G→A substitution at position −75 in the promoter of the *apo AI* gene is common in the general population. A recent meta-analysis has shown that the minor allele *A* is associated with mildly elevated apoA1 levels in healthy individuals (Juo et al. 1999). Ordovas and co-workers (2002a) found a significant gene–diet interaction associated with the *APOAI G-75A* polymorphism in women. In female carriers of the *A* allele, higher dietary polyunsaturated fatty acid (PUFA) intakes were associated with higher HDL cholesterol concentrations (Ordovas et al. 2002a). In men the situation was more complex because of interactions with smoking and alcohol drinking.

In a small study of 58 male subjects with atorvastatin (40 mg/day) or placebo in a crossover design, atorvastatin was less effective in carriers of the *apoA1 −75A* allele ($n=15$) in lowering triglycerides, both in the fasting and in the postprandial state (Ordovas et al. 1999). Larger studies are needed to confirm these results.

5.5
Apolipoprotein A-IV

Apolipoprotein A-IV (apoA4) is a 46-kDa glycoprotein synthesized by intestinal enterocytes and incorporated into the surface of chylomicrons, VLDL and HDL. Presumably apoA4 plays a role in intestinal lipid absorption, modulates enzymes involved in lipoprotein metabolism and serves as a saturation signal (Tso et al. 1999). Polymorphisms of apoA4 may modulate the physical properties of chylomicrons and, ultimately, the intestinal dietary lipid absorption (Weinberg 2002). The *apoA4 His360Glu* polymorphism has been associated with increased postprandial hypertriglyceridaemia and a reduced low-density lipoprotein response to dietary cholesterol. However, this response is modulated depending on total fat intake and its qualitative composition (MUFA/PUFA ratio). Again it should be noted that study findings are far from conclusive and appear to be at odds at times, because most of these studies were underpowered for detection of subtle gene effects. Furthermore, study protocols varied with respect to the populations and dietary modification and only a few have addressed the potentially complex interactions of *apoA4* alleles with gender and life-style (evidence summarized by Weinberg 2002). In a study with 144 participants, the *apoA4 His360Glu* polymorphism had no significant effect on cholesterol lowering with statin therapy (Ordovas et al. 1995).

5.6
Apolipoprotein B

Apolipoprotein B (apoB) is the only apolipoprotein of LDL and is responsible for receptor-mediated LDL uptake. Thus, mutations and polymorphisms of the *apoB* gene may modulate the lipid response to statins. Familial defective apolipoprotein B-100 (FDB) is a group of autosomal dominantly inherited disorders in which the cellular uptake of LDL from the blood is diminished due to mutations within the apolipoprotein B-100 (apoB-100) receptor binding domain (Fischer et al. 1999). Out of several point mutations of the putative receptor binding domain of apoB-100, only three have so far been proven to produce binding-defective apoB-100. The most frequent one is *apoB-100 (arg3500→gln)* (Soria et al. 1989). Hypercholesterolaemia is in general less severe in FDB compared to LDL receptor deficiency (März et al. 1993; Myant et al. 1993). We found that the residence time of LDL apoB-100 was prolonged fourfold in homozygous FDB, but the production rate of LDL apoB-100 was approximately half of normal. This resulted from an enhanced removal of apo E-containing LDL precursors by LDL receptors, which may be up-regulated as a consequence of the decreased flux of LDL-derived cholesterol into hepatocytes (Schaefer et al. 1997). The availability of apo E for the receptor-mediated removal of remnant particles may also explain why FDB patients, homozygous or heterozygous, respond to statins as well as individuals with other types of hypercholesterolaemia.

Numerous polymorphisms have been identified at the *apo B* locus and many of them are associated with differences in baseline lipids levels (Bentzen et al. 2002). Among these, a (silent) polymorphic *XbaI* site has been examined extensively. The physiological role of the *apo B XbaI* polymorphism in codon 2488 in exon 26 is still unclear. In most studies, carriers of the *XbaI* cutting site had moderately increased LDL cholesterol. The polymorphism alters plasma lipid concentrations and LDL catabolism even though it does not alter the amino acid sequence of apoB. In one controlled dietary low- and high-fat intervention study in 44 healthy middle-aged subjects, absence of the *apoB XbaI* restriction site (X^-/X^-) was associated with a greater increase in LDL cholesterol during the high-fat diet phase of the study (Rantala et al. 2000). The same authors also investigated the *apo B EcoRI* polymorphism in exon 29, which changes the amino acid sequence of apoB, but whose functional role is unclear. The R^- allele has been associated with high plasma total cholesterol concentrations and coronary heart disease in some studies, but not in all. In their dietary intervention study, plasma LDL cholesterol concentrations increased during the high-fat diet by as much as 59±10% in $R-/R-$ subjects, compared to 39±6% in $R+/R-$ and 26±2% in $R+/R+$ subjects.

The apoB MspI polymorphism is located in the same exon as the XbaI restriction site, but causing a substitution of arginine by glutamine. In the same study, the $M+/M+$ genotype (homozygous presence of the MspI restriction site in codon 3611) was also more responsive (41±3% increase in LDL cholesterol) than the $M+/M-$ genotype (27±10% increase) (Rantala et al. 2000). In their meta-analysis of other published studies, the authors found a significant effect of the *EcoRI* and *MspI* polymorphisms, but not of the XbaI polymorphism, in dietary intervention (Rantala et al. 2000). The final conclusion was that determination of *apoB* gene polymorphisms does not now add much clinical value to dietary counselling at present.

One study addressing the impact of this polymorphism on the response to lovastatin treatment (20 or 40 mg/day; $n=211$) was negative (Ojala et al. 1991). There are a few studies investigating the role of polymorphisms of the *apoB* gene in modulating the response to fibrates. Although the *apoB XbaI* and signal peptide insertion/deletion polymorphisms were associated with different baseline levels of LDL cholesterol, they did not influence the response to fibrate therapy (Aalto-Setala et al. 1991; Hayashi et al. 1998).

5.7
Apolipoprotein E

The apolipoprotein E (apoE) polymorphism determines the greatest fraction (around 5%) of the population variance of LDL cholesterol among known variants of genes related to lipoprotein metabolism. In humans, there are three common alleles designated *ε2*, *ε3*, *ε4*, giving rise to three homozygous (designated *22*, *33*, *44*) and three heterozygous genotypes (designated *32*, *42*, *43*) (for review see Mahley and Huang 1999; Mahley and Rall 2000). This polymorphism of *apoE* affects the concentration of LDL by modifying the expression of hepatic LDL receptors. *ApoE4* enhances the catabolism of remnants by virtue of its preferential association with triglyceride-rich lipoproteins and stronger binding to lipoprotein receptors. Consequently, hepatic LDL receptors are down-regulated and LDL levels increase. For this reason, *apo E4* is associated with increased LDL cholesterol and atherosclerosis. The *ε2* allele exerts an opposite effect on lipoprotein levels. *ApoE2* is defective in binding to lipoprotein receptors. This decreases the flux of remnant-derived cholesterol into the liver, up-regulates hepatic LDL receptor and lowers LDL cholesterol. Ultimately, *apoE2* may confer protection against the development of vascular disease. For yet unknown reasons, however, one out of 20 *apoE22* homozygotes develops type III hyperlipoproteinaemia, a highly atherogenic disorder characterized by accumulation of excessive amounts of cholesterol-rich remnant lipoproteins derived from the partial catabolism of chylomicrons and very-low-density lipoproteins.

There are many studies coping with the impact of *apoE* genotypes on lipid response to diet, but their results are far from conclusive. Although the baseline total and LDL cholesterol levels were higher among *apoE4* carriers and lower among *apoE2* carriers compared with *apoE3* homozygotes, the plasma lipid response to dietary intervention did not differ significantly across *apoE* genotypes in most of these studies (see study summary by Rubin and Berglund 2002). Most were post hoc studies involving small numbers of subjects in which modest effects might be difficult to detect in the presence of confounders. In the Quebec Heart Health survey, a gene–nutrient interaction was reported for alcohol intake and apoE polymorphism in women. In women with an *E43* genotype, alcohol consumption intensified the expected increase in LDL cholesterol and, paradoxically, the decrease in HDL cholesterol, associated with increasing body mass index (Lussier-Cacan et al. 2002).

Studies with a metabolic challenge have generally been more successful in replicating effects across *apoE* genotypes. In the largest study of the effect of *apoE* polymorphism on the response of plasma cholesterol to various dietary interventions, involving 395 mostly normolipidaemic subjects, the authors concluded that *apoE* genotype may affect the cholesterol response to dietary saturated fat and cafestol, but that the effects were small. Thus, knowledge of the *apoE* genotype by itself may be of little use in the identification of subjects who respond to diet (Weggemans et al. 2001).

Similarly, reports on the effects of the *apoE* polymorphism on the efficacy of hypolipidaemic drugs are conflicting (Table 2). Despite a majority of publications describing a lower cholesterol reduction in *apoE4* carriers (Korhonen et al. 1999; Ordovas et al. 1995; Kuivenhoven et al. 1998; Ordovas et al. 2000; Ballantyne et al. 2000; Pedro-Botet et al. 2001), there are several negative reports (Sanllehy et al. 1998; Ojala et al. 1991; Gerdes et al. 2000). In view of the fact that the *apoE* polymorphism is a strong predictor of baseline LDL cholesterol, it is surprising that there is such a weak interaction, if any at all, between the *apoE* genotype and the change in the LDL cholesterol concentration with statin treatment. In the early 1990s, the interaction of the *apoE* polymorphism with the efficacy of statin treatment in familial hypercholesterolaemia was investigated in at least nine studies (reviewed by Thompson et al. 2002). Again, of those early studies, only about half confirmed that statins were less efficacious in *apoE4* carriers.

ApoE genotype was determined in 328 out of 730 patients in the atorvastatin arm of a multicentre study (Pedro-Botet et al. 2001). There was considerable interindividual variability for the change in LDL lowering, ranging from almost 0% to 60% in men and women. Men carrying the *apoE2* allele had a significantly higher mean LDL cholesterol response (−44%) than *apoE3* homozygotes (−37%) and *apoE4* carriers (−34%); $p=0.01$ for *apoE* group by treatment interaction. No such gene–treatment interaction was noted in women with average decreases in LDL cholesterol of 34%, 39% and 34%, respectively (Pedro-Botet et al. 2001). The mechanisms responsible for these effects have not been elucidated. In a subgroup of the Scandinavian Simvastatin Survival Study (4S), patients with coronary disease with high absorption (high basal cholestanol:cholesterol ratio) and low synthesis of cholesterol respond less to HMGCoA reductase inhi-

Table 2 Studies investigating the impact of the apolipoprotein E polymorphism on the LDL-lowering response to statin treatment

Gene	Variant	Clinical phenotype	Intervention	Design	Reported effect	Number of subjects	Significance	First Author, year
apoE	apoE2 (Cys Cys)[a]	Hyper-cholesterolaemia	Atorvastatin 10 mg	Double-blind parallel group (subgroup analysis for atorvastatin)	LDL lowering more effective in male ε2 allele carriers (but not in females)	195 Men	0.01	Pedro-Botet 2001 (see Davidson 1997 for study design)
	apoE3 (Cys Arg)	LDL ≥160 mg/dl	Lovastatin 20 mg			133 Women		
	apoE4(Arg Arg)	TG <400 mg/dl	Placebo					
apoE	ε4 Carriers	MI survivors	Simvastatin 20–40 mg	Double-blind	ε4 carriers equally responsive as non-ε4 carriers			
	non-ε4 Carriers	Total cholesterol 5.5–8.0 mmol/l, TG ≤2.5 mmol/l	Placebo	Parallel group		966	NS	Gerdes 2000
	(ε4 Carriers defined as ε42, ε43, ε44)			(Subgroup analysis for Danish + Finnish)			NS	
apoE	apoE 2/3	CHD patients LDL 115–190 mg/dl	Fluvastatin 40 mg	Double-blind	LDL lowering most effective in subjects with E3/3 genotype	320	0.01	Ballantyne 2000
	apoE 3/3	TG <300 mg/dl	Placebo (plus cholestyramine if LDL >160 mg/dl)	Parallel group		(10/102/49)[b]		
	apoE 4/3 and 4/4					(12/103/44)[b]		

Table 2 (continued)

Gene	Variant	Clinical phenotype	Intervention	Design	Reported effect	Number of subjects	Significance	First Author, year
apoE	ε4 carriers	Hyperlipidaemic subjects treated in lipid clinic	Pravastatin	Open label	Less effective LDL lowering in ε4 carriers	142 (Analysed as subgroups 14–48 subjects)	<0.05 for pravastatin	Drmanac 2001
	non-ε4 Carriers (ε4 Carriers defined as ε42, ε43, ε44)		Lovastatin Simvastatin	Retrospective				
apoE	apoE2 (Cys Cys)[a]	Hypercholesterolaemic outpatients No diabetics	Atorvastatin Pravastatin 20 mg	Open label	No significant effect on lipid response	401 (56% Female)	NS	Pena 2002
	apoE3 (Cys Arg)			195 Primary health care physicians				
	apoE4 (Arg Arg)	TG≤4.5 mmol/l						

[a] ApoE2, E3 and E4 haplotypes (also designated as ε2, ε3 and ε4 alleles) defined by amino acid coding in exon 3 codon 112 and exon 4 codon 158 of the apoE gene. Note: the nomenclature apoE2 or ε2 has historical reasons, to distinguish phenotyping based on the apoE protein or genotyping assays for apoE determination.

[b] In brackets number of subjects with apoE 2/3, apoE 3/3, apoE 4/3 and 4/4 genotypes in the fluvastatin (upper bracket) and placebo (lower bracket) groups

bition by statins (Miettinen et al. 1998). Presumably intestinal cholesterol absorption is also related to *apoE* phenotype, as it is related to bile acid and cholesterol synthesis (Gylling et al. 1992; Kesaniemi et al. 1987); so that the *apo E* polymorphism may exert its effect on the response to statins via modulating intestinal cholesterol absorption. Supportive data have been reported in study of 19 patients with refractory heterozygous familial hypercholesterolaemia. Again, *apoE4* was more common in poor responders to atorvastatin. The authors concluded that poor responders to statins have a low basal rate of cholesterol synthesis that may be secondary to a genetically determined increase in cholesterol absorption, possibly mediated by *apoE4*. If so, statin responsiveness could be enhanced by reducing dietary cholesterol or inhibiting cholesterol absorption (O'Neill et al. 2001).

In a substudy of Scandinavian Simvastatin Survival Study (4S), Gerdes et al. (2000) found that the risk of death or a major coronary event in survivors of myocardial infarction (MI) was related to the *apoE* genotype. They analysed 5.5 years of follow-up data of 966 Danish and Finish myocardial infarction survivors enrolled in 4S and found that MI survivors with the *apoE4* allele were at nearly twice the risk of death, and that treatment with simvastatin abolished the excess mortality. They concluded that the effect of *apoE4* may involve mechanisms unrelated to serum lipoproteins because (a) baseline lipid levels did not differ between *apoE* genotypes, and (b) *apoE4* carriers and patients with other genotypes were equally responsive to simvastatin treatment in terms of LDL cholesterol lowering (Gerdes et al. 2000), which contrasts with the differential effect of *apoE* genotype observed in men receiving atorvastatin (Petro-Botet et al. 2001). It should be noted that both reports are retrospective subgroup findings and not prespecified analyses in the study protocols of the respective trials. In a subgroup analysis of the Lipoprotein and Coronary Atherosclerosis Study (LCAS), *apoE4* carriers were less responsive in LDL cholesterol lowering to fluvastatin. However, neither baseline lipids nor clinical outcome differed between *apoE* genotype (Ballantyne et al. 2000).

The *apoE* genotype, which has shown to influence plasma cholesterol level (see statins), had no effect on the hypolipidaemic efficacy of colestipol (Korhonen et al. 1999). The reports of response to fibrates in relation to the *apoE* locus are conflicting (Manttari et al. 1991; Nemeth et al. 1994 and 1995; Yamada et al. 1997; Sanllehy et al. 1998). It is important to note that all of these studies are very small ($n= 63-230$) and therefore their power to detect an effect of the *apoE* genotype was low. In a larger study, Brisson and co-workers (2002) investigated the interaction of several genetic variants, including *apoE* polymorphism, with the lipid response to fenofibrate in 292 hypertriglyceridaemic subjects. Overall *apoE2* carriers were most responsive in lowering non-HDL cholesterol (defined as total cholesterol minus HDL cholesterol, reflecting the cholesterol fraction associated with apoB containing lipoproteins).

Two *apoE* promoter polymorphisms (among other variants of genes involved in lipid metabolism) have been examined in subjects randomized to treatment with atorvastatin ($n=56$) or bezafibrate ($n=60$) (Garcia-Otin et al. 2002). Subjects on atorvastatin showed greater reductions in total cholesterol, LDL cholesterol and non-HDL cholesterol if carriers of a *T* at position −491 of the *apo E* promoter (*T+* subjects) compared to those homozygous for *A* at that position (*T−* subjects). In contrast, *T+* subjects treated with bezafibrate were less responsive to reductions in triglyceride concentrations. No effect was observed for the *apoC3 C3238G 3'utr* and the *LPL D9N* and *N291S* gene variants.

Taken together it appears that the *apoE4* allele is associated with an enhanced response (of LDL lowering) to dietary interventions, but a reduced response to statin-induced LDL cholesterol lowering (Ordovas and Mooser 2002). However, at most, the effect of *apoE* on the LDL response to statins is clinically modest and not observed in a large outpatient study of 400 hypercholesterolaemic patients treated with pravastatin (Pena et al. 2002).

5.8
Cholesteryl Ester Transfer Protein

Cholesteryl ester transfer protein (CETP) is involved in reverse cholesterol transport and several polymorphisms with a functional impact on plasma HDL cholesterol and triglycerides have been identified (Yamashita et al. 2000a). CETP mediates the transfer of neutral lipids between lipoproteins and plays a central role in HDL metabolism. CETP transfers cholesteryl esters associated with HDL to triglyceride-rich lipoproteins, facilitating the clearance of cholesteryl esters from plasma (Vaughan et al. 2000). Although the potential contribution of CETP to reverse cholesterol transport suggests an anti-atherogenic mode of action, knowledge of the physiological role of CETP in lipoprotein metabolism remains incomplete. The overall effect of CETP on atherogenesis may vary depending on both metabolic context and molecular variation in the *CETP* gene (Tall 1995). Interestingly, CETP deficiency is associated with elevated HDL-C, but may, paradoxically, increase the risk for CHD (Yamashita et al. 2000b).

The *B2* allele of the *Taq1B* polymorphism of the *CETP* gene, a silent base exchange in nucleotide 277 of the first intron, has been associated with decreased CETP activity and increased HDL cholesterol (Kuivenhoven et al. 1998; Ordovas et al. 2000). Several other single nucleotide polymorphisms (SNPs) in the *CETP* gene have been associated with interindividual variation in CETP plasma concentrations, HDL cholesterol levels and risk of cardiovascular disease (Agellon et al. 1990; Agerholm-Larsen et al. 2000a, b). The Taq1B polymorphism has also been shown to serve as a marker of lipoprotein response to dietary intervention (Dullaart et al. 1997; Wallace et al. 2000). Other studies have demonstrated a link between the *CETP I405V* gene polymorphism and HDL-C but not with response to diet (Gudnason et al. 1999; Friedlander et al. 2000). Sample size and known confounders influencing HDL-C (i.e. smoking status, level of exercise, alcohol consumption) are the most probable explanation for such conflicting data.

In the REGRESS study, pravastatin therapy slowed the progression of coronary atherosclerosis in *B1B1* CETP TaqIB carriers, but not in *B2B2* carriers who represented 16% of the patients (Kuivenhoven et al. 1998). This effect was independent of the degree of lipid lowering, which was not significantly different across all three *CETP TaqIB* genotypes (*B1B1*, *B1B2*, *B2B2*) (Kuivenhoven et al. 1998). Since this finding was the result of a retrospective analysis, it awaits replication in other studies. It was not observed in WOSCOPS (West of Scotland Coronary Prevention Study) (Freeman et al. 2000). REGRESS was an angiography-based trial in men with pre-existing coronary disease, whereas WOSCOPS was a primary prevention study in men with elevated LDL cholesterol. The different populations and primary endpoints of these studies are possible reasons for the inconsistent results.

SNP haplotypes are more informative than SNPs considered separately (Knoblauch et al. 2002). We investigated the impact of several genetic variants and haplotypes of *CETP* on lipid-lowering response in 103 dyslipidaemic patients treated with several statin agents for primary and secondary prevention. Statin doses were titrated according to current guidelines to reach LDL cholesterol levels of 130 mg/dl and 100 mg/dl, respectively. Nine single nucleotide polymorphisms (SNPs) were identified from the literature or by sequencing the *CETP* gene in a reference population of ethnic diversity. Strong associations between variants in the *CETP* gene and baseline CETP mass and activity were found. Furthermore, most of the nine SNPs showed a significant association with baseline levels of HDL cholesterol and triglycerides. No significant association was observed between individual *CETP* SNPs and the response to statin treatment. However, when the SNPs were organized into haplotypes, we not only confirmed the associations with CETP mass and activity and baseline lipid levels, but also identified *CETP* haplotypes that significantly predicted the lipid response to treatment with a statin (even after adjusting for multiple comparisons): patients with two copies of a particular CETP haplotype showed the largest increase in HDL-C levels, those without this haplotype the lowest increase in HDL-C levels, and patients with one copy had an intermediate response. A similar inverse relationship was observed for the decrease in triglycerides. There was no significant interaction between any of the *CETP* gene variants with baseline LDL cholesterol or cholesterol response after statin therapy (Winkelmann et al. 2002). In conclusion, in our study haplotypes were better predictors of the individual response to treatment with a statin than a single SNP.

5.9
Hepatic Triglyceride Lipase

Hepatic triglyceride lipase (HTGL) or hepatic lipase (*HL*, the HL gene symbol is *LIPC*) catalyses the hydrolysis of triglycerides of HDL and remnant lipoproteins like IDL. Further, it is involved in their uptake in the liver. Whether HL is pro- or antiatherogenic is still a matter of debate (Santamarina-Fojo et al. 1998). Recently, a $C \rightarrow T$ polymorphism at position −514 (−480) in the promoter of the *HL* gene has been described which is in complete linkage disequilibrium with three other polymorphic sites within the promoter (G-250A, T-710C, A-763G) (Guerra et al. 1997) and is associated with HDL particle size (Couture et al.

2000). The common *C* allele is associated with higher HL activity and an atherogenic lipid profile, characterized by lower levels of HDL_2 cholesterol and dense LDL particles (Zambon et al. 1999). A gene-nutrient interaction of the *C-514T* HL gene polymorphism was observed in the Framingham study. Carriers of a *TT* genotype had the highest baseline HDL levels, but showed an impaired adaptation to higher animal fat intake (Ordovas et al. 2002b).

Zambon et al. (2001) treated 49 dyslipidaemic men with elevated ($\geq$125 mg/dl) apoB levels and established CAD with 40 mg daily of lovastatin and colestipol and analysed their lipid response depending on the HL *C-514T* genotype. Subjects with a HL *−514 CC* genotype had the highest baseline HL activity and the largest absolute and relative decrease in HL activity (18% decrease compared to 9% and 5% in *CT* and *TT* carriers). They also showed the largest increase in HDL cholesterol, particularly in HDL_2 cholesterol, and in LDL buoyancy as a measure of LDL particle size (Zambon et al. 2001). Parallel to the changes in lipid profile, the *CC* homozygous subjects had a significantly better angiographic outcome. To our knowledge the angiographic findings of that study have not yet been replicated by other studies. Other polymorphisms within the coding region of the *HL* gene are known to influence the activity of the lipase (Nie et al. 1998) and should be investigated for functional and clinical effects during lipid-lowering intervention.

One study in 198 type 2 diabetics replicated the baseline associations between *HL C-514T* gene polymorphism and HL activity, but did not find any interaction of this gene variant and the lipid lowering potency of 10 mg and 80 mg atorvastatin in a randomized double-blind placebo-controlled parallel group protocol. A daily dose of 10 mg atorvastatin lowered HL activity to a similar degree in male and female *CC* and *CT* carriers (10%–13%). The higher dose of 80 mg resulted in an even greater reduction. Thus, atorvastatin treatment resulted in a dose-dependent decrease in HL activity, regardless of sex or the HL gene promoter variant (Berk-Planken et al. 2003).

5.10
Lipoprotein Lipase

Lipoprotein lipase (LPL) is the rate-limiting enzyme in the hydrolysis of triglyceride-rich lipoproteins, i.e. chylomicrons and VLDL. The genetic variants of LPL associated with elevated triglycerides and low HDL include the Asn291Ser (or N291S) and Asp9Asn (or D9N) polymorphisms, and the Ser447Stop (or S447X) polymorphism, a C→G nucleotide exchange at position 1595 that results in a stop codon and truncation of the LPL protein by two amino acids (Zhang et al. 1995; Wittrup et al. 1999). Another intronic polymorphism of *LPL*, the intron 8 *HindIII LPL* polymorphism, has been associated with differences in LDL cholesterol, but not with triglycerides and HDL in females (Larson et al. 1999). There is strong evidence that the latter two polymorphisms, *S447X* and *HindIII*, are in significant linkage disequilibrium (Humphries et al. 1998a).

The *LPL Asp9Asn* polymorphism was associated with lower HDL, a significantly lower decrease in total and LDL cholesterol upon treatment with pravastatin and a significantly higher rate of angiographic progression (in the placebo group only) in the REGRESS trial. Although the lipid-lowering effect of pravastatin was attenuated in patients carrying the Asp9Asn variant, angiographic progression of coronary atherosclerosis was prevented in the pravastatin group regardless of the presence or absence of the *Asp9Asn* polymorphism (Jukema et al. 1996). The *LPL HindIII* polymorphism predicted non-responsiveness to statin treatment (lovastatin, plus cholestyramine, if needed for aggressive LDL lowering to levels <100 mg/dl) in the Post-CABG trial, both in the moderate and aggressive treatment arm (as evidenced by a significantly higher rate of saphenous graft worsening in LPL Hind III *2/2* homozygotes), while no such association was observed for the Asn291Ser and *Asp9Asn* polymorphisms (Taylor et al. 1999). Finally, the REGRESS study group has published a gene–environment interaction between the *Ser447Stop* variant of *LPL* and the use of beta-blockers. In presence of the *447 stop codon*, LPL activity and HDL cholesterol were significantly lower in users of beta-blockers, while such associations were absent in the patients not taking beta-blockers (Groenemeijer et al. 1997).

As always, caution should be used when assessing such findings. These results were obtained in subgroups of large placebo-controlled randomized trials studying the lipid-lowering response of statins. The genetic analysis was not a prespecified endpoint, but a retrospective analysis. None of the reports corrected for multiple testing. Therefore, these studies should be taken as exploratory hypothesis-generating studies that need to be replicated.

5.11
Lipoprotein (a)

Lipoprotein (a) (Lp(a) consists of two components: a LDL particle and apolipoprotein (a) which are linked by a disulfide bridge. Apo(a) reveals a genetically determined size polymorphism resulting from a variable number of plasminogen kringle IV-type repeats (Utermann 1999; Hobbs and White 1999). Lp(a) levels are amazingly stable over the lifetime of an individual, indicating a strong genetic component. Lp(a) is considered an independent risk factor for coronary artery disease (Marcovina et al. 1998). However, Lp(a) was not an independent predictor of CHD risk in the Quebec Cardiovascular study (Cantin et al. 2002) or in the Strong Heart study (Wang et al. 2002).

Diets rich in saturated fatty acids consistently lower plasma Lp(a) concentration. However, any benefit of lowering elevated Lp(a) is counteracted by the ability of saturated fats to raise LDL levels. Subjects with low to moderate alcohol intake have been shown to have lower Lp(a) concentrations than teetotallers (Puckey and Knight 1999).

The effect of statins on Lp(a) is controversial (Klausen et al. 1993; Maerz et al. 1994). In a study of 51 FH patients treated with 40 mg pravastatin daily, an increase in Lp(a) was observed which was greatest in patients with the low-molecular-weight apo(a) phenotypes (Klausen et al. 1993). In another double-blinded study of 391 hypercholesterolaemic subjects treated with atorvastatin (10 mg daily) or simvastatin (20 mg daily), plasma levels of apo(a) fragments were not modified by either statin. A minor reduction in Lp(a) plasma levels was observed on treatment with atorvastatin (6% variation ; $p<0.001$) and simvastatin (0.02% variation; $p=0.048$) (Gonbert et al. 2002). Such marginal findings have most probably no clinical relevance, especially since they were not replicated in other large statin intervention studies.

5.12
Cholesterol-7α-Hydroxylase

Cholesterol-7α-hydroxylase (CYP7) activity seems to be inversely correlated with plasma cholesterol levels (Cohen et al. 1999) and bile acids in the intestine seem to activate the pregnane X nuclear receptor (PXR), which subsequently represses *CYP7A1* gene expression (Goodwin et al. 2002). Resins such as cholestyramine and colestipol impede the recycling of bile acids by trapping them in the lumen of the intestine (Grundy et al. 1971). As a consequence, the hepatic conversion of cholesterol to bile acid is increased by up-regulation of CYP7 (Reihner et al. 1989), the rate-limiting enzyme of bile acid synthesis.

There exists at least one common polymorphism within the regulatory region of the *CYP7* gene (*C-278A* [Wang et al. 1998] or *A-204C* [Couture et al. 1999]). Depending on the population studied, the *C-278A* polymorphism accounted for 1%–15% of the variation of LDL cholesterol (Wang et al. 1998; Couture et al. 1999). The effect of this SNP on the regulation of *CYP7* has not been evaluated in detail. It is, therefore, difficult to predict whether it will influence the lipid-lowering effect of bile acid sequestrants or HMG-CoA reductase inhibitors.

5.13
Peroxisome Proliferator-Activated Receptors

Peroxisome proliferator-activated receptors (PPARs) are nuclear receptors activated by fatty acids and derivatives (Berger and Moller 2002; see also the chapter by Vidal-Puig and Abel, this volume). PPARα is predominantly expressed in tissues that metabolize high amounts of fatty acids (liver, kidney, heart, and muscle), where it stimulates their oxidative degradation (Auboeuf et al. 1997). PPARγ is mainly found in adipose tissues, where it promotes adipocyte differentiation and lipid storage (Auwerx 1999). PPARβ/δ is expressed ubiquitously and seems to play a role in the control of adipogenesis. PPARα mediates the hypolipidaemic action of fibrates, and PPARγ is the molecular target of glitazone antidiabetics (Corton et al. 2000; Barbier et al. 2002).

Several SNPs in the *PPARα* gene have been published recently: a *G/A* transversion in intron 3, *R131Q*, and *L162V* (Vohl et al. 2000; Flavell et al. 2000; Sapone et al. 2000). In all studies, the frequency of the minor allele was lower than 10%. There was no evidence that the mutations within the coding region of

PPARα have a major role in type 2 diabetes, although they might have a borderline impact on LDL cholesterol levels (Vohl et al. 2000; Flavell et al. 2000).

No polymorphisms have been described within the PPAR responsive elements (*PPRE*) of the promoters of LPL, apo CIII, apo AI and apo AII that might directly influence the binding of these transcription factors. On the other hand, there are several possible polymorphisms in the target genes of PPARα that might interact with the action of fibrates, e.g. *LPL D9N, N291S* and *S447X*. Carriers of the truncated LPL variant, *S447X*, which is associated with higher plasma LPL activity, might have greater benefit, whereas carriers of *LPL 9N* and *291S*, who have lower plasma LPL activity, might have a smaller benefit from fibrate therapy, but this has not yet been proven experimentally.

In a study of 71 dyslipidaemic subjects randomized to gemfibrozil (600 mg daily) or placebo, the HDL_2 response to gemfibrozil was modulated by the *PPARα L162V* gene polymorphism. *PPARα 162L* homozygotes showed a 5% increase in HDL_2 cholesterol compared to a 50% increase in *162V* carriers ($p=0.03$) (Bossé et al. 2002).

In the SENDCAP study, bezafibrate-treated *V162* allele carriers (13 patients) showed a twice as much lowering of total cholesterol (−0.90 vs −0.42 mmol/l, $p=0.04$) and non-HDL-C (−1.01 vs −0.50 mmol/l, $p=0.04$) than *L162* allele homozygotes (109 patients) (Flavell et al. 2000). As bezafibrate is not PPARα specific but also interacts with PPARγ and PPAR β/δ, the effects of the *V162* variation might even be greater in the case of other, more specific fibrates. However, the *Leu to Val* change at position *162* of the *PPARα* gene was not associated with a difference in lipid-lowering effectiveness in a study of 96 lipid clinic subjects prescribed fibrates (Puckey and Knight 2001) and in another study with fenofibrate described below (Brisson et al. 2002).

Brisson et al. (2002) observed in 292 hypertriglyceridaemic subjects treated with fenofibrate that the *LPL D9N* (= low LPL) and *PPARγ P12A* mutations did not affect fenofibrate lipid-lowering action. Neither was the *PPARα L162V* polymorphism associated with any significant difference in lipid-lowering efficacy of fenofibrate in this study. In contrast, only the *LPL P207L* (LPL-null) variant was significantly associated with residual post-treatment hypertriglyceridaemia. Furthermore, *apoE2* carrier status was most strongly associated non-HDL cholesterol response to fenofibrate, both in the entire cohort and in the subgroups with the simultaneous presence of the *PPARα 162V* or *LPL 207L* mutations (Brisson et al., 2002)

6
Non-lipid Pathway Genes

6.1
Stromelysin-1

Recently, a functional *5A/6A* polymorphism has been described within the *stromelysin-1 promoter* (Ye et al. 1995). Stromelysin-1 is a member of the metalloproteinases that degrade extracellular matrix (see the chapter by Henney, this volume). In situ hybridization and histopathological studies suggest that stromelysin-1 activity is important in connective tissue remodelling associated with atherogenesis and plaque rupture. Patients homozygous for the *6A* allele showed greater progression of angiographic disease than those with other genotypes (Ye et al. 1995). In the REGRESS study (Regression Growth Evaluation Study), patients within the placebo group with the *5A6A* or *6A6A* genotype had more clinical events than patients with the *5A5A* genotype. In the pravastatin group, the risk of clinical events in patients with *5A6A* or *6A6A* genotypes was lower compared with placebo (deMaat et al. 1999). Similar data were obtained for the incidence of repeat angioplasty. These beneficial changes were independent of the effects of pravastatin on lipid levels, raising the possibility that pravastatin exerts pleiotropic effects on stromelysin-1 expression or activity. Until now there have been two studies, one with gemfibrozil (LOCAT) (Humphries et al. 1998) and the REGRESS study conducted with pravastatin (deMaat et al. 1999), suggesting that the *stromelysin-1 promoter* polymorphism confers a genotype-specific response to medication.

7
Limitations and Conclusions

Today, a physician's selection of drug treatment and dosage is usually based upon empirical averages obtained from clinical trials, but not upon the individual who will take the drug. This practice of undifferentiated treatment leads to a lesser degree of efficacy and increased toxicity (Lazarou et al. 1998). Thus, the promise of the post-genomic era is that fast and inexpensive gene-measurement technologies will allow variation at the DNA level to be incorporated into algorithms designed to identify populations and individuals at risk and tailor drug dosing for an optimal drug response. However, according to a recent study of the *apoE* polymorphism and its contribution to the variability in quantitative measures of lipid metabolism, such expectations may be unrealistic (Stengard JH et al. 2002). Another study of genetic variants known to affect HDL cholesterol (*TaqIB CETP* polymorphism, *C-514T* polymorphism of hepatic lipase, and *S447X* polymorphism of lipoprotein lipase) concluded that only 2.5% of the variance in HDL cholesterol could be explained by those variants (Talmud et al. 2002). The same may apply to predictability of individual drug response: It may not be possible to identify either a particular genetic variant or a particular subset of variants (haplotypes, SNP patterns across genes located on different chromosomes) that are specific or sensitive enough to identify an individual at risk. Complexity research has shown that it is extremely difficult to explain an individual outcome of a highly interactive system in terms of the behaviour of a particular state or certain subsets (Sole et al. 2000): it may even be theoretically impossible (Axelrod et al. 2000). Although the functional effects of polymorphic drug targets are under study by many groups through the world, this area is plainly not yet mature enough to provide clear-cut recommendations for the choice of drug and the drug dosage in the individual (Ingelman-Sundberg 2001). We would like to close with a citation from Maitland-van der Zee et al. (2002), who summarized their own review concerning the impact of genetic polymorphisms on the response to HMG-CoA reductase inhibitors as follows: "At present, no single polymorphism has been identified that renders statin treatment ineffective, based on clinical outcomes. Therefore, results from large-scale population studies are needed to complement results from clinical trials and small-scale studies in selected populations".

References

Aalto-Setala K, Kontula K, Manttari M et al (1991) DNA polymorphisms of apolipoprotein B and AI/CIII genes and response to gemfibrozil treatment. Clin Pharmacol Ther 50:208–214

Agellon LB, Quinet EM, Gillette TG et al (1990)Organization of the human cholesteryl ester transfer protein gene. Biochemistry 29:1372–1376

Agerholm-Larsen B, Nordestgaard BG, Steffensen R et al (2000a) Elevated HDL cholesterol is a risk factor for ischemic heart disease in white women when caused by a common mutation in the cholesteryl ester transfer protein gene. Circulation 101:1907–1912

Agerholm-Larsen B, Tybjaerg-Hansen A, Schnohr P et al (2000b) Common cholesteryl ester transfer protein mutations, decreased HDL cholesterol, and possible decreased risk of ischemic heart disease: The Copenhagen City Heart Study. Circulation 102:2197–2203

Aguilar-Salinas SA, Barnett H, Schonfeld G (1998) Metabolic modes of action of statins in the hyperlipoproteinemias. Atherosclerosis 141:203–207

Altman RB, Klein TE (2002) Challenges for biomedical informatics and pharmacogenomics. Annu Rev Pharmacol Toxicol 42:113–133

Anonymous (2000) Pharmacogenomics. Nature Biotechnol 18[Suppl.]:IT40-IT42.

Auboeuf D, Rieusset J, Fajas L et al (1997) Tissue distribution and quantification of the expression of mRNAs of peroxisome proliferator-activated receptors and liver X receptor-alpha in humans: no alteration in adipose tissue of obese and NIDDM patients. Diabetes 46:1319–27

Auwerx J (1999) PPARγ, the ultimately thrifty gene. Diabetologia 42:1033–1049

Axelrod R, Cohen MD (2000) Harnessing complexity. New York: Basic Books

Bailey DS, Bondar A, Furness LM (1998) Pharmacogenomics – it's not just pharmacogenetics. Curr Opin Biotechnol 9:595–601

Ballantyne CM, Herd JA, Stein EA et al (2000) Apolipoprotein E genotypes and response of plasma lipids and progression-regression of coronary atherosclerosis to lipid-lowering drug therapyJ Am Coll Cardiol 36:1572–8

Barbier O, Torra IP, Duguay Y et al (2002) Pleiotropic actions of peroxisome proliferator-activated receptors in lipid metabolism and atherosclerosis. Arterioscler Thromb Vasc Biol 22:717–726

Beaird SL (2000) HMG-CoA reductase inhibitors: asssessing differences in drug interactions and safety profiles. J Am Pharm Assoc 40:637–644

Bennett MK, Lopez JM, Sanchez HB et al (1995) Sterol regulation of fatty acid synthase promoter. Coordinate feedback regulation of two major lipid pathways. J Biol Chem 270:25578–25583

Bentzen J, Jørgensen T, Fenger M (2002) The effect of six polymorphisms in the apolipoprotein B gene on parameters of lipid metabolism in a Danish population. Clin Genet 61:126–134

Berger J, Moller DE (2002) The mechanisms of action of PPARs. Annu Rev Med 53:409–435

Berk-Planken IIL, Bootsma AH, Hoogerbrugge N et al, on behalf of the DALI study group (2003) Atorvastatin dose-dependently decreases hepatic lipase activity in type 2 diabetes -effect of sex and the LIPC promoter variant. Diabetes Care 26:427–432

Bottorff M, Hansten P (2000) Long-term safety of hepatic hydoxylmethyl glutaryl coenzyme A reductase inhibitors. Arch Intern Med 160:2273–2280

Bouly M, Masson D, Gross B et al (2001) Induction of the phospholipid transfer protein gene accounts for the high density lipoprotein enlargement in mice treated with fenofibrate. J Biol Chem 276:25841–25847

Bossé Y, Pascot A, Dumont M et al (2002) Influences of the PPARγ L162 V polymorphism on plasma HDL_2 cholesterol response of abdoiminally obese men treated with gemfibrozil. Genet Med 4:311–315

Brisson D, Ledoux D, Bossé Y et al (2002) Effect of apolipoprotein E, peroxisome proliferator-activated receptor alpha and lipoprotein lipase gene mutations on the ability of fenofibrate to improve lipid profiles and reach clinical guideline targets among hypertriglyceridemic patients. Pharmacogenetics 12:313–320

Brorholt-Petersen JU, Jensen HK, Raungaard B et al (2001) LDL-receptor gene mutations and the hypocholesterolemic response to statin therapy. Clin Genet 59:397–405

Brown MS, Goldstein JL (1997) The SREBP pathway: regulation of cholesterol metabolism by proteolysis of a membrane-bound transcription factor. Cell 89:331–340

Brown WV (2001) What are the priorities for managing cholesterol effectively. Am J Cardiol. 88 (Suppl. F): 21F-24F

Bucher HC, Griffith LE, Guyatt HG (1999) Systematic review on the risk and benefit of different cholesterol-lowering interventions Arterioscler Thromb Vasc Biol 19:187–195

Cantin B, Lamarche B, Després JP et al (2002) Does correction of the friedewald formula using lipoprotein(a) change our estimation of ischemic heart disease risk? The Quebec Cardiovascular Study. Atherosclerosis 163:261–267

Cardon LR, Idury RM, Harris TJR et al (2000) Testing drug response in the presence of genetic information: sampling issues for clinical trials. Pharmacogenetics 10:503–510

Carson PE, Flanagan CL, Ickes CE et al (1956) Enzymatic deficiency in primaquine sensitive erythrocytes. Science 124:484–485

Caslake MJ, Packard CJ, Gaw A et al (1993) Fenofibrate and LDL metabolic heterogeneity in hypercholesterolemia. Arterioscler Thromb 13:702–711

Chaves FJ, Real JT, Garcia-Garcia AB et al (2001) Genetic diagnosis of familial hypercholesterolemia in a South European outbreed population: influence of low-density lipoprotein (LDL) receptor gene mutations on treatment response to simvastatin in total, LDL, and high-density lipoprotein cholesterol. J Clin Endocrinol Metab 86:4926–4932

Chinetti G, Lestavel S, Bocher V et al (2001) PPAR-alpha and PPAR-gamma activators induce cholesterol removal from human macrophage foam cells through stimulation of the ABCA1 pathway. Nat Med 7:53–58

Cohen JC (1999) Contribution of cholesterol 7-alpha-hydroxylase to the regulation of lipoprotein metabolism. Curr Opin Lipidol 10:303–307

Corton JC, Anderson SP, Stauber A (2000) Central role of peroxisome proliferator-activated receptors in the actions of peroxisome proliferators. Annu Rev Pharmacol Toxicol 40:491–518

Couture P, Brun LD, Szots F et al (1998) Association of specific LDL receptor gene mutations with differential plasma lipoprotein response to simvastatin in young French

Canadians with heterozygous familialhypercholesterolemia. Arterioscler Thromb Vasc Biol 18:1007–1012

Couture P, Otvos JD, Cupples LA et al (1999) Association of the A-204C polymorphism in the cholesterol 7alpha- hydroxylase gene with variations in plasma low density lipoprotein cholesterol levels in the Framingham Offspring Study. J Lipid Res 40:1883–1889

Couture P, Otvos JD, Cupples LA et al (2000) Association of the C-514T polymorphism in the hepatic lipase gene with variations in lipoprotein subclass profiles—the Framingham Offspring study. Arterioscler Thromb Vasc Biol 20:815–822

Davidson M, McKenney J, Stein E, Schrott M, Bakker-Arkema R, Fayyad R, Black D, for the Atovvastatin Study Group I (1997) Comparison of one-year efficacy and safety of atovvastatin versus lovastatin in primary hypercholesterolemia. Am J Cardiol 70:1475–1481

Davidson MH (2001) Introduction: utilization of surrogate markers of atherosclerosis for the clinical development of pharmaceutical agents. Am J Cardiol 87 [Suppl A]:1A–7A

Diabetes Atherosclerosis Intervention Study Investigators (2001) Effect of fenofibrate on progression of coronary-artery disease in type 2 diabetes: the Diabetes Atherosclerosis Intervention Study, a randomised study. Lancet 357:905–910

Desager JP, Horsmans Y, Vandenplas C et al (1996) Pharmacodynamic activity of lipoprotein lipase and hepatic lipase, and pharmacokinetic parameters measured in normolipididaemic subjects receiving ciprofibrate (100 or 200 mg/day) or micronised fenofibrate (200 mg/day) therapy for 23 days. Atherosclerosis 124(Suppl.): S65-S73

Després JP (2001) Increasing high-density lipoprotein cholesterol: an update of fenofibrate. Am J Cardiol 88 (Suppl. N):30N-36 N

Destenaves B, Thomas F (2000) New advances in pharmacogenomics. Curr Opin Chem Biol 4:440–444

Dullaart RP, Hoogenberg K, Riemens SC et al (1997) Cholesteryl ester transfer protein gene polymorphism is a determinant of HDL cholesterol and of the lipoprotein response to a lipid-lowering diet in type 1 diabetes. Diabetes 46:2082–2087

Ericsson CG, Hamsten A, Nilsson J et al (1996) Angiographic assessment of effects of bezafibrate on progression of coronary artery disease in young male postinfarction patients. Lancet 347:849–853

Evans WE, Johnson JA (2001) Pharmacogenomics: the inherited basis for interindividual differences in drug response. Annu Rev Genomics Hum Genet 2:9–39

Fan YM, Laaksonen R, Janatuinen T et al (2001) Effects of pravastatin therapy on serum lipids and coronary reactivity are not associated with SREBP cleavage-activating protein polymorphism in healthy young men. Clin Genet 60:319–321

Feher MD, Webb JC, Patel DD et al (1993) Cholesterol-lowering drug therapy in a patient with receptor-negative homozygous familial hypercholesterolaemia. Atherosclerosis 103:171–180

Fischer E, Scharnagl H, Hoffmann MM et al (1999) Mutations in the Apolipoprotein (apo) B-100 Receptor-binding region: Detection of apo B-100 (Arg3500–>Trp) Associated with Two New Haplotypes and Evidence That apo B-100 (Glu3405–>Gln) Diminishes Receptor-mediated Uptake of LDL. Clin Chem 45:1026–1038

Flavell DM, Pineda Torra I, Jamshidi Y et al (2000) Variation in the PPARalpha gene is associated with altered function in vitro and plasma lipid concentrations in Type 2 diabetic subjects. Diabetologia 43:673–680

Freeman DJ, Wilson V, McMahon AD et al (2000)A polymorphism of the Cholesteryl Ester Transfer Protein (CETP) gene predicts cardiovascular events in the West of Scotland Coronary Prevention Study (WOSCOPS). Atherosclerosis 151:91

Frenette PS (2001) Locking a leukocyte integrin with statins. N Engl J Med 345:1419–1421

Frick MH, Syvänne M, Nieminen MS et al, for the Lopid Coronary Angiohraphy Trial (LOCAT) (1997) Prevention of the angiographic progression of coronary and vein-graft atherosclerosis by gemfibrozil after coronary bypass surgery in men with low levels of HDL cholesterol. Circulation 96:2137–2143

Friedlander Y, Leitersdorf E, Vecsler R et al (2000) The contribution of candidate genes to the response of plasma lipids and lipoproteins to dietary challenge. Atherosclerosis 152:239–248

Fruchart JC, Brewer HB, Leitersdorf E (1998) Consensus for the use of fibrates in the treatment of dyslipoproteinemia and coronary heart disease. Am J Cardiol 81:912–917

Fruchart JC, Duriez P, Staels B (1999) Peroxisome proliferator-activated receptor-alpha activators regulate genes governing lipoprotein metabolism, vascular inflammation and atherosclerosis. Curr Opin Lipidol 10:245–257

Fruchart JC (2001) Peroxisome proliferator-activated receptor-α activation and high-density lipoprotein metabolism. Am J Cardiol 88 (Suppl. N):24N-29 N

García-Otín AL, Civeira F, Aristegui R et al, on behalf of the ATOMIX Study Group (2002) Allelic polymorphism 491A/T in apo E gene modulates the lipid-lowering response in combined hyperlipidemia treatment. Eur J Clin Invest 32:421–428

Gerdes LU, Gerdes C, Kervinen K et al (2000) The apolipoprotein epsilon4 allele determines prognosis and the effect on prognosis of simvastatin in survivors of myocardial infarction : a substudy of the Scandinavian simvastatin survival study. Circulation 101:1366–1371

Ginsberg HN (1998) Effects of statins on triglyceride metabolism. Am J Cardiol 81[Suppl 4A]:32–35

Gonbert S, Malinsky S, Sposito AC et al (2002) Atorvastatin lowers lipoprotein(a) but not apolipoprotein(a) fragment levels in hypercholesterolemic subjects at high cardiovascular risk. Atherosclerosis 164:305–311

Goodwin B, Redinbo MR, Kliewer SA (2002) Regulation of CYP3A gene transcription by the pregnane X receptor. Annu Rev Pharmacol Toxicol 42:1–23

Groenemeijer B, Hallman MD, Reymer PWA et al, on behalf of the REGRESS study group (1997) Genetic variant showing a positive interaction with ß-blocking agents with a beneficial influence on lipoprotein lipase activity, HDL cholesterol, and triglyceride levels in coronary artery disease patients. Circulation 95:2628–2635

Grundy SM, Ahrens EH, Jr., Salen G (1971) Interruption of the enterohepatic circulation of bile acids in man: comparative effects of cholestyramine and ileal exclusion on cholesterol metabolism. J Lab Clin Med 78:94–121

Gudnason V, Kakko S, Nicaud V et al (1999) Cholesteryl ester transfer protein gene effect on CETP activity and plasma high-density lipoprotein in European populations. The EARS Group. Eur J Clin Invest 29:116–128

Guerra R, Wang J, Grundy SM et al (1997) A hepatic lipase (LIPC) allele associated with high plasma concentrations of high density lipoprotein cholesterol. Proc Natl Acad Sci U S A 94:4532–4537

Guérin M, Bruckert E, Dolphin PJ et al (1996) Fenofibrate reduces plasma cholesteryl ester transfer from HDL to VLDL and normalizes the atherogenic, dense LDL profile in combined hyperlipidemia. Arterioscler Thromb Vasc Biol. 16:763–772

Gylling H, Miettinen TA (1992) Cholesterol absorption and synthesis related to low density lipoprotein metabolism during varying cholesterol intake in men with different apoE phenotypes. J Lipid Res 33:1361–1371

Hayashi K, Kurushima H, Kuga Y et al (1998) Comparison of the effect of bezafibrate on improvement of atherogenic lipoproteins in Japanese familial combined hyperlipidemic patients with or without impaired glucose tolerance. Cardiovasc Drugs Ther 12:3–12

Heart Protection Study Collaborative Group (2002) MRC/BHF Heart Protection Study of cholesterol lowering with simvastatin in 20 536 high-risk individuals: a randomised placebocontrolled. Lancet 360:7–22

Heath KE, Gudnason V, Humphries SE et al (1999) The type of mutation in the low density lipoprotein receptor gene influences the cholesterol-lowering response of the HMG-CoA reductase inhibitor simvastatin in patients with heterozygous familial hypercholesterolaemia. Atherosclerosis 143:41–54

Heath KE, Gahan M, Whittall RA et al (2001) Low-density lipoprotein receptor gene (LDLR) world-wide website in familial hypercholesterolaemia: update, new features and mutation analysis. Atherosclerosis 154:243–246

Herman RJ. Drug interactions and the statins (1999) CMAJ 161:1281–6

Hobbs HH, White AL (1999) Lipoprotein(a): intrigues and insights. Curr Opin Lipidol 10:225–236

Hodis H (1999) Triglyceride-rich lipoprotein remnant particles and risk of atherosclerosis. Circulation 99:2852–2854

HoogerbruggeN, Jansen H (1999) Atorvastatin increases low-density lipoprotein size and enhances high-density lipoprotein cholesterol concentration in male, but not in female patients with familial hypercholesterolemia. Atherosclerosis 146:167–174

Humphries SE, Luong LA, Talmud PJ et al (1998) The 5A/6A polymorphism in the promoter of the stromelysin-1 (MMP-3) gene predicts progression of angiographically determined coronary artery disease in men in the LOCAT gemfibrozil study. Lopid Coronary Angiography Trial. Atherosclerosis 139:49–56

Humphries SE, Nicaud V, Margalef J et al (1998a) Lipoprotein lipase gene variation is associated with a paternal history of premature coronary artery disease and fasting and postprandial plasma triglycerides: the European Atherosclerosis Research Study (EARS). Arterioscler Thromb Vasc Biol 18:526–34

Ichihara K, Satoh K (2002) Disparity between angiographic regression and clincial event rates with hydrophobic statins. Lancet 359:2195–2198

Ingelman-Sundberg M (2001) Pharmacogeneitcs: an opportunity for a safer and more efficient pharmacotherapy. J Intern Med 250:186–200

Iwaki K, Nakajima T, Ota N et al (1999) A common Ile796Val polymorphism of the human SREBP cleavage-activating protein (SCAP) gene. J Hum Genet 44:421–422

Jukema WJ, van Boven AJ, Groenemeijer B et al, on behalf of the REGRESS study group (1996) The ASP$_9$ASN mutation in the lipoprotein lipase gene is associated with increased progression of coronary atherosclerosis. Circulation 94:1913–1918

Juo SH, Wyszynski DF, Beaty TH et al (1999) Mild association between the A/G polymorphism in the promoter of the apolipoprotein A-I gene and apolipoprotein A-I levels: a meta-analysis. Am J Med Genet 82:235–241

Kafatos FC (2001) The future of genomics. Mol Aspect Med 22:101–111

Kalow W (1956) Familial incidence of low pseudocholinesterase level. Lancet 211:576–577

Kalow W (2001) Pharmacogenetics, pharmacogenomics, and pharmacobiology. Clin Pharmacol Ther 70:1–4

Karayan L, Qiu S, Betard C et al (1994) Response to HMG CoA reductase inhibitors in heterozygous familial hypercholesterolemia due to the 10-kb deletion ("French Canadian mutation") of the LDL receptor gene. Arterioscler Thromb 14:1258–1263

Kesaniemi YA, Ehnholm C, Miettinen TA (1987) Intestinal cholesterol absorption efficiency in man is related to apoprotein E phenotype. J Clin Invest 80:578–581

Klausen IC, Gerdes LU, Meinertz H et al (1993) Apolipoprotein(a) polymorphism predicts the increase of Lp(a) by pravastatin in patients with familial hypercholesterolaemia treated with bile acid sequestration. Eur J Clin Invest 23:240–5

Knoblauch H, Bauerfeind A, Krähenbühl C et al (2002) Common haplotypes in five genes in.uence genetic variance of LDL and HDL cholesterol in the general population. Hum Mol Genet 11:1477–1485

Knopp RH (1999) Drug treatment of lipid disorders. N Engl J Med 341:498–511

Korhonen T, Hannuksela ML, Seppanen S et al (1999) The effect of the apolipoprotein E phenotype on cholesteryl ester transfer protein activity, plasma lipids and apolipoprotein A I levels in hypercholesterolaemic patients on colestipol and lovastatin treatment. Eur J Clin Pharmacol 54:903–910

Kuivenhoven JA, Jukema JW, Zwinderman AH et al (1998) The role of a common variant of the cholesteryl ester transfer protein gene in the progression of coronary atherosclerosis. N Engl J Med 338:86–93

Lamb RG, Koch JC, Bush SR (1993) An enzymatic explanation of the differential effects of oleate and gemfibrozil on cultured hepatocyte triacylglycerol and phosphatidylcholine biosynthesis and secretion. Biochim Biophys Acta 1165:299–305

LaRosa JC, He J, Vupputuri S (1999) Effect of statins on risk of coronary disease—a meta-analysis of randomized controlled trials. JAMA 282:2340–2346

Larson I, Hoffmann MM, Ordovas JM et al (1999) The lipoprotein lipase HindIII polymorphism: association with total cholesterol and LDL cholesterol, but not with HDL and triglycerides in 342 females. Clin Chem 45:7;963–968

Lazarou J, Pomeranz BH, Corey PN (1998) Incidence of adverse drug reactions in hospitalized patients. A meta-analysis of prospective studies. JAMA 279:1200–1205

Leitersdorf E, Eisenberg S, Eliav O et al (1993) Genetic determinants of responsiveness to the HMG-CoA reductase inhibitor fluvastatin in patients with molecularly defined heterozygous familial hypercholesterolemia [In Fredrickson DS. Dyslipoproteinemia – from phenotypes to genotypes... a remarkable quarter century]. Circulation 87 (Suppl III): 35–44

Linder MW, Prough RA, Valdes R, Jr (1997) Pharmacogenetics: a laboratory tool for optimizing therapeutic efficiency. Clin Chem 43:254–266

Lopez JM, Bennett MK, Sanchez HB et al (1996) Sterol regulation of acetyl coenzyme A carboxylase: a mechanism for coordinate control of cellular lipid. Proc Natl Acad Sci U S A 93:1049–1053

Lu AY (1998) Drug-metabolism research challenges in the new millenium: individual variability in drug therapy and drug safety. Drug Metab Dispos 26:1217–1222

Lussier-Cacan S, Bolduc A, Xhignesse M et al (2002) Impact of alcohol intake on measures of lipid metabolism depends on context defined by gender, body mass index, cigarette smoking, and apolipoprotein E genotype. Arterioscler Thromb Vasc Biol 22:824–831

de Maat MP, Jukema JW, Ye S et al (1999) Effect of the stromelysin-1 promoter on efficacy of pravastatin in coronary atherosclerosis and restenosis. Am J Cardiol 83:852–856

Maerz W, Grutzmacher P, Paul D et al (1994) Effects of lovastatin (20–80 mg daily) on lipoprotein fractions in patients with severe primary hypercholesterolemia. Int J Clin Pharmacol Ther 32:92–97

März W, Baumstark M, Scharnagl H et al (1993) Accumulation of 'small dense' low density lipoproteins in a homozygous patient with familial defective apolipoprotein B-100 results from heterogenous interaction of LDL subfractions with the LDL receptor. J Clin Invest 92:2922–2933

März W, Winkelmann BR (2002) HMG CoA reductase inhibition in the treatment of atherosclerosis: effects beyond lipid lowering. J Kardiol 9:284–294

Mahley RW, Huang Y (1999) Apolipoprotein E: from atherosclerosis to Alzheimer's disease and beyond. Curr Opin Lipidol 10:207–217

Mahley RW, Rall SC (2000) Apolipoprotein E: far more than a lipid transport protein. Annu Rev Genomics Hum Genet 1:507–537

Maitland-van der Zee AH, Klungel OH, Stricker BHC et al (2002) Genetic polymorphisms: importance for response to HMG-CoA reductase inhibitors. Atherosclerosis 163:213–222

Manttari M, Koskinen P, Ehnholm C et al (1991) Apolipoprotein E polymorphism influences the serum cholesterol response to dietary intervention. Metabolism 40:217–221.

Marcovina SM, Koschinsky ML (1998) Lipoprotein(a) as a risk factor for coronary artery disease. Am J Cardiol 82:57U-66U

McLeod HL, Evans WE (2001) Pharmacogenomics: unlocking the human genome for better drug therapy. Annu Rev Pharmacol Toxicol 41:101–121

Miettinen TA, Gylling H, Strandberg T et al (1998) Baseline serum cholestanol as predictor of recurrent coronary events in subgroup of Scandinavian simvastatin survival study. Finnish 4S Investigators. Br Med J 316:1127–1130

Milosavjevic D, Griglio S, Le Naour G et al (2001)Preferential reduction of very low density lipoprotein-1 particle number by fenofibrate in type IIB hyperlipidemia: consequences for lipid accumulation in human monocyte-derived macrophages. Atherosclerosis 155:251–260

Miserez AR, Muller PY, Barella L et al (2002) Sterol-regulatory element-binding protein (SREBP)-2 contributes to polygenic hypercholesterolaemia. Atherosclerosis 164:15–26

Muller PJ, Miserez AR (2001) Mutations in the gene encoding sterol-regulatory element-binding protein-2 in hypercholesterolaemic subjects. Atherosclerosis Supplements 2:69

Murphy MP (2000) Current pharmacogenomic approaches to clinical drug development. Pharmacogenomics 1:115–123

Myant NB (1993) Familial defective apolipoprotein B-100: a review, including some comparisons with familial hypercholesterolaemia [published erratum appears in Atherosclerosis 1994;105:253]. Atherosclerosis 104:1–18

Nakajima T, Ota N, Kodama T et al (1999) Isolation and radiation hybrid mapping of a highly polymorphic CA repeat sequence at the SREBP cleavage-activating protein (SCAP) locus. J Hum Genet 44:350–351

Nemeth A, Dinya E, Audikovszky M et al (1994) Effect of Gevilon therapy and its relation to E polymorphism. Orv Hetil 135:735–741

Nemeth A, Szakmary K, Kramer J et al (1995) Apolipoprotein E and complement C3 polymorphism and their role in the response to gemfibrozil and low fat low cholesterol therapy. Eur J Clin Chem Clin Biochem 33:799–804

Nie L, Niu S, Vega GL et al (1998) Three polymorphisms associated with low hepatic lipase activity are common in African Americans. J Lipid Res 39:1900–1903

Ojala JP, Helve E, Ehnholm C et al (1991)Effect of apolipoprotein E polymorphism and XbaI polymorphism of apolipoprotein B on response to lovastatin treatment in familial and non-familial hypercholesterolaemia. J Intern Med 230:397–405

O'Neill FH, Patel DD, Knight BL et al (2001) Determinants of variable response to statin treatment in patients with refractory familial hypercholesterolemia. Arterioscler Thromb Vasc Biol 21:832–837

Ordovas JM, Lopez-Miranda J, Perez-Jimenez F et al (1995) Effect of apolipoprotein E and A-IV phenotypes on the low density lipoprotein response to HMG CoA reductase inhibitor therapy. Atherosclerosis 113:157–166

Ordovas JM, Vargas C, Santos A et al (1999) The G/A promoter polymorphism at the apoAI gene locus predicts individual variability in fasting and postprandial responses to the HMG CoA reductase inhibitor atorvastatin. Circulation 100(Suppl I):I-239

Ordovas JM, Cupples LA, Corella D et al (2000) Association of cholesteryl ester transfer protein-TaqIB polymorphism with variations in lipoprotein subclasses and coronary heart disease risk: the Framingham study. Arterioscler Thromb Vasc Biol 20:1323–1329

Ordovas JM, Mooser V (2002) The apoE locus and the pharmacogenetics of lipid response. Curr Opin Lipidol 13:113–117

Ordovas JM, Corella D, Cupples LA et al (2002a) Polyunsaturated fatty acids modulate the effects of the APOA1 G-A polymorphism on HDL-cholesterol concentrations in a sex-specific manner: the Framingham Study. Am J Clin Nutr 75:38–46

Ordovas JM, Corella D, Demissie S et al (2002b) Dietary fat intake determines the effect of a common polymorphism in the hepatic lipase gene promoter on high-density lipoprotein metabolism—evidence of a strong dose effect in this gene-nutrient interaction in the Framingham study. Circulation 106:2315–2321

Osborne TF (2000) Sterol regulatory element binding protein (SREBPs): Key regulators of nutritional homeostasis and insulin action. J Biol Chem 275:32379–32382

Pedro-Botet JP, Schaefer EJ, Bakker-Arkema RG et al (2001) Apolipoprotein E genotype affects plasma lipid response to atorvastatin in a gender specific manner. Atherosclerosis 58:183–193

Pena R, Lahoz C, Mostaza JM (2002) Effect of apoE genotype on the hypolipidaemic response to pravastatin in an outpatient setting. J Int Med 251:518–525

Puckey L, Knight B (1999) Dietary and genetic interactions in the regulation of plasma lipoprotein(a). Curr Opin Lipidol 10:35–40

Puckey L, Knight B (2001) Variation at position 162 of peroxisome proliferator-activated receptor α does not influence the effect of fibrates on cholesterol or triacylglycerol concentrations in hyperlipidaemic subjects. Pharmacogenetics 11:619–624

Rantala M, Rantala TT, Savolainen MJ et al (2000) Apolipoprotein B gene polymorphisms and serum lipids: meta-analysis of the role of genetic variation in responsiveness to diet. Am J Clin Nutr 71:713–24

Reihner E, Bjorkhem I, Angelin B et al (1989) Bile acid synthesis in humans: regulation of hepatic microsomal cholesterol 7 alpha-hydroxylase activity. Gastroenterology 97:1498–1505

Rubin J, Berglund L (2002) Apolipoprotein E and diets: a case of gene-nutrient interaction? Curr Opin Lipidol 13:25–32

Rubins HB, Robins SJ, Collins D et al, for the Veterans Affairs High-Density Lipoprotein Cholesterol Intervention Trial Study Group (1999) Gemfibrozil for the secondary prevention of coronary heart disease in men with low levels of high-density lipoprotein cholesterol. N Engl J Med 341:410–418

Rusnak JM, Kisabeth RM, Herbert DP et al (2001) Pharmacogenomics: a clinicians primer on emerging technologies for improved patient care. Mayo Clin Proc 2001;76:299–309

Rust S, Rosier M, Funke H et al (1999) Tangier disease is caused by mutations in the gene encoding ATP-binding cassette transporter 1. Nat Genet 22:352–355

Salek L, Lutucuta S, Ballantyne CM et al (2002) Effects of SREBF-1a and SCAP polymorphisms on plasma levels of lipids, severity, progression and regression of coronary atherosclerosis and response to therapy with fluvastatin. J Mol Med 80:737–744

Sanllehy C, Casals E, Rodriguez-Villar C et al (1998) Lack of interaction of apolipoprotein E phenotype with the lipoprotein response to lovastatin or gemfibrozil in patients with primary hypercholesterolemia. Metabolism 47:560–565

Santamarina-Fojo S, Haudenschild C, Amar M (1998) The role of hepatic lipase in lipoprotein metabolism and atherosclerosis. Curr Opin Lipidol 9:211–219

Sapone A, Peters JM, Sakai S et al (2000) The human peroxisome proliferator-activated receptor alpha gene: identification and functional characterization of two natural allelic variants. Pharmacogenetics 10:321–333

Schaefer JR, Scharnagl H, Baumstark MW et al (1997) Homozygous familial defective apolipoprotein B-100. Enhanced removal of apolipoprotein E-containing VLDLs and decreased production of LDLs. Arterioscler Thromb Vasc Biol 17:348–353

Schoonjans K, Peinado-Onsurbe J, Lefebvre AM et al (1996) PPARalpha and PPARgamma activators direct a distinct tissue-specific transcriptional response via a PPRE in the lipoprotein lipase gene. Embo J 15:5336–48

De la Serna G, Cadarso C (1999) Fenofibrate decreases plasma fibrinogen, improves lipid profile, and reduces uricemia. Clin Pharmacol Ther 66;2:166–172

Serruys PW, de Feyter P, Macaya C et al (2002) Fluvastatin for prevention of cardiac events following successful first percutaneous coronary intervention: a randomized controlled trial. JAMA 287:3215–3222

Shah PK, Kaul S, Nilsson J et al (2001) Exploiting the vascular protective effects of high-density lipoprotein and its apolipoproteins – an idea whose time for testing is coming, part I. Circulation 104:2376–2383

Shepherd J, Blauw GJ, Murphy MB et al (2002) Pravastatin in elderly individuals at risk of vascular disease (PROSPER): a randomised controlled trial. Lancet 360:1623–1630.

Sica DA, Gehr TWB (2002) Rhabdomyolysis and statin therapy: relevance to the elderly. Am J Geriat Cardiol 11:48–55

Sijbrands EJ, Lombardi MP, Westendorp RG et al (1998) Similar response to simvastatin in patients heterozygous for familial hypercholesterolemia with mRNA negative and mRNA positive mutations. Atherosclerosis 136:247–254

Sijbrands EJG, Westendorp RGJ, Defesche JC et al (2001) Mortality over two centuries in large pedigree with familial hypercholesterolaemia: family tree mortality study. Brit Med J 322:1019–1023

Sole R, Goodwin B (2000) Signs of life: how complexity pervades biology. New York: Basic Books

Soria LF, Ludwig EH, Clarke HR et al (1989) Association between a specific apolipoprotein B mutation and familial defective apolipoprotein B-100. Proc Natl Acad Sci U S A 86:587–91

Staffa JA, Chang J, Green L (2002) Cerivastatin and reports of fatal rhabdomyolysis [letter]. N Engl J Med 346:539–540

Staels B, Peinado-Onsurbe J, Auwerx J (1992) Down-regulation of hepatic lipase gene expression and activity by fenofibrate. Biochim Biophys Acta 1123:227–230

Staels B, Vu-Dac N, Kosykh V et al (1995) Fibrates downregulate apolipoprotein C-III expression independent of induction of peroxisomal acyl coenzyme A oxidase. J. Clin. Invest 95:705–712

Staels B, Dallongeville J, Auwerx J et al (1998a) Mechanism of action of fibrates on lipid and lipoprotein metabolism. Circulation. 98:2088–2093

Staels B, Koenig W, Habib A et al (1998b) Activation of human aortic smooth-muscle cells is inhibited by PPARα but not by PPARγ activators. Nature. 393:790–793

Stengard JH, Clark AG, Weiss KM et al (2002) Contributing of 18 additional DNA sequence variations in the gene encoding apolipoprotein E to explaining variation in quantitative measures of lipid metabolism. Am J Hum Genet 71:501–517

Tall A (1995) Plasma lipid transfer proteins. Annu Rev Biochem 64:235–257

Talmud PJ, Hawe E, Robertson K et al (2002) Genetic and environmental determinants of plasma high density lipoprotein cholesterol and apolipoprotein AI concentration in healthy middle-aged men. Ann Hum Genet 66:111–124

Tanaka E (1999) Update: genetic polymorphism of drug metabolizing enzymes in humans. J Clin Pharm Ther 24:323–329

Taylor KD, Scheuner MT, Rotter JI et al (1999) Genetic test to determine non-responsiveness to statin drug treatment. Cedars Sinai Medical Center, Los Angeles, U.S.A. US Patent PCT/US00/18308, EP 1228241

The BIP study group (2000) Secondary prevention by raising HDLcholesterol and reducing triglycerides in patients with coronary artery disease—the Bezafibrate Infarction Prevention (BIP) Study. Circulation 102:21–27

Thompson GR, ÓNeill F, Seed M (2002) Why some patients respond poorly to statins and how this might be remedied. Eur Heart J 23:200–206

Tso P, Liu M, Kalogeris TJ (1999) The role of apolipoprotein A-IV in food intake regulation. J Nutr 129:1503–1506

Tu AY, AlbersJJ (1999) DNA sequences responsible for reduced promoter activity of human phospholipid transfer protein by fibrate. Biochem Biophys Res Commun 264:802–807

Ucar M, Mjörndal T, Dahlqvist R (2000) HMG-CoA reductase inhibitors and myotoxicity. Drug Safety 22:441–457

Ulrich S, Hingorani AD, Martin J (2000) What is the optimal age for starting lipid lowering treatment? a mathematical model. Brit Med J 320:1134-1140

Utermann G (1999) Genetic architecture and evolution of the lipoprotein(a) trait. Curr Opin Lipidol 10:133-41

Van Hout BA, Simoons ML (2001) Cost-effectiveness of HMG coenzyme reductase inhibitors – whom to treat? Eur Heart J 22:751-761

Vaughan CJ, Gotto AM, Jr, Basson CT (2000) The evolving role of statins in the management of atherosclerosis. J Am Coll Cardiol 35:1-10

Vedie B, Jeunemaitre X, Megnien JL et al (2001) A new DNA polymorphism in the 5' untranslated region of the human SREBP- 1a is related to development of atherosclerosis in high cardiovascular risk population. Atherosclerosis 154:589-597

Vergopoulos A, Knoblauch H, Schuster H (2002) DNA testing for familial hypercholesterolemia – improving disease recognition and patient care. Am J Pharmacogenomics 2:253-262

Vesell ES (2000) Advances in pharmacogentics and pharmacogenomics. J Clin Pharmacol 40:930-938

Vogel F (1959) Moderne Probleme in der Humangenetik. Ergeb Inn Med Kinderheilkd 12:52-125

Vohl MC, Lepage P, Gaudet D et al (2000) Molecular scanning of the human PPARa gene. Association of the l162v mutation with hyperapobetalipoproteinemia. J Lipid Res 41:945-952

Vohl MC, Szots F, Lelièvre M et al (2002) Influence of LDL receptor gene mutation and apo E polymorphism on lipoprotein response to simvastatin treatment among adolescents with heterozygous familial hypercholesterolemia. Atherosclerosis 160:361-368

Vuorio AF, Ojala JP, Sarna S et al (1995) Heterozygous familial hypercholesterolaemia: the influence of the mutation type of the low-density-lipoprotein receptor gene and PvuII polymorphism of the normal allele on serum lipid levels and response to lovastatin treatment. J Intern Med 237:43-48

Wallace AJ, Humphries SE, Fisher RM et al (2000) Genetic factors associated with response of LDL subfractions to change in the nature of dietary fat. Atherosclerosis 149:387-394

Wang J, Freeman DJ, Grundy SM et al (1998) Linkage between cholesterol 7alpha-hydroxylase and high plasma low- density lipoprotein cholesterol concentrations. J Clin Invest 101:1283-1291

Wang W, Hu D, Lee ET et al (2002) Lipoprotein(a) in American Indians is low and not independently associated with cardiovascular disease: The Strong Heart Study. Ann Epidemiol 12:107-114

Weggemans RM, Zock PL, Ordovas JM et al (2001) Apoprotein E genotype and the response of serum cholesterol to dietary fat, cholesterol and cafestol. Atherosclerosis 154:547-555

Weinberg RB (2002) Apolipoprotein A-IV polymorphisms and diet-gene interactions. Curr Opin Lipidol 13:125-134

Wieczorek SJ, Tsongalis GJ (2001) Pharmacogenomics: will it change the field of medicine. Clin Chim Acta 308:1-8

Winkelmann BR, Stack CB, Stephens JC et al (2002) Haplotypes of CETP predict clinical response to statins. [Asia Pacific Scientific Forum: the Genomics Revolution: Bench to Bedside to Community and 42nd Annual Conference on Cardiovascular Disease Epidemiology and Prevention. Honolulu, Hawaii, April 23–26, 2002]. Circulation 105:e96(abstract 51 page 10)

Witte JS, Elston RC, Cardon LR (2000) On the relative sample size required for multiple comparisons. Stat Med 69:369-372

Wittrup HH, Tybjaerg-Hansen A, Nordestgaard BG (1999) Lipoprotein lipase mutations, plasma lipids and lipoproteins, and risk of ischemic heart disease. A meta-analysis. Circulation 99:2901–2907

Yamada M. (1997) Influence of apolipoprotein E polymorphism on bezafibrate treatment response in dyslipidemic patients. J Atheroscler Thromb 4:40–44

Yamashita S, Hirano K, Sakai N et al (2000a) Molecular biology and pathophysiological aspects of plasa cholesteryl ester transfer protein. Biochim Biophys Acta 1529:257–275

Yamashita S, Maruyama T, Hirano KI et al (2000b) Molecular mechanisms, lipoprotein abnormalities and atherogenicity of hyperalphalipoproteinemia. Atherosclerosis 152:271–285

Ye S, Watts GF, Mandalia S et al (1995) Preliminary report: genetic variation in the human stromelysin promotor is associated with progression of coronary atherosclerosis. Br Heart J 73:209–215

Zhang Q, Cavanna J, Winkelmann BR et al (1995) Common genetic variants of lipoprotein lipase that relate to lipid transport in patients with premature coronary artery disease. Clin Genet 48:293–298

Zambon A, Hokanson JE, Brown BG et al (1999) Evidence for a new pathophysiological mechanism for coronary artery disease regression : hepatic lipase-mediated changes in LDL density. Circulation 99:1959–64

Zambon A, Deeb SS, Brown BG et al (2001) Common hepatic lipase gene promoter variant determines clinical response to intensive lipid lowering treatment. Circulation 103:792–798

The Genetic Basis of Essential Hypertension and Its Implications for Treatment

K. M. O'Shaughnessy[1] · M. R. Wilkins[2]

[1] Clinical Pharmacology Unit, Level 6, Addenbrooke's Centre for Clinical Investigation, Addenbrooke's Hospital, Cambridge, CB2 2QQ, UK
e-mail: kmo22@medschl.cam.ac.uk

[2] Faculty of Medicine, Imperial College of Science, Technology and Medicine, Hammersmith Campus, Du Cane Road, London, W12 0NN, UK
e-mail: m.wilkins@ic.ac.uk

1	Introduction	151
2	Insight from Single Genes with Large Effects on Blood Pressure	151
2.1	Genetic Mutations Altering Renal Ion Channels and Transporters	151
2.1.1	Liddle's Syndrome	151
2.1.2	Recessive Pseudohypoaldosteronism Type 1 (PHA1)	152
2.1.3	Gitelman's and Bartter's Syndromes	153
2.2	Genetic Mutations in the Mineralocorticoid Receptor	153
2.2.1	Hypertension Exacerbated by Pregnancy	153
2.2.2	Autosomal Dominant Pseudohypoaldosteronism Type 1	153
2.3	Genetic Mutations That Affect Mineralocorticoid Hormone Levels	154
2.3.1	Glucocorticoid-Remediable Aldosteronism	154
2.3.2	Apparent Mineralocorticoid Excess	154
2.3.3	Defective Aldosterone Synthesis	154
2.4	Other Mendelian Forms of Hypertension	155
2.4.1	Pseudohypoaldosteronism Type II	155
2.4.2	PPARγ Mutations	155
2.4.3	Hypertension with Brachydactyly	155
3	Genome-Wide Studies in Human Hypertension	156
3.1	Framingham Heart Study	156
3.2	Icelandic deCode study	158
3.3	UK BRIGHT Study	158
4	Candidate Gene Studies in Human Hypertensive Populations	159
4.1	Angiotensin-Converting Enzyme	159
4.2	Angiotensinogen	160
4.3	The Epithelial Sodium Channel	161
4.4	Adducin	161
4.5	The G-Coupling Protein GNβ3	162
4.6	Nitric Oxide Synthase	162
5	Animal Genetic Models of Hypertension	163
5.1	Quantitative Trait Loci for Blood Pressure	163
5.2	Strategies for Refining QTLs	164
5.2.1	Congenic Strains	164
5.2.2	Consomic Strains	164

5.3 Strategies for Defining Causative Genes 165
5.3.1 Gene Expression Profiling. 165
5.3.2 Knockouts 166
5.3.3 Mouse Mutagenesis Studies 166
5.4 Rodent Versus Human Loci 166

6 **The Genetics of Target Organ Damage** 166
6.1 Cardiac Hypertrophy 167
6.2 Stroke 167
6.3 Renal Disease 168

7 **Genotype Versus Phenotype** 168

8 **Conclusion** 170

References 171

Abstract Hypertension is an asymptomatic condition but a major risk factor for cardiovascular events and stroke. Blood pressure exhibits a skewed normal distribution in the general population with no natural hypertensive threshold. Decisions regarding who to treat and with what drugs are based upon morbidity and mortality data in large population studies, the presence/absence of co-existing disease and cost. Because of variation between individuals in their response to antihypertensive drugs, patients are frequently exposed to a number of different drugs before a suitable agent (or combination of agents) is found. This increases the potential for adverse drug reactions and/or poor compliance. A few patient characteristics can be used to help predict their blood pressure response to a drug such as age, race and perhaps renin levels. Genetic factors also influence the level of blood pressure in an individual, susceptibility to target organ damage and the response to antihypertensive drugs. Genes with a large influence on blood pressure have been identified for rare familial forms of hypertension, but these account for a very small fraction of the general hypertensive population. Their elucidation has helped define pathophysiological pathways and suggests new biochemical factors for further genetic studies or drug targeting. The hunt is on for genes which influence blood pressure in a much larger proportion of the population. The broader vision is that knowledge of a patient's genotype coupled with epidemiological and clinical data can help in tailoring therapy to the individual patient. How useful it turns out to be will ultimately depend upon whether blood pressure is regulated by a relatively small number of genes with significant effects or a large number of genes with very small effects.

Keywords Hypertension · Candidate genes · Animal models · Target organ damage

1
Introduction

It is widely accepted that Pickering won the debate with Robert Platt over the inheritance of high blood pressure. By the middle-half of the twentieth century, it was confidently assumed that many genes involved in the regulation of blood pressure would be identified: perhaps as many as 20 or 30. Yet, half a century on, we have made relatively poor progress towards identifying any genes that influence blood pressure variation in any significant proportion of the population. In fact, it is an irony that Platt's single-gene hypothesis has been proved right for a very small number of rare familial hypertension syndromes, and an understanding of the molecular basis of these syndromes has thus far been the real success of research in this area.

Nonetheless it remains a firm belief that essential hypertension is a genetically heterogenous condition and that elucidating the genes involved will allow the subdivision of the condition into distinct molecular subtypes and identify new therapeutic targets. It is also recognized that patients with hypertension vary in their tolerance of the condition – that some have significant hypertension without much evidence of target organ damage. Rather than relying solely on population risk factor scores to assess an individual's need for medicines, it is hoped that genetic information will provide valuable insight into the type of hypertension and risk of complications for each patient and permit a more individually tailored approach to treatment.

2
Insight from Single Genes with Large Effects on Blood Pressure

Linkage analysis in families with an extreme phenotype (e.g. severe hypertension or hypotension evident at an early age) coupled to sequence analysis of candidate genes within the regions of linkage has identified over a dozen genes with a large effect on blood pressure (Lifton et al. 2001). In most cases the genes encode known drug targets or proteins already implicated in the pathogenesis of hypertension; indeed, this knowledge was instructive in choosing the candidates to sequence. In practically all cases to date, the mutated gene products disrupt, directly or indirectly, sodium homeostasis by the kidney (Fig. 1).

2.1
Genetic Mutations Altering Renal Ion Channels and Transporters

2.1.1
Liddle's Syndrome

This is an autosomal dominant condition in which hypertension is associated with hypokalaemic alkalosis and suppressed plasma renin and aldosterone levels. Linkage analysis in Liddle's original kindred identified a mutation in the

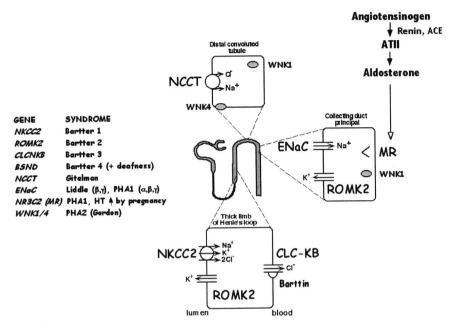

Fig. 1 Summary of known single gene mutations with major effects on blood pressure. (Adapted from Karet 2003)

gene encoding the β subunit of the epithelial sodium channel (ENaC) of the cortical collecting tubule of the kidney (Shimkets et al. 1994). This is a site of action of amiloride, an effective treatment for the condition. Subsequently, mutations have been found elsewhere in this gene and in the gene for the γ subunit of this channel (Hansson et al. 1995a, b). The mutations all result in a gain-of-function. It seems that the clearance of the normal channel from the membrane is dependent upon the PPPXY sequence in the cytoplasmic tail. The mutations affect the motif either as missense mutations within it or by the introduction of premature stop codons that delete the entire C-terminus (containing the PPPXY motif). Channels lacking an intact PPPXY motif show reduced interaction with, and endocytosis by, clathrin-coated pits, thus prolonging their half-life at the cell surface. The increased abundance of these channels leads to avid sodium reabsorption independent of aldosterone.

2.1.2
Recessive Pseudohypoaldosteronism Type 1 (PHA1)

Loss-of-function mutations in any of the three subunits making up ENaC lead to a neonatal salt wasting condition with hyperkalaemia and metabolic acidosis that requires life-long sodium supplementation and treatment of hyperkalaemia (Chang et al. 1996, Strautnieks et al. 1996).

2.1.3
Gitelman's and Bartter's Syndromes

These are normotensive-hypotensive, autosomal recessive conditions characterized by hypokalaemic alkalosis that result from mutations in four different genes, two of which encode protein targets for commonly used diuretics. Gitelman's syndrome is caused by loss-of-function mutations in the gene for the thiazide-sensitive Na-Cl co-transporter of the distal convoluted tubule (Simon et al. 1996a). The characteristics of the condition are typical of thiazide use in that in addition to hypokalaemia, serum magnesium is low and urinary calcium excretion reduced. One variant of Bartter's syndrome is due to loss-of-function mutations in the apical Na-K-2Cl co-transporter of the ascending limb (TAL) of Henle, the site of action of furosemide (Simon et al. 1996b). Two other molecular variants of Bartter's syndrome have been reported: one due to loss-of-function mutations in the ATP-sensitive K^+ channel, ROMK, and the other due to mutations in *CLCNKB*, which impairs the function of a Cl^- channel in TAL (Simon et al. 1996c, 1997). Given the phenotype of patients with mutations in ROMK and *CLCNKB*, it has been suggested that pharmacological antagonism of these channels may offer a novel approach to the treatment of hypertension.

2.2
Genetic Mutations in the Mineralocorticoid Receptor

2.2.1
Hypertension Exacerbated by Pregnancy

A missense mutation, *MR S810L*, in the ligand binding domain of the mineralocorticoid receptor has been identified in a family in which hypertension was evident below the age of 20 and exacerbated by pregnancy (Geller et al. 2000). The mutation alters the steric requirements for activation of the receptor such that steroids lacking 21-hydroxl groups, such as progesterone, which normally bind but do not activate the receptor and the antagonist spironolactone become potent agonists.

2.2.2
Autosomal Dominant Pseudohypoaldosteronism Type 1

Heterozygous loss-of-function mutations in the mineralocorticoid receptor result in severe neonatal salt wasting with hypotension, hyperkalaemia and metabolic acidosis, despite markedly elevated aldosterone levels (Hanukoglu 1991; Geller et al. 1998). Interestingly, once through the neonatal period, heterozygous patients are able to survive on a normal salt-rich diet.

2.3
Genetic Mutations That Affect Mineralocorticoid Hormone Levels

2.3.1
Glucocorticoid-Remediable Aldosteronism

Glucocorticoid-remediable aldosteronism (GRA) is an autosomal dominant trait arising from the inheritance of a chimeric gene formed from the 5' end of 11β hydroxylase and the 3' end of aldosterone synthase (Lifton et al. 1992a, b). This causes ectopic expression of the aldosterone synthase gene in the adrenal zona fasciculata, where it is regulated by adrenocorticotropic hormone (ACTH) rather than angiotensin II. The synthesis of aldosterone, therefore, becomes linked to cortisol secretion and is maintained at higher than normal levels, leading to hypertension, hypokalaemia and metabolic alkalosis. It responds to treatment with low-dose dexamethasone, which suppresses ACTH levels.

2.3.2
Apparent Mineralocorticoid Excess

Apparent mineralocorticoid excess (AME) arises when corticosteroids other than aldosterone activate the mineralocorticoid receptor. One of these is cortisol itself, which has similar affinity to aldosterone at the mineralocorticoid receptor. Cortisol circulates in concentrations 1000-fold higher than aldosterone but is prevented from accessing the receptor in the kidney by 11β-hydroxysteroid dehydrogenase-2 (11β-HSD2). This enzyme converts cortisol to cortisone, which does not activate the mineralocorticoid receptor. Homozygous loss-of-function mutations in 11β-HSD2 permits cortisol to access and stimulate the mineralocorticoid receptor, leading to hypertension and hypokalaemia (Mune et al. 1995). The same phenotype can be produced by over-production of cortisol and by inhibition of the enzyme, e.g. by glycyrrhetinic acid, a metabolite of liquorice. Inherited deficiencies in 11β-hydroxylase and 17α-hydroxylase impair cortisol production and divert steroid synthesis into the over-production of 21-hydroxylated steroids, which exhibit a high affinity for the mineralocorticoid receptor (Kagimoto et al. 1988; White et al. 1991).

2.3.3
Defective Aldosterone Synthesis

Mutations resulting in the loss of aldosterone synthase and deficiencies in 21-hydroxylase impair aldosterone synthesis, resulting in salt wasting and hypotension.

2.4
Other Mendelian Forms of Hypertension

2.4.1
Pseudohypoaldosteronism Type II

Pseudohypoaldosteronism type II (PHAII) is hypertension associated with hyperkalaemia and as a heritable trait has been mapped to at least three genetic loci. Mutations in two genes encoding serine-threonine kinases have been reported (Wilson et al. 2001). Those in *WNK1* are large deletions in intron 1 that increase expression of the enzyme while those in *WNK4* are missense mutations in two highly conserved domains. Both kinases localize to distal nephron segments and are thought to be involved in Na and Cl transport. A recent study suggests that the normal function of *WNK4* is to inhibit expression of the thiazide-sensitive Na-Cl co-transporter; missense mutations that abrogate the kinase function prevent this action, leading to gain-of-function (Wilson et al. 2002). The responsible gene on chromosome *1q31–42* has not been identified and a fourth genetic locus has also been suggested (Mansfield et al. 1997).

2.4.2
PPARγ Mutations

Dominant-negative missense mutations in the peroxisome-proliferator activated receptor subtype gamma (PPARγ) have been identified in three patients with diabetes, acanthosis nigricans and hypertension (Barroso et al. 1999). The mechanism by which disruption of PPARγ function can cause hypertension is unclear but one intriguing possibility is through regulation of serum- and glucocorticoid kinase activity (sgk) (Pearce 2001). This enzyme is a major transcriptional target for aldosterone and stimulates ENaC membrane expression and activity in response to the mineralocorticoid. It also lies in the insulin signalling pathway, providing a biochemical link between insulin signalling, aldosterone and sodium homeostasis.

2.4.3
Hypertension with Brachydactyly

There is a familial condition in which hypertension is associated with abnormal skeletal development in the hand and wrist. It has been mapped to chromosome *12p12.2–11.2* (Schuster et al. 1996), but the gene has not been identified. However, the pathophysiology of the hypertension is unusual in that it appears to be neurogenic in origin and possibly associated with abnormal vascular loops abutting the brainstem.

3
Genome-Wide Studies in Human Hypertension

A number of genome-wide linkage studies have been published in the last few years and the largest of these are summarized in Table 1. These are difficult studies and there is much debate over the best study design, sample size and statistical approach. Other complicating factors include ethnic mix and the availability of pre-treatment blood pressures. It is not surprising perhaps that the reproducibility of loci between study populations is poor.

There are statistical problems with most if not all the published studies. Of concern is the issue of repetitive testing. A p value of <0.05 is not appropriate when several hundred tests are being performed for a typical genome-wide set of markers. The appropriate level is still debated, although Lander and Kruglyak in 1995 provided stringent definitions: they set suggestive and significant linkage log of the odds ratio (LOD) thresholds at 2.2 and 3.6, corresponding to a $p<0.0007$ and $p<0.00002$ respectively. In a study of over 30 such genome scans, Altmuller and colleagues found that the majority did not achieve the Lander and Kruglyak threshold for genome-wide significance (Altmuller et al. 2001). The single most important useful factor that emerged from this analysis was the use of a large sample size drawn from a single ethnic group. The problems of combining data across ethnic groupings is exemplified by the recent meta-analysis from the Family Blood Pressure Program (Province et al. 2003).

To date, just two published hypertension scans have reported a locus that achieved genome-wide significance. The first is the chromosome *17q* locus from the Framingham Heart Study (see Fig. 2) and the second is on chromosome *18q* in the deCode study population from Iceland. The final analysis of the UK BRIGHT study has not been published, although it is reported to show a locus that reaches genome-wide significance on chromosome *6q*, with three further loci on *2q*, *5q* and *9q* (Caulfield et al. 2002).

3.1
Framingham Heart Study

The discovery of one of the disease genes for Gordon's syndrome (*WNK4*) under the chromosome *17q* locus is interesting, although it must be pointed out that *17q* is actually very gene-rich. Further refinement may be possible with fine mapping, but the suggestion of *WNK4* as an explanation for the association raises an important issue of how liberal one needs to be in identifying candidate genes. If the *WNK4* kinase had not been highlighted by work on a rare monogenic form of hypertension it is likely to have remained an orphan kinase for some time. In fact, we still know so little about how the WNK kinases function that it seems unlikely that *WNK4* would have been identified as a candidate gene by most researchers.

Table 1 Genome-wide linkage studies in human hypertension

Study	Ethnicity	Families ascertained	Study variable[a]	Population size	Loci reaching genome-wide significance[b]	Most significant loci	Comments
Anquig Study, Xu et al. 1999	Chinese	Sib pairs (extreme discordant)	QT	207 Sib pairs	None	>2 for 5 loci on: ch 3,11, 15–17	Untreated isolated rural population
NHLBI, Hunt et al. 2002	White American	Extended families	Both	2,959 Subjects in 402 families	None	Max LOD for SBP 3.3 on ch 6. No LOD >2 for DBP	Subjects obese (mean BMI 27.8) BP recorded pretreatment
Framingham Heart Study, Levy et al. 2000	White American	Extended families	QT	1,702 Subjects in 332 families	LOD 4.7 for SBP on ch 17q	Max LOD for DBP 2.1 ch 17q and 18p	
Krushkal et al. 1999	White Americans	Sib pairs (discordant)	QT	69 Sib pairs	None	4 had a p>0.01 on ch 2,5,6 & 15	
Province et al. 2003	Mixed (Asian, African-American and White American)	Both	Both	6,245 Subjects (in sib pairs and families)	None	No locus reached LOD >2. Max was 1.08 over ch 10 (reached 2.15 in Asian subset)	This meta-analysis did not refine hits from ethnic-specific analyses
Quebec Family Study, Rice et al. 2000	White American	Sib pairs	QT	226 Random and 109 obese sib pairs	None	2.7 for SBP on 5p, 2–2.5 on 2p, 7q, 8q and 19p. No DBP loci had LOD >1.5	Ch 1p, 7q and 5p QTLs predominantly reflect BP.
Kristjansson et al. 2002	White Northern Europeans (Icelanders)	Extended families	AP	490 Subjects in 120 families	LOD of 4.6 for ch 18q	No other loci had LOD <1.5	A population with a founder effect?
San Antonio Heart Study, Atwood et al. 2001	Mexican-Americans	Extended families	QT	440 Subjects in 10 families	None	Max LOD of 2.8 on ch 21, ~2 on ch 7, 18 & 18	Pulse pressure used though same hits with SBP
Shanghai study, Zhu et al. 2001	Chinese	Sib pairs	AP	283 Sib pairs (replicated in a further 637)	None	Max LOD of 2.24 over 2q	

[a] QT, BP treated as continuous quantitative trait or AP, affected phenotype (hypertensive)—discontinuous.
[b] Lander and Kruglyak criterion of LOD 3.6.

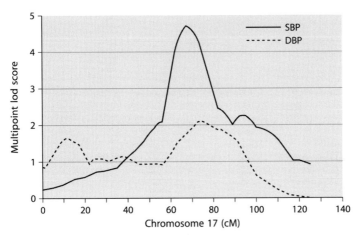

Fig. 2 Multipoint LOD score map for chromosome 17, taken from the genome-wide scan of the Framingham Heart Study. (From Levy et al. 2000, with permission)

3.2
Icelandic deCode study

The putative locus on chromosome *18* represents one of a number of positive genome-wide scans that have been reported from the Icelandic deCode database. This island population is unusual for being geographically isolated until relatively recently and subject to several bottlenecks in its millennium-long history. Founder populations such as this have been widely heralded as being more useful than outbred ones in linkage-disequilibrium mapping, but both modelling and recent haplotype analysis within Finnish and Sardinian populations suggests that this is not the case (Eaves et al. 2000). Founder effects also imply that the disease mutations identified may be distinct from those operating in other populations. We should be partly reassured that this locus coincides with one reported for human familial orthostatic *hypo*tension (DeStefano et al. 1998) and quantitative trait loci (see the chapter by Winkelmann et al., this volume) reported from both mouse and rat models (Stoll et al. 2000; Wright et al. 1999).

3.3
UK BRIGHT Study

The UK BRIGHT Study of White Northern Europeans, using over 1500 sib pairs, has reported a LOD score of 2.2 for chromosome *6q*, with smaller effects from three further loci on *2q, 5q* and *9q* (Caulfield et al. 2002). The *5q* locus is a potentially interesting locus containing several obvious candidate genes including the β_2 (*ADRB2*) and α_{1b} (*ADRA1B*) adrenoceptor genes. The *ADRB2* gene has attracted a lot of attention, since its common variants, *R16G* and *Q27E*, clearly impact β2-adrenoceptor function (Liggett 1997) and it may be a susceptibility gene for other polygenic diseases such as asthma. But the current literature on whether the *ADRB2* locus is the explanation for the hypertension locus in this region is divided (Bray et al. 2000; Tomaszewski et al. 2002).

4
Candidate Gene Studies in Human Hypertensive Populations

Genome-wide searches represent a huge commitment in terms of time and resources. Not surprisingly, other simpler approaches have been very popular. In fact, association studies are the most published type of genetic study and the accelerated discovery program for single nucleotide polymorphisms (SNPs) and availability of high throughput chip-based genotyping will probably see this increase further in the next few years. However, their popularity does disguise major shortcomings with much of the published literature (Anonymous 1999). The power of many studies is a frequent concern, as is the risk of admixing (population stratification) within the samples chosen, which probably explain the poor reproducibility of many association studies (Dahlman et al. 2002).

In fact, given the reality that the influence of any single polymorphism is likely to be small in a complex disease such as hypertension, sample sizes of many thousands of patients are needed to reliably isolate them. Consider, for example, that the association of the P12A PPARγ polymorphism in another complex disease, type 2 diabetes, was finally confirmed using a sample over 3,000 subjects (Altshuler et al. 2000). The individual risk was just 1.25, which compares closely to the influence of the angiotensinogen M235T polymorphism in hypertension (see below). But it should be emphasized that these small effects can have large attributable risks for a population as a whole if the particular polymorphism and disease are common. In the case of P12A PPARγ, some 25% of the population risk of type 2 diabetes can be attributed to it (Altshuler et al. 2000). The following reviews the better-tested candidate genes in hypertension.

4.1
Angiotensin-Converting Enzyme

The importance of angiotensin-converting enzyme (ACE) to many areas of cardiovascular biology has made it the focus for a considerable effort to document the extent and functional importance of polymorphisms within the gene. The strongest and most consistent association has been between an *Alu* repeat in intron 16 (referred to as I when present and D when absent) and the levels of circulating serum ACE; in fact, some 50% of the population variance in serum ACE levels has been attributed to it (Rigat et al. 1990). *DD* homozygotes have higher levels that those with the *II* genotype. However, moving from this intermediate phenotype to the complex one of blood pressure or essential hypertension has been much more problematic. Association and linkage has been claimed and refuted extensively (Jeunemaitre et al. 1992), although significantly the two largest studies including the Framingham cohort have reported positive associations or linkage (O'Donnell et al. 1998; Fornage et al. 1998). Curiously both studies could only detect significant effects for men and the nature of this sex-specific effect of the ACE gene is, so far, unexplained. Follow-up analysis of the Rochester heart Study Cohort showed just how much the effect of ACE was dependent on other variables such as age, height and weight as well as sex (Turner et al. 1999). This highlights the importance of gene–gene and gene–environment interactions in regulating (and minimizing) the influence of putative hypertensive alleles. Besides the *Alu* variant, a further 78 SNPs have been identified within the ACE gene (Rieder et al. 1999). Seventeen of these are in linkage disequilibrium with the *Alu* variant, and no less than 13 haplotypes were identified. Bearing in mind this was based on just 22 individuals, the level of genomic diversity that might be expected at the population level is daunting but in keeping with extensive SNP analyses that have been carried out in other genes (Nickerson et al. 1998).

4.2
Angiotensinogen

After the ACE gene, angiotensinogen (AGT) is amongst the most extensively studied of the candidate genes. The first genetic data implicating angiotensinogen in essential hypertension appeared from a combined study cohort of French and Mid-Western American Caucasians (Jeunemaitre et al. 1992b). This provided evidence of linkage in a total of 379 sib pairs, and association was subsequently found amongst just 2 of the 15 SNPs reported: *M235T* and *T174M*. These two SNPs were actually in complete linkage disequilibrium, so it was not clear which (if either) of these alleles actually conferred the risk. It is clear, however, that angiotensinogen levels are affected by the *M235T* polymorphism and the results of Winkelman et al. (1999) are typical, showing a stepwise rise according to the number of T235 alleles present (14.8±3.9, 15.7±5.1 and 17.3±4.7 nmol/l with 0, 1 or 2 alleles, respectively).

The association and linkage of the angiotensinogen gene with essential hypertension has been replicated in a number of populations including Hutterites (Hegele et al. 1994), African Caribbeans (Caulfield et al. 1994, 1995) and Japanese (Hata et al. 1994). The association with the *M235T* and *T174M* polymorphisms has proved to be less robust, and studies on rural Chinese populations have consistently failed to replicate linkage to AGT or association with *M235T* or other SNPs (Niu et al. 1999a, b, c). There are substantial differences in the population frequencies of the incriminating angiotensinogen SNPs that may partly explain the contradictory findings, although the highest frequencies are ironically in Canadian aborigines where essential hypertension is rare (Hegele et al. 1998).

It is also clear that M235T is in tight linkage to a promoter polymorphism, G(-6)A, which suggests that the increased tissue expression of AGT seen with the *235T* allele reflects increased gene transcription (Hegele et al. 1994). However, a recent study using human AGT transgenes expressed in the mouse has caste doubt on this explanation (Cvetkovic et al. 2002). Nevertheless, it is likely that the *235T* and *−6A* alleles actually represent the haplotype of the ancestral AGT gene whose effects are now being reported in essential hypertension. Its contribution is very modest with ~3% of the total variation in systolic pressure attributable to it (Hegele et al. 1994). Put another way, the odds ratio for the *235T* allele was put at 1.22 in a Japanese meta-analysis (Kato et al. 1999), which although small is slightly higher than the estimate from the NHLBI Family Program Study (Province et al. 2000). This should be a sobering finding if (as seems likely) this represents the largest single gene contribution in essential hypertension, and puts into stark relief the scale of the task ahead for studying other minority gene effects.

4.3
The Epithelial Sodium Channel

The amiloride-sensitive epithelial sodium channel (ENaC) is an obvious candidate gene, especially for salt-sensitive hypertension. A number of polymorphisms have been identified within the β and γ subunits of this channel, and some of these cause amino acid substitutions, although their effects on channel activity are marginal when the variants are expressed in Xenopus oocytes (Persu et al. 1998). However, these effects on basal channel activity are misleading, since there is evidence that ENaC is regulated in vivo by protein kinase C-dependent phosphorylation of the *T594* site. This probably explains why the striking effects of the *T594M* substitution are only obvious in EBV-transformed lymphocytes when they are stimulated with cAMP (Cui et al. 1997).

The *T549M* variant has been studied in several cohorts and in the largest, a UK cohort of African descent, was present in 8% of hypertensives vs 2% in matched normotensives (Baker et al. 1998). Work in other African-descent populations has not confirmed the association, but was almost certainly underpowered to do so (Tiago et al. 2001). Subsequent work with the UK South London cohort has also shown that the blood pressure of hypertensive *T594M* carriers were particularly sensitive to amiloride, although the study was unblinded (Baker et al. 2002). Further work is needed on the quantitative importance of this variant, but since its expression is restricted to populations of African descent, it cannot contribute to salt-sensitive hypertension in other populations, such as the Japanese (Matsubara et al. 2002).

4.4
Adducin

Adducin, a cytoskeletal protein, first emerged as a candidate from work on the Milan strain of hypertensive rat (Bianchi et al. 1994). However, a plausible mechanism to explain how adducins might alter Na reabsorption within the nephron only emerged with the discovery of their effect on Na-K-ATPase activity, to which they bind with high affinity (Ferrandi et al. 1999). Linkage to hypertension was initially reported in an Italian cohort of sib pairs and supported by positive association with the *G460W* polymorphism in α-adducin (Cusi et al. 1997). Yet even within Italian cohorts, subsequent association studies have been inconsistent, with the same polymorphism appearing to be a more robust predictor of response to diuretic therapy rather than hypertension itself (Glorioso et al. 1999). A comprehensive series of association and linkage studies carried out on US-based White, African-American and Asian cohorts have also failed to replicate the Italian findings, casting serious doubt over the importance of α-adducin and its *G460W* polymorphism in particular (Schork et al. 2000; Bray et al. 2000; Ranade et al. 2000).

4.5
The G-Coupling Protein GNβ3

Studies on lymphocytes and fibroblasts from hypertensive patients highlighted enhanced signal transduction through G_i-coupled proteins (Siffert et al. 1995), which led to the identification of a novel variation, C825T, in exon 10 of the *GNβ3* gene (Siffert et al. 1998). It was novel because C825T did not affect the donor or acceptor sites, yet was able to cause splice variation from an in-frame deletion within the adjacent exon 9. The resulting propeller structure is thought to confer a dominant gain-of-function on the mutated protein. A positive association of hypertension with the *825T* allele has been reported in a large Caucasian European cohort (Siffert et al. 1998). This has been verified in other populations (Schunkert et al. 1998; Benjafield et al. 1998), although there are notably exceptions such as the Japanese (Ishikawa et al. 2000). This may be relevant, since markers of obesity and fat distribution show stronger associations with the *825T* allele than hypertension itself (Siffert et al. 1999; Hegele et al. 1999). This suggests that any impact on hypertension of the *825T* allele may be indirect.

4.6
Nitric Oxide Synthase

The role of endothelial nitric oxide synthase (eNOS or NOS3), and its product nitric oxide, in regulating blood vessel tone has made this gene an obvious target for investigation (see the chapter by Huang, this volume). This was strengthened by the discovery that its targeted disruption in the mouse elevated blood pressure (Huang et al. 1995), but initial attempts to link the eNOS locus to hypertension were not successful in either European (Bonnardeaux et al. 1995) or Australian sib pair collections (Takami et al. 1999). The discovery of potentially functional variants within exon 7 (*R298G*) (Tesauro et al. 2000) and the 5' flanking region of eNOS (*T-758C*) has lead to subsequent reports of positive association with these SNPs in both essential hypertension (Miyamoto et al. 1998; Hydman et al. 2002) and pregnancy-induced hypertension (Arngrimsson et al. 1997). Once again, these findings are by no means universal (Kajiyama et al.

2000) or apparently reproducible within the same study population (Arngrimsson et al. 1999). It is relevant that a large population study in the UK has recently found no evidence that plasma $(NO)_x$ levels are influenced by either the exon 7 or promoter SNPs; neither was the systolic or diastolic BP correlated to plasma $(NO)_x$ (Jeerooburkhan et al. 2001).

5
Animal Genetic Models of Hypertension

The use of animal models of hypertension to define genes influencing blood pressure has developed in parallel with the human studies, and their attractions are very persuasive. They are employed to identify genetic loci, test the biological importance of candidate genes, explore the effects of genetic background and investigate therapeutic strategies. The vast majority of work in this area has used inbred rat strains, although the more complete genomic map for the mouse as well as the opportunity to knock out genes in this species has prompted the appearance of studies in mice as well (Sugiyama et al. 2001).

The ten most widely used laboratory rat strains have been comprehensively reviewed recently by Rapp (2000). There is striking heterogeneity in the hypertensive phenotype across the various strains, especially with regards their salt sensitivity and propensity for end-organ damage. This has afforded the opportunity to explore the genetic basis of susceptibility to morbidity from elevated blood pressure as part of the hypertensive phenotype.

5.1
Quantitative Trait Loci for Blood Pressure

The paradigm for dissecting the genetic basis for hypertension in these rodent strains follows a well-trodden path. The first step involves crossing the strain with a normotensive reference strain then crossing the F1 offspring either as a backcross [(AXB)F1 X B] or an intercross [(AXB)F1X (AXB)F1]. The F2 offspring are then genotyped for a panel of markers scattered throughout the rodent genome with the object of identifying markers that co-segregate with blood pressure (or some other hypertension marker) in the F2 offspring (Lander and Botstein 1989). The result is the identification of genomic regions that may carry the genes responsible for the hypertensive phenotype, quantitative trait loci (QTLs).

Using this approach, blood pressure QTLs have been identified on virtually every rodent chromosome, except chromosomes 6, 11 and 15 (Rapp 2000). The locus is usually referred to by the chromosome number, so that the locus on chromosome 1, for example, is designated *BP1*. Many of these loci are actually shared across strains, and the ones that are most robust in this respect map to chromosome 2, 10 and 18 (Dominiczak et al. 2000).

Candidate genes for blood pressure have been identified within each locus but at present no blood pressure QTL, let alone gene, has met stringent criteria for identification. The fact that several of the QTLs have regions of conserved synteny on human chromosomes is intriguing, in particular the recent work on the rodent *BP10* locus (Rapp et al. 1998; Jacob et al. 1991), and the homologous segment of human chromosome 17 that is now known to contain the monogenic hypertension gene *WNK4*.

5.2
Strategies for Refining QTLs

5.2.1
Congenic Strains

It is important to realize, however, that QTLs are defined statistically and often cover large areas of a chromosome; a typical size of tens of centiMorgans might encompass ~1,000 genes. It is, in fact, simply not feasible to genotype enough F2 offspring with a dense enough marker set to refine these QTLs to regions that would make positional cloning possible. This is the first technical impasse in moving from co-segregation studies to a QTL and eventually a candidate gene (Mackay 2001). One solution has been to create congenic rat strains, in which the polygenic host strain is essentially converted to a monogenic model of hypertension (Nadeau and Frankel 2000). In practice, F1 offspring are successively backcrossed onto the recipient strain, so as to achieve offspring that differ from the recipient genome only in the region of the QTL of interest (Jeffs et al. 2000). Congenic strains have been generated for nearly all the blood pressure QTLs to demonstrate transfer of the phenotype with the QTL. In some cases, this has refined the locus and excluded a candidate gene. In other cases the accompanying flanking DNA may include more than one QTL. Continuing the backcross may refine the congenic interval further and confine it to region small enough to positionally clone (~1 cM or about a megabase of DNA). There are examples where the phenotype disappears; such results highlight the importance of the background (recipient) genome, which may even completely suppress any measurable effect of the congenic on blood pressure.

5.2.2
Consomic Strains

Another approach to the problems of mapping QTLs has been the use of chromosome substitution strains of animals or consomics (Nadeau et al. 2000). These provide panels of animals with a common recipient genetic background that differ at a single chromosome. It is a simple matter to compare the consomic strain with its background strain to identify whether a particular chromosome carries QTLs for a give phenotype. In fact, the consomic approach has been successfully used to demonstrate the pressor effects of chromosome 13 from the Dahl rat (Cowley et al. 2001) as well as the pressor effect of the Y chromosome in the SHRSP strain (Negrin et al. 2001). The consomic strains are useful in backcrosses to localize a QTL within a chromosome without the confounding effects of other co-segregating QTLs. Consomics are also a valuable resource for rapidly generating congenic strains, but they are currently being employed for a much more ambitious exercise in functional cardiovascular genomics.

The Physgen program (http://pga.mcw.edu) has been initiated to study the functional impact of single chromosome substitution of a number of standard hypertensive rat strains. Currently, a half-dozen or so chromosomes in the salt-sensitive Dahl strain have been individually replaced with chromosomes from the Brown-Norway strain. It is expected that up to 15 new strains will follow annually for the duration of the program. The animals are being exposed to one of several physiological challenges including salt-loading and their responses quantified using a panel of more than 200 phenotypic variables. This project should give a much better understanding of the functional genomics of the QTLs carried on each chromosome. They may not necessarily help identify the hypertension genes themselves, but could identify genes that modulate their effect. Some of these genes could potentially represent novel antihypertensive drug targets.

5.3
Strategies for Defining Causative Genes

5.3.1
Gene Expression Profiling

Expression profiling using high-density microarrays can be used to detect differences in the abundance of transcribed genes from tissues of congenic vs recipient animals. These differences (ignoring of course gene–gene interactions) must reflect the small genomic differences between the animals (i.e. the congenic segment), allowing candidate genes to be rapidly identified. The proof of concept for this was first shown for the Cd36 gene product in the SHR rat, although this was actually an insulin-resistance not a blood pressure QTL (Pravenec et al. 1999; Aitman et al. 1999; Collison et al. 2000). This strategy using tissues that have small but well-defined genomic differences bypasses the laborious conventional approach to positional cloning using YAC or BAC contigs. Nevertheless, it is not a quick fix and success with expression profiling is still limited. It has been successfully applied to a number of other phenotypes, but often in non-mammalian genomes (Wayne and McIntyre 2002). None of the blood pressure QTLs themselves have been dissected to date by profiling, although advances seem certain in the next few years (Pravenec et al. 2002).

5.3.2
Knockouts

Mice in which specific genes have been genetically engineered to disrupt or enhance their expression are powerful tools for testing the role of candidate genes on blood pressure (and target-organ response) and new therapeutic strategies. These models can reveal the importance of genes in development, gene–gene and gene–environment interactions and novel phenotypes. For example, recent studies of the *sgk1* knock-out mouse have unmasked the importance of sgk1 in the regulation of sodium balance and blood pressure in sodium-depleted states and has raised its profile as a potential causal gene in some forms of pseudohypoaldosteronism (Wulff et al. 2002).

5.3.3
Mouse Mutagenesis Studies

Other strategies are also available in animals for dissecting complex disease traits that could be applied in future to hypertension, such as ENU mutagenesis to provide artificial allelic variation (Nadeau and Frankel 2000). These provide single-gene models that avoid the difficult dissection of the QTLs in inbred animal strains.

5.4
Rodent Versus Human Loci

It is clear that localization of candidate loci within the human genome is in its early stages compared to the mapping success of QTLs in rodents. But advances in comparative genomics may alter this balance rapidly in the very near future, because, despite large-scale chromosomal differences between humans and rodents, they often conserve the order of their genes over substantial genomic distances. With the arrival of good draft sequences for the rat and mouse genomes, it will be relatively easy to map refined rodent QTLs onto the human genome (Stoll and Jacob 2001). These candidate regions could then be subjected to LD mapping using an appropriate high-density SNP map. This approach is likely to be especially important for mapping more modest gene effects, which will be too small to reach significance in a genome-wide scan without enormous sample sizes being available.

6
The Genetics of Target Organ Damage

As intriguing as the genetic basis of hypertension are the genes that influence cardiac hypertrophy and failure, renal failure and stroke. These represent the end-organ processes that cause the mortality and morbidity associated with hypertension, and underlying genetic factors can operate independently and cooperatively with genes that regulate blood pressure to cause cardiovascular disease. Identifying these susceptibility genes could have immense clinical importance in predicting the types of end-organ damage a hypertensive patient will be at risk of developing (Turner and Boerwinkle 2000).

6.1
Cardiac Hypertrophy

In rodent studies, QTLs on chromosomes 2, 14, 17 and X have been reported to influence heart weight independent of blood pressure (Rapp 2000). Several genes are now known to influence cardiac mass in humans (see the chapter by Marian, this volume), but their interaction with blood pressure is not well understood.

There is interest in the role of the renin–angiotensin system and cardiac hypertrophy, and several studies have investigated the role of ACE genotype in the pathogenesis of cardiac hypertrophy and its response to inhibition of angiotensin II activity. Two small studies have reported a greater reduction in cardiac mass with ACE inhibition in hypertensive patients homozygous for deletion of the *Alu* repeat (*DD* genotype) compared to the *ID* and *II* variants (Sasaki et al. 1996; Khono et al. 1999). One study found no effect of genotype on the response of cardiac hypertrophy to angiotensin II receptor (AT1) blockade given during a fitness training program (Myerson et al. 2001).

6.2
Stroke

Genetic susceptibility to stroke has been explored in rodent models and humans. Stroke has been reported to be influenced by genomic regions on chromosomes 1, 4 and 5 (Rapp 2002). A candidate for the chromosome 5 locus is the *Nppa* gene, which encodes the precursor for atrial natriuretic factor (Shimkets et al. 1999). A number of molecular variants of this gene have been identified in SHR and SHRSP rats, one of which may be functionally relevant (Rubattu et al. 1999a). However, the ANP gene has been excluded in another study of SHRs. Similarly, a *G664A* polymorphism in exon 1 of the human ANP gene has been associated with an increased risk of stroke in one study in Caucasian subjects (Rubattu et al. 1999b), but there was no association with another polymorphism of this gene in Japanese (Kato et al. 2002). Among other candidates that have been investigated in detail in animal models are genes encoding the NOS isoforms (see the chapter by Huang, this volume).

A number of conditions in which stroke occurs are inherited in a classical mendelian pattern (Hassan and Markus 2000). One of the best examples is cerebral autosomal dominant arteriopathy with subcortical infarcts and leukoencephalopathy (CADASIL), which is associated with the *NOTCH3* gene (Joutel et al. 1996). Although hypertension is not a feature of this disorder, it shares a similar vascular pathology. A genome-wide scan of Icelandic pedigrees with two or more members who have suffered an ischaemic or haemorrhagic stroke has reported significant linkage to chromosome 5 (*5q12*), suggesting the presence of an as yet unidentified stroke-susceptibility gene. A number of candidate genes have also be reported, alone or in combination, to be associated with an increased risk of stroke; among them, the *Alu* variant of the ACE gene, the *C677T* polymorphism of the methylenetetrahydrofolate reductase gene, the *M235T* allele of angiotensinogen, the *A1166C* allele of the angiotensin 1 receptor gene, the *APOE4* allele and the Leiden V variant (Seinio et al.1998; Takami et al. 2001; Szolnoki et al. 2002; Sierra et al. 2002). Interestingly, the study with the largest subject number found no association with the ACE gene polymorphism, emphasizing the importance of large subject numbers for association studies (Zee et al. 1999).

6.3
Renal Disease

African-Americans are more likely to develop end-stage renal disease than Caucasian Americans, even allowing for socio-economic factors. Linkage between markers on chromosome 10 and non-diabetic renal disease in this population has attracted attention, particularly because a human homologue of the rodent renal failure gene, *Rf1*, is located on this chromosome (Freedman et al. 2002). *Rf1* is one of two genes linked to renal disease in genetic studies of the fawn-hooded rat (Brown et al. 1996). Located on rat chromosome 1, it explains 40% of the genetic variance of renal impairment in this strain, independent of blood pressure. The second gene, *Rf2*, maps to a locus that also influences blood pressure in several rat populations. Data to date suggest that a region near marker D10S677 contributes to susceptibility to renal failure in African-Americans; this marker does not align precisely with the human homolog of *Rf1* and it is not clear if this gene or one nearby accounts for the linkage at present (Freedman et al. 2002).

7
Genotype Versus Phenotype

The rational use of antihypertensive drugs is currently based upon clinical outcome data in large populations, the presence or absence of other diseases and cost considerations. The level of hypertension together with data on co-existing cardiovascular risk factors is used to generate a risk score, an estimate of the 10-year risk of myocardial infarction and stroke, for each patient. Allowances are made for pre-existing target organ damage and providing the risk reaches an agreed level (e.g. 20% over 10 years), the patient receives treatment. Calculating individual risk scores based solely on these clinical phenotype criteria has its limitations. First, the algorithm used to generate the risk score is based upon data from a large Caucasian population (Framingham) and attempts to modify it for other ethnic groups (for example, to take into account the greater propensity of African-Americans for stroke rather than myocardial infarction) have so far been disappointing. Second, the calculation is not applicable to patients already on treatment for hypertension or hyperlipidaemia. Third, it is much more valuable to be able to predict the development of complications rather than wait for their appearance before making the decision to treat.

Until recently, there was little evidence that any one class of antihypertensive drug offered benefit in terms of reduction in myocardial infarction and stroke over any other. So the choice of drug for each patient is based on a clinical assessment of the risk of the drug causing an adverse event in that patient (e.g. gout with a thiazide, asthma with a beta-blocker, etc.) and the cost of the drug. This is unsatisfactory. It is well known that patients vary in their response to drugs, not only in terms of adverse effects but also therapeutic effect. The present practice is little better than a best guess approach to treatment; patients may end up rotated through several different treatments in an attempt to bring blood pressure under control and in an asymptomatic condition, this can contribute to poor compliance. It is recognized that African-Americans are more responsive to diuretics and calcium channel blockers and less responsive to beta-blockers and angiotensin-converting enzyme inhibitors than Caucasians. There is also support for the idea that older patients respond better to diuretic and calcium antagonists than younger patients. Some advocate the measurement of plasma renin activity to assist the choice of treatment, but others argue that this is simply a surrogate for age and race.

Incorporating genetic information into the risk factor calculation has the potential to improve its predictive power and can assist in the choice and dose of therapeutic agent with a view to maximizing the possibility of response and minimizing adverse effects. However, the present application of genotype data is restricted. No one would dispute the value of recognizing patients with Liddle's syndrome or GRA with respect to finding the most appropriate treatment. However, single gene mutations account for only a small fraction of patients with hypertension and genetic screening for these disorders is not indicated at present; rather it is reserved for patients for whom there is a high index of suspicion of the diagnosis (e.g. early onset, hypokalaemia, family history of premature stroke). It is too early to advocate screening for the T594M variant in the ENaC in African-Americans to select out patients who might benefit from amiloride or to use ACE genotype to predict the response of hypertensive patients with cardiac hypertrophy to ACE inhibitors.

The Genetics of Hypertension Associated Treatment (GenHAT) study is an attempt to address this (Arnett et al. 2002). It is a genetic study coupled to the Antihypertensive and Lipid Lowering Treatment to Prevent Heart Attack Trial (ALLHAT) (The ALLHAT Officers and Coordinators for the ALLHAT Collaborative Research Group 2002). A total of 42,411 high-risk hypertensive subjects aged 55 years and above have been randomized to chlorthalidone (a diuretic), lisinopril (and ACE inhibitor), doxazosin (an alpha-blocker) or amlodipine (a calcium antagonist), with a view to reducing blood pressure below 140/90 and the combination of coronary heart disease death and non-fatal myocardial in-

Table 2 Polymorphisms that will be examined for association with response to treatment in the Genetics of Hypertension Associated Treatment (GenHAT) study (Arnett et al. 2002)

Genetic variant	Genotypes	Assumption
AGT-6	AA, AG, GG	Lowest relative risk of CHD associated with *AA* homozygotes treated with an ACE inhibitor
ACE I/D	II, ID, DD	Lowest relative risk of CHD associated with *DD* homozygotes treated with an ACE inhibitor
AT-1	AA, AC, CC	Lowest relative risk of CHD associated with *AC/CC* genotypes treated with an ACE inhibitor
β2 Receptor	3.7/3.7, 3.7/3.4, 3.4/3.4	Lowest relative risk of CHD associated with *3.4/3.4* homozygotes treated with chlorthalidone
α-Adducin	Gly/Gly, Gly/Trp, Trp/Trp	Lowest relative risk of CHD associated with *Gly/Trp* and *Trp/Trp* genotypes treated with chlorthalidone
LPL	H1H1, H1H2, H2H2	Lowest relative risk of CHD associated with *H1H1* and *H1H2* genotypes treated with doxazosin

farction. Patients will be genotyped for polymorphisms in six genes (Table 2) to test the possibility that certain variations are associated with a better response to one of the study drugs. It is the largest study of its kind and expects to report in 2003 or 2004.

Whether or not the genetic screening of hypertensive patients will become more widespread will depend upon the genetic model of the disease. If there are relatively few genes with significant effects, be it on blood pressure or susceptibility to complications such as stroke, then early recognition will be important. If there are a number of genes, each with a very small effect, then genetic screening will be much less useful. Moreover, with progressive lowering of the ideal blood pressure for patients, i.e. stricter treatment target blood pressure goals, it becomes less likely that identifying a single gene will obfuscate the need for several drugs to achieve an acceptable reduction.

8
Conclusion

To summarize, the current state of knowledge on the genes involved in essential hypertension, one simply needs to update the remarks made by Pierre Corvol and his group in 1999 in relation to the AGT and ENaC genes (Corvol et al. 1999). Thus:

1. The majority (if not all) linkage or association studies still lack adequate statistical power.
2. The results are still heavily influenced by ethnicity.
3. The definition of the hypertensive phenotype frequently lacks precision.
4. Proving a causal link between a molecular variant and hypertension remains as difficult and elusive – and without a yardstick.

5. Rodent genetic studies have still not come of age.
6. Most molecular variants will probably carry low attributable risks at the population level or a low individual effect at the individual level.
7. It is still impossible to predict with any certainty individual responses to salt restriction or specific classes of antihypertensive agent based on a patient's genotype.

Despondent as this may seem, this field is in good company and this critique could be equally made against any of the other complex diseases. On a positive note, large hypertensive cohorts are being recruited and the use of more informative intermediate phenotypes, particularly amongst patients with low-renin hypertension, may address some of these concerns in the coming years.

References

Aitman TJ, Glazier AM, Wallace CA et al (1999)Identification of Cd36 (Fat) as an insulin-resistance gene causing defective fatty acid and glucose metabolism in hypertensive rats [see comments]. Nat Genet 21:76–83

Altmuller J, Palmer LJ, Fischer G et al (2001) Genomewide scans of complex human diseases: true linkage is hard to find. Am J Hum Genet 69:936–950

Altshuler D, Hirschhorn JN, Klannemark M et al (2000) The common PPARgamma Pro12Ala polymorphism is associated with decreased risk of type 2 diabetes. Nat Genet 26:76–80

Anonymous (1999) Freely associating. Nat Genet 22(1):1–2

Arnett DK, Boerwinkle E, Davis BR et al (2002) Pharmacogentic Approaches to hypertension therapy: design and rationale for the Genetic of Hypertension Associated Treatment (GenHAT) study. The Pharmacogenetics J 2:309–317

Arngrimsson R, Hayward C, Nadaud S et al (1997) Evidence for a familial pregnancy-induced hypertension locus in the eNOS-gene region. Am J Hum Genet 61:354–362

Arngrimsson R, Sigurard TS, Frigge ML et al (1999) A genome-wide scan reveals a maternal susceptibility locus for pre-eclampsia on chromosome 2p13. Hum Mol Genet 8:1799–1805

Baker EH, Dong YB, Sagnella GA et al (1998) Association of hypertension with T594 M mutation in beta subunit of epithelial sodium channels in black people resident in London. Lancet 351:1388–1392

Baker EH, Duggal A, Dong Y et al (2002) Amiloride, a specific drug for hypertension in black people with T594 M variant? Hypertension 40:13–17

Benjafield AV, Jeyasingam CL, Nyholt DR et al (1998) G-protein beta3 subunit gene (GNB3) variant in causation of essential hypertension. Hypertension 32:1094–1097

Bergman S, Key BO, Kirk KA et al (1996) Kidney disease in the first-degree relatives of African-Americans with hypertensive end-stage renal disease. Am J Kidney Dis 27:341–346

Bianchi G, Tripodi G, Casari G et al (1994) Two point mutations within the adducin genes are involved in blood pressure variation. Proc Natl Acad Sci U S A 91:3999–4003

Bonnardeaux A, Nadaud S, Charru A et al (1995) Lack of evidence for linkage of the endothelial cell nitric oxide synthase gene to essential hypertension. Circulation 91:96–102

Bray MS, Krushkal J, Li L et al (2000) Positional genomic analysis identifies the beta(2)-adrenergic receptor gene as a susceptibility locus for human hypertension. Circulation 101:2877–2882

Bray MS, Li L, Turner ST et al (2000) Association and linkage analysis of the alpha-adducin gene and blood pressure. Am J Hypertens 13:699–703

Brown DM, Provoost AP, Daly MJ et al (1996) Renal disease susceptibility and hypertension are under independent genetic control in the fawn-hooded rat. Nat Genet 12:44–51

Caulfield M, Lavender P, Farrall M et al (1994) Linkage of the angiotensinogen gene to essential hypertension. N Engl J Med 330:1629–1633

Caulfield M, Lavender P, Newell-Price J et al (1995) Linkage of the angiotensinogen gene locus to human essential hypertension in African Caribbeans. J Clin Invest 96:687–692

Caulfield M, Pembroke J, Dominiczak A et al (2002) The MRC British Genetics of Hypertension Study: genome-wide results. J Hum Hypertens 2002;16:896

Collison M, Glazier AM, Graham D et al (2000) Cd36 and molecular mechanisms of insulin resistance in the stroke-prone spontaneously hypertensive rat. Diabetes 49:2222–2226

Corvol P, Persu A, Gimenez-Roqueplo AP et al (1999) Seven lessons from two candidate genes in human essential hypertension: angiotensinogen and epithelial sodium channel. Hypertension 33:1324–1331

Cowley AW, Jr., Roman RJ, Kaldunski ML et al (2001) Brown Norway chromosome 13 confers protection from high salt to consomic Dahl S rat. Hypertension 37:456–461

Cui Y, Su YR, Rutkowski M et al (1997) Loss of protein kinase C inhibition in the beta-T594 M variant of the amiloride-sensitive Na+ channel. Proc Natl Acad Sci USA 94:9962–9966

Cusi D, Barlassina C, Azzani T et al (1997) Polymorphisms of alpha-adducin and salt sensitivity in patients with essential hypertension. Lancet 349:1353–1357

Cvetkovic B, Keen HL, Zhang X et al (2002) Physiological significance of two common haplotypes of human angiotensinogen using gene targeting in the mouse. Physiol Genomics

Dahlman I, Eaves IA, Kosoy R et al. (2002) Parameters for reliable results in genetic association studies in common disease. Nat Genet 30:149–150

DeStefano AL, Baldwin CT, Burzstyn M et al. (1998) Autosomal dominant orthostatic hypotensive disorder maps to chromosome 18q. Am J Hum Genet 63:1425–1430

Dominiczak AF, Negrin DC, Clark JS et al (2000). Genes and hypertension: from gene mapping in experimental models to vascular gene transfer strategies. Hypertension 35:164–172

Eaves IA, Merriman TR, Barber RA et al (2000) The genetically isolated populations of Finland and Sardinia may not be a panacea for linkage disequilibrium mapping of common disease genes. Nat Genet 25:320–323.

Ferrandi M, Salardi S, Tripodi G et al. (1999) Evidence for an interaction between adducin and Na(+)-K(+)-ATPase: relation to genetic hypertension. Am J Physiol 277:H1338-H1349

Fornage M, Amos CI, Kardia S et al (1998) Variation in the region of the angiotensin-converting enzyme gene influences interindividual differences in blood pressure levels in young Caucasian males. Circulation 97:1773–1779

Freedman BI, Rich SS, Yu Hongrun et al (2002) Linkage heterogeneity of end-stage renal disease on human chromosome 10. Kidney Internat 62:770–774

Glorioso N, Manunta P, Filigheddu F et al (1999) The role of alpha-adducin polymorphism in blood pressure and sodium handling regulation may not be excluded by a negative association study. Hypertension 34:649–654

Hasson A, Lansbury A, Catto AJ et al (2002) Angiotensin converting enzyme insertion/deletion genotype is associated with leukoaraiosis in lacunar syndromes. J Neurol Neurosurg Psychiatry 72:343–346

Hata A, Namikawa C, Sasaki M et al (1994) Angiotensinogen as a risk factor for essential hypertension in Japan. J Clin Invest 93:1285–1287

Hegele RA, Anderson C, Young TK et al (1999) G-protein beta3 subunit gene splice variant and body fat distribution in Nunavut Inuit. Genome Res 9:972–977

Hegele RA, Brunt JH, Connelly PW (1994) A polymorphism of the angiotensinogen gene associated with variation in blood pressure in a genetic isolate. Circulation 90:2207–2212

Hegele RA, Harris SB, Hanley AJ et al (1998) -6A promoter variant of angiotensinogen and blood pressure variation in Canadian Oji-Cree. J Hum Genet 43:37–41

Huang PL, Huang Z, Mashimo H et al (1995) Hypertension in mice lacking the gene for endothelial nitric oxide synthase. Nature 377:239–242

Hunt SC, Ellison RC, Atwood LD et al (2002) Genome scans for blood pressure and hypertension: the National Heart, Lung, and Blood Institute Family Heart Study. Hypertension 40:1–6

Hyndman ME, Parsons HG, Verma S et al (2002) The T-786->C mutation in endothelial nitric oxide synthase is associated with hypertension. Hypertension 39:919–922

Ishikawa K, Imai Y, Katsuya T et al (2000) Human G-protein beta3 subunit variant is associated with serum potassium and total cholesterol levels but not with blood pressure. Am J Hypertens 13:140–145

Jacob HJ, Lindpaintner K, Lincoln SE et al (1991)Genetic mapping of a gene causing hypertension in the stroke- prone spontaneously hypertensive rat. Cell 67:213–224

Jeerooburkhan N, Jones LC, Bujac S et al (2001) Genetic and environmental determinants of plasma nitrogen oxides and risk of ischemic heart disease. Hypertension 38:1054–1061

Jeffs B, Clark JS, Anderson NH et al (1997) Sensitivity to cerebral ischaemic insult in a rat model of stroke is determined by a single genetic locus. Nat Genet 16:364–367

Jeffs B, Negrin CD, Graham D et al (2000) Applicability of a "speed" congenic strategy to dissect blood pressure quantitative trait loci on rat chromosome 2. Hypertension 35:179–187.

Jeunemaitre X, Lifton RP, Hunt SC et al (1992) Absence of linkage between the angiotensin converting enzyme locus and human essential hypertension. Nat Genet 1:72–75

Jeunemaitre X, Soubrier F, Kotelevtsev YV et al (1992) Molecular basis of human hypertension: role of angiotensinogen. Cell 71:169–180

Kajiyama N, Saito Y, Miyamoto Y et al (2000) Lack of association between T-786->C mutation in the 5'-flanking region of the endothelial nitric oxide synthase gene and essential hypertension. Hypertens Res 23:561–565

Karet F (2003) Monogenic tubular salt and acid transporter disorders. J Nephrol 15 [Suppl 6]:S57–S68

Kato N, Sugiyama T, Morita H et al (1999) Angiotensinogen gene and essential hypertension in the Japanese: extensive association study and meta-analysis on six reported studies. J Hypertens 17:757–763

Khono M, Yokokawa K, Minami M et al (1999) Association between angiotensin-converting enzyme gene polymorphisms and regression of left ventricular hypertrophy in patients treated with angiotensin-converting enzyme inhibitors. Am J Med 106; 544–549.

Klag MJ, Whelton PK, Randall BL et al (1997) End-stage renal disease in African-American and Caucasian men. 16-year MRFIT findings. J Am Med Assoc 277:1293–1298

Kostulas K, Huang W-X, Crisby M et al (1999) An angiotensin-converting enzyme gene polymorphism suggests a genetic distrinction between ischaemic stroke and carotid stenosis. Eur J Clin Invest 29:478–483

Kristjansson K, Manolescu A, Kristinsson A et al (2002) Linkage of essential hypertension to chromosome 18q. Hypertension 39:1044–1049

Krushkal J, Ferrell R, Mockrin SC et al (1999) Genome-wide linkage analyses of systolic blood pressure using highly discordant siblings. Circulation 9:1407–1410

Lander ES, Botstein D (1989) Mapping mendelian factors underlying quantitative traits using RFLP linkage maps. Genetics 121:185–199

Levy D, DeStefano AL, Larson MG et al (2000) Evidence for a gene influencing blood pressure on chromosome 17. Genome scan linkage results for longitudinal blood pressure phenotypes in subjects from the framingham heart study. Hypertension 36:477–483

Liggett SB (1997) Polymorphisms of the beta2-adrenergic receptor and asthma. Am J Respir Crit Care Med 156:S156–S162

Mackay TF (2001) The genetic architecture of quantitative traits. Annu Rev Genet 35:303–339

Matsubara M, Metoki H, Suzuki M et al (2002) Genotypes of the betaENaC gene have little influence on blood pressure level in the Japanese population. Am J Hypertens 15:189–192

Miyamoto Y, Saito Y, Kajiyama N et al (1998) Endothelial nitric oxide synthase gene is positively associated with essential hypertension. Hypertension 32:3–8

Myerson SG, Montgomery HE, Whittingham M et al (2001) Left ventricular hypertrophy with exercise and ACE gene insertion/deletion polymorphism: a randomised controlled trial with losartan. Circulation 103:226–230

Nadeau JH, Frankel WN (2000) The roads from phenotypic variation to gene discovery: mutagenesis versus QTLs. Nat Genet 25:381–384

Nadeau JH, Singer JB, Matin A et al (2000) Analysing complex genetic traits with chromosome substitution strains. Nat Genet 24:221–225

Negrin CD, McBride MW, Carswell HV et al (2001) Reciprocal consomic strains to evaluate y chromosome effects. Hypertension 37:391–397

Nickerson DA, Taylor SL, Weiss KM et al (1998) DNA sequence diversity in the 9.7-kb region of the human lipoprotein lipase gene. Nat Genet 19:233–240

Niu T, Chen C, Yang J et al (1999) Blood pressure and the T174 M and M235T polymorphisms of the angiotensinogen gene. Ann Epidemiol 9:245–253

Niu T, Xu X, Cordell HJ et al (1999) Linkage analysis of candidate genes and gene-gene interactions in chinese hypertensive sib pairs. Hypertension 33:1332–1337

Niu T, Yang J, Wang B et al (1999) Angiotensinogen gene polymorphisms M235T/T174 M: no excess transmission to hypertensive Chinese. Hypertension 33:698–702

Nothnagel M, Ott J (2002) Statistical gene mapping of traits in humans–hypertension as a complex trait: is it amenable to genetic analysis? Semin Nephrol 22:105–114

O'Donnell CJ, Lindpaintner K, Larson MG et al (1998) Evidence for association and genetic linkage of the angiotensin-converting enzyme locus with hypertension and blood pressure in men but not women in the Framingham Heart Study. Circulation 97:1766–1772

Persu A, Barbry P, Bassilana F et al (1998) Genetic analysis of the beta subunit of the epithelial Na+ channel in essential hypertension. Hypertension 32:129–137

Pravenec M, Zidek V, Simakova M et al (1999) Genetics of Cd36 and the clustering of multiple cardiovascular risk factors in spontaneous hypertension. J Clin Invest 103:1651–1657

Price DA, Crook ED (2002) Kidney disease in African Americans: genetic considerations. J Natl Med Assoc 94 (8 Suppl):16S-27S

Province MA, Boerwinkle E, Chakravarti A et al (2000) Lack of association of the angiotensinogen-6 polymorphism with blood pressure levels in the comprehensive NHLBI Family Blood Pressure Program. National Heart, Lung and Blood Institute. J Hypertens 18:867–876

Province MA, Kardia SLR, Ranade K et al (2003) A Meta-Analysis of Genome-Wide Linkage Scans for Hypertension:The National Heart,Lung and Blood Institute Family Blood Pressure Program. Am J Hypertens 16:144–147

Ranade K, Hsuing AC, Wu KD et al (2000) Lack of evidence for an association between alpha-adducin and blood pressure regulation in Asian populations. Am J Hypertens 13:704–709

Rapp JP (2000) Genetic analysis of inherited hypertension in the rat. Physiol Rev 80:135–172

Rapp JP, Garrett MR, Deng AY (1998) Construction of a double congenic strain to prove an epistatic interaction on blood pressure between rat chromosomes 2 and 10. J Clin Invest 101:1591–1595

Reich DE, Schaffner SF, Daly MJ et al(2002) Human genome sequence variation and the influence of gene history, mutation and recombination. Nat Genet 32:135–142

Rice T, Rankinen T, Province MA et al (2000) Genome-wide linkage analysis of systolic and diastolic blood pressure: the Quebec Family Study. Circulation 102:1956–1963

Rieder MJ, Taylor SL, Clark AG et al (1999) Sequence variation in the human angiotensin converting enzyme. Nat Genet 22:59–62

Rigat B, Hubert C, Alhenc-Gelas F et al (1990) An insertion/deletion polymorphism in the angiotensin I-converting enzyme gene accounting for half the variance of serum enzyme levels. J Clin Invest 86:1343–1346

Rubattu S, Lee-Kirsch MA, dePaolis P et al (1999) Altered structure, regulation, and function of the gene encoding the atrial natriuretic peptide in the stroke-prone spontaneously hypertensive rat. Circ Res 85:900–905

Sasaki M, Oki T, Luchi A et al (1996) Relationship between the angiotensin converting enzyme gene polymorphism and the effects of enalapril on left ventricular hypertrophy and impaired diastolic filling in essential hypertension: M-mode and pulsed-Doppler echocardiographic studies. J Hypertens 14:1403–1408

Schork NJ, Chakravarti A, Thiel B et al (2000) Lack of association between a biallelic polymorphism in the adducin gene and blood pressure in Caucasians and African Americans. Am J Hypertens 13:693–698.

Schunkert H, Hense HW, Doring A et al (1998) Association between a polymorphism in the G protein beta3 subunit gene and lower renin and elevated diastolic blood pressure levels. Hypertension 32:510–513

Seino Y, Ikeda U, Maeda Y et al (1998) Angiotensin-converting enzyme gene polymorphism and plasminogen activator inhibitor 1 levels in subjects with cerebral infarction. J Thrombosis and Throbolysis 5:263–267

Sierra C, Coca A, Gómez-Angelats E et al (2002) Renin-angiotensin system genetic polymorphisms and cerebral white matter lesions in essential hypertension. Hypertension 39:343–347.

Siffert W, Forster P, Jockel KH et al (1999) Worldwide ethnic distribution of the G protein beta3 subunit 825T allele and its association with obesity in Caucasian, Chinese, and Black African individuals. J Am Soc Nephrol 10:1921–1930

Siffert W, Rosskopf D, Moritz A et al (1995) Enhanced G protein activation in immortalized lymphoblasts from patients with essential hypertension. J Clin Invest 96:759–766

Siffert W, Rosskopf D, Siffert G et al (1998) Association of a human G-protein beta3 subunit variant with hypertension. Nat Genet 18:45–48

Stephens JC, Schneider JA, Tanguay DA et al (2001) Haplotype Variation and Linkage Disequilibrium in 313 Human Genes. Science 293:489–493

Stoll M, Jacob HJ. (2001) Genetic rat models of hypertension: relationship to human hypertension. Curr Hypertens Rep 3:157–164

Stoll M, Kwitek-Black AE, Cowley AW et al (2000) New target regions for human hypertension via comparative genomics. Genome Res 10:473–482

Sugiyama F, Churchill GA, Higgins DC et al (2001) Concordance of murine quantitative trait loci for salt-induced hypertension with rat and human loci. Genomics 71:70–77

Takami S, Imai Y, Katsuya T et al (2000) Gene polymorphism of the renin-angiotensin system associates with risk for lacunar infarction. The Ohasama Study. Am J Hypertension 13:121–127

Takami S, Wong ZY, Stebbing M et al (1999) Linkage analysis of endothelial nitric oxide synthase gene with human blood pressure. J Hypertens 17:1431–1436

Tesauro M, Thompson WC, Rogliani P et al (2000) Intracellular processing of endothelial nitric oxide synthase isoforms associated with differences in severity of cardiopulmonary diseases: cleavage of proteins with aspartate vs. glutamate at position 298. Proc Natl Acad Sci U S A 97:2832–2835

The ALLHAT Officers and Coordinators for the ALLHAT Collaborative Research Group (2002) Major outcomes in high-risk hypertensive patients randomized to angiotensin-converting enzyme inhibitor or calcium channel blocker vs diuretic: The Antihypertensive and Lipid-Lowering Treatment to Prevent Heart Attack Trial (ALLHAT). JAMA 288: 3039–3042

Province MA, Kardia SLR, Ranade K et al (2003). A meta-analysis of genome-wide linkage scans for hypertension:The National Heart, Lung and Blood Institute Family Blood Pressure Program. Am J Hypertens 16:144–147.

Tiago AD, Nkeh B, Candy GP et al (2001) Association study of eight candidate genes with renin status in mild-to-moderate hypertension in patients of African ancestry. Cardiovasc J S Afr 12:75–80

Tiret L, Poirier O, Nicaud V et al (2002) Heterogeneity of linkage disequilibrium in human genes has implications for association studies of common diseases. Hum Mol Genet 11:419–429

Tomaszewski M, Brain NJ, Charchar FJ et al (2002) Essential hypertension and beta2-adrenergic receptor gene: linkage and association analysis. Hypertension 40:286–291

Tournier-Lasserve E (2002) New players in the genetics of stroke. N Engl J Med 347:1711–1712

Turner ST, Boerwinkle E (2000) Genetics of hypertension, target-organ complications, and response to therapy. Circulation 102:IV40-IV45

Turner ST, Boerwinkle E, Sing CF (1999) Context-dependent associations of the ACE I/D polymorphism with blood pressure. Hypertension 34:773–778

Wayne ML, McIntyre LM (2002) Combining mapping and arraying: An approach to candidate gene identification. Proc Natl Acad Sci U S A 99:14903–14906

Wilson FH, Disse-Nicodeme S, Choate KA et al (2001) Human Hypertension Caused by Mutations in WNK Kinases. Science 293:1107–1112

Wilson FH, Kahle KT, Sabath E et al (2002) Molecular pathogenesis of inherited hypertension with hyperkalemia: The Na–Cl cotransporter is inhibited by wild-type but not mutant WNK4. Proc Natl Acad Sci 100:680–684

Winkelmann BR, Russ AP, Nauck M et al (1999) Angiotensinogen M235T polymorphism is associated with plasma angiotensinogen and cardiovascular disease. Am Heart J 137:698–705

Wright FA, O'Connor DT, Roberts E et al. (1999) Genome Scan for Blood Pressure Loci in Mice. Hypertension 34:625–630

Wulff P, Vallon V, Huang DY et al (2002) Impaired renal Na^+ retention in the *sgk1*-knockout mouse. J Clin Invest 110:1263–1268

Xu X, Rogus JJ, Terwedow HA et al (1999) An extreme-sib-pair genome scan for genes regulating blood pressure. Am J Hum Genet 64:1694–1701

Zee RYL, Ridker PM, Stampfer MJ et al (1999) Prospective evauation of the angiotensin-converting enzyme insertion/deletion polymorphism and the risk of stroke. Circulation 99:340–343

Zhu DL, Wang HY, Xiong MM et al (2001) Linkage of hypertension to chromosome 2q14-q23 in Chinese families. J Hypertens 19:55–61

Genetic Predisposition to Cardiac Hypertrophy

A. J. Marian

Section of Cardiology, Baylor College of Medicine, One Baylor Plaza, 519D, Houston, TX 77030, USA
e-mail: amarian@bcm.tmc.edu

1	Introduction	178
2	Cardiac Hypertrophy in Mendelian Genetic Disorders	179
2.1	Hypertrophic Cardiomyopathy as a Genetic Model of the Cardiac Hypertrophic Response	179
2.1.1	Genetic Basis of HCM	180
2.1.2	Causal Genes and Mutations	181
2.1.3	Impact of Causal Mutations on Cardiac Hypertrophy	182
2.1.4	Modifier Genes	183
2.1.5	Impact of Modifier Genes on Cardiac Hypertrophy	185
2.1.6	Gene Expression in HCM	186
2.1.7	Pathogenesis of HCM	187
2.1.8	Reversal and Attenuation of Cardiac Phenotypes in HCM	188
2.2	Cardiac Hypertrophy in Trinucleotide Repeat Syndromes	189
2.3	Cardiac Hypertrophy in Noonan and Leopard Syndromes	190
2.4	Cardiac Hypertrophy in Genetic Metabolic Disorders	191
3	Cardiac Hypertrophy in Non-Mendelian Genetic Disorders	192
3.1	Cardiac Hypertrophy in Mitochondrial Disorders	192
3.2	Single-Nucleotide Polymorphisms and Cardiac Hypertrophy	193
4	Conclusions	194
	References	195

Abstract Cardiac hypertrophy is the common response of the heart to a number of physiological and pathological stimuli. Regardless of the etiology, the magnitude and nature of the response is influenced by both genetic and nongenetic factors. Mutations in single genes underlie hypertrophic cardiomyopathy (HCM) but there is remarkable variability in the phenotypic expression of HCM. Both the genetic background, referred to as the modifier genes, and probably the environmental factors affect the severity of the phenotype. Cardiac hypertrophy is also a common feature of a variety of other mendelian and non-mendelian genetic disorders, including trinucleotide repeat syndromes, Noonan and leopard syndromes, inherited metabolic disorders and mitochondrial myopathies. Furthermore, single nucleotide polymorphisms also affect cardiac hypertrophic response and modulate regression of cardiac hypertrophy in response to pharmacological and nonpharmacological interventions. Deciphering the molecular pathogenesis of cardiac hypertrophic response could provide for the opportunity to identify new therapeutic targets. Moreover, it may permit early genetic diagnosis and risk stratification, independent of and prior to development of the clinical phenotypes, and so the implementation of preventive and therapeutic measures in those at risk and ultimately individualization of pharmacological (pharmacogenetics) and nonpharmacological therapy.

Keywords Genetics · Hypertrophy · Cardiomyopathy · Polymorphism · Mutations

1
Introduction

Cardiac hypertrophy is the ubiquitous response of the heart to all forms of stimuli. Adult cardiac myocytes are considered terminally differentiated and unable to proliferate. Therefore, the cardiac response to a stimulus is restricted primarily to an increase in the size of myocytes but not their number. Cardiac and myocyte hypertrophy develop as a consequence of physiological as well as pathological stimuli, regardless of whether the stimulus is intrinsic, such as a defect in a structural protein, or extrinsic, such as an increased load. The primary purpose of cardiac and myocyte hypertrophy is adaptive, aimed at reducing stress applied per unit of the working myofibril. The ensuing gross phenotype is increased cardiac muscle mass; however, there are differences in the morphological, molecular, structural and functional phenotypes between cardiac hypertrophy secondary to physiological stimuli and that are caused by pathological stimuli.

Similarly, the clinical consequences of cardiac hypertrophy also depend upon the underlying cause. As such, cardiac hypertrophy resulting from physiological stimuli such as exercise is an adaptive response that has beneficial effects. The so-called athlete's heart is characterized by an increase in left ventricular dimension, wall thickness and cardiac mass and has a normal function. In contrast, hypertrophy caused by pathological stimuli such as hypertension or mutations in contractile proteins is considered maladaptive and is associated with increased cardiovascular mortality and morbidity (Levy et al. 1990). In addition to myocyte hypertrophy, the hypertrophic process often encompasses an increase in the number of cardiac fibroblasts and extracellular matrix protein content, which could contribute to the undesirable consequences of cardiac hypertrophy.

Regardless of whether the primary stimulus is an acquired condition or a genetic defect, expression of the hypertrophic phenotype is determined by the complex interactions of genetic and nongenetic factors. In acquired forms of cardiac hypertrophy such as in valvular disease or in physiological states such as in athletes' heart genetic factors are important determinants of the magnitude of hypertrophic response. Similarly, in genetic forms of cardiac hypertrophy such as in hypertrophic cardiomyopathy (HCM), which is considered a classic single-gene disorder, expression of cardiac hypertrophy is affected not only by the causal mutation but also by the non-causal genetic and nongenetic factors (Marian 2001a). Thus, cardiac hypertrophy is neither strictly genetic nor entirely acquired or environmental but is rather a consequence of complex interplay between genetic and nongenetic factors.

Epidemiological studies in monozygotic and dizygotic twins provided the initial evidence for the role of genetic factors in determining cardiac size (Adams et al. 1985; Verhaaren et al. 1991; Landry et al. 1985; Harshfield et al. 1990; Bielen et al. 1990). The results showed cardiac size and left ventricular mass were less variable in subjects with an identical genetic background. A significant part of the influence of the genetic background on cardiac size is indirect because of shared genetic factors influencing body size, blood pressure and gender (Adams et al. 1985; Bielen et al. 1990; Post et al. 1997), which are themselves major determinants of cardiac size. Nevertheless, genetic factors impose a direct and independent influence on cardiac mass corrected for age, gender, body size and blood pressure (Post et al.1997; Bielen et al.1990; Schunkert et al. 1999). Genetic factors accounted for a discernible, albeit small part of the variability of left ventricular mass in the Framingham Heart and Framingham Offspring Studies, which comprise relatively homogenous Caucasian populations and which are expected to have less variance than a nonhomogenous population (Post et al. 1997). Genetic factors not only contribute to cardiac size in the general population but also contribute significantly to the development of cardiac hypertrophy, independent of known risk factors for cardiac hypertrophy (Schunkert et al. 1999). In this chapter, the impact of genetic factors on the development of cardiac hypertrophy in conditions with simple mendelian and non-mendelian inheritance will be discussed.

2
Cardiac Hypertrophy in Mendelian Genetic Disorders

Cardiac hypertrophy is the predominant phenotype in several genetic disorders with simple mendelian inheritance. Hypertrophic cardiomyopathy (HCM) is the prototype of this group, which has been studied extensively during the last decade, resulting in the unraveling of its molecular genetic basis.

2.1
Hypertrophic Cardiomyopathy as a Genetic Model
of the Cardiac Hypertrophic Response

HCM is a primary disease of the myocardium characterized by left ventricular hypertrophy in the absence of an increased external load, i.e., unexplained cardiac hypertrophy. HCM is a relatively common disease with an estimated prevalence of approximately 1:500 in young individuals (Maron et al. 1995). The prevalence may be higher in older subjects because of age-dependent penetrance of the causal mutations (Niimura et al. 1998).

Cardiac hypertrophy is asymmetric in approximately two-thirds of cases and the interventricular septum is the predominant site of involvement; hence the term asymmetric septal hypertrophy. In the remainder, cardiac hypertrophy is symmetric, and rarely hypertrophy is restricted to apex of the heart (apical HCM). Morphologically, the left ventricular cavity is small and left ventricular ejection fraction, a measure of global systolic function, is preserved. However, more sensitive indices of myocardial function show impaired contraction and relaxation (Nagueh et al. 2001). Diastolic function is commonly impaired, leading to increased left ventricular end diastolic pressure and thus symptoms of heart failure.

Myocyte hypertrophy, myocyte disarray and interstitial fibrosis are the most common pathological phenotypes, and disarray is considered the pathological hallmark (Maron and Roberts, 1979; Maron et al. 1981). Disarray often comprises more than 20%–0% of the myocardium as opposed to less than 5% in normal hearts. Myocyte disarray is more prominent in the interventricular septum, but scattered myocyte disarray is often present throughout the myocardium (Maron et al. 1981). Cardiac hypertrophy, interstitial fibrosis and myocyte disarray are considered major determinants of risk of sudden cardiac death (SCD), mortality and morbidity in patients with HCM (Shirani et al. 2000; Spirito et al. 2000; Varnava et al. 2001a, b).

The clinical manifestations of HCM are variable. Overall, HCM is considered a relatively benign disease with an annual mortality rate of less than 1% in adults (Cannan et al. 1995; Maron et al. 2000). The majority of patients are asymptomatic or mildly symptomatic. The main symptoms comprise those of heart failure, secondary to an elevated ventricular filling pressure, along with chest pain, palpitations and infrequently syncope. Cardiac arrhythmias, in particular atrial fibrillation and nonsustained ventricular tachycardia, are relatively common but Wolff-Parkinson-White syndrome is uncommon. SCD is the most dreadful presentation of HCM and is tragic since it often occurs as the first manifestation of HCM in young, asymptomatic and apparently healthy individuals (Maron et al. 1996; McKenna et al. 1981). HCM is the most common cause of SCD in young competitive athletes (Maron et al. 1996).

2.1.1
Genetic Basis of HCM

HCM is a genetic disease with an autosomal dominant mode of inheritance. It is caused primarily by mutations in contractile sarcomeric proteins (Marian and Roberts 2001). It is a familial disease in approximately two-thirds of the cases and in the remainder it is considered sporadic, but these also have a genetic basis and are caused by de novo mutations in sarcomeric proteins. A de novo mutation will be transmitted to the offspring of the index case with sporadic HCM. A founder effect in HCM is uncommon, which suggests the majority of mutations have occurred independently (Watkins et al. 1992a; Watkins et al. 1993, 1995a). The morphological phenotype of HCM, defined as cardiac hypertrophy in the absence of an external load, can also develop as a consequence of mutations in mitochondrial genes. In such cases, the mode of inheritance is matrilinear.

2.1.2
Causal Genes and Mutations

Approximately 12 years ago, Dr. Seidman's group made the seminal discovery of an *R403Q* mutation in the β-myosin heavy chain (MyHC) in a family with HCM (Geisterfer-Lowrance et al. 1990), which led to elucidation of the molecular genetic basis of HCM. Since then, a large number of mutations in 11 different genes, all encoding contractile sarcomeric proteins (Table 1), have been identified (Marian and Roberts 2001). This has led to the notion that HCM is primarily a disease of contractile sarcomeric proteins (Thierfelder et al. 1994). Overall, the identified causal genes and mutations account for approximately two-thirds of all HCM cases (Arad et al. 2002; Marian and Roberts 2001). Mutations in genes encoding β-MyHC (*MYH7*), myosin binding protein-C (*MYBPC3*) and cardiac troponin T (*TNNT2*) account for the vast majority of known mutations (Marian and Roberts 2001). *MYH7* mutations account for approximately 35%–50% of all HCM cases (Seidman 2000). Over 100 different mutations in *MYH7* have been identified, and the vast majority are missense mutations. Codons 403 and 719 are considered hot spots for mutations (Anan et al. 1994; Dausse et al. 1993). There is a propensity for mutations to be localized to the globular head of the

Table 1 Genetic causes of hypertrophic cardiomyopathy caused by mutant sarcomeric proteins

Gene	Symbol	Locus	Frequency	Mutations
β-Myosin heavy chain	MYH7	14q12	~35%	Most common, predominantly missense mutations
Myosin binding protein-C	MYBPC3	11p11.2	~20%	Probably 2nd most common, predominantly splice junction and insertion/deletion mutations
Cardiac troponin T	TNNT2	1q32	~20%	Probably 3rd most common, mostly missense
α-Tropomyosin	TPM1	15q22.1	~5%	Missense mutations
Cardiac troponin I	TNNI3	19p13.2	~5%	Missense and deletion mutations
Essential myosin light chain	MYL3	3p21.3	<5%	Missense mutations
Regulatory myosin light chain	MYL2	12q23–24.3	<5%	Missense and truncation mutations
Cardiac α-actin	ACTC	15q11	<5%	Missense mutations
Titin	TTN	2q24.1	<5%	Missense mutation
α-Myosin heavy chain	MYH6	14q1	Rare	Missense and rearrangement mutations
Cardiac troponin C	TNNC1	3p21.3–3p14.3	Rare	Missense mutation

myosin molecule. However, missense, deletion and insertion/deletion mutations in the rod and tail regions have also been described (Tesson et al. 1998; Nakajima-Taniguchi et al. 1995; Marian et al. 1992; Cuda et al. 1996). The frequency of each *MYH7* mutation is relatively low and a founder effect is uncommon.

Mutations in *MYBPC3* and *TNNT2* account for approximately 20% and 15% of all HCM cases, respectively (Niimura et al. 1998; Erdmann et al. 2001; Marian, 2001b; Seidman and Seidman 1998). More than 40 different mutations in the *MYBPC3* have been identified and again the frequency of each mutation is relatively low (Bonne et al. 1995; Niimura et al. 1998; Carrier et al. 1997; Watkins et al. 1995b; Erdmann et al. 2001). However, unlike mutations in *MYH7*, which are mostly missense mutations, the majority of mutations in *MYBPC3* are deletion/insertion or splice junction mutations (Erdmann et al. 2001). Deletion/insertion mutations are expected to result in a frame shift or truncation of the MyBP-C protein, leading either to severe structural and functional defects in the protein or immediate degradation of the expressed mutant proteins.

Mutations in *TNNT2* are also a relatively common cause of HCM and over 20 mutations have been identified (Marian and Roberts 2001). The vast majority of the mutations are missense mutations and codon 92 is considered a hot spot for mutations (Thierfelder et al. 1994; Forissier et al. 1996). Deletion mutations involving splice donor sites have been described that could lead to truncated proteins (Thierfelder et al. 1994). Collectively, mutations in *MYH7*, *MYBPC3*, and *TNNT2* account for approximately two-thirds of all known HCM mutations. The remainder can be explained by mutations in α-tropomyosin (*TPM1*) (Thierfelder et al. 1994; Coviello et al. 1997; Karibe et al. 2001; Watkins et al. 1995a), cardiac troponin I (*TNNI3*) (Kimura et al. 1997; Kokado et al. 2000), cardiac troponin C (*TNNC1*) (Hoffmann et al. 2001), titin (*TTN*) (Satoh et al. 1999), cardiac α-actin (*ACTC*) (Mogensen et al. 1999; Olson et al. 2000) and essential and regulatory light chains (*MYL3* and *MYL2*, respectively) (Poetter et al. 1996; Flavigny et al. 1998)

2.1.3
Impact of Causal Mutations on Cardiac Hypertrophy

Collective data indicate that the causal mutations exhibit highly variable clinical, electrocardiographic and echocardiographic manifestations and no particular phenotype is mutation-specific (Marian 2001b). Nonetheless, genotype–phenotype correlation studies suggest that different mutations affect the magnitude of cardiac hypertrophy, an important determinant of risk of SCD, to different degrees (Spirito et al. 2000). A theme emerges from the comparison of the phenotypes of the three most common genes for HCM. In general, mutations in *MYH7* are associated with an earlier onset, more extensive hypertrophy and a high incidence of SCD (Charron et al. 1998b) . In contrast, *MYBPC3* mutations are generally associated with a late onset and relatively mild hypertrophy and a low incidence of SCD (Niimura et al. 1998; Charron et al. 1998b; Erdmann et al. 2001). Mild cardiac hypertrophy, a high incidence of SCD and more extensive disarray characterize the phenotype of HCM caused by mutations in *TNNT2* (Varnava et al. 2001b; Watkins et al. 1995c).

These are, however, generalizations. Despite the overall benign nature of mutations in the myosin-binding protein C (MyBP-C) protein, significant variability exists and malignant mutations in the *MYBPC3* gene have been described (Erdmann et al. 2001). This is also the case for other causal genes for HCM. Mutations in the α-tropomyosin gene, which are generally associated with a benign phenotype and mild left ventricular hypertrophy, also have been associated with a high incidence of SCD (Karibe et al. 2001). Mutations in essential and regulatory myosin light chains have been associated with mid-cavity obstruction in HCM and skeletal myopathy in some patients (Poetter et al. 1996) but not in others (Flavigny et al. 1998). Mutations in titin (Satoh et al. 1999) and α-actin (Mogensen et al. 1999; Olson et al. 2000) are uncommon and have been observed in a small number of families and so here it is difficult to comment on any particular pattern of effect.

A fundamental tenet of all genetic disorders, including HCM, regardless of the causal genes and mutations and regardless of the degree of variability of clinical phenotypes, is the age dependence of penetrance. Causal mutations are present from the formation of the single-cell embryo; however, the clinical phenotype is often absent until the third or fourth decade of life. The reasons for age-dependent penetrance of causal mutations in HCM remain unknown. The clinical implication of the age dependence of penetrance is that a normal physical examination and clinical testing at an early age do not effectively exclude the presence of the disease-causing mutation. This is particularly the case for HCM caused by mutations in MyBP-C protein, since the phenotype often develops in the fifth or sixth decade of life (Maron et al. 2001).

2.1.4
Modifier Genes

The phenotypic expression of HCM, whether it is the magnitude of hypertrophy or the risk of SCD, varies significantly among affected individuals, including family members who share identical mutations. Such variability indicates that factors other than the causal mutation contribute to the phenotype. These factors include the genetic background of the individual on which the mutation is expressed and nongenetic factors.

The genetic background will include modifier genes that are not in themselves causative but influence the development of the phenotype. The identity of modifier genes for HCM and the magnitude of their effects remain largely unknown. Given the complexity of molecular genetics and biology of cardiac hypertrophy, a large number of genes and their functional variants are expected to be involved, each exerting only a modest effect. The ensuing phenotype is the result of complex genotype–genotype and genotype–environment interactions. Table 2 lists several SNPs and genes that have been implicated as modifiers in HCM.

Table 2 Selected SNPs associated with cardiac hypertrophy

Gene	Symbol	Locus	Polymorphism	Results
Angiotensin-I converting enzyme-1	ACE	17q23	I/D	DD is associated with higher risk of SCD in HCM (Marian et al. 1993) DD is associated with severity of hypertrophy in HCM (Lechin et al. 1995; Tesson et al. 1997) DD is more common in HCM patients (Pfeufer et al. 1996; Yoneya et al. 1995) Frequency of DD genotype unchanged in HCM (Yamada et al. 1997) No association with indices of hypertrophy in HCM (Osterop et al. 1998; Yamada et al. 1997) No association between I/D genotypes and cardiac mass in Framingham Heart Study subjects (Lindpaintner et al. 1996) DD is associated with left ventricular mass in athletes (Diet et al. 2001; Montgomery et al. 1997)
Angiotensinogen	AGT	1q42	−6G/A	No association with indices of hypertrophy in HCM (Brugada et al. 1997; Yamada et al. 1997)
			T174M	235T allele more common in HCM (Ishanov et al. 1997)
			M235T	Frequency of T174M and M235T unchanged in HCM (Yamada et al. 1997) 235T allele is associated with left ventricular hypertrophy in endurance athletes (Karjalainen et al. 1999)
Angiotensin-II receptor 1	AGTR1	3q21-q25	1166A/C	No association with indices of hypertrophy I HCM (Brugada et al. 1997) C allele is associated is associated with severity of hypertrophy in HCM (Brugada et al. 1997)
Chymase	CMA1	14q11.1	1625A/G	No changes in frequency in HCM (Brugada et al 1997)
Bradykinin B2 receptor	BDKRB2	14q32.1-q32.2	−412C/G T21M	T21M was found in HCM cases but not in controls (Erdmann et al. 1998)
Aldosterone synthase	CYP11B2	8q21-q22	−344T/C	No association with indices of hypertrophy (Patel et al. 2000)
Endothelin-1	EDN1	6p24.1	8002G/A	A allele is associated with severity of hypertrophy (Brugada et al. 1997)
G protein β3 subunit	GNB3	12p13	825C/T	825T is associated with left ventricular mass in hypertensives (Semplicini et al. 2001; Poch et al. 2000)
Tumor necrosis factor α	TNF	6p21.3	−308G/A	A allele is associated with severity of hypertrophy in HCM (Patel et al. 2000)
Insulin-like growth factor 2	IGF2	11p15.5	820G/A	No association with indices of hypertrophy in HCM (Patel et al. 2000)

Table 2 (continued)

Gene	Symbol	Locus	Polymorphism	Results
Transforming growth factor β1	TGFB1	19p13.2	−509C/T	No association with indices of hypertrophy in HCM (Patel et al. 2000)
Interleukin-6	IL6	7p21	−174G/C	No association with indices of hypertrophy in HCM (Patel et al. 2000)
Peroxisome proliferator-activated receptor α	PPARA	22q13.31	Intron 7 G/C	C allele is associated with left ventricular mass in hypertensives (Jamshidi et al. 2002)
Platelet activating factor acetylhydrolase	PLA2G7	6p21.2-p12	994G/T (V279T)	T allele is more common in HCM and is associated with increased left ventricular dimension and decreased function (Yamada et al. 2001)

2.1.5
Impact of Modifier Genes on Cardiac Hypertrophy

The ACE-1 gene was the first gene implicated as a potential modifier of human HCM (Marian et al. 1993). ACE-1 is a transmembrane-ectopeptidase that catalyzes the conversion of angiotensin-I to angiotensin-II and inactivates bradykinin. Angiotensin-II and bradykinin are potent agents with opposing effects on cardiac growth and cellular hyperplasia (Yamazaki et al. 1999). ACE-1 is up-regulated in pressure overload-induced cardiac hypertrophy and in heart failure (Schunkert et al. 1993). Furthermore, inhibition of ACE-1 induces regression of cardiac hypertrophy independent of load and prevents dilatation and remodeling of the ventricle after myocardial infarction (Mathew et al. 2001). Plasma levels of ACE-1 are under tight genetic control and vary significantly among individuals. ACE-1 gene has over 28 different polymorphisms including an insertion (I)/deletion (D) polymorphism, due to the presence or absence of a 287-base pair Alu repeat in intron 16. The I/D polymorphism has been associated with variation in plasma, cellular and tissue levels of ACE-1 and has been studied extensively (Rigat et al. 1990). The influence of the I/D genotype on plasma ACE-1 levels is co-dominant, so that subjects with the *DD* genotype have the highest, those with the *ID* an intermediary, and those with the *II* genotype the lowest plasma and tissue levels (Rigat et al. 1990).

Several studies have explored a potential modifier role for the ACE-1 I/D polymorphism on expression of cardiac phenotypes, particularly cardiac hypertrophy, in HCM (Marian et al. 1993; Lechin et al. 1995; Tesson et al. 1997; Pfeufer et al. 1996; Yoneya et al. 1995; Yamada et al. 1997; Osterop et al. 1998). The initial report suggested an association between the ACE-1 I/D genotypes and the risk of SCD (Marian et al. 1993). The *DD* genotype was found to be more common in HCM families with a high incidence of SCD, compared to those with a low incidence of this outcome (Marian et al. 1993). Subsequent studies showed an association between the I/D genotype and the severity of cardiac hypertrophy (Tesson et al. 1997; Lechin et al. 1995). Indices of cardiac hypertrophy, such as interventricular septal thickness, left ventricular mass index (indexed to body surface area) and a semi-quantitative index of left ventricular hypertrophy, referred to as the Wigle score, were greater in HCM patients with the *DD* genotype (Tesson et al. 1997; Lechin et al. 1995). The observed association followed a gradient consistent with the biological effect of the I/D variants on plasma and tissue levels of ACE (*DD>ID>II*) (Lechin et al. 1995). However, the overall impact of ACE-1 I/D genotype on expression of cardiac hypertrophy was relatively small, accounting for 3%–5% of the variability. The impact of the I/D genotype on expression of cardiac hypertrophy was greater in affected members of a single family, accounting for approximately 10%–15% of the variability (Lechin et al. 1995). These results have been corroborated in several additional studies (Tesson et al. 1997; Osterop et al. 1998; Yamada et al. 1997). In addition, an interaction between the modifying effect of the ACE-1 I/D genotype and the underlying causal mutation has been reported, suggesting the presence of a significant modifying effect in those with the *R403Q* mutation in the β-MyHC protein but not in others (Tesson et al. 1997).

Variants of several other genes have been implicated as having modifying effects on cardiac phenotypes in HCM (Table 2). For example, variants of endothelin-1 and tumor necrosis factor-α have been associated with severity of cardiac hypertrophy (Brugada et al. 1997; Patel et al. 2000). However, the results of association studies have been inconsistent, perhaps because of the small sample size of studies, differences in population characteristics and the presence of confounders that are frequently encountered in SNP-association studies (discussed in Sect. 7.4.2).

2.1.6
Gene Expression in HCM

The heart is a complex organ with a number of cellular components involved in the hypertrophic process. Accordingly, the cardiac hypertrophic response entails changes in expression of a large number of genes. Expression profiling studies have identified changes in the abundance of a variety of genes including those encoding contractile sarcomeric proteins, cytoskeletal proteins, ion channels, intracellular signaling transducers, proteins maintaining the redox state of the myocardium, as well as genes involved in transcriptional and translation machinery (Lim et al. 2001a; Hwang et al. 2002). The best-known and perhaps the most common up-regulated genes in HCM are the markers of secondary cardiac hypertrophy such as skeletal α-actin, isoforms of myosin light chain, and brain natriuretic factor, which are activated in pressure overload-induced (secondary) cardiac hypertrophy. Significant up-regulation of expression of heat shock 70-kD protein 8 (*HSPA8*), also known as *HSP73*, is also notable. This gene is a member of *HSP70* multigene family that encode for proteins with chaperoning roles for nascent polypeptides, facilitating their correct folding, translocation, and degradation (Tavaria et al. 1995). The expression of several other genes encoding proteins with unknown functions such as sarcosin (Taylor et al. 1998) and slimmer (Brown et al. 1999), are also up-regulated in HCM. Up-regulation of *NDUFB10*, which encodes for the first enzyme complex in the electron transport chain of mitochondria (Loeffen et al. 1998), and *HSPA8*, together with increased levels of many ribosomal proteins is consistent with the increase in protein synthesis and mitochondria function in hypertrophic states and signifies their role as potential modulators of cardiac phenotype in HCM. The diversity of molecular phenotype in HCM is in accord with the diversity of pathological and clinical phenotypes that encompass not only myocyte hypertrophy and disarray, but also interstitial fibrosis, thickening of the media of intramural coronary arteries, and arrhythmias. Increased expression of the markers of secondary cardiac hypertrophy supports the hypothesis that hypertrophy in HCM is a secondary phenotype and common pathways are involved in induction of cardiac hypertrophy in genetic and nongenetic forms.

2.1.7
Pathogenesis of HCM

The broad view of the pathogenesis of HCM is that the genetic mutation leads to a dysfunctional protein, which impairs myocyte contractile function. The hypertrophy, disarray and fibrosis are secondary changes, arising out of an attempt by the heart to overcome the contractile impairment and reduce myocyte stress.

The majority of causal mutations in HCM are missense mutations coding mutant proteins that incorporate and assemble into myofibrils and sarcomeres, albeit sometimes inefficiently. Once incorporated into myofibrils, mutant sarcomeric proteins induce a diverse array of functional defects including altered Ca^{+2} sensitivity of myofibrils, reduced ATPase activity and on occasions, sarcomere dysgenesis (Marian and Roberts 2000). In addition, impaired calcium trafficking in myofibrils has been implicated as a major initial defect (Fatkin et al. 2000). Identification of deletion or truncation mutations that abolish the stop codon and/or the polyadenylation signal or encode truncated proteins that are likely to degrade immediately after translation (Marian et al. 1992; Rottbauer et al. 1997) have raised the possibility of haplo-insufficiency. Such mutations could function as so-called null-alleles altering the stoichiometry of the sarcomeric proteins.

These initial defects would each be expected to impair cardiac myocyte function, resulting in increased mechanical and biochemical stress on these cells. The final common pathway would include the activation of stress-responsive intracellular signaling kinases, which activate the transcription machinery leading to cardiac hypertrophy, interstitial fibrosis and other histological and clinical phenotypes of HCM (Marian 2000a).

The greater hypertrophy of the left ventricle (high-pressure chamber) and its absence in the right ventricle (low-pressure chamber), despite equal expression of mutant MyHC protein in both ventricles, supports the idea that the hypertrophy is compensatory. Furthermore, functional abnormalities occur early and prior to structural changes and cardiac hypertrophy. Accordingly, tissue Doppler velocities of myocardial contraction and relaxation are reduced in human subjects with HCM, causing mutations in the absence of discernible cardiac hypertrophy (Nagueh et al. 2001). Similarly, myocardial contraction and relaxation velocities are reduced in β-MyHC-Q403 transgenic rabbits prior to the development of cardiac hypertrophy or interstitial fibrosis (Nagueh et al. 2000). Studies in adult cardiac myocytes also show impairment of function prior to the development of discernible sarcomere or myofibrillar disarray (Marian et al. 1997; Rust et al. 1999). Skeletal myotubes and muscle fibers, isolated from the skeletal muscles of patients with HCM, show reduced force generation in the absence of structural abnormality (Lankford et al. 1995; Malinchik et al. 1997). Myocytes isolated from the hearts of transgenic mice expressing a mutant α-MyHC protein show impaired mechanical performance (Kim et al. 1999). Collectively these results suggest the functional impairment precedes the structural changes and development of cardiac hypertrophy in HCM. This is an important concept since potentially these secondary phenotypes are reversible.

2.1.8
Reversal and Attenuation of Cardiac Phenotypes in HCM

It must be noted that the current pharmacological interventions in human patients are empirical and none have been shown to induce regression of cardiac hypertrophy, fibrosis or disarray, major predictors of mortality and morbidity in HCM (Shirani et al. 2000; Spirito et al. 2000). Understanding the molecular pathogenesis of HCM and the development of several genetically engineered animal models have provided the opportunity to test the effects of pharmacological interventions targeted to specific pathways involved in the pathogenesis of HCM. Current technology, however, does not permit correction of the causal mutations; therefore, the emphasis of pharmacological interventions has been on blockade of the intermediary phenotypes such as signaling kinases and the products of the modifier genes.

Two recent studies have shown significant attenuation of cardiac phenotypes in transgenic animal models expressing mutant sarcomeric proteins known to cause HCM in humans (Lim et al. 2001b; Patel et al. 2001). The first study examined blockade of angiotensin-II receptor 1 in cardiac troponin T-Q92 transgenic mice that exhibit myocyte disarray and interstitial fibrosis; the result was a reduction in interstitial collagen volume by approximately 50% to normal levels (Lim et al. 2001b). Expression levels of collagen $\alpha 1$ (I) mRNA and TGF-$\beta 1$ protein, a known mediator of pro-fibrotic effects of angiotensin II, were also reduced significantly (Lim et al. 2001b). Normalization of interstitial collagen content, a major predictor of SCD in HCM (Shirani et al. 2000), through blockade of angiotensin-II supports a modifying role of the renal-angiotensin system (RAAS) in HCM. The results also raise the possible use of inhibitors of the RAAS in treatment of patients with HCM, a proposal that merits testing in human patients.

Recent data suggest HMG-CoA reductase inhibitors could block signaling kinases involved in the pathogenesis of cardiac hypertrophy and thus become attractive agents for the treatment of cardiac hypertrophy and fibrosis in the pathological states (Oi et al. 1999; Park and Galper 1999; Su et al. 2000). The effects of simvastatin, a pleiotropic HMG-CoA reductase inhibitor, have been tested on cardiac structure and function in β-MyHC-Q403 transgenic rabbits. These transgenic rabbits exhibit significant cardiac hypertrophy, increased interstitial fibrosis and cardiac myocyte disarray and fully recapitulate the phenotype of human HCM (Marian et al. 1999). A 37% reduction in left ventricular mass, a 20% reduction in septal and posterior wall thickness and ~ 50% in collagen volume fraction were recorded. In addition, indices of left ventricular filling pressure were improved significantly.

There has been significant controversy regarding the utility of calcineurin inhibitors in treatment and prevention of cardiac hypertrophy in a variety of pathological conditions. Pre-treatment with diltiazem, an L-type Ca^{+2} channel blocker, prevented the exaggerated cardiac hypertrophic response to inhibitors of calcineurin. The results suggest altered calcium handling in the hearts of α-MyHC-Q403$^{+/-}$ mice (Fatkin et al. 2000).

2.2
Cardiac Hypertrophy in Trinucleotide Repeat Syndromes

Trinucleotide repeat syndromes are a group of genetic disorders caused by expansion of naturally occurring GC-rich triplet repeats in genes (Table 3). The group comprises more than ten different diseases, including myotonic muscular dystrophy (DM), several neurodegenerative disorders such as Huntington's disease, and fragile site syndromes (Cummings and Zoghbi 2000). In several triplet repeat syndromes, such as DM, cardiac involvement is common and is a major

Table 3 Genetic causes of cardiac hypertrophy in triplet repeats syndromes and Noonan syndrome

Gene	Symbol	Locus	Frequency	Mutations
Frataxin (Friedreich ataxia)	FRDA	9q13	Rare	Expansion of GAA repeats in intron 1
Myotonin protein kinase (Myotonic dystrophy)	DMPK	19q13	Uncommon	Expansion of CTG repeats in 3'-UTR
	DMWD	19q13	Uncommon	Probably loss of function mutations
Protein tyrosine phosphatase, non-receptor type 11	PTPN11	12q24	Rare	Noonan or Leopard syndrome Most missense mutations in N-SH2 and PTP domains

determinant of morbidity and mortality (Marian 2000b; Cummings and Zoghbi 2000). The phenotype includes cardiac hypertrophy as a primary phenotype, i.e., in the absence of an increased external load, but dilated cardiomyopathy may also occur.

DM is an autosomal dominant disorder and the most common form of muscular dystrophy in adults (approximately 1:8,000 in the North American population). It commonly manifests as progressive degeneration of muscles and myotonia, cardiomyopathy, conduction defects, male pattern baldness, infertility, premature cataracts, mental retardation and endocrine abnormalities (Korade-Mirnics et al. 1998). Cardiac involvement is common, usually apparent as cardiac conduction abnormalities and less frequently as cardiomyopathy (Phillips and Harper 1997). Mutations in two genes have been associated with DM. Expansion of GCT trinucleotide repeats in the 3' untranslated region of dystrophia myotonica protein kinase (*DMPK*) is one recognized association. The number of CTG repeats in normal individuals varies between 5 and 37 repeats. It expands from 50 to more than several thousands in patients with DM (Korade-Mirnics et al. 1998). Expansion of the repeats could interfere with DMPK transcription, RNA processing, and/or translation resulting in decreased levels of DMPK protein. The length of CTG repeats often correlates with severity of clinical (including cardiac) phenotype. The second gene implicated in DM is gene 59 or *DMWD*, which is located immediately upstream of *DMPK* (Cummings and Zoghbi 2000). The mechanisms by which mutations in *DMWD* could cause DM remain unknown but loss of function has been implicated (Cummings and Zoghbi 2000).

Friedreich ataxia (FRDA) is an autosomal recessive neurodegenerative disease caused by expansion of GAA repeat sequences in the intron of *FRDA* (Palau 2001). FRDA is a multi-system disorder that primarily involves central and peripheral nervous system and less frequently manifests as cardiomyopathy and occasionally as diabetes mellitus. The encoded protein is frataxin, which is a soluble mitochondrial protein with 210 amino acids. (Palau 2001). Cardiac involvement could manifest as either dilated or hypertrophic cardiomyopathy. The severity of clinical manifestations of Friedreich ataxia also correlates with the size of the repeats (Bit-Avragim et al. 2001).

2.3
Cardiac Hypertrophy in Noonan and Leopard Syndromes

Noonan syndrome is an uncommon autosomal dominant disorder characterized by dysmorphic facial features, hypertrophic cardiomyopathy, pulmonic stenosis, mental retardation and bleeding disorders. Leopard syndrome (*l*entigines, *e*lectrocardiographic conduction abnormalities, *o*cular hypertelorism, *p*ulmonic stenosis, *a*bnormal genitalia, *r*etardation of growth, and *d*eafness) is an allelic variant of the Noonan syndrome. Pulmonic stenosis and HCM are the primary cardiac phenotypes and endocardial and myocardial fibroelastosis have also been reported. The responsible gene for Noonan and Leopard syndromes, in approximately half of the cases, is protein-tyrosine phosphatase, the non-receptor type 11 (*PTPN11*) gene (Tartaglia et al. 2001, 2002), located on chromosome *12q24.1*. Mutations in *PTPN11* are missense mutations located in interacting portions of the amino-terminal src-homology 2 (N-SH2) and protein tyrosine phosphatase (PTP) domains (Tartaglia et al. 2001, 2002). The proposed mechanism for the pathogenesis of Noonan syndrome is a gain-of-function effect on the phosphotyrosine phosphatase domains. The other genes responsible for Noonan syndrome remain unknown.

2.4
Cardiac Hypertrophy in Genetic Metabolic Disorders

A phenotype grossly similar to that of HCM has been observed in a variety of metabolic diseases (Table 4). Metabolic diseases in which cardiac involvement is not a primary phenotype will be mentioned and those with predominance of hypertrophic response will be briefly discussed.

Refsum disease is an autosomal recessive disorder characterized clinically by a tetrad of retinitis pigmentosa, peripheral neuropathy, cerebellar ataxia, and elevated protein levels in the cerebrospinal fluid. Electrocardiographic abnormalities are common but cardiac hypertrophy and heart failure are uncommon. Refsum disease is caused by mutations in the gene encoding phytanoyl-CoA hydroxylase (*PAHX* or *PHYH*) (Mihalik et al. 1997), which leads to enzymatically inactive protein (Jansen et al. 1997). Consequently, phytanic acid, an unusual branched-chain fatty acid, accumulates in tissues and body fluids.

Glycogen storage disease type II (Pompe disease) is an autosomal recessive disorder caused by deficiency of alpha-1,4-glucosidase (acid maltase) in the liver and muscles, leading to storage of glycogen in lysosomal membranes. Its clinical manifestations are remarkable for hypertrophic cardiomyopathy with heart failure, conduction defects and muscular hypotonia (Raben et al. 2002). The causal gene encodes acid maltase, which when mutated leads to a deficiency of acid al-

Table 4 Genetic causes of selected mitochondrial and metabolic cardiac hypertrophy

Gene	Symbol	Locus	Frequency	Mutations
Phytanoyl-Co A hydroxylase	PAHX (PHYH)	10pter-p11.2	Rare	Refsum disease
α-1,4 Galactosidase	GLA	Xq22	Rare	Mutations lead to deficiency of acid α-galactosidase
AMP-activated protein kinase, γ2 regulatory subunit	PRKAG2	7q35-q36	Uncommon	Missense and insertion mutations
Acyl-CoA Dehydrogenase	ACADL	2q34-q45	Rare	Fatty acid beta oxidation
Mitochondrial DNA	MTTI	Mitochondrial	Rare	tRNA Isoleucine and tRNA glycine
Mitochondrial DNA	MTTI	Mitochondrial	Rare	Kearns-Sayre Syndrome

pha-glucosidase. High-protein diet and recombinant acid alpha-glucosidase have been used effectively for treatment of this disorder (Amalfitano et al. 1999).

A form of glycogen-storage disease is caused by mutations in the *PRKAG2* gene, which encodes the γ2 regulatory subunit of AMP-activated protein kinase (AMPK). Mutations in *PRKAG2* lead to a phenotype of HCM, conduction defects and Wolff-Parkinson-White syndrome (Gollob et al. 2001; Blair et al. 2001). There is significant variability in expression of the disease as in some families the predominant phenotype is pre-excitation and conduction abnormalities (Gollob et al. 2001), and cardiac hypertrophy is present in a minority of patients (Gollob et al. 2001). In others, early cardiac hypertrophy predominates and pre-excitation may be present (Blair et al. 2001).

Cardiac hypertrophy also has been described in patients with mucopolysaccharidosis, Niemann-Pick disease, Gaucher disease, hereditary hemochromatosis and CD36 deficiency. A comprehensive review of metabolic cardiomyopathies, including mitochondrial cardiomyopathies can be found in Guertl et al. (2000).

3
Cardiac Hypertrophy in Non-Mendelian Genetic Disorders

Cardiac hypertrophy also has been observed in a variety of genetic disorders with a non-mendelian inheritance, such as mitochondrial myopathies, which exhibit a matrilineal transmission. In addition, SNPs are important determinants of the cardiac hypertrophic response to external stimuli, such as pressure overload, and internal stimuli such as causal mutations.

3.1
Cardiac Hypertrophy in Mitochondrial Disorders

Mitochondrial DNA is a circular double-stranded genome of approximately 16.5 kb, which codes for 13 polypeptides of the respiratory chain complexes I, III, IV, and V subunits, 28 ribosomal RNAs, and 22 tRNAs. Mutations in mitochondrial oxidative phosphorylation pathways often result in a complex phenotype involving multiple organs, including the heart (Simon and Johns 1999). Cardiac involvement can lead to hypertrophy as well as dilatation. Each mitochondrion has multiple copies of its own DNA and each cell contains thousands of mitochondrial DNA. Therefore, mutations result in a significant degree of heteroplasmy, which increases over time as mitochondria multiply. In general, approximately 80%–90% of mitochondrial DNA need to mutate in order to affect mitochondrial function and lead to a clinical phenotype (Williams 2000).

Kearns-Sayre syndrome (KSS) is a mitochondrial disease caused by sporadically occurring mutations in mitochondrial DNA (Ashizawa and Subramony 2001). It is characterized by a triad of progressive external ophthalmoplegia, pigmentary retinopathy and cardiac conduction defects (Ashizawa and Subramony 2001). The classic cardiac abnormality in KSS is conduction defects, however, dilated and hypertrophic cardiomyopathies are also often observed, but a lower frequency.

An example of mitochondrial disease caused by mutations in nuclear DNA is L-carnitine deficiency, which can lead to hypertrophic cardiomyopathy, and more commonly, to heart failure due to dilated cardiomyopathy (Guertl et al. 2000). Carnitine is an important component of fatty acid metabolism and necessary for the entry of long-chain fatty acids into mitochondria. Systemic carnitine deficiency can arise by inadequate dietary intake or defective synthesis, uptake and transport or decreased tubular reabsorption. Mutations in chromosomal genes encoding solute carrier family 22, member 5 (*SLC22A5*) or *OCTN2* transporter impair transport of carnitine to mitochondria and cause systemic carnitine deficiency. Several other enzymes are involved in the transfer and metabolism of carnitine, including carnitine mitochondrial carnitine palmitoyltransferase I (CATI), located in the outer mitochondrial membrane and translocase (*SLC25A20*), located in the inner membrane. They mediate the process of esterification and transfer of long-chain fatty acids into mitochondria. Mutations in *CAT1* and translocase lead to defective carnitine uptake and decreased tissue levels of carnitine. The phenotype is characterized by skeletal myopathy, congestive heart failure, as well as abnormalities of the central nervous system and liver. It rarely causes hypertrophic cardiomyopathy (Guertl et al. 2000). Treatment with high doses of oral carnitine alleviates the symptoms.

Mutations in acyl-CoA dehydrogenase also impair mitochondrial fatty acid oxidation and could lead to hypertrophic cardiomyopathy (Guertl et al. 2000; Kelly and Strauss 1994). The clinical manifestations are remarkable for hypertrophic cardiomyopathy with diminished systolic function, fasting hypoglycemia, inadequate ketotic response to hypoglycemia, hepatic dysfunction, skeletal myopathy and sudden death (Guertl et al. 2000; Kelly and Strauss 1994). The majority of medium-chain acyl-CoA dehydrogenase deficiency is caused by substitution of glutamic acid for lysine in the mutant protein, while the molecular genetic basis of short-chain acyl-CoA dehydrogenase deficiency is more heterogeneous.

3.2
Single-Nucleotide Polymorphisms and Cardiac Hypertrophy

Single-nucleotide polymorphisms (SNPs) are major determinants of interindividual variations in disease susceptibility and clinical phenotypes such as variation in cardiac hypertrophic response. Indeed, the influence of genetic background, i.e., SNPs, on cardiac hypertrophic response has been shown previously through epidemiological and family studies in humans (Adams et al. 1985; Schunkert et al. 1999) and experimental data in animals (Sebkhi et al. 1999; Innes et al. 1998). However, individual SNPs that affect clinical phenotype are largely unknown. Completion of the final sequence of the human genome along with the development of SNP and haplotype maps is expected to accelerate the pace of discoveries of genetic determinants of quantitative traits such as cardiac size and the hypertrophic response to stimuli.

Unlike monogenic disorders such as HCM, whereby application of genetic linkage techniques have led to the successful identification of a large number of causal genes and mutations, conventional linkage techniques have limited utility in mapping susceptibility SNPs for complex traits. The strength of linkage disequilibrium studies, as opposed to conventional linkage analysis, is in their ability to detect modest effects from modifier genes and SNPs on clinical phenotype. Each gene contains multiple SNPs that cooperatively regulate its expression and affect the function of the encoded protein. Multiple potentially functional SNPs and variable linkage disequilibrium across the genome (Dawson et al. 2002) means that large numbers of SNPs and haplotypes have to be typed. For example, the human gene encoding the ACE-1 has a large number of SNPs that collectively affect ACE-1 levels and its function (Zhu et al. 2001). Given that several SNPs in each gene are in linkage disequilibrium, genotyping for a fraction of all of the SNPs of a gene may be sufficient to construct the main haplotypes of a gene (Daly et al. 2001; Reich et al. 2001). Nevertheless, comprehensive analysis of SNPs in each candidate susceptibility gene is often necessary. In addition, given the diversity of factors that contribute to a complex trait, the impact of each SNP on biological and clinical phenotype is modest and decreases further for more distant phenotypes. Furthermore, the effects of SNPs on the phenotype of interest may be additive, synergistic, or subtractive. Epistatic interactions between SNPs or genes, epigenetic regulation and environmental factors can also affect the impact of SNPs on gene expression and protein function. Given these complexities, the results of association studies should be interpreted in the context of population characteristics, the design of the study, sample size, exploratory hypothesis testing, the biological plausibility of the association, the functional significance of the SNPs, the strength of the association and the presence of genetic and biological gradients (Marian 2001c). Because of the inherent weaknesses of association studies with SNPs, results must be considered provisional until confirmed in repeat studies and through experimentation (Lander and Schork 1994).

SNPs in multiple genes have been associated with cardiac size and indices of cardiac hypertrophy in a variety of pathological states. While none have yet been validated and the results have been inconsistent, several SNPs that are considered to have biologically plausible effects on cardiac hypertrophy have been listed in Table 2.

4
Conclusions

Cardiac hypertrophy is a common response of the heart to all forms of stimuli. Advances in molecular genetic studies have elucidated the genetic basis of cardiac hypertrophy arising from genetic defects with simple mendelian inheritance, such as HCM. Similarly, it has also become evident that genetic variations, primarily SNPs, affect cardiac hypertrophic response not only in complex traits and acquired conditions but also in single-gene disorders. Regardless of the underlying causes, cardiac hypertrophy is a complex phenotype, which results from intricate interactions between genetics, epigenetic and nongenetic factors. Understanding the genetic basis of the cardiac hypertrophic response could provide for better diagnosis, risk stratification and treatment of this ubiquitous cardiac phenotype, which is a major predictor of morbidity and mortality.

Acknowledgements. This work is supported in part by a grant from the National Heart, Lung, and Blood Institute, Specialized Centers of Research (P50-HL42267-01) and R01-HL68884.

References

Abchee A, Marian AJ (1997) Prognostic significance of beta-myosin heavy chain mutations is reflective of their hypertrophic expressivity in patients with hypertrophic cardiomyopathy. J Investig Med. 45:191-196

AdamsTD, Yanowitz FG, FisherAG et al (1985) Heritability of cardiac size: an echocardiographic and electrocardiographic study of monozygotic and dizygotic twins. Circulation 71:39-44

Amalfitano A, McVie-Wylie AJ, Hu H. et al (1999) Systemic correction of the muscle disorder glycogen storage disease type II after hepatic targeting of a modified adenovirus vector encoding human acid-alpha-glucosidase. Proc Natl Acad Sci. USA 96:8861-8866

Anan R, Greve G, Thierfelder L et al (1994) Prognostic implications of novel beta cardiac myosin heavy chain gene mutations that cause familial hypertrophic cardiomyopathy. J Clin Invest 93:280-285

Arad M, Seidman JG, Seidman CE (2002) Phenotypic diversity in hypertrophic cardiomyopathy. Hum Mol Genet 11:2499-2506

Ashizawa T,Subramony SH (2001) What is Kearns-Sayre syndrome after all? Arch. Neurol. 58:1053-1054

Bielen E, Fagard R, AmeryA (1990) Inheritance of heart structure and physical exercise capacity: a study of left ventricular structure and exercise capacity in 7-year-old twins. Eur Heart J 11:7-16

Bit-Avragim N, Perrot A., Schols L et al (2001) The GAA repeat expansion in intron 1 of the frataxin gene is related to the severity of cardiac manifestation in patients with Friedreich's ataxia. J Mol Med 78:626-632

Blair E, Redwood C, Ashrafian H et al (2001) Mutations in the gamma(2) subunit of AMP-activated protein kinase cause familial hypertrophic cardiomyopathy: evidence for the central role of energy compromise in disease pathogenesis. Hum Mol Genet 10:1215-1220

Bonne G, Carrier L, Bercovici J et al (1995) Cardiac myosin binding protein-C gene splice acceptor site mutation is associated with familial hypertrophic cardiomyopathy. Nat. Genet. 11:438-440

Brown S, McGrath, MJ, Ooms LM et al (1999) Characterization of two isoforms of the skeletal muscle LIM protein 1, SLIM1. Localization of SLIM1 at focal adhesions and the isoform slimmer in the nucleus of myoblasts and cytoplasm of myotubes suggests distinct roles in the cytoskeleton and in nuclear-cytoplasmic communication. J Biol Chem 274:27083-27091

Brugada R, Kelsey W, Lechin M et al (1997) Role of candidate modifier genes on the phenotypic expression of hypertrophy in patients with hypertrophic cardiomyopathy. J Investig Med 45:542–551

Cannan CR, Reeder GS, Bailey KR et al (1995) Natural history of hypertrophic cardiomyopathy. A population-based study, 1976 through 1990. Circulation 92:2488–2495

Carrier L, Bonne G, Bahrend E et al (1997) Organization and sequence of human cardiac myosin binding protein C gene (MYBPC3) and identification of mutations predicted to produce truncated proteins in familial hypertrophic cardiomyopathy. Circ Res 80:427–434

Charron P, Dubourg O, Desnos M et al (1998a) Clinical features and prognostic implications of familial hypertrophic cardiomyopathy related to the cardiac myosin-binding protein C gene. Circulation 97:2230–2236

Charron P, Dubourg O, Desnos M et al (1998b) Genotype-phenotype correlations in familial hypertrophic cardiomyopathy. A comparison between mutations in the cardiac protein-C and the beta-myosin heavy chain genes. Eur Heart J 19:139–145

Coviello DA, Maron BJ, Spirito P et al (1997) Clinical features of hypertrophic cardiomyopathy caused by mutation of a "hot spot" in the alpha-tropomyosin gene. J Am Coll Cardiol 29:635–640

Cuda G, Perrotti N, Perticone F et al (1996) A previously undescribed de novo insertion-deletion mutation in the beta myosin heavy chain gene in a kindred with familial hypertrophic cardiomyopathy. Heart 76:451–452

Cummings CJ, Zoghbi HY (2000) Trinucleotide repeats: mechanisms and pathophysiology. Annu Rev Genomics Hum Genet 1:281–328

Daly MJ, Rioux, D, Schaffner SF et al (2001) High-resolution haplotype structure in the human genome. Nat Genet 29:229–232

Dausse E, Komajda M, Fetler L et al(1993) Familial hypertrophic cardiomyopathy. Microsatellite haplotyping and identification of a hot spot for mutations in the beta-myosin heavy chain gene. J Clin Invest 92:2807–2813

Dawson E, Abecasis GR, Bumpstead S et al(2002) A first-generation linkage disequilibrium map of human chromosome 22. Nature 418:544–548

Derchi G, Bellone P, Chiarella F et al (1992) Plasma levels of atrial natriuretic peptide in hypertrophic cardiomyopathy. Am J Cardiol 70:1502–1504

Diet F, Graf C, Mahnke N et al (2001) ACE and angiotensinogen gene genotypes and left ventricular mass in athletes. Eur J Clin Invest 31:836–842

Epstein ND, Cohn GM, Cyran F et al (1992) Differences in clinical expression of hypertrophic cardiomyopathy associated with two distinct mutations in the beta-myosin heavy chain gene. A 908Leu-Val mutation and a 403Arg-Gln mutation. Circulation 86:345–352

Erdmann J, Hegemann N, Weidemann A et al (1998) Screening the human bradykinin B2 receptor gene in patients with cardiovascular diseases: identification of a functional mutation in the promoter and a new coding variant (T21 M). Am J Med Genet 80:521–525

ErdmannJ, Raible J, Maki-AbadiJ et al (2001) Spectrum of clinical phenotypes and gene variants in cardiac myosin-binding protein C mutation carriers with hypertrophic cardiomyopathy. J Am Coll Cardiol 38:322–330

Fananapazir L, Epstein ND (1994) Genotype-phenotype correlations in hypertrophic cardiomyopathy. Insights provided by comparisons of kindreds with distinct and identical beta-myosin heavy chain gene mutations. Circulation 89:22–32

Fatkin D, McConnell BK, Mudd JO (2000) An abnormal Ca(2+) response in mutant sarcomere protein-mediated familial hypertrophic cardiomyopathy. J Clin Invest 106:1351–1359

Flavigny J, Richard P, Isnard R et al (1998) Identification of two novel mutations in the ventricular regulatory myosin light chain gene (MYL2) associated with familial and classical forms of hypertrophic cardiomyopathy. J Mol Med 76:208–214

Forissier JF, Carrier L, Farza H et al (1996) Codon 102 of the cardiac troponin T gene is a putative hot spot for mutations in familial hypertrophic cardiomyopathy. Circulation 94:3069–3073

Geisterfer-Lowrance AA, Kass S, Tanigawa G et al (1990) A molecular basis for familial hypertrophic cardiomyopathy: a beta cardiac myosin heavy chain gene missense mutation. Cell 62:999–1006

Gollob MH, Green MS, Tang AS et al (2001) Identification of a gene responsible for familial Wolff-Parkinson-White syndrome. N Engl J Med 344:1823–1831

Guertl B, Noehammer C, Hoefler G (2000) Metabolic cardiomyopathies. Int J Exp Pathol 81:349–372

Harshfield GA, Grim CE, Hwang C et al (1990) Genetic and environmental influences on echocardiographically determined left ventricular mass in black twins. Am J Hypertens 3:538–543

Hasegawa K, Fujiwara H, Doyama K et al (1993) Ventricular expression of brain natriuretic peptide in hypertrophic cardiomyopathy. Circulation 88:372–380

Hasegawa K, Fujiwara H, Koshiji M et al (1996) Endothelin-1 and its receptor in hypertrophic cardiomyopathy. Hypertension 27:259–264

Ho CY, Lever HM, DeSanctis R et al (2000) Homozygous mutation in cardiac troponin T : implications for hypertrophic cardiomyopathy [In Process Citation]. Circulation 102:1950–1955

Hoffmann B, Schmidt-Traub H, Perrot A et al(2001) First mutation in cardiac troponin C, L29Q, in a patient with hypertrophic cardiomyopathy. Hum Mutat 17:524

Hwang JJ, Allen PD, Tseng GC et al (2002) Microarray gene expression profiles in dilated and hypertrophic cardiomyopathic end-stage heart failure. Physiol Genomics 10:31–44

Innes BA, McLaughlin M.G, Kapuscinski MK et al (1998) Independent genetic susceptibility to cardiac hypertrophy in inherited hypertension. Hypertension 31:741–746

Ishanov A, Okamoto H, Yoneya K et al (1997) Angiotensinogen gene polymorphism in Japanese patients with hypertrophic cardiomyopathy. Am Heart J 133:184–189

Jamshidi Y, Montgomery HE, Hense HW et al (2002) Peroxisome proliferator–activated receptor alpha gene regulates left ventricular growth in response to exercise and hypertension. Circulation 105:950–955

Jansen GA, Ofman R, FerdinandusseS et al (1997) Refsum disease is caused by mutations in the phytanoyl-CoA hydroxylase gene. Nat Genet 17:190–193

Jeschke B, Uhl K, Weist B et al (1998) A high risk phenotype of hypertrophic cardiomyopathy associated with a compound genotype of two mutated beta-myosin heavy chain genes. Hum Genet 102:299–304

Karibe A, TobacmanLS, Strand J et al (2001) Hypertrophic cardiomyopathy caused by a novel alpha-tropomyosin mutation (V95A) is associated with mild cardiac phenotype, abnormal calcium binding to troponin, abnormal myosin cycling, and poor prognosis. Circulation 103:65–71

Karjalainen J, Kujala UM, Stolt A et al (1999) Angiotensinogen gene M235T polymorphism predicts left ventricular hypertrophy in endurance athletes. J Am. Coll. Cardiol 34:494–499

KellyDP, StraussAW(1994) Inherited cardiomyopathies. N EnglJ Med 330:913–919

Kim SJ, IizukaK, Kelly RA et al (1999) An alpha-cardiac myosin heavy chain gene mutation impairs contraction and relaxation function of cardiac myocytes. Am J Physiol 276: H1780-H1787

Kimura A, Harada H, ParkJE et al (1997) Mutations in the cardiac troponin I gene associated with hypertrophic cardiomyopathy. Nat Genet 16:379–382

Kokado H, Shimizu M, Yoshio H et al (2000) Clinical features of hypertrophic cardiomyopathy caused by a Lys183 deletion mutation in the cardiac troponin I gene. Circulation 102:663–669

Korade-Mirnics Z, Babitzke P, Hoffman E (1998) Myotonic dystrophy: molecular windows on a complex etiology. Nucleic Acids Res 26:1363–1368.
Lander ES, Schork NJ (1994) Genetic dissection of complex traits. Science 265:2037–2048
Landry F, Bouchard C, Dumesnil J (1985) Cardiac dimension changes with endurance training. Indications of a genotype dependency. JAMA 25477–80
Lankford EB, Epstein ND, Fananapazir L et al (1995) Abnormal contractile properties of muscle fibers expressing beta-myosin heavy chain gene mutations in patients with hypertrophic cardiomyopathy. J Clin Invest 95:1409–1414
Lechin M, Quinones MA, Omran A et al (1995) Angiotensin-I converting enzyme genotypes and left ventricular hypertrophy in patients with hypertrophic cardiomyopathy. Circulation 92:1808–1812
Levy D, Garrison RJ, Savage DD et al (1990) Prognostic implications of echocardiographically determined left ventricular mass in the Framingham Heart Study. N Engl J Med 322:1561–1566
Li RK, Li G, Mickle DA et al (1997) Overexpression of transforming growth factor-beta1 and insulin-like growth factor-I in patients with idiopathic hypertrophic cardiomyopathy. Circulation 96:874–881
Lim DS, Lutucuta S, Bachireddy P et al (2001b) Angiotensin II blockade reverses myocardial fibrosis in a transgenic mouse model of human hypertrophic cardiomyopathy. Circulation 1–3:789–791
Lim DS, Roberts R, Marian AJ (2001a) Expression profiling of cardiac genes in human hypertrophic cardiomyopathy: insight into the pathogenesis of phenotypes. J Am. Coll. Cardiol 38:1175–1180
Lindpaintner K, Lee M, Larson MG et al (1996) Absence of association or genetic linkage between the angiotensin-converting-enzyme gene and left ventricular mass. N. Engl J Med 334:1023–1028
Loeffen JL, Triepels RH, van den Heuvel LP (1998). cDNA of eight nuclear encoded subunits of NADH:ubiquinone oxidoreductase: human complex I cDNA characterization completed. Biochem Biophys Res Commun 253:415–422
Malinchik S, Cuda G, Podolsky RJ et al (1997) Isometric tension and mutant myosin heavy chain content in single skeletal myofibers from hypertrophic cardiomyopathy patients. J Mol Cell Cardiol 29667–676
Marian AJ (2000b) Genetics for Cardiologists. London: REMEDICA Publishing, pp 1–53
Marian AJ (2000a) Pathogenesis of diverse clinical and pathological phenotypes in hypertrophic cardiomyopathy. Lancet 355:58–60
Marian AJ (2001a) Modifier genes for hypertrophic cardiomyopathy. Curr Opin Cardiol
Marian AJ (2001b) On genetic and phenotypic variability of hypertrophic cardiomyopathy: nature versus nurture. J Am Coll Cardiol 38:331–334
Marian AJ (2001c) On genetics, inflammation, and abdominal aortic aneurysm: can single nucleotide polymorphisms predict the outcome? Circulation 103:2222–2224
Marian AJ, Boerwinkle E (2002) "Into thin air" and the genetics of complex traits. Circulation 106:768–769
Marian AJ, Mares A, Jr, Kelly DP (1995) Sudden cardiac death in hypertrophic cardiomyopathy. Variability in phenotypic expression of beta-myosin heavy chain mutations. Eur Heart J 16:368–376
Marian AJ, Roberts R (2000) Molecular pathogenesis of cardiomyopathies. In: Sperelakis N, Kurachi Y, Terzic A, Cohen M (eds) Heart Physiology and Pathophysiology. Academic Press, San Diego, pp 1045–1063
Marian AJ, Roberts R (2001) The molecular genetic basis for hypertrophic cardiomyopathy. J Mol Cell Cardiol 33:655–670
Marian AJ, Wu Y, Lim DS et al (1999) A transgenic rabbit model for human hypertrophic cardiomyopathy. J Clin Invest 104:1683–1692

Marian AJ, Yu QT, Mares A, Jr et al (1992) Detection of a new mutation in the beta-myosin heavy chain gene in an individual with hypertrophic cardiomyopathy. J Clin Invest 90:2156–2165

Marian AJ, Yu QT, Workman R et al (1993) Angiotensin-converting enzyme polymorphism in hypertrophic cardiomyopathy and sudden cardiac death. Lancet 342:1085–1086

Marian AJ, Zhao G, Seta Y et al (1997) Expression of a mutant (Arg92Gln) human cardiac troponin T, known to cause hypertrophic cardiomyopathy, impairs adult cardiac myocyte contractility. Circ Res 81:76–85

Maron BJ, Anan TJ, Roberts WC (1981) Quantitative analysis of the distribution of cardiac muscle cell disorganization in the left ventricular wall of patients with hypertrophic cardiomyopathy. Circulation 63:882–894

Maron BJ, Gardin JM, Flack JM et al (1995) Prevalence of hypertrophic cardiomyopathy in a general population of young adults. Echocardiographic analysis of 4111 subjects in the CARDIA Study. Coronary Artery Risk Development in (Young) Adults. Circulation 92:785–789

Maron BJ, Niimura H, Casey SA et al (2001) Development of left ventricular hypertrophy in adults in hypertrophic cardiomyopathy caused by cardiac myosin-binding protein C gene mutations. J Am Coll Cardiol 38:315–321

Maron BJ, Olivotto I, Spirito P et al (2000) Epidemiology of hypertrophic cardiomyopathy-related death: revisited in a large non-referral-based patient population. Circulation 102:858–864

Maron BJ, RobertsWC (1979) Quantitative analysis of cardiac muscle cell disorganization in the ventricular septum of patients with hypertrophic cardiomyopathy. Circulation 59:689–706

Maron BJ, Shirani J, Poliac LC et al (1996) Sudden death in young competitive athletes. Clinical, demographic, and pathological profiles. JAMA 276:199–204

Mathew J, Sleight P, Lonn E (2001) Reduction of cardiovascular risk by regression of electrocardiographic markers of left ventricular hypertrophy by the angiotensin-converting enzyme inhibitor ramipril. Circulation 104:1615–1621

McKenna W, Deanfield J, Faruqui (1981) Prognosis in hypertrophic cardiomyopathy: role of age and clinical, electrocardiographic and hemodynamic features. Am Cardiol 47:538

Mihalik SJ, Morrell JC, Kim D et al (1997) Identification of PAHX, a Refsum disease gene. Nat Genet 17185–189

Mogensen J, Klausen IC, Pedersen AK (1999) Alpha-cardiac actin is a novel disease gene in familial hypertrophic cardiomyopathy. J Clin Invest 103:R39-R43

Montgomery HE, Clarkson P, Dollery CM (1997) Association of angiotensin-converting enzyme gene I/D polymorphism with change in left ventricular mass in response to physical training. Circulation 96:741–747

Nagueh SF, Bachinski L, Meyer D et al (2001) Tissue Doppler imaging consistently detects myocardial abnormalities in patients with familial hypertrophic cardiomyopathy and provides a novel means for an early diagnosis prior to an independent of hypertrophy. Circulation 104:128–130

Nagueh SF, Kopelen HA, Lim DS (2000) Tissue Doppler imaging consistently detects myocardial contraction and relaxation abnormalities, irrespective of cardiac hypertrophy, in a transgenic rabbit model of human hypertrophic cardiomyopathy. Circulation 102:1346–1350

Nakajima-Taniguchi C, Matsui H, Eguchi N (1995) A novel deletion mutation in the beta-myosin heavy chain gene found in Japanese patients with hypertrophic cardiomyopathy. J Mol Cell Cardiol 27:2607–2612

Niimura H, Bachinski LL, Sangwatanaroj S et al (1998) Mutations in the gene for cardiac myosin-binding protein C and late-onset familial hypertrophic cardiomyopathy. N. Engl. J Med 338:1248–1257

Oi S, Haneda T, Osaki J et al (1999) Lovastatin prevents angiotensin II-induced cardiac hypertrophy in cultured neonatal rat heart cells. Eur J Pharmacol 376:139–148

Olson TM, DoanTP, Kishimoto NY et al (2000) Inherited and de novo mutations in the cardiac actin gene cause hypertrophic cardiomyopathy. J Mol Cell Cardiol 32:1687–1694

Osterop AP, Kofflard MJ, Sandkuijl LA et al (1998) AT1 receptor A/C1166 polymorphism contributes to cardiac hypertrophy in subjects with hypertrophic cardiomyopathy. Hypertension 32:825–830

Palau F (2001) Friedreich's ataxia and frataxin: molecular genetics, evolution and pathogenesis. Int J Mol Med 7:581–589

Park HJ, Galper JB (1999) 3-Hydroxy-3-methylglutaryl CoA reductase inhibitors up-regulate transforming growth factor-beta signaling in cultured heart cells via inhibition of geranylgeranylation of RhoA GTPase. Proc Natl Acad Sci USA 96:11525–11530

Patel R, Lim DS, Reddy D et al (2000) Variants of trophic factors and expression of cardiac hypertrophy in patients with hypertrophic cardiomyopathy. J Mol Cell Cardiol 32:2369–2377

Patel R, Nagueh SF, Tsybouleva N et al (2001) Simvastatin induces regression of cardiac hypertrophy and fibrosis and improves cardiac function in a transgenic rabbit model of human hypertrophic cardiomyopathy. Circulation 104:317–324

Pfeufer A, Osterziel KJ, Urata H et al (1996) Angiotensin-converting enzyme and heart chymase gene polymorphisms in hypertrophic cardiomyopathy. Am J Cardiol 78:362–364

Phillips MF, Harper PS (1997) Cardiac disease in myotonic dystrophy. Cardiovasc Res 33:13–22

Poch E, Gonzalez D, Gomez-Angelats E et al (2000) G-Protein beta(3) subunit gene variant and left ventricular hypertrophy in essential hypertension. Hypertension 35:214–218

Poetter K, Jiang H, Hassanzadeh S et al (1996) Mutations in either the essential or regulatory light chains of myosin are associated with a rare myopathy in human heart and skeletal muscle. Nat Genet 13:63–69

Post WS, Larson MG, Myers RH et al (1997) Heritability of left ventricular mass: the Framingham Heart Study. Hypertension 30:1025–1028

Raben N, Plotz P, Byrne BJ (2002) Acid alpha-glucosidase deficiency (glycogenosis type II, Pompe disease). Curr Mol Med 2:145–166

Reich DE, Cargill M, Bolk S et al (2001) Linkage disequilibrium in the human genome . Nature 411:199–204

Rigat B, Hubert C, Alhenc-Gelas F et al (1990) An insertion/deletion polymorphism in the angiotensin I-converting enzyme gene accounting for half the variance of serum enzyme levels. J Clin Invest 86:1343–1346

Rottbauer W, Gautel M, Zehelein J et al (1997) Novel splice donor site mutation in the cardiac myosin-binding protein-C gene in familial hypertrophic cardiomyopathy. Characterization Of cardiac transcript and protein. J Clin Invest 100:475–482

Rust EM, Albayya FP, Metzger JM (1999) Identification of a contractile deficit in adult cardiac myocytes expressing hypertrophic cardiomyopathy-associated mutant troponin T proteins. J Clin Invest 103:1459–1467

Satoh M, Takahashi M, Sakamoto T et al (1999) Structural analysis of the titin gene in hypertrophic cardiomyopathy: identification of a novel disease gene. Biochem. Biophys. Res Commun 262:411–417

Schunkert H, Brockel U, Hengstenberg C et al (1999) Familial predisposition of left ventricular hypertrophy. J Am Coll Cardiol 33:1685–1691

Schunkert H, Jackson B, Tang SS et al (1993) Distribution and functional significance of cardiac angiotensin converting enzyme in hypertrophied rat hearts. Circulation 87:1328–1339

Sebkhi A, Zhao L, Lu L et al (1999) Genetic determination of cardiac mass in normotensive rats: results from an F344xWKY cross. Hypertension 33:949–953

Seidman CE. Hypertrophic Cardiomyopathy: from man to mouse. J Clin Invest 106:S9-S13. 11-15-2000. Ref Type: Generic

Seidman CE, Seidman JG (1998) Molecular genetic studies of familial hypertrophic cardiomyopathy. Basic Res Cardiol 93, Suppl 3:13–16

Semplicini A, Siffert W, Sartori M et al (2001). G protein beta3 subunit gene 825T allele is associated with increased left ventricular mass in young subjects with mild hypertension. Am J Hypertens 14:1191–1195

Shirani J, PickR, Roberts WC et al (2000) Morphology and significance of the left ventricular collagen network in young patients with hypertrophic cardiomyopathy and sudden cardiac death. J Am Coll Cardiol 35:36–44

Simon DK, Johns DR (1999) Mitochondrial disorders: clinical and genetic features. Annu Rev Med 50:111–127

Spirito P, Bellone P, Harris KM et al (2000) Magnitude of left ventricular hypertrophy and risk of sudden death in hypertrophic cardiomyopathy. N Engl J Med 342:1778–1785

Su SF, Hsiao CL, Chu CW et al (2000). Effects of pravastatin on left ventricular mass in patients with hyperlipidemia and essential hypertension. Am J Cardiol 86:514–518

Tartaglia M, KalidasK, Shaw A et al (2002). PTPN11 mutations in Noonan syndrome: molecular spectrum, genotype-phenotype correlation, and phenotypic heterogeneity. Am. J Hum Genet 70:1555–1563

Tartaglia M, Mehler EL, Goldberg R et al (2001) Mutations in PTPN11, encoding the protein tyrosine phosphatase SHP-2, cause Noonan syndrome. Nat Genet 29:465–468

Tavaria M, Gabriele T, Anderson RL et al (1995) Localization of the gene encoding the human heat shock cognate protein, HSP73, to chromosome 11. Genomics 29:266–268

Taylor A, Obholz K, Linden G et al (1998). DNA sequence and muscle-specific expression of human sarcosin transcripts. Mol Cell Biochem. 183:105–112

Tesson F, DufourC, Moolman JC et al (1997). The influence of the angiotensin I converting enzyme genotype in familial hypertrophic cardiomyopathy varies with the disease gene mutation. J Mol Cell Cardiol 29:831–838

Tesson F, Richard P, Charron P et al (1998) Genotype-phenotype analysis in four families with mutations in beta-myosin heavy chain gene responsible for familial hypertrophic cardiomyopathy. Hum Mutat 12:385–392

Thierfelder L, Watkins H, MacRae C et al (1994) Alpha-tropomyosin and cardiac troponin T mutations cause familial hypertrophic cardiomyopathy: a disease of the sarcomere. Cell 77:701–712

Varnava AM, Elliott PM, Baboonian C et al (2001a) Hypertrophic cardiomyopathy: histopathological features of sudden death in cardiac troponin t disease. Circulation 104:1380–1384

Varnava AM, Elliott PM, Mahon N et al (2001b) Relation between myocyte disarray and outcome in hypertrophic cardiomyopathy. Am J Cardiol 88:275–279

Verhaaren HA, Schieken RM, Mosteller M et al (1991). Bivariate genetic analysis of left ventricular mass and weight in pubertal twins (the Medical College of Virginia twin study). Am J Cardiol 68:661–668

Watkins H, Anan R, Coviello DA et al (1995a) A de novo mutation in alpha-tropomyosin that causes hypertrophic cardiomyopathy. Circulation 91:2302–2305

Watkins H, Conner D, Thierfelder L et al (1995b) Mutations in the cardiac myosin binding protein-C gene on chromosome 11 cause familial hypertrophic cardiomyopathy. Nat Genet 11:434–437

Watkins H, McKenna WJ, Thierfelder L et al (1995c). Mutations in the genes for cardiac troponin T and alpha-tropomyosin in hypertrophic cardiomyopathy. N Engl J Med 332:1058–1064

Watkins H, Rosenzweig A, Hwang DS et al (1992b) Characteristics and prognostic implications of myosin missense mutations in familial hypertrophic cardiomyopathy. N. Engl J Med 326:1108–1114

Watkins H, Thierfelder L, Anan R et al (1993) Independent origin of identical beta cardiac myosin heavy-chain mutations in hypertrophic cardiomyopathy. Am J Hum Genet 53:1180–1185

Watkins H, Thierfelder L, Hwang DS et al (1992a) Sporadic hypertrophic cardiomyopathy due to de novo myosin mutations. J Clin Invest 90:1666–1671

Williams RS (2000) Canaries in the coal mine: mitochondrial DNA and vascular injury from reactive oxygen species. Circ Res 86:915–916

Yamada Y, Ichihara S, Fujimura T et al (1997) Lack of association of polymorphisms of the angiotensin converting enzyme and angiotensinogen genes with nonfamilial hypertrophic or dilated cardiomyopathy. Am J Hypertens 10:921–928

Yamada Y, Ichihara S, Izawa H et al (2001) Association of a G994 ->T (Val279 ->Phe) polymorphism of the plasma platelet-activating factor acetylhydrolase gene with myocardial damage in Japanese patients with nonfamilial hypertrophic cardiomyopathy. J Hum Genet 46:436–441

Yamazaki T, Komuro I, Yazaki Y (1999) Role of the renin-angiotensin system in cardiac hypertrophy. Am J Cardiol 83:53H-57H

Yoneya K, Okamoto H, Machida M et al (1995) Angiotensin-converting enzyme gene polymorphism in Japanese patients with hypertrophic cardiomyopathy. Am Heart J 130:1089–1093

Zhu X, Bouzekri N, SouthamL et al (2001) Linkage and association analysis of angiotensin I-converting enzyme (ACE)-gene polymorphisms with ACE concentration and blood pressure. Am J Hum Genet 68:1139–1148

Genetic Determinants of Susceptibility, Prognosis and Treatment in Heart Failure

T. Stanton[1] · J. M. C. Connell[2] · J. J. V. McMurray[3]

[1] Division of Cardiovascular and Medical Sciences, Western Infirmary,
University of Glasgow, Glasgow, G11 6NT, UK
e-mail: ts39u@clinmed.gla.ac.uk
[2] MRC Blood Pressure Unit, Division of Cardiovascular and Medical Sciences,
Western Infirmary, University of Glasgow, Glasgow, G11 6NT, UK
e-mail: jmcc1m@clinmed.gla.ac.uk
[3] Clinical Research Initiative in Heart Failure, Department of Medicine,
University of Glasgow, West Medical Building, Glasgow, G12 8QQ, UK
e-mail: j.mcmurray@bio.gla.ac.uk

1	Heart Failure Incidence and Prevalence.	204
2	Types of Heart Failure.	205
2.1	Familial Idiopathic Dilated Cardiomyopathy	205
2.2	Non-familial Idiopathic Dilated Cardiomyopathy	206
3	Negative Studies and Result Interpretation.	206
4	Polymorphisms Affecting Susceptibility to Heart Failure.	207
4.1	Angiotensin-Converting Enzyme.	207
4.2	Beta-1 Adrenoceptor.	208
4.3	Endothelin Type-A Receptor	209
4.4	Platelet-Activating Factor acetylhydrolase	210
4.5	Aldosterone Synthase	210
4.6	Transforming Growth Factor-β1	210
4.7	Tumour Necrosis Factor α.	211
5	Polymorphisms Affecting Prognosis in Heart Failure.	211
5.1	Angiotensin-Converting Enzyme.	211
5.2	Angiotensin-II Type 1 Receptor	212
5.3	Beta-1 Adrenoceptor.	212
5.4	Beta-2 Adrenoceptor.	212
5.5	Endothelin Type-A Receptor	213
5.6	Adenosine Monophosphate Deaminase 1	214
6	Polymorphisms Affecting Response to Heart Failure Treatment.	214
6.1	Diuretics	214
6.2	Beta-Blockers.	215
6.3	ACE Inhibitors.	216
6.4	Angiotensin-II Type I Receptor Blockers	217
6.5	Digoxin.	217
7	Summary.	218
References		218

Abstract Heart failure is a major burden on public health resources and the prevalence of the condition is rising as the population ages and with the increasing success of salvage strategies for myocardial infarction. The emphasis in management has changed over the last 20 years from the use of predominantly inotropic agents to drugs that regulate neurohumoral activity. This has reaped rewards in terms of improving reducing mortality but much remains to be done. Now, in common with the approach taken to other cardiovascular diseases, there is considerable interest in the use of genetic information to refine and personalize the treatment of heart failure, in the expectation that this will bring about further benefits to the patient and a more rational use of resources. The genetic information that may be of use in this respect is very wide ranging, encompassing factors that predispose to the disease (through atheroma, hypertension, resistance to oxidative damage, etc.), and genetic traits that influence prognosis, the ability to metabolize drugs and drug targets. This review will focus on studies in humans that have investigated the relationship between genotypes in candidate genes and susceptibility to heart failure, prognosis and response to treatment. In general these studies have been conducted on relatively small patient groups and as such have not yielded conclusive results which have changed clinical practice. Nonetheless, the studies provide encouragement and a basis for further investigations in larger, well-characterized populations.

Keywords Heart failure · Polymorphisms · Candidate genes

1
Heart Failure Incidence and Prevalence

Heart failure is a worldwide public health problem associated with high morbidity and mortality. The overall prevalence of clinically identified heart failure is estimated to be 3–20 cases per 1,000 of the population, rising with age to greater than 100 cases per 1,000 of the population in those over 65 years of age. The overall annual incidence of clinically overt heart failure in middle-aged men and women is approximately 0.1%–0.2%, again rising with age to 2%–3% in those over 85 years. Following diagnosis, the 5-year mortality rate is approximately 60% (in comparison, average 5-year survival for men and women with all cancers in the US at time of sampling was 50%). In industrialized countries, heart failure admission rates are rising steadily and in the early 1990s the cost of managing heart failure was estimated at 1%–2% of the total health care budget. This is almost certain to continue to rise as increased survival after acute myocardial infarction and increased longevity in the Western world leads to an increase in the overall prevalence of heart failure (McMurray and Stewart 2000).

Given the medical and socioeconomic importance of heart failure, there has been a great deal of effort expended on defining the factors involved in both the development and progression of the disease. Genetic factors have been found to play an important part. This review will aim to summarize the major genetic variations associated with heart failure and how they may provide an insight

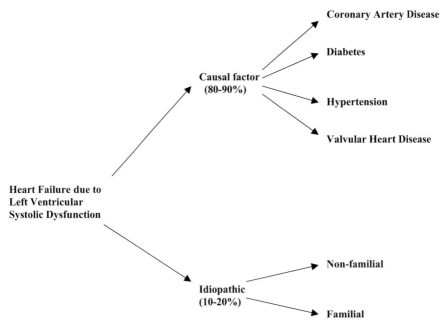

Fig. 1 Causes of left ventricular systolic dysfunction

into a patient's prognosis. We will also discuss how genetics may affect response to treatment and the opportunities for newer treatments such as gene therapy.

2
Types of Heart Failure

Heart failure has been defined as "a complex pathophysiologic condition that arises when myocardial performance is insufficient to adequately supply blood to other organs" (Fig. 1). This chapter will focus primarily on heart failure due to left ventricular systolic dysfunction, the most common and best understood form of heart failure. It is estimated that roughly 80%–90% of heart failure is due either to coronary artery disease, hypertension, valvular heart disease or diabetes (Fig. 2). The remaining 10%–20% is due to idiopathic dilated cardiomyopathy, which can be subdivided into familial and non-familial forms (Seidman and Seidman et al 2001).

2.1
Familial Idiopathic Dilated Cardiomyopathy

It is thought familial idiopathic dilated cardiomyopathy (IDCM) may represent 20%–30% of the total number of cases of idiopathic dilated cardiomyopathy (Komajda et al. 1999). As in familial hypertrophic cardiomyopathy, autosomal

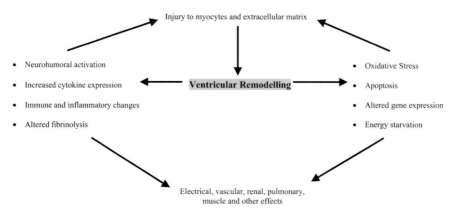

Fig. 2 Causes of left ventricular systolic dysfunction

dominant transmission predominates but other modes of inheritance have also been reported, such as X-linked cardiomyopathy. Genetic abnormalities in the cardiac actin, dystrophin (X-linked cardiomyopathy), lamin A/C and delta-sarcoglycan genes (Charron and Komajda 2001) have been identified, all of which encode structural proteins in the cytoskeleton or sarcomere. In total and to date, five different genes and ten other loci have been identified as being responsible for IDCM. Often in any single family, a unique genetic alteration causing IDCM within one of the genes mentioned above is responsible.

2.2
Non-familial Idiopathic Dilated Cardiomyopathy

Non-familial IDCM is thought to be the result of a complex interaction between genetic factors and viral, immunological, toxic and other stimuli. It is more common than familial IDCM and has thus far been the focus of research aimed at identifying possible biological triggers for disease expression. Investigators have selected candidate genes on the basis of their plausible biological impact (e.g. angiotensin-converting enzyme, beta-1 adrenoceptor). The major studies reporting an association between these genes and IDCM susceptibility and prognosis are summarized below.

3
Negative Studies and Result Interpretation

The use of genetics to define heart failure susceptibility and prognosis is a relatively new exercise. The first reports of an association between a common polymorphism and a cardiovascular disease was in 1992 when the *D* allele of the angiotensin-converting enzyme (ACE) gene was identified as a risk factor for myocardial infarction (Cambien et al. 1992). Since then there has been a raft of studies reporting associations between polymorphisms of candidate genes and cardiovascular disease, including heart failure. Most of the studies discussed in this review demonstrate a positive association between a candidate gene and heart failure.

There are reasons, however, why each report has to be critically reviewed. Many of the studies involved small numbers of patients, in racially homogenous populations, and contain little or no information on the cause or severity of heart failure. Given the relative paucity of negative association studies reported, it is evident that there is a publication bias for positive studies.

One of the few negative association studies to be published was conducted by Tiret et al. (2000). These investigators screened 403 patients with IDCM and 401 controls for polymorphisms in the *ACE*, angiotensinogen, angiotensin-II type 1 receptor (*ATRG*), aldosterone synthase (*CYP11B2*), tumour necrosis factor (TNF), transforming growth factor beta1 (TGF-β1), endothelial nitric oxide synthase (NOS3) and brain natriuretic peptide (BNP) genes. No association with IDCM was found. While the association between genetic variation and heart failure seems intuitive, it is important to remain sceptical until larger studies are undertaken.

Almost all of the studies detailed below investigate a survivor cohort of CHF patients at a fixed point in time. Most are candidate gene studies which aim to detect a change in genotype frequency compared to a control population. If a particular genotype is of increased frequency (e.g. *ACE DD*) compared to a control population, then the conclusion is often drawn that this genotype is a risk factor for CHF. Alternatively, it is possible this genotype (e.g. *ACE DD*) is protective in CHF populations and patients with CHF and the contrasting genotype (e.g. *ACE II*) die prior to the date of study, thus skewing the results. None of the studies detailed below address this concern and the conclusions reported are those of the investigators themselves.

4
Polymorphisms Affecting Susceptibility to Heart Failure

4.1
Angiotensin-Converting Enzyme

One of the most thoroughly researched genetic variants is the ACE insertion (*I*) / deletion (*D*) polymorphism. This polymorphism is located in intron 16 of the *ACE* gene (i.e. non-coding), where there is either the presence (*I*) or absence (*D*) of a 287-base pair sequence. Individuals homozygous for the *D* allele have been shown to have higher serum levels of ACE and of ACE activity (Tiret et al. 1992) and an increased frequency of this genotype has been reported in patients with coronary artery disease (Cambien et al. 1992).

The *ACE I/D* polymorphism has been studied in patients with end-stage heart failure by Raynolds et al. (1993). Genotype frequency was compared between patients with end-stage heart failure due to either coronary heart disease or idiopathic dilated cardiomyopathy and controls with normally functioning hearts.

The *ACE DD* genotype comprised 35.7% and 39.2% of the IDCM and ischaemic cardiomyopathy groups, respectively, compared to only 24.0% of the normal controls (p=0.008 for both groups). This finding suggested that the *ACE* genotype may be of pathological significance in both idiopathic and ischaemic cardiomyopathy, with the *D* allele conferring a disadvantage.

This association has been refuted in three subsequent studies, all of which showed no association between *ACE* genotype and patients with dilated cardiomyopathy, either idiopathic or ischaemic (Montgomery et al. 1995; Sanderson et al. 1996; Vancura et al. 1999).

4.2
Beta-1 Adrenoceptor

The beta-1 adrenoceptor (β1-AR) is a key regulator of myocyte function. It is the predominant beta-adrenoreceptor found in the myocardium. The benefits of beta-blockers in the treatment of heart failure are now established as a result of studies such as CIBIS-II (CIBIS-II investigators 1999), MERIT-HF (MERIT-HF Investigators 1999), and COPERNICUS (Packer et al. 2001). Two polymorphisms in the coding region of the β1-AR have been identified and linked with heart failure. The *Gly49Ser* variation (*A* for *G* at nucleotide 145) is located in the extracellular domain of the receptor but there is no evidence to suggest it alters receptor function. In contrast, the *Arg389Gly* polymorphism (*C* for *G* at nucleotide position 1165) is located within the carboxy terminal tail portion of this hepta-helical G-protein-coupled receptor, specifically in a highly conserved region critical for G-protein coupling and so intracellular signalling. In vitro studies (Mason et al. 1999) have indicated that the *Arg389* variant has higher basal adenylyl cyclase activity and this effect is greatly magnified when bound by agonist with a subsequent increase in G-protein coupling.

Podlowski et al. (2000) investigated the prevalence of the *Gly49Ser* variation in 37 patients with IDCM patients and 40 controls. *Gly49* allele was found in 6 (16.2%) patients with IDCM and none of the controls. Although the number of patients studied was very small, the investigators suggested that this variation may increase susceptibility to IDCM.

Börjesson et al. (2000) examined the effect of the *Gly49Ser* polymorphism on survival in heart failure. They studied 184 patients with chronic heart failure and 77 controls. Each group was followed up for 5 years. Interestingly, these investigators also found that the allele frequency of the *Gly49* variant was more common in heart failure patients (p=0.019). However, patients *without* this variation had a significantly poorer 5-year survival, (risk ratio, 2.34, p=0.003). This was despite the greater use of beta-blockers and ACE inhibitors in these patients. Thus, while the *Gly49* polymorphism may confer susceptibility to heart failure, it may also be associated with altered beta-1 adrenoceptor function, resulting in myocardial protection in heart failure patients.

The prevalence of the *Arg389Gly* genotype was investigated in 600 patients with NYHA II or III heart failure in a substudy of MERIT-HF by White et al.

(2002). Genotype frequencies were *Arg/Arg* 51.3%, *Arg/Gly* 40.2% and *Gly/Gly* 8.5%, which do not significantly differ from that shown in normal populations (Mason 1999). The investigators concluded that the *Arg389Gly* genotype did not confer increased susceptibility to heart failure.

The interaction of the *β1-AR Arg389Gly* and α2c-adrenoceptor (α_{2C}-AR) *Del322-325* polymorphisms has also been investigated in patients with cardiomyopathy. As stated above, the *β1-AR Arg389Gly* variation affects receptor activity when bound by norepinephrine (NE). The α_{2C}-AR *Del322-325* is a common coding polymorphism resulting in the loss of four consecutive amino acids in a G-protein-coupling domain. It is associated with interruption of the normal autoinhibitory negative feedback loop governing NE release from sympathetic nerve endings and so enhanced presynaptic release of NE. Small et al. (2002) studied the interaction of these two polymorphisms in a biracial group of patients who either had IDCM or ischaemic cardiomyopathy. In an African-American population subset, they found that individuals homozygous for both α_{2C}-*Del322-325/Del322-325* (↑ NE release) and *β1-Arg389/Arg389* (↑ response to NE) variants were at a substantially increased risk of heart failure (odds ratio, 10.11). No relationship was found in the larger Caucasian subset. While the study contains relatively small numbers, the fact that this finding has a biochemical basis strengthens its validity. It also highlights the value of studying haplotypes rather than isolated SNPs.

4.3
Endothelin Type-A Receptor

Plasma endothelin-1 (ET-1) concentrations and left ventricular endothelin-A (ET_A) receptor densities are increased in patients with dilated cardiomyopathy suggesting that the endothelin system is involved in the pathophysiology of heart failure (Zolk O et al. 1999). ET-1 is produced by cardiac myocytes and fibroblasts as well as endothelial cells and exerts its main effects through ET_A receptors (Herrmann et al. 2001). It is a potent vasoconstrictor and has both anti-natriuretic and anti-diuretic properties. ET-1 also augments the vasoconstrictor effects of vasopressin, the renin-angiotensin-aldosterone system and the sympathetic nervous system (Petrie 1999). It is of interest, therefore, to know whether variation in the genes encoding ET-1 or the endothelin type A or type B receptors may affect susceptibility or prognosis in heart failure.

Charron et al (1999) examined five different polymorphisms in the *ET-1*, ET_A receptor and ET_B receptor genes in 433 French patients with IDCM and 400 age and sex-matched controls. They found that a +1,363 cytosine(*C*) / thymine (*T*) polymorphism in the non-translated part of exon 8 of the endothelin type-A receptor had a relationship with IDCM. Although the functional significance of this polymorphism is unknown, the authors found that individuals homozygous for the *T* allele (*TT*) were significantly more frequent in the IDCM group than in the control group (13.8% vs. 7.8%, *p*=0.045). No significant differences between the IDCM and control groups were found for the other polymorphisms. It was suggested that this variant was a genetic risk factor for IDCM.

4.4
Platelet-Activating Factor acetylhydrolase

Plasma platelet-activating factor (PAF) acetylhydrolase acts as a key defence against oxidative stress by hydrolysing PAF and oxidized phospholipids. A polymorphism of the gene encoding PAF acetylhydolase causes a G (M allele) to T (m allele) change at nucleotide 994 in exon 9 which encodes the catalytic domain. The m allele has much diminished PAF acetylhydolase activity compared to the wild-type M allele (Stafforini et al. 1996).

Ichihara et al. (1998) proposed that individuals with the m allele may be exposed to more oxidative stress and therefore be more susceptible to dilated cardiomyopathy. They compared 122 Japanese patients with IDCM and 226 healthy controls. The frequency of the m allele was higher in the IDCM group than controls (3.2% vs 1.8%). The investigators suggested that this mutation was a risk factor for dilated cardiomyopathy.

4.5
Aldosterone Synthase

The gene encoding aldosterone synthase (*CYP11B2*) has been linked to hypertension (Davies et al. 1999). Aldosterone controls sodium balance and intravascular volume, which are important in CHF. It may also promote changes in the cardiac extra-cellular matrix.

A polymorphism in the *CYP11B2* gene (*−344 C/T*) is associated with increased aldosterone secretion (Hengstenberg et al. 2000). Kupari et al. (1998) examined the influence of this polymorphism on left ventricular (LV) mass and function in a group of 84 healthy individuals. They found that *CC* homozygotes had on average a 28% larger end-diastolic volume (LVEDD) and a 21% greater LV mass than *TT* homozygotes. These investigators have suggested that this polymorphism may well play a part in determining LV structure.

4.6
Transforming Growth Factor-β1

Transforming growth factor-β1 is a regulatory cytokine which inhibits the proliferation of many cell types, including smooth muscle, epithelial, and endothelial cells (Wang et al. 1997). Elevated TGF-β1 gene expression has been measured in ventricular biopsies from patients with IDCM (Li et al.1997). Variations in the *TGF-β1* gene alter the plasma concentration of TGF-β1 and so may be important in heart failure (Grainger et al. 1999).

Holweg et al. (2001) investigated two polymorphisms, *Leu10Pro* and *Arg25-Pro*, responsible for changes in the signalling sequence of the TGF-β1 protein.

Individuals homozygous for *Arg25* have been shown, both in vitro and in vivo, to have higher plasma TGF-β1 levels. Two hundred and fifty-three heart transplant recipients (109 due to IDCM and 144 due to ischaemic heart disease, IHD) and 94 controls were genotyped. The individuals receiving transplants because of IDCM had a different *TGF-β1* genotype distribution from the other two groups. Patients with IDCM had a higher frequency of the *Pro10* allele when compared to the both those with HF due to IHD ($p=0.04$) and healthy controls ($p=0.02$). The other two groups (IHD and control) did not differ in their genotype frequency. All three groups had a similar genotype frequency for *Arg25Pro*. As a result, these investigators suggested that the *Leu10Pro TGF-β1* variation is associated with end-stage heart failure due to IDCM.

4.7
Tumour Necrosis Factor α

Tumour necrosis factor α (TNF-α) is a cytokine with negative inotropic effects. Plasma and myocardial TNF concentrations are increased in heart failure (Levine at al. 1990). A change from guanine to adenosine at position -308 in the promoter region of the gene causes formation of the *TNF2* allele, which is associated with a six- to sevenfold increase in inducible TNF-α gene transcription (Wilson et al. 1997).

Densem et al. (2002) investigated the frequency of the *TNF2* allele in 175 heart transplant recipients (69 with non-ischaemic cardiomyopathy and 106 with ischaemic myocardial dysfunction) and 212 controls. The *TNF2* allele was more frequent in those with non-ischaemic myocardial dysfunction, suggesting that this polymorphism may play a role in susceptibility for IDCM.

5
Polymorphisms Affecting Prognosis in Heart Failure

Genetic changes may affect not only an individual's susceptibility to heart failure, but also his or her prognosis once heart failure is established. Once again, candidate gene studies have been the main method of investigation.

5.1
Angiotensin-Converting Enzyme

The effect of *ACE* genotype on prognosis was examined by Andersson et al. (1996). They followed 193 patients with IDCM for 5 years compared with a group of 77 aged-matched, healthy controls. While genotype frequencies did not vary between the heart failure and control groups, 5-year survival data showed that heart failure patients with the *DD* genotype fared significantly worse compared to the *II/ID* genotypes (49% vs 72%, $p=0.001$). The *DD* genotype was therefore suggested to be a marker of poor prognosis in heart failure.

5.2
Angiotensin-II Type 1 Receptor

Given the importance of the renin-angiotensin-aldosterone system in heart failure, other genetic variants affecting prognosis have been sought. A polymorphism in the 3' untranslated region of the angiotensin-II type 1 receptor gene (*ATRG*) consisting of an *A* or *C* variant (*A1166C*) has been identified. The *C* mutant allele has been associated with ischaemic heart disease and hypertension (Tiret et al. 1994).

Andersson et al. (1999) studied this polymorphism and the *ACE I/D* polymorphism, and their relationship in 194 patients with IDCM. Patients with the *DD* ACE genotype and the *C* ATRG allele had lower LV ejection fraction, higher LV mass and overall higher mortality. This finding suggests that these genotypes may interact to confer a worse prognosis in heart failure.

5.3
Beta-1 Adrenoceptor

The *Arg389Gly* polymorphism of the β1-AR is in an area critical for G-protein coupling and intracellular signalling. The *Arg389* allele has been shown, in vitro, to be associated with higher adenylyl cyclase activity and G-protein coupling (Mason et al. 1999). As stated above, the other known β1-AR polymorphism, *Gly49Ser*, is not currently known to alter receptor function.

The role of these polymorphisms has been investigated by Wagoner et al. (2002), who examined the response to exercise of 263 patients with CHF. Patients homozygous for the *Gly389* genotype had lowest peak VO_2 and exercise times (Arg389, 17.7±0.4 ml/kg per minute and 9.6±0.3, respectively, compared to *Gly389*, 14.5±0.6 ml/kg per minute and 7.0 ±0.5, respectively). In a subsequent haplotype analysis, two haplotypes displayed the most divergent peak VO_2; homozygous *Gly389/Ser49* and homozygous *Arg389/Gly49* carriers (14.4±0.5 vs 18.2±0.8 ml/kg per minute). No effect of either polymorphism was seen on heart rate.

A substudy of the MERIT-HF trial also examined the *Arg389Gly* polymorphism in CHF (White et al. 2002). Six hundred patients with CHF were genotyped and followed for 1 year. Of these 600, 155 (26%) had reached the combined endpoint of death or hospitalization at the end of the year. In contrast to Wagoner's study, no effect of genotype was observed. It should be noted that all patients were taking the beta-1 selective blocker, metoprolol CR/XL, and whether this negated any genotype effect is uncertain.

5.4
Beta-2 Adrenoceptor

Both beta-1 (β1-AR) and beta-2 adrenoceptors (β2-AR) are involved in cardiac inotropy and chronotropy. In the non-failing heart, the β1-AR subtype predominates, representing 70%–80% of the total beta-adrenoceptor population. As the heart fails, there is selective down-regulation of the β1-AR subtype, such that the $\beta1$:$\beta2$ ratio comes closer to 50:50 (Port and Bristow 2001). β2-AR polymorphisms affect both blood pressure, (*Arg16Gly*, Hoit et al. 2000) and resistance artery function, (*Glu27Gln*, Cockcroft et al. 2000) and these effects could be important in heart failure.

Liggett et al. (1998) studied 259 patients with NYHA class II–IV heart failure (due to ischaemic heart failure or IDCM) and followed them up for 1 year. They were genotyped for the *Thr164Ile* polymorphism in β2-AR, the *Thr* to *Ile* switch at amino acid 164 in the fourth transmembrane-spanning domain, conferring decreased binding affinity for catecholamines and defective G-protein coupling (Green at al. 1993). No patients were homozygous for the *Ile164* mutation and only 4% ($n=10$) were heterozygous. The 1-year survival for these heterozygotes was 42%, compared to 76% for the wild type. Despite the small number of patients studied, the investigators suggested that this polymorphic variant may alter heart failure prognosis.

In addition to studying β1-AR, Wagoner et al. (2000) have investigated the relationship between polymorphic forms of the β2-AR and exercise capacity, studying 232 patients with either ischaemic heart failure or IDCM in heart failure. As well as *Thr164Ile*, they looked at two other alleles causing abnormal receptor-effector coupling of the β2 receptor: *Arg16Gly* and *Gln27Glu*. The *Ile164* and *Gly16* variant forms are associated with decreased receptor function compared to wild type, while the *Glu27* change is associated with enhanced function. Perhaps not surprisingly, patients with the less active receptor forms (*Ile164*, *Gly16* and a combination of *Gln27* and *Gly16*) had depressed exercise performance (as determined by peak VO$_2$ during cardiopulmonary exercise testing). The authors suggested these polymorphisms exerted a functional effect in heart failure.

5.5
Endothelin Type-A Receptor

Herrmann et al. (2001) genotyped 125 patients with IDCM patients for six polymorphisms of the *ET-1*, *ET$_A$* and *ET$_B$* genes. Unfortunately, the +1,363 C/T polymorphism previously shown to be associated with heart failure susceptibility was not studied. Herrmann found that a *H323H (C/T)* polymorphism in exon 6 of the *ET$_A$* gene was linked with a shorter 2-year survival time after diagnosis. The odds ratio for carriers of the *ET$_A$ T* allele dying within 2 years of diagnosis compared to non-carriers was 5.5 ($p=0.013$). The influence of this change remained significant even when echocardiographic measurements, age and NYHA classification were taken into account. This polymorphism may therefore affect prognosis in IDCM.

5.6
Adenosine Monophosphate Deaminase 1

Adenosine monophosphate deaminase (AMPD) is a key enzyme involved in adenine nucleotide catabolism, which produces adenosine. A $C{\rightarrow}T$ transition in codon 12 in exon 2 in the *AMPD1* gene results in a non-sense mutation predicting a severely truncated AMPD peptide. It is associated with reduced activity of the enzyme in skeletal muscle and increased adenosine levels (Morasaki et al. 1992). Approximately 20% of both African-Americans and Caucasians are heterozygous for the variant allele.

Loh et al. (1999) hypothesized that heart failure patients with variant exon 2 *AMPD1* allele may have improved survival compared to the wild type. In a study involving 132 heart failure patients and 91 controls, individuals heterozygous ($n=20$) and homozygous ($n=1$) for the variant allele survived longer (odds ratio, 8.6) and had delayed progression of symptoms compared to wild type. The polymorphism was thus deemed to be cardioprotective in heart failure patients is spite of the small numbers involved.

6
Polymorphisms Affecting Response to Heart Failure Treatment

The concept that an individual's response to treatment may vary depending on genotype is one which has garnered interest in recent years. The following section details some of the genetic polymorphisms that may alter response to some of the most common drug treatments for heart failure. Importantly, most of the studies listed were not conducted in groups with heart failure and include small population numbers.

6.1
Diuretics

Loop diuretics such as furosemide are commonly used in heart failure. They decrease intracellular sodium and thus intracellular calcium, causing relaxation of arterial and venous vascular smooth muscle, thereby lowering ventricular preload and afterload.

Manunta et al. (1998) investigated how a polymorphism in the α-adducin gene (*G460W*) affected the response to furosemide. The α-adducin gene variants are thought to affect renal tubular sodium reabsorption (Cusi et al. 1997). One hundred and eight hypertensive patients were given a single oral dose of furosemide 25 mg. Patients carrying one or two copies of the variant *W* allele had a smaller plasma renin and fractional sodium excretion increase after furosemide. Patients with the *W* allele also had a less steep negative pressure–natriuresis relationship suggesting that the variant allele causes an increased propensity to renal tubular sodium reabsorption.

6.2
Beta-Blockers

Beta-blockers are now established as a mainstay of heart failure treatment as a result of large clinical trials with these agents (CIBIS-II, MERIT-HF, COPERNICUS). Studies to investigate what determines the often varied response to these agents are thus of great clinical importance.

McNamara et al. (2001) studied the influence of the *ACE I/D* polymorphism in heart failure. They followed up 328 patients with LV systolic dysfunction to assess the impact of the *D* allele on transplant-free survival. Survival was reduced in patients carrying the *D* allele. The *D* allele has been previously suggested to be deleterious in conditions such as myocardial infarction (Cambien 1992), LV hypertrophy (Schunkert 1994) and hypertrophic cardiomyopathy (Marian 1993). The investigators took the opportunity to study the effect of treatment with beta-blockers. If patients were not on beta-blockers at the time of entry into the trial, the adverse impact of the *D* allele was increased, whereas the deleterious effect of the *D* allele was abolished if patients were already receiving beta-blocker therapy. For example, for *DD* homozygotes, 1-year survival was 67% if not on a beta-blocker compared to 86% if on beta-blocker therapy. A possible pharmacogenetic interaction between the *ACE I/D* polymorphism and beta-blocker therapy for heart failure was suggested. The mechanism for this remains unclear and has not, as yet, been replicated in other studies.

De Groote et al. (2001) investigated the effect of two known β1-adrenoceptor polymorphisms (*Gly49Ser* and *Arg389Gly*) on the response to beta-blockade in heart failure. One hundred and fifty-one patients with the condition underwent echocardiography and cardiopulmonary exercise testing before and after beta-blockade treatment (bisoprolol or carvedilol). Neither polymorphism had any effect on echocardiographic parameters before or after beta-blocker treatment. Beta-blockade, however, reduced peak VO_2 in response to exercise in those carrying the *G389* allele. It was concluded that exercise capacity was reduced after beta-blockade in heart failure patients carrying the *G389* variant.

The *Arg389Gly* polymorphism and its possible effect on response to beta-blockade (atenolol and bisoprolol) has also been investigated by O'Shaughnessy et al. (2000). They found that heart rate and blood pressure fell by the same amount irrespective of genotype after 4 weeks of beta-blocker treatment in 147 hypertensive patients.

As beta-adrenoreceptors exert their intracellular effects via G-protein coupling, Jia et al. (1999) examined the effect of a polymorphism (*GNAS1*), determined by the presence (+) or absence (−) of a *Fok*I restriction enzyme site in the G-protein α-subunit, on the response to beta-blockade. Again, hypertensive patients (n=114) were studied. Patients were given beta-blockers and then classified into good or poor responders dependent on their fall in mean arterial pressure. A good response was defined as a fall greater than 15 mmHg and a poor one, a fall less than 11 mmHg. They found that good responders were more likely to carry the *FokI*+ allele than poor responders (62.5% vs 41.7%). It is thus possible that the *GNAS1* locus influences response to beta-blockade.

6.3
ACE Inhibitors

ACE inhibitors improve survival, decrease morbidity, relieve symptoms, and increase exercise capacity in patients with symptomatic heart failure or isolated LV systolic dysfunction (Garg and Yusuf 1995). The possibility that variation in the *ACE* gene may alter response and thus benefit is of considerable interest.

O'Toole et al. (1998) investigated the effect of *ACE I/D* polymorphism in 34 heart failure patients randomized to 6 weeks of either lisinopril (10 mg once daily) or captopril (25 mg three times daily) in a double-blind crossover study. The change in ambulatory 24-h mean arterial pressure (MAP) and glomerular filtration rate (GFR) were recorded. *DD* homozygotes had a significantly smaller fall in MAP on captopril treatment than *ID* or *II* subjects. There was a significant relation between *ACE* genotype and fall in MAP in patients on lisinopril. Genotype did not significantly affect the change in GFR on either drug, but there was some evidence for a greater fall in *II* patients. The authors acknowledge that their results are not conclusive but suggest they support the possibility of an interaction between *ACE* genotype and response to ACE inhibitors in heart failure.

Todd et al. (1995) evaluated whether or not *ACE* genotype affected the fall in serum ACE activity after enalapril (10 mg once daily). A total of nine healthy individuals of each genotype (*DD*, *ID*, *II*) had ACE activity measured before and after dosing. Throughout the study, the serum ACE activity of the *DD* group was consistently higher than that of the *II* group. However, the fall in serum ACE activity was significantly greater in the *DD* group than the *II* group at 2, 4, and 6 but not 24 h. Genotype did not appear to influence fall in mean arterial pressure.

Despite long-term ACE inhibitor therapy, up to 20% of heart failure patients still have an elevated serum aldosterone concentration, termed "aldosterone escape" (MacFadyen et al. 1999). Aldosterone levels are an important prognostic indicator in heart failure (Pitt et al. 1999) and aldosterone escape seems to have important clinical consequences. Cicoira et al. (2001) investigated the relationship between *ACE I/D* genotype and aldosterone escape in 132 patients with heart failure receiving long-term (>6 months) ACE inhibitor treatment. Thirteen patients out of 132 (10%) were labelled as having aldosterone escape. A significantly higher proportion of the escape group had the *DD* genotype than patients with suppressed aldosterone (62% vs 24%), again raising the possibility that *ACE I/D* genotype may modulate the neurohormonal response to ACE inhibition.

A polymorphism in the ATRG (*A166C*) has also been shown to influence response to the ACE inhibitor perindopril. Benetos et al. (1996) showed that carriers of the *C* allele had a threefold greater reduction in carotid-femoral pulse wave velocity (a measure of aortic stiffness) when given perindopril compared to *AA* homozygotes. Once again a genetic influence on the response to ACE inhibitor therapy was suggested.

A more recent study by Tiago et al. (2002) investigated the influence of *RAAS* genotypes on response to medical treatment in a group of 107 NYHA II–IV patients of African ancestry with IDCM. These patients were newly diagnosed and were initiated on medical treatment consisting of furosemide, digoxin and trandolapril. A polymorphism in the aldosterone synthetase gene, *CYP11B2* (discussed above), predicted that those carrying the −344C allele had greater subsequent improvement in LV ejection fraction (23±6% improved to 35±14% for *CC/CT*; 25±7% improved to 29±12% for *TT* individuals). *ACE* and angiotensinogen genotypes had no effect. The results of this study, however, fail to make clear whether this genotype alters response to therapy or the natural course of IDCM.

6.4
Angiotensin-II Type I Receptor Blockers

Both the ELITE II (Pitt et al. 2000) and Val-He FT (Baruch et al. 1999) trials evaluated the efficacy of angiotensin II receptor blockers on morbidity and mortality in patients with symptomatic heart failure. While they have not replaced ACE inhibitors as treatment of choice in heart failure, they are a practical alternative in patients with ACE inhibitor-induced cough.

The *A1166C* polymorphism in the *ATR* gene and its effect on response to the angiotensin-II type I receptor blocker, losartan, was investigated by Miller et al. (1999). In this study of 66 healthy men, those carrying the *C* allele had lower baseline GFRs, renal plasma flow (ERPF) and renal blood flow (RBF) than *AA* homozygotes. Losartan increased GFR and decreased MAP in the *AC/CC* group but not in the *AA* group. The fall in aldosterone was also less in the *AA* group after losartan. These results suggest that this mutation is important in determining response to angiotensin-II type I receptor blockers.

Kurland et al. (2001) investigated what influence the ATRG gene *A1166C*, the ACE *I/D* and the angiotensinogen (*T174M* and *M235T*) polymorphisms may have on response to an angiotensin-II type I receptor blocker (irbesartan) and a beta-blocker (atenolol). These investigators found that the greatest reduction in diastolic blood pressure with losartan was seen in ACE *II* patients. No interaction was seen for with the other genotypes or in the response to atenolol.

6.5
Digoxin

The place of digoxin in the treatment of heart failure remains controversial. The DIG study (Digitalis Investigation Group 1997) investigated the effect of digoxin in patients with heart failure who were in sinus rhythm. While no effect was seen on all cause mortality, digoxin decreased the number of hospitalizations for worsening heart failure by 28% and the combined risk of deaths and hospitalizations for heart failure by 25%. It probably still has an add-on role in patients already taking diuretics, ACE inhibitors and beta-blockers.

Plasma concentrations obtained after orally administered digoxin have been shown to be dependent on a polymorphism in the multidrug-resistance (*MDR*)-1 gene. This gene encodes an integral membrane protein, P-glycoprotein, which is present in organ systems that influence drug absorption (intestine), distribution (central nervous system and leukocytes), and elimination (liver and kidney). Patients homozygous for the *T* allele of the *C3435T* polymorphism in exon 26 demonstrated significantly lower duodenal MDR-1 expression and the highest digoxin plasma levels after 5 days of digoxin 0.25 mg o.d. (Hoffmeyer et al. 2000).

7
Summary

Given the socioeconomic burden of heart failure, a better understanding of the factors governing susceptibility, prognosis and response to treatment is necessary. In this regard, the genetic basis of inter-individual variation is poorly understood. A better appreciation of how key genes interact with the environment in determining cardiac function will be important. The studies conducted to date have involved small population numbers with heterogeneous types of heart failure. While these have given us clues as to possible candidate genes, larger and more refined studies are required before any firm conclusions can be reached.

As discussed elsewhere in this book, a considerable effort is being placed in mapping genetic variation, in the form of single nucleotide polymorphisms and haplotypes across the human genome and this new knowledge will be applied to the management of heart failure as with other cardiovascular diseases. This will require the development of new technologies to enable high-throughput genotyping and the bioinformatics support to interpret the data. There will be no substitute for careful and well-designed, hypothesis-driven studies in appropriate patient groups. Nonetheless, the use of genetic information to refine our current management of heart failure patients, together with the prospect of gene and stem cell therapy, herald an exciting future for the treatment of heart failure.

References

Andersson B, Blange I, Sylven C (1999) Angiotensin-II type 1 receptor gene polymorphism and long-term survival in patients with idiopathic congestive heart failure. Eur J Heart Fail. 1:363–369

Andersson B, Sylven C (1996) The DD genotype of the angiotensin-converting enzyme gene is associated with increased mortality in idiopathic heart failure. J Am Coll Cardiol. 28:162–167

Baruch L, Anand I, Cohen IS et al (1999) Augmented short- and long-term hemodynamic and hormonal effects of an angiotensin receptor blocker added to angiotensin converting enzyme inhibitor therapy in patients with heart failure. Vasodilator Heart Failure Trial (V-HeFT) Study Group. Circulation. 99:2658–2664

Benetos A, Gautier S, Ricard S et al (1996) Influence of angiotensin-converting enzyme and angiotensin II type 1 receptor gene polymorphisms on aortic stiffness in normotensive and hypertensive patients. Circulation. 94:698–703

Borjesson M, Magnusson Y, Hjalmarson A et al (2000) A novel polymorphism in the gene coding for the beta(1)-adrenergic receptor associated with survival in patients with heart failure. Eur Heart J. 21:1853–1858

Cambien F, Poirier O, Lecerf L et al (1992) Deletion polymorphism in the gene for angiotensin-converting enzyme is a potent risk factor for myocardial infarction. Nature 359:641–644

Charron P, Komajda M (2001) Are we ready for pharmacogenomics in heart failure? Eur J Pharm 417:1–9.

Charron P, Tesson F, Poirier O et al (1999) Identification of a genetic risk factor for idiopathic dilated cardiomyopathy. Involvement of a polymorphism in the endothelin receptor type A gene. CARDIGENE group. Eur Heart J 20:1587–1591

CIBIS-II Investigators and Committee (1999) The cardiac insufficiency bisoprolol study II. Lancet 353:9–13

Cicoira M, Zanolla L, Rossi A et al (2001) Failure of aldosterone suppression despite angiotensin-converting enzyme (ACE) inhibitor administration in chronic heart failure is associated with ACE DD genotype. J Am Coll Cardioll. 37:1808–1812

Cockcroft JR, Gazis AG, Cross DJ et al (2000) Beta(2)-adrenoceptor polymorphism determines vascular reactivity in humans. Hypertension 36:371–375

Cusi D, Barlassina C, Azzani T et al (1997) Polymorphisms of alpha-adducin and salt sensitivity in patients with essential hypertension. Lancet 349(9062): 353–1357

Davies E, Holloway CD, Ingram MC et al (1999) Aldosterone excretion rate and blood pressure in essential hypertension are related to polymorphic differences in the aldosterone synthase gene. Hypertension 33:703–707

De Groote P, Lamblin N et al (2001) Relationship between beta-1 adrenergic receptor polymorphisms and the response to beta-blockers in patients with heart failure. Circ 102 Suppl. II:661.

Densem CG, Hutchinson IV, Yonan N et al (2002) Tumour necrosis factor alpha gene polymorphism: a predisposing factor to non-ischaemic myocardial dysfunction? Heart. 87:153–155

Digitalis Investigation Group (1997) The effect of digoxin on mortality and morbidity in patients with heart failure. N Engl J Med 336:525–533.

Garg R, Yusuf S (1995) Overview of randomized trials of angiotensin-converting enzyme inhibitors on mortality morbidity in patients with heart failure. Collaborative Group on ACE Inhibitor Trials. JAMA. 273:1450–6.

Grainger DJ, Heathcote K, Chiano M et al (1999) Genetic control of the circulating concentration of transforming growth factor type beta1. Hum Mol Genet. 8:93–97

Green SA, Cole G, Jacinto M et al (1993) A polymorphism of the human beta 2-adrenergic receptor within the fourth transmembrane domain alters ligand binding and functional properties of the receptor. J Biol Chem. 268(31):23116–23121

Hengstenberg C, Holmer SR, Mayer B et al (2000) Evaluation of the aldosterone synthase (CYP11B2) gene polymorphism in patients with myocardial infarction. Hypertension 35(3):704–709

Herrmann S, Schmidt-Petersen K, Pfeifer J et al (2001) A polymorphism in the endothelin-A receptor gene predicts survival in patients with idiopathic dilated cardiomyopathy. Eur Heart J 22:1948–1953

Hoffmeyer S, Burk O, von Richeter O et al (2000) Functional polymorphisms of the human multidrug-resistance gene: multiple sequence variations and correlation of one allele with P-glycoprotein expression and activity in vivo. Proc Natl Acad Sci USA 97:3473–3478

Hoit BD, Suresh DP, Craft L et al (2000) Beta 2-adrenergic receptor polymorphisms at amino acid 16 differentially influence agonist-stimulated blood pressure and peripheral blood flow in normal individuals. Am Heart J 139:537–542.

Holweg CT, Baan CC, Niesters HG et al (2001) TGF-beta1 gene polymorphisms in patients with end-stage heart failure. J Heart Lung Transplant. 20:979–984

Ichihara S, Yamada Y, Yokota M (1998) Association of a G994->T missense mutation in the plasma platelet-activating factor acetylhydrolase gene with genetic susceptibility to nonfamilial dilated cardiomyopathy in Japanese. Circulation. 98:1881–1885

Isner JM (2002) Myocardial gene therapy. Nature 415:234–9

Jia H, Hingorani AD, Sharma P et al (1999) Association of the G(s)alpha gene with essential hypertension and response to beta-blockade. Hypertension. 34:8–14

Kirshenbaum LA, de Moissac D (1997) The bcl-2 gene product prevents programmed cell death of ventricular myocytes. Circulation. 96:1580–1585

Komajda M, Charron P, Tesson F (1999) Genetic aspects of heart failure. Eur J H Fail 1:121–126

Kupari M, Hautanen A, Lankinen L et al (1998) Associations between human aldosterone synthase (CYP11B2) gene polymorphisms and left ventricular size, mass, and function. Circulation. 97:569–575

Kurland L, Melhus H, Karlsson J et al (2001) Angiotensin converting enzyme gene polymorphism predicts blood pressure response to angiotensin II receptor type 1 antagonist treatment in hypertensive patients. J Hypertens. 19:1783–1787

Levine B, Kalman J, Mayer L et al (1990) Elevated circulating levels of tumor necrosis factor in severe chronic heart failure. N Engl J Med.323:236–241

Li RK, Li G, Mickle DA et al (1997) Overexpression of transforming growth factor-beta1 and insulin-like growth factor-I in patients with idiopathic hypertrophic cardiomyopathy. Circulation. 96:874–881

Liggett SB, Wagoner LE, Craft LL et al (1998) The Ile164 beta2-adrenergic receptor polymorphism adversely affects the outcome of congestive heart failure. J Clin Invest 102:1534–1539

Loh E, Rebbeck TR, Mahoney PD et al (1999) Common variant in AMPD1 gene predicts improved clinical outcome in patients with heart failure. Circulation 99:1422–1425

MacFadyen RJ, Lee AF, Morton JJ et al (1999) How often are angiotensin II and aldosterone concentrations raised during chronic ACE inhibitor treatment in cardiac failure? Heart 82:57–61

Manunta P, Cusi D, Barlassina C et al (1998) Alpha-adducin polymorphisms and renal sodium handling in essential hypertensive patients. Kidney Int 53:1471–1478

Mason DA, Moore JD, Green SA et al (1999) A gain-of-function polymorphism in a G-protein coupling domain of the human beta1-adrenergic receptor. J Biol Chem 274(18):12670–1264

Matsui T, Li L, del Monte F et al (1999) Adenoviral gene transfer of activated phosphatidylinositol 3'-kinase and Akt inhibits apoptosis of hypoxic cardiomyocytes in vitro. Circulation 100:2373–2379

MacGowan GA, McNamara DM. (2002) New molecular insights into heart failure and cardiomyopathy: potential strategies and therapies. Ir J Med Sci 171(2):99–104

McMurray JJ, Stewart S (2000) Epidemiology, aetiology and prognosis of heart failure. Heart 83:596–602.

McNamara DM, Holubkov R, Janosko K et al (2001) Pharmacogenetic interactions between beta-blocker therapy and the angiotensin-converting enzyme deletion polymorphism in patients with congestive heart failure. Circulation.103:1644–1648

MERIT-HF Investigators. (1999) Effect of metoprolol CR/XL in chronic heart failure: Metoprolol CR/XL Randomised Intervention Trial in Congestive Heart Failure (MERIT-HF). Lancet 353:2001–7.

Miller JA, Thai K, Scholey JW (1999) Angiotensin II type 1 receptor gene polymorphism predicts response to losartan and angiotensin II. Kidney Int 56:2173–2180

Miyamoto MI, del Monte F, Schmidt U et al (2000) Adenoviral gene transfer of SERCA2a improves left-ventricular function in aortic-banded rats in transition to heart failure. Proc Natl Acad Sci USA. 97:793–798

Montgomery HE, Keeling PJ, Goldman JH et al (1995) Lack of association between the insertion/deletion polymorphism of the angiotensin-converting enzyme gene and idiopathic dilated cardiomyopathy. J Am Coll Cardiol 2:1627–1631

Morisaki T, Gross M, Morisaki H et al (1992) Molecular basis of AMP deaminase deficiency in skeletal muscle. Proc Natl Acad Sci USA 89(14):6457–6461

O'Shaughnessy KM, Fu B, Dickerson C et al (2000) The gain-of-function G389R variant of the beta1-adrenoceptor does not influence blood pressure or heart rate response to beta-blockade in hypertensive subjects. Clin Sci (Lond) 99:233–238

O'Toole L, Stewart M, Padfield P et al (1998) Effect of the insertion/deletion polymorphism of the angiotensin-converting enzyme gene on response to angiotensin-converting enzyme inhibitors in patients with heart failure. J Cardiovasc Pharmacol 32:988–994

Packer M, Coats AJ, Fowler MB et al (2001) Effect of carvedilol on survival in severe chronic heart failure. N Engl J Med 344:1651–1658

Pitt B, Poole-Wilson PA, Segal R et al (2000) Effect of losartan compared with captopril on mortality in patients with symptomatic heart failure: randomised trial–the Losartan Heart Failure Survival Study ELITE II. Lancet. 355:1582–1587

Pitt B, Zannad F, Remme WJ et al (1999) The effect of spironolactone on morbidity and mortality in patients with severe heart failure. Randomized Aldactone Evaluation Study Investigators. N Engl J Med 341:709–717

Podlowski S, Wenzel K, Luther HP et al (2000) Beta1-adrenoceptor gene variations: a role in idiopathic dilated cardiomyopathy? J Mol Med 78:87–93

Port JD, Bristow MR (2001) Altered beta-adrenergic receptor gene regulation and signaling in chronic heart failure. J Mol Cell Cardiol. 33:887–905

Raynolds MV, Bristow MR, Bush EW et al (1993) Angiotensin-converting enzyme DD genotype in patients with ischaemic or idiopathic dilated cardiomyopathy. Lancet 342:1073–1075

Rusnak JM, Kisabeth RM, Herbert DP et al (2001) Pharmacogenomics: a clinician's primer on emerging technologies for improved patient care. Mayo Clin Proc 76:299–309

Sanderson JE, Young RP, Yu CM et al (1996) Lack of association between insertion/deletion polymorphism of the angiotensin-converting enzyme gene and end-stage heart failure due to ischaemic or idiopathic dilated cardiomyopathy in the Chinese. Am J Cardiol 77:1008–1010

Seidman JG, Seidman C (2001) The genetic basis for cardiomyopathy: from mutation identification to mechanistic paradigms. Cell 104:557–567

Small KM, Wagoner LE, Levin AM et al (2002) Synergistic polymorphisms of beta1- and alpha2c-adrenergic receptors and the risk of congestive heart failure. N Engl J Med. 347(15):1135–1142

Stafforini DM, Satoh K, Atkinson DL et al (1996) Platelet-activating factor acetylhydrolase deficiency. A missense mutation near the active site of an anti-inflammatory phospholipase. J Clin Invest 97(12):2784–2791

Tiago AD, Badenhorst D, Skudicky D et al (2002) An aldosterone synthase gene variant is associated with improvement in left ventricular ejection fraction in dilated cardiomyopathy. Cardiovasc Res 54(3):584–589

Tiret L, Bonnardeaux A, Poirier O et al (1994) Synergistic effects of angiotensin-converting enzyme and angiotensin-II type 1 receptor gene polymorphisms on risk of myocardial infarction. Lancet 344:910–913

Tiret L, Mallet C, Poirier O et al (2000) Lack of association between polymorphisms of eight candidate genes and idiopathic dilated cardiomyopathy. The CARDIGENE study. J Am Coll Cardiol 35:29–35

Tiret L, Rigat B, Visvikis S et al. (1992) Evidence, from combined segregation and linkage analysis, that a variant of the angiotensin I-converting enzyme (ACE) gene controls plasma ACE levels. Am J H Gen 51:197–205

Todd GP, Chadwick IG, Higgins KS et al (1995) Relation between changes in blood pressure and serum ACE activity after a single dose of enalapril and ACE genotype in healthy subjects. Br J Clin Pharmacol 39:131–134

Vancura V, Hubacek J, Malek I et al (1999) Does angiotensin-converting enzyme polymorphism influence the clinical manifestation and progression of heart failure in patients with dilated cardiomyopathy? Am J Cardiol 83:461–462

Wagoner LE, Craft LL, Zengel P et al (2002) Polymorphisms of the beta(1)-adrenergic receptor predict exercise capacity in heart failure. Am Heart J 144(5):840–846

Wagoner LE, Craft LL, Singh B et al (2000) Polymorphisms of the beta(2)-adrenergic receptor determine exercise capacity in patients with heart failure. Circ Res 86:834–40

Wang XL, Liu SX, Wilcken DEl (1997) Circulating transforming growth factor beta 1 and coronary artery disease. Cardiovasc Res.34(2):404–410

Weig HJ, Laugwitz KL, Moretti A et al (2000) Enhanced cardiac contractility after gene transfer of V2 vasopressin receptors in vivo by ultrasound-guided injection or transcoronary delivery. Circulation 101:1578–1585

White HL, de Boer RA et al (2002) An evaluation of the beta-1 adrenergic receptor Arg389Gly polymorphism in patients with heart failure: A MERIT-HF substudy. EHJ Suppl 1(1): 26.

Wilson AG, Symons JA, McDowell TL et al (1997) Effects of a polymorphism in the human tumor necrosis factor alpha promoter on transcriptional activation. Proc Natl Acad Sci USA 94(7):3195–3199

Zolk O, Quattek J, Sitzler G et al (1999) Expression of endothelin-1, endothelin-converting enzyme, and endothelin receptors in chronic heart failure. Circulation 99(16):2118–2123

The Genetics of Cardiac Channelopathies: Implications for Therapeutics

D. M. Roden

Division of Clinical Pharmacology, Vanderbilt University School of Medicine, 532 Robinson Research Building, Nashville, TN 37232, USA
e-mail: Dan.Roden@Vanderbilt.edu

1	Introduction	224
2	Long QT-Related Arrhythmias	224
2.1	Causes of QT Prolongation	224
2.2	The Congenital Long QT Syndromes	227
2.3	Acquired Long QT Syndrome, Including Drug-Induced	228
2.4	Basic Electrophysiological Mechanisms in Long QT-Related Arrhythmias	228
2.5	Implications of New Genetic Information from the Congenital Syndrome	230
2.6	DNA Variants Associated with Long QT-Related Arrhythmias	231
2.6.1	Mutations	231
2.6.2	Polymorphisms	232
2.6.3	Other Candidates	233
3	Loss of Sodium Channel Function Is Also Arrhythmogenic	234
3.1	A congenital Arrhythmia Syndrome Caused by Loss of Sodium Channel Function	234
3.2	Other Evidence that Reduced Sodium Current Is Arrhythmogenic	235
4	Implications for Therapy	236
	References	237

Abstract Reviews of clinical cases have led to identification of risk factors predisposing individual patients to proarrhythmia. However, in an individual patient, these risk factor predictions are imperfect and the reaction has been generally termed unpredictable (or idiosyncratic). As a result, we and others have suggested that a variable genetic context may contribute to risk of proarrhythmia on exposure to specific drugs. This research is most advanced in studies of drug-induced long QT syndrome, and it is on this area that this review will focus.

Keywords Arrhythmias · Long QT syndromes · Ion channels

1
Introduction

Drugs have been used to control abnormalities of cardiac rhythm for centuries. Digitalis entered widespread clinical use for heart failure and arrhythmias in the mid-eighteenth century, and quinidine in the early twentieth century. However, the mechanisms whereby these derivatives of naturally occurring plant products modulate cardiac rhythm remained obscure until the last several decades. Similarly, the idea that drugs might not only control cardiac rhythm but also exacerbate them, or create entirely new arrhythmias (the phenomenon of proarrhythmia), is a relatively new one, coinciding with the development of electronic systems capable of recording cardiac rhythm for long periods of time (hours or days). This technological capability, in turn, has lead to delineation of specific syndromes of proarrhythmia, driven by specific electrocardiographic patterns. In parallel, molecular mechanisms of drug action have been increasingly well understood. Taken together, these clinical and basic electrophysiological data have led to the delineation of specific mechanisms underlying the phenomenon of drug-induced arrhythmias (Roden 1994, 1998a).

We and others have suggested that a variable genetic context may contribute to risk of proarrhythmia on exposure to specific drugs. This research is most advanced in studies of drug-induced long QT syndrome. As outlined in Table 1, multiple rare genetic causes of arrhythmias (in the absence of other heart disease) have now been recognized. These genes have become initial candidates for modulators of drug-response phenotype, although the extent to which variations in these disease genes might contribute to variable responses to drug therapy (or other exogenous stressors) remains unknown.

2
Long QT-Related Arrhythmias

2.1
Causes of QT Prolongation

When QT interval on the surface electrocardiogram is markedly prolonged, a morphologically distinctive and potentially lethal form of ventricular tachycardia, termed torsades de pointes, can ensue (Schwartz et al. 1999). The commonest causes of torsades de pointes are the congenital long QT syndromes, therapy with certain antiarrhythmic drugs (Fig. 1A), therapy with a range of drugs prescribed for noncardiovascular indications, hypokalemia, and bradycardia. QT interval prolongation is also a feature of many forms of heart disease, including congestive heart failure and cardiac hypertrophy, although torsades de pointes in the absence of a usual precipitator is unusual. However, QT prolongation in these settings has been found to be a marker of increased mortality (Barr et al. 1994; Spargias et al. 1999; Algra et al. 1991), through mechanisms that are not

Table 1 Genotype–phenotype correlations in congenital arrhythmia syndromes

Eponym	Subtype	Inheritance	Disease gene	Ionic current changed	Frequency	Conditions under which arrhythmia occurs	Clinical arrhythmia	ECG characteristics
Romano Ward syndrome (long QT syndrome)	LQT1	AD	KCNQ1 (KvLQT1)	↓I_{Ks}	40% of LQT syndromes	Exercise; emotional fright; swimming; diving	Torsades de pointes	↑T wave amplitude
	LQT2	AD	KCNH2 (HERG)	↓I_{Kr}	40% of LQT syndromes	Exercise; sudden loud noise	Torsades de pointes	Low-amplitude notched T wave
	LQT3	AD	SCN5A	Plateau I_{Na}	10% of LQT syndromes	Rest		Long isoelectric ST segment; peaked T wave
	LQT4	AD	Ankyrin B					
	LQT5	AD	minK (KCNE1)	↓I_{Ks}				
	LQT6	AD	MiRP1 (KCNE2)	↓I_{Kr} (?)				
	LQT7	AD	KCNJ2	↓I_{K1}				
Andersen's syndrome							Bidirectional VT	
Jervell-Lange-Neilson syndrome (long QT syndrome + deafness)	JLN1	AR	KCNQ1	↓↓I_{Ks}		Exercise; emotional fright; swimming; diving	Torsades de pointes	↑T wave amplitude
	JLN2	AR	KCNE1	↓↓I_{Ks}				
Brugada syndrome	BS1	AD	SCN5A	↓I_{Na}		Rest	Torsades de pointes VF	↑T wave amplitude ↑J point V1–V3 (see Fig. 1)
	BS2	AD						↑J point V1–V3 (see Fig. 1)

Table 1 (continued)

Eponym	Subtype	Inheritance	Disease gene	Ionic current changed	Frequency	Conditions under which arrhythmia occurs	Clinical arrhythmia	ECG characteristics
Familial conduction system disease		AD	SCN5A	↓I$_{Na}$			Bradycardia	
Polymorphic ventricular tachycardia with short QT interval		AD	RYR2			Exercise	Polymorphic VT with exercise	
		AR	Calsequestrin					

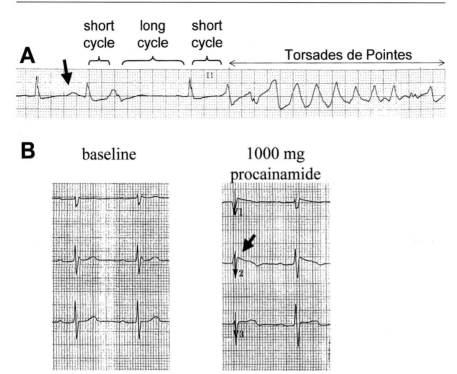

Fig. 1 Drug-induced ECG changes. **A** Torsades de pointes due to sotalol therapy. The series of cycle length changes (short-long-short) prior to the onset of the arrhythmia is typical, as is a very long QT interval, indicated by the *arrow*. **B** Brugada syndrome exposed by procainamide challenge. The baseline ECG shows slight J point elevation in lead *V2*, but this is of a saddleback configuration and not diagnostic. After drug challenge, the changes in *V2* are diagnostic. Note also the long PR interval, indicating conduction slowing typical of the Brugada syndrome and evident even at baseline

yet well understood, and may well include a contribution from torsades de pointes.

2.2
The Congenital Long QT Syndromes

One form of the congenital long QT syndrome (LQTS), inherited in an autosomal recessive fashion and associated with deafness, was described in 1957 by Jervell and Lange-Neilson (Jervell and Lange-Nielsen 1957). In a single kindred, four out of six children had congenital deafness, marked QT interval prolongation, and suffered sudden death. The mechanism underlying the sudden death was unknown and asystole was discussed as one possible cause. Parents of children with the Jervell-Lange-Neilson (JLN) syndrome usually have normal electrocardiograms. In 1963, Ward (Ward 1964) and Romano (Romano et al. 1963) separately described the much more common variant of the long QT syndrome, with autosomal dominant inheritance, QT prolongation, normal hearing, and an increased risk for syncopal episodes and sudden death. These events generally occurred with adrenergic stimulation, typically diving into cold water, hearing a loud noise (such as an alarm clock or a fire alarm), or with severe emotional stress, although exceptions were recognized.

2.3
Acquired Long QT Syndrome, Including Drug-Induced

Around the same time, the technique of continuous online monitoring was applied to understand the phenomenon of quinidine syncope. This is an uncommon reaction to quinidine, in which patients abruptly lose consciousness, often after receiving only one or two doses of the drug. Since quinidine is a relatively potent vasodilator, it was assumed these reactions were vascular in origin. However, Selzer and Wray (Selzer and Wray 1964) documented an unusual polymorphic ventricular tachycardia, later termed torsades de pointes (Dessertenne 1966), as the cause of quinidine syncope. Analyses of patients with drug-induced and congenital LQTS frequently reveals other clinical risk factors, including hypokalemia, underlying bradycardia, QT prolongation prior to drug administration, and underlying heart diseases such as congestive heart failure or hypertrophy (Kay et al. 1983; Roden et al. 1986; Houltz et al. 1998). Further, women are 2–3 times more commonly represented then men (Makkar et al. 1993). Interestingly, the same female preponderance of symptoms is seen in congenital LQTS families in whom obligate mutation carriers can be identified (Locati et al. 1998). The fact that this unusual arrhythmia occurs in both a congenital syndrome and a drug-induced form, and the similarities in risk factors between the two conditions, support the idea that DNA variants in congenital LQTS disease genes are candidates for modulating the drug-induced phenotype.

2.4
Basic Electrophysiological Mechanisms in Long QT-Related Arrhythmias

The duration of the QT interval on the surface electrocardiogram reflects the duration of action potentials in the ventricle (Fig. 2). The prolongation of the QT interval on the surface electrocardiogram therefore implies prolongation of action potential in at least some regions of the ventricle. While ventricular depolarization is a rapid event driven by rapid inward movement of sodium through sodium channels, the repolarization process is much slower and much more complex, reflecting a balance between maintained inward currents that prolong repolarization and time-dependent outward currents that promote repolarization. The major inward currents during the plateau phase are carried by calcium and, perhaps, sodium channels, whereas the major outward currents are carried by potassium channels.

Ion currents are generated by pore-forming proteins, termed ion channels, which, upon an appropriate stimulus such as a change in voltage or binding of a

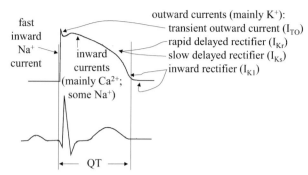

Fig. 2 Relationship between the surface ECG (*bottom*) and individual ventricular action potentials, and the currents that generate them (*top*). At a first approximation, the QT interval is an estimate of action potential duration (APD), although in fact there is considerable heterogeneity among APDs. This likely arises from heterogeneity in expression or function of the genes underlying the individual currents

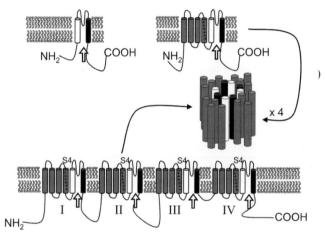

Fig. 3 Basic ion channel structures. The most primitive channel was probably a two-membrane-spanning segment potassium channel of the type shown in the *upper left*. The *open arrows* indicate the segment of the protein that generates the ion-conducting pore. Addition of four further membrane spanning segments in the course of evolution, including the S4 segment with evenly spaced positive charges, resulted in the structure of a voltage-gated potassium channel shown at the *upper right*. Potassium channels assemble as tetramers to generate pore-forming structures. Sodium and calcium channel genes generate the structure shown *below*, which likely arose by reduplication of potassium channel genes; Na^+ and Ca^{2+} channel proteins generate a pore-forming structure without a requirement for other proteins. Nevertheless, for all three types of channels (Na^+, K^+, and Ca^{2+}), ancillary proteins, the products of different genes, are often required to generate currents observed in cardiac myocytes

ligand, open a pore to allow specific ions to move with their electrochemical gradients into or out of cells (Fig. 3). Most evidence suggests that these channels include not only pore-forming proteins, but also ancillary function-modifying subunits, and likely other regulatory proteins. Mutations in one of three genes account for the vast majority of congenital LQTS (Table 1): *KvLQT1* (also termed

KCNQ1) and *HERG* (*KCNH2*), encoding the α-subunits underlying two specific potassium currents, I_{Ks} and I_{Kr}, respectively, and *SCN5A*, the gene underlying the cardiac sodium channel (Curran et al. 1995; Wang et al. 1995, 1996). The two other disease genes are *KCNE1* (*minK*) a function modifying subunit for *KvLQT1* (Splawski et al. 1997b), and perhaps *HERG*, and *KCNE2* (and *MiRP1*), a function modifying subunit for *HERG* (Abbott et al. 1999). Mutations in *KvLQT1*, *HERG*, *KCNE1*, and *KCNE2* reduce I_{Ks} and/or I_{Kr}, resulting in prolonged action potentials (Abbott et al. 1999; Sanguinetti et al. 1995, 1996; Bianchi et al. 2000; Splawski et al. 1997b; Chouabe et al. 1997). By contrast, mutations in *SCN5A* result in increased sodium current during the plateau phase of the action potential, thereby accounting for action potential prolongation (Bennett et al. 1995). At least one sodium channel *LQT3* mutant has been reported to result in a secondary increase in inward calcium current as its mechanism of QT prolongation (Wehrens et al. 2000).

2.5
Implications of New Genetic Information from the Congenital Syndrome

The state of the art of the congenital LQTS is an area of rapid scientific evolution and has been subject of numerous in-depth recent views (Ackerman 1998; Ackerman and Clapham 1997; Keating and Sanguinetti 2001; Priori et al. 1999a; Roden and Spooner 1999). One area of active research is the emerging association between specific genotypes and certain clinical features such as the details of the ECG abnormalities, conditions under which syncope occurs, or response to exercise stress (Priori et al. 1998; Wilde et al. 1998; Ackerman et al. 1999; Locati et al. 1998; Zareba et al. 1998; Moss et al. 2000, 2002; Zhang et al. 2000; Schwartz et al. 2001; Kimbrough et al. 2001). This genetic heterogeneity extends into the fundamental mechanisms, where multiple functional defects in the encoded proteins have now been described. These include protein misfolding with altered trafficking (Sanguinetti et al. 1996; Zhou et al. 1998; Ficker et al. 2000; Kupershmidt et al. 2002), altered gating of channels that traffic normally (Sanguinetti et al. 1996), and altered selectivity of channels that traffic normally (Lees-Miller et al. 2000).

Another important result of the cloning of the LQTS disease genes was the recognition that virtually all drugs that cause Torsades de Pointe block I_{Kr}, the current resulting from expression of *HERG* (Woosley et al. 1993; Yang et al. 2001; Roden 2000; Mitcheson et al. 2000). This applies to not only antiarrhythmic agents, the commonest recognized cause of drug-induced arrhythmias, but also to a wide range of drugs developed for noncardiovascular indications, including certain antihistamines, antibiotics, and antipsychotics (http//:www.torsades.org). Even a small predisposing risk of torsades de pointes may profoundly affect the balance between risk and benefit with such agents, and this issue has therefore become one of great concern in the drug industry and the drug regulatory communities (Haverkamp et al. 2000; Anderson et al. 2002; De Ponti et al. 2002).

One aspect of this research that is particularly important for drug-associated long QT-related arrhythmias is the finding that when genetic diagnostic approaches have been applied to kindreds with LQTS, incomplete penetrance has been observed (Priori et al. 1999b). That is, mutation carriers with entirely normal phenotypes have now been described, and penetrance as low as 25% (or lower) has been reported in specific kindreds. A natural question, then, is whether such individuals are at increased risk for QT prolongation during drug challenge, a hypothesis whose testing is described further below.

Along these same lines, a second group of mutation carriers with near-normal phenotypes are JLN parents. JLN is now recognized to arise when a child inherits two abnormal I_{Ks} alleles, either *KCNQ1* or *KCNE1* (Schulze-Bahr et al. 1997; Splawski et al. 1997a; Neyroud et al. 1997). Because I_{Ks} is important in establishing normal endolymph flow in the inner ear, its complete absence leads to congenital deafness. JLN children can inherit two abnormal alleles because of consanguinity or compound heterozygosity, and both mechanisms have been reported; this mechanistic insight establishes that JLN parents are obligate mutation carriers, and indeed sudden death under emotional stress has been reported (Splawski et al. 1997a).

2.6
DNA Variants Associated with Long QT-Related Arrhythmias

2.6.1
Mutations

It is well recognized that the development of torsades de pointes on exposure to a QT-prolonging drug may indicate underlying congenital long QT syndrome (Roden et al. 1986; Donger et al. 1997). Following cloning of the LQTS disease gene, anecdotes began to emerge of individuals or small kindreds with minimal QT prolongation in the absence of drug, but who developed torsades de pointes on exposure to a QT prolonging agent. Because of the possibility of a genetic contributor to risk for drug-induced torsades, we began collecting DNA samples from affected individuals in the early 1990s. As in other series, two-thirds of patients were women. Heart failure was present in 25%, hypokalemia in 18%, and no identifiable risk factors in 11%. Antiarrhythmics, primarily quinidine and sotalol, were the culprit drugs in 77%. With the identification of the disease genes in the congenital long QT syndrome, we have now been in a position to screen these samples for DNA variants that might predispose to the drug-induced form. In a database of 98 patients, nine mutations were identified (Yang et al. 2002; Sesti et al. 2000) (Fig. 4). Each mutant cDNA was then transfected into tsa-201 or CHO cells, and voltage clamp techniques used to characterize mutant channel function. Each of the six potassium channel defects (in *KvLQT1*, *HERG*, and *MiRP1*) were found to reduce K^+ current (a QT-prolonging effect) or to increase sensitivity of the encoded channel to blocking drugs. By contrast,

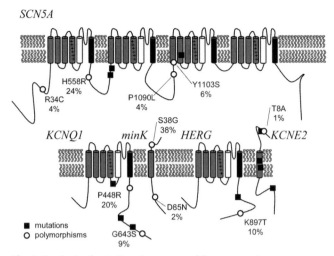

Fig. 4 Hundreds of mutations in any one of five genes, whose products are shown here, can result in the congenital long QT syndrome (LQTS). The *solid squares* show mutations that have been identified in individuals with drug-associated LQTS. The *open circles* represent nonsynonymous DNA polymorphisms, with minor allele frequencies shown

the channels encoded by three sodium channel mutants behaved near normally, and therefore may represent functionally unimportant mutations or rare polymorphisms. Thus, we concluded that approximately 6% of patients with the drug-associated long QT syndrome carry mutations in congenital LQTS disease genes that predispose them to torsades de pointes. Whether these patients truly have subclinical congenital long QT syndrome or more simply mutations predisposing to drug-induced arrhythmias may be simply a matter of semantics. We cannot entirely exclude the sodium channel lesions as causative, but at this point we further conclude that in vitro or other functional characterization should be undertaken prior to labeling a specific mutation as contributing to a drug-induced arrhythmia phenotype. This lesson very likely extends to other pharmacogenetic phenotypes.

2.6.2
Polymorphisms

In the course of these studies, we and others have identified nonsynonymous coding region polymorphisms in the LQTS disease genes (Fig. 4), with minor allele frequencies from 1.5% to 25% (Iwasa et al. 2001; Yang et al. 2002). The polymorphism D85N in KCNE1 was present in 8% of our drug-induced cases, and 2%–4% of a range of controls; we have not yet encountered a homozygote for this variant. In vitro characterization of I_{Ks} generated by co-expression of *KCNQ1* and wild-type or variant *KCNE1* demonstrated subtle changes in I_{Ks} gating with the variant. Incorporation of these subtle changes into a model in which action potentials are reconstructed by a computer model incorporating cardiac ion currents, exchangers, and elements controlling intracellular calcium homeostasis, demonstrated an increased susceptibility of N85 *KCNE1* action potentials to generate arrhythmogenic early after depolarizations, particularly following exposure to an I_{Kr} blocking drug (Wei et al. 1999). These in vitro and in silico experiments, combined with the clinical association study, have suggested to us that N85 is a polymorphism predisposing to drug-induced QT prolongation and torsades de pointes. More recently, the polymorphism Y1103S in SCN5A has been identified in African-American populations only, and the minor allele (*Y*) appears to be a risk factor for an arrhythmia phenotype that may include sensitivity to dug challenge (Splawski et al. 2002). Prospective studies are required to establish the value of these and other polymorphisms in predicting drug responses.

Two other polymorphisms, K897T in *HERG*, and T8A in *KCNE2* (Abbott et al. 1999; Sesti et al. 2000), also alter the magnitude and/or drug sensitivity of the encoded channel. However, for K897T we were unable to demonstrate any difference in allele frequencies among patients with drug-induced arrhythmias and controls, although another study has reported that this polymorphisms associates with longer QT intervals among women (Pietila et al. 2002). The minor allele frequency for T8A is so low that association studies could not be undertaken. The polymorphism with the commonest minor allele, H558R in *SCN5A*, also does not alter sodium current to prolong action potentials although, as described below, it may play a role in mutations that result in loss of sodium channel function. Other polymorphisms have not yet been characterized in vitro. In the course of screening our drug-induced database, we also identified P448R in *KvLQT1* as a mutation associated with drug-induced arrhythmias. However, this so-called mutation arose in an individual of Japanese origin, and P448R is in fact a relatively common polymorphism in that population, but not in Caucasians or African-Americans. Two other polymorphisms have been identified in the five LQTS disease genes in Japanese populations only.

2.6.3
Other Candidates

Other genes whose products contribute to control of normal repolarization such as other potassium channel genes or calcium channel genes are also candidate genes for drug-induced phenotype. Multiple lines of evidence suggest that abnormal intracellular calcium homeostasis is a major contributor to arrhythmias, so polymorphisms in genes controlling this process such as the ryanodine release channel (Laitinen et al. 2001) or the calcium-dependent uptake pump are also candidate genes. Since sympathetic activation may play a role in the development of arrhythmias related to mutations in *KvLQT1* and *HERG*, we screened our drug-induced database for recognized polymorphisms in the genes encoding the β_1 and β_2 adrenergic receptors, and found no frequency differences between patients and controls (Kanki et al. 2002).

In summary, our clinical studies have now identified DNA variants—mutations or polymorphisms—in 10%–15% of patients with drug-associated QT prolongation and torsades de pointes. An emerging understanding of the physiology and pathophysiology of action potential control suggests that multiple redundant mechanisms participate in this process. Thus, in ordinary conditions, there is considerable reserve in the repolarization process. Individual lesions predisposing to torsades de pointes such as hypokalemia, bradycardia, and heart failure each reduce this reserve, usually by reducing outward potassium current function or expression (Roden 1998b). Subclinical mutations in genes encoding cardiac potassium channels can now be considered risk factors in the same fashion. In this situation of reduced repolarization reserve, baseline QT interval may be near normal, but superposition of a further challenge such as an I_{Kr} blocking drug, profound hypokalemia, or profound bradycardia may then bring out the full-blown torsades de pointes syndrome. Further studies will be required to more fully define the candidate DNA variants that may contribute to this reduced repolarization reserve and establish, prospectively, that pre-prescription genotyping might result in a decreased risk for this potentially fatal reaction. This framework for considering multiple environmental and genetic inputs may obviously apply to other rare drug-associated adverse effects. The key to unraveling these will be a clear understanding of the underlying biology and of the drug targets.

3
Loss of Sodium Channel Function Is Also Arrhythmogenic

3.1
A congenital Arrhythmia Syndrome Caused by Loss of Sodium Channel Function

Sodium channel lesions causing LQTS result in increased inward current during the plateau phase of the action potential. However, lesions reducing sodium channel availability may also be arrhythmogenic. A starting point in considering this possibility is an unusual electrocardiographic syndrome associated with sudden death, described by Brugada and Brugada (Brugada and Brugada 1992). The Brugada syndrome has a number of parallels to the long QT syndrome: both have an unusual baseline electrocardiographic phenotype, which, in some individuals, may always be manifest, whereas in other individuals it may come and go (Alings and Wilde 1999). In both settings, challenge with drug may elicit the full-blown phenotype (Fig. 1B). For the long QT syndrome, drug challenge with a QT prolonging drug may produce this effect. For the Brugada syndrome, sodium channel-blocking drugs (such as flecainide, procainamide, and some tricyclic antidepressants) may elicit the phenotype. Finally, both the Brugada syndrome and the long QT syndrome are associated with potentially life-threatening ventricular arrhythmias: torsades de pointes in the long QT syndrome and ventricular fibrillation in the Brugada syndrome.

Because sodium channel blockers can elicit or exacerbate the clinical Brugada syndrome phenotype, *SCN5A* (encoding the cardiac sodium channel) became a candidate gene and *SCN5A* mutations have now been identified in individuals and in kindreds with the Brugada syndrome (Chen et al. 1998). Like the congenital long QT syndrome, these mutations produce multiple functional effects in vitro, including abnormal gating (Wang et al. 2000; Viswanathan et al. 2001b; Dumaine et al. 1999), abnormal trafficking (Baroudi et al. 2002; Valdivia et al. 2002), and generation of truncated proteins (Chen et al. 1998), each of which results in decreased sodium channel availability. In addition, as with the congenital long QT syndrome, other kindreds have been linked to non-*SCN5A* regions in the genome, although these disease genes have not yet been identified (Weiss et al. 2002). Unlike the congenital long QT syndrome, the manifest Brugada syndrome appears commoner in men, and indeed until the advent of HIV infection was the commonest cause for sudden death syndromes in young men described in certain regions of Southeast Asia (Nademanee et al. 1997; Vatta et al. 2002). Loss of function mutations in *SCN5A* not only produce the Brugada phenotype, but others as well: isolated conduction system disease (Tan et al. 2001), some cases of the sudden infant death syndrome (Ackerman et al. 2001), and mixed (e.g. LQT3 + Brugada) phenotypes (Bezzina et al. 1999; Veldkamp et al. 2000).

3.2
Other Evidence that Reduced Sodium Current Is Arrhythmogenic

From a pharmacogenetic context, the development of the Brugada ECG phenotype on exposure to sodium channel-blocking agents may suggest an increased risk for sudden death during therapy with such drugs. Along these lines, a landmark clinical trial, the Cardiac Arrhythmia Suppression Trial (CAST), demonstrated in the late 1980s that therapy with flecainide and related sodium channel-blocking drugs unexpectedly (at the time) increased mortality in patients convalescing from a myocardial infarction (CAST Investigators 1989). Further analysis of the CAST database showed that the subset of patients treated with sodium channel-blocking drugs who were at greatest risk for death were those with clinical characteristics suggesting ongoing myocardial ischemia (Akiyama et al. 1991). Thus, the CAST result suggests that sodium channel-blocking drug therapy and recurrent myocardial ischemia together promote the development of lethal ventricular arrhythmias, and in vitro studies further support this contention. One major effect of ischemia is to reduce sodium channel availability, through mechanisms that are now being worked out (Pu et al. 1998). Thus, taken together, Brugada syndrome genetics and drug response, the CAST result, and in vitro and in vivo studies of myocardial ischemia all raise the possibility that patients with subclinical reduction-of-function variations in the sodium channel gene may be phenotypically entirely ordinary, until challenged with the sodium channel-blocking drug, myocardial ischemia or their combination, when arrhythmia risk would be enhanced. We have preliminary evidence that while the H558R polymorphism in *SCN5A* does not affect basal sodium current;

it may modulate the in vitro phenotype under certain pathophysiological study conditions (Viawanathan et al. 2001a). While epidemiological studies do support the contention that risk factors for sudden death include a family member who has died suddenly (Jouven et al. 1999; Friedlander et al. 1998), the ideal database in which to test this hypothesis would be the patients who participated in CAST. Unfortunately, the study was done at a time when a sensitivity to pharmacogenetic issues was not well developed, and the opportunity has now been lost. As ongoing studies pay more attention to the issue of genetic factors in modulating to these phenotypes or drug response, such approaches may prove more fruitful in the future.

4
Implications for Therapy

The idiosyncratic nature of many arrhythmias has raised the possibility that the complex biological context in which they develop may be modulated by genetic factors. An extension of this concept is that individuals who are phenotypically unremarkable in the baseline state may nevertheless display an increased susceptibility to disease or aberrant responses to drugs because of such genetic influences. Our initial studies with drug-induced long QT syndrome as a model support the concept of such an increased susceptibility, and we can identify reduced repolarization reserve in 10%–15% of patients. Important lessons that we have learned in the course of these studies are that the identification of a nonsynonymous coding region (or any) polymorphism does not, itself, necessarily imply that the resulting phenotype will be abnormal, and that there is in ion channel genes, as in virtually all other genes, a striking potential for interethnic variability in DNA variants. In our studies of drug-induced arrhythmias, we have used molecular and genetic results, notably in the congenital long QT syndrome, that have implicated specific gene products in the generation of arrhythmias. These genes, then, become initial candidates for modulating drug-response phenotype. Nevertheless, as a more complete picture of the complex genetic and cellular biology of arrhythmias is developed, further candidates emerge. Importantly, such candidates are not necessarily disease genes for congenital arrhythmia or other syndromes. As our identification and functional characterization of rare mutations and, more importantly, from a public health point of view, common polymorphisms continues, a more complete picture of arrhythmia susceptibility may emerge. Although DNA variants appearing to increase risk have been identified, it is important that these results have only been obtained in limited populations, examined retrospectively. The concept of preprescription genotyping to establish risk is very appealing, but much more work will be required to delineate which DNA variants ought to be included in such an evaluation. Importantly, prospective trials will be required to truly establish the value, in terms of increased efficacy and reduced toxicity of drug therapy, as well as in cost-effectiveness terms, of any such intervention. An important consequence of this work has been the identification of the HERG channel as the target for most drugs associated with torsades de pointes and it is now becoming routine to screen for HERG activity in all new drug entities. Finally, as these studies push forward our understanding of basic molecular mechanisms in arrhythmias, an important outcome will be the development of new and better targets for drug intervention.

Acknowledgements. This work was supported by grants from the United Public Health Service (HL46681 and HL65962).

References

Abbott GW, Sesti F, Splawski I et al (1999) MiRP1 Forms I_{Kr} Potassium Channels with HERG and Is Associated with Cardiac Arrhythmia. Cell 97:175–187

Ackerman MJ (1998) The long QT syndrome: ion channel diseases of the heart. Mayo Clinic Proceedings 73:250–269

Ackerman MJ, Clapham DE (1997) Mechanisms of disease—Ion channels: Basic science and clinical disease. N Engl J Med 336:1575–1586

Ackerman MJ, Siu BL, Sturner WQ et al (2001) Postmortem molecular analysis of SCN5A defects in sudden infant death syndrome. JAMA 286:2264–2269

Ackerman MJ, Tester DJ, Porter CJ (1999) Swimming, a gene-specific arrhythmogenic trigger for inherited long QT syndrome. Mayo Clin Proc 74:1088–1094

Akiyama T, Pawitan Y, Greenberg H et al (1991) Increased risk of death and cardiac arrest from encainide and flecainide in patients after non-Q-wave acute myocardial infarction in the Cardiac Arrhythmia Suppression Trial. Am J Cardiol 68:1551–1555

Algra A, Tijssen JGP, Roelandt JRTC et al (1991) QTc prolongation measured by standard 12-lead electrocardiography is an independent risk factor for sudden death due to cardiac arrest. Circulation 83:1888–1894

Alings M, Wilde A (1999) "Brugada" syndrome: clinical data and suggested pathophysiological mechanism. Circulation 99:666–673

Anderson ME, Al-Khatib SM, Roden DM et al (2002) Proceedings of an expert panel conference on repolarization changes: Cardiac Repolarization: current knowledge, critical gaps, and new approaches to drug development and patient management. Am Heart J 144:769–81

Baroudi G, Acharfi S, Larouche C et al (2002) Expression and Intracellular Localization of an SCN5A Double Mutant R1232 W/T1620 M Implicated in Brugada Syndrome. Circ Res 90:E11-E16

Barr CJ, Naas A, Freeman M et al (1994) QT dispersion and sudden unexpected death in chronic heart failure. Lancet 343:327–329

Bennett PB, Yazawa K, Makita N et al (1995) Molecular mechanism for an inherited cardiac arrhythmia. Nature 376:683–685

Bezzina C, Veldkamp MW, van Den Berg MP et al (1999) A single Na (+) channel mutation causing both long-QT and Brugada syndromes. Circ Res 85:1206–1213

Bianchi L, Priori SG, Napolitano C et al (2000) Mechanisms of I (Ks) suppression in LQT1 mutants. Am J Physiol Heart Circ Physiol 279:H3003-H3011

Brugada P, Brugada J (1992) Right bundle branch block, persistent ST segment elevation and sudden cardiac death: a distinct clinical and electrocardiographic syndrome. A multicenter report. J Am Coll Cardiol 20:1391–1396

CAST Investigators (1989) Preliminary Report: Effect of encainide and flecainide on mortality in a randomized trial of arrhythmia suppression after myocardial infarction. N Engl J Med 321:406–412

Chen QY, Kirsch GE, Zhang DM et al (1998) Genetic basis and molecular mechanism for idiopathic—ventricular fibrillation. Nature 392:293–296

Chouabe C, Neyroud N, Guicheney P et al (1997) Properties of KvLQT1 K+ channel mutations in Romano-Ward and Jervell and Lange-Nielsen inherited cardiac arrhythmias. EMBO Journal 16:5472–5479

Curran ME, Splawski I, Timothy KW et al (1995) A molecular basis for cardiac arrhythmia: HERG mutations cause long QT syndrome. Cell 80:795–803

De Ponti F, Poluzzi E, Cavalli A et al (2002) Safety of non-antiarrhythmic drugs that prolong the QT interval or induce torsade de pointes: an overview. Drug Saf 25:263–286

Dessertenne F (1966) La tachycardie ventriculaire à deux foyers opposés variables. Arch Mal Coeur 59:263–272

Donger C, Denjoy I, Berthet M et al (1997) KVLQT1 C-terminal missense mutation causes a forme fruste long-QT syndrome. Circulation 96:2778–2781

Dumaine R, Towbin JA, Brugada P et al (1999) Ionic mechanisms responsible for the electrocardiographic phenotype of the Brugada syndrome are temperature dependent. Circ Res 85:803–809

Ficker E, Thomas D, Viswanathan PC et al (2000) Novel characteristics of a misprocessed mutant HERG channel linked to hereditary long QT syndrome. Am J Physiol Heart Circ Physiol 279:H1748-H1756

Friedlander Y, Siscovick DS, Weinmann S et al (1998) Family history as a risk factor for primary cardiac arrest. Circulation 97:155–160

Haverkamp W, Breithardt G, Camm AJ et al (2000) The potential for QT prolongation and proarrhythmia by non-antiarrhythmic drugs: clinical and regulatory implications. Report on a Policy Conference of the European Society of Cardiology. Eur Heart J 21:1216–1231

Houltz B, Darpo B, Edvardsson N et al (1998) Electrocardiographic and clinical predictors of Torsades de Pointes induced by almokalant infusion in patients with chronic atrial fibrillation or flutter. A prospective study. PACE 21:1044–1057

Iwasa H, Kurabayashi M, Nagai R et al (2001) Multiple single-nucleotide polymorphisms (SNPs) in the Japanese population in six candidate genes for long QT syndrome. J Hum Genet 46:158–162

Jervell A, Lange-Nielsen F (1957) Congenital deaf-mutism, functional heart disease with prolongation of the Q-T interval and sudden death. Am Heart J 54:59–68

Jouven X, Desnos M, Guerot C et al (1999) Predicting sudden death in the population: the Paris Prospective Study I. Circulation 99:1978–1983

Kanki H, Yang P, Xie HG et al (2002) Polymorphisms in β-adrenergic receptor genes in patients with the acquired long QT syndrome. J Cardiovasc Electrophysiol 252–256

Kay GN, Plumb VJ, Arciniegas JG et al (1983) Torsades de pointes: The long-short initiating sequence and other clinical features: Observations in 32 patients. J Am Coll Cardiol 2:806–817

Keating MT, Sanguinetti MC (2001) Molecular and cellular mechanisms of cardiac arrhythmias. Cell 104:569–580

Kimbrough J, Moss AJ, Zareba W et al (2001) Clinical implications for affected parents and siblings of probands with long-QT syndrome. Circulation 104:557–562

Kupershmidt S, Yang T, Chanthaphaychith S et al (2002) Defective human Ether-a-go-go-related gene trafficking linked to an endoplasmic reticulum retention signal in the C terminus. J Biol Chem 277:27442–27448

Laitinen PJ, Brown KM, Piippo K et al (2001) Mutations of the Cardiac Ryanodine Receptor (RyR2) Gene in Familial Polymorphic Ventricular Tachycardia. Circulation 103:485–490

Lees-Miller JP, Duan Y, Teng GQ et al (2000) Novel gain-of-function mechanism in K(+) channel-related long-QT syndrome: altered gating and selectivity in the HERG1 N629D mutant. Circ Res 86:507–513

Locati EH, Zareba W, Moss AJ et al (1998) Age- and sex-related differences in clinical manifestations in patients with congenital long-QT syndrome: findings from the International LQTS Registry. Circulation 97:2237–2244

Makkar RR, Fromm BS, Steinman RT et al (1993) Female gender as a risk factor for torsades de pointes associated with cardiovascular drugs. JAMA 270:2590–2597

Mitcheson JS, Chen J, Lin M et al (2000) A structural basis for drug-induced long QT syndrome. Proc Natl Acad Sci U S A 97:12329–12333

Moss AJ, Zareba W, Hall WJ et al (2000) Effectiveness and limitations of beta-blocker therapy in congenital long-QT syndrome. Circulation 101:616–623

Moss AJ, Zareba W, Kaufman ES et al (2002) Increased risk of arrhythmic events in long-QT syndrome with mutations in the pore region of the human ether-a-go-go-related gene potassium channel. Circulation 105:794–799

Nademanee K, Veerakul G, Nimmannit S et al (1997) Arrhythmogenic marker for the sudden unexplained death syndrome in Thai men. Circulation 96:2595–2600

Neyroud N, Tesson F, Denjoy I et al (1997) A novel mutation in the potassium channel gene KVLQT1 causes the Jervell and Lange-Nielsen cardioauditory syndrome. Nature Genetics 15:186–189

Pietila E, Fodstad H, Niskasaari E et al (2002) Association between HERG K897T polymorphism and QT interval in middle-aged Finnish women. J Am Coll Cardiol 40:511–514

Priori SG, Barhanin J, Hauer RN et al (1999a) Genetic and molecular basis of cardiac arrhythmias: impact on clinical management. Study group on molecular basis of arrhythmias of the Working Group on Arrhythmias of the European Society of Cardiology. Eur Heart J 20:174–195

Priori SG, Napolitano C, Schwartz PJ (1999b) Low penetrance in the long-QT syndrome: clinical impact. Circulation 99:529–533

Priori SG, Napolitano C, Schwartz PJ et al (1998) The risk of sudden death as first cardiac event in asymptomatic patients with the long QT syndrome. Circulation 98:I-777 (abstract)

Pu JL, Balser JR, Boyden PA (1998) Lidocaine Action On Na+ Currents In Ventricular Myocytes From The Epicardial Border Zone Of The Infarcted Heart. Circ Res 83:431–440

Roden DM (1994) Risks and benefits of antiarrhythmic therapy. N Engl J Med 331:785–791

Roden DM (1998a) Mechanisms and management of proarrhythmia. Am J Cardiol 82:49I-57I

Roden DM (1998b) Taking the idio out of idiosyncratic—predicting torsades de pointes. PACE 21:1029–1034

Roden DM (2000) Point of View: Acquired Long QT Syndromes and the Risk of Proarrhythmia. J Cardiovasc Electrophysiol 11:938–940

Roden DM, Spooner PM (1999) Inherited Long QT Syndromes: A Paradigm for Understanding Arrhythmogenesis. J Cardiovasc Electrophysiol 10:1664–1683

Roden DM, Woosley RL, Primm RK (1986) Incidence and clinical features of the quinidine-associated long QT syndrome: implications for patient care. Am Heart J 111:1088–1093

Romano C, Gemme G, Pongiglione R (1963) Aritmie cardiache rare in eta' pediatrica. Clin Pediatr 45:656–683

Sanguinetti MC, Curran ME, Spector PS et al (1996) Spectrum of HERG K$^+$ channel dysfunction in an inherited cardiac arrhythmia. Proc Natl Acad Sci 93:2208–2212

Sanguinetti MC, Jiang C, Curran ME et al (1995) A mechanistic link between an inherited and an acquired cardiac arrhythmia: *HERG* encodes the I_{Kr} potassium channel. Cell 81:299–307

Schulze-Bahr E, Wang Q, Wedekind H et al (1997) KCNE1 mutations cause Jervell and Lange-Nielsen syndrome. Nature Genetics 17:267–268

Schwartz PJ, Priori SG, Napolitano C (1999) Long QT Syndrome. 3rd:615–640

Schwartz PJ, Priori SG, Spazzolini C et al (2001) Genotype-Phenotype Correlation in the Long-QT Syndrome : Gene-Specific Triggers for Life-Threatening Arrhythmias. Circulation 103:89–95

Selzer A, Wray HW (1964) Quinidine syncope, paroxysmal ventricular fibrillations occurring during treatment of chronic atrial arrhythmias. Circulation 30:17–17

Sesti F, Abbott GW, Wei J et al (2000) A common polymorphism associated with antibiotic-induced cardiac arrhythmia. Proc Natl Acad Sci U S A 97:10613–10618

Spargias KS, Lindsay SJ, Kawar GI et al (1999) QT dispersion as a predictor of long-term mortality in patients with acute myocardial infarction and clinical evidence of heart failure. Eur Heart J 20:1158–1165

Splawski I, Timothy KW, Tateyama M et al (2002) Variant of SCN5A sodium channel implicated in risk of cardiac arrhythmia. Science 297:1333–1336

Splawski I, Timothy KW, Vincent GM et al (1997a) Molecular basis of the long QT syndrome associated with deafness. N Engl J Med 336:1562–1567

Splawski I, Tristanti-Firouzi M, Lehmann MH et al (1997b) Mutations in the *hminK* gene cause long QT syndrome and suppress I_{Ks} function. Nature Genetics 17:338–340

Tan HL, Bink-Boelkens MT, Bezzina CR et al (2001) A sodium-channel mutation causes isolated cardiac conduction disease. Nature 409:1043–1047

Valdivia CR, Ackerman MJ, Tester DJ et al (2002) A novel SCN5A arrhythmia mutation, M1766L, with expression defect rescued by mexiletine. Cardiovasc Res 55:279–289

Vatta M, Dumaine R, Varghese G et al (2002) Genetic and biophysical basis of sudden unexplained nocturnal death syndrome (SUNDS), a disease allelic to Brugada syndrome. Hum Mol Genet 11:337–345

Veldkamp MW, Viswanathan PC, Bezzina C et al (2000) Two distinct congenital arrhythmias evoked by a multidysfunctional Na(+) channel. Circ Res 86:E91-E97

Viswanathan PC, Benson DW, Balser JR (2001a) An *SCN5A* mutation causing isolated conduction disease is modulated by a common polymorphism. Circulation 104:II-134 (abstract)

Viswanathan PC, Bezzina CR, George AL et al (2001b) Gating-dependent mechanisms for flecainide action in SCN5A-linked arrhythmia syndromes. Circulation 104:1200–1205

Wang DW, Makita N, Kitabatake A et al (2000) Enhanced Na(+) Channel Intermediate Inactivation in Brugada Syndrome. Circ Res 87:E37-E43

Wang Q, Curran ME, Splawski I et al (1996) Positional cloning of a novel potassium channel gene: *KVLQT1* mutations cause cardiac arrhythmias. Nature Genetics 12:17–23

Wang Q, Shen J, Li Z, Timothy K et al (1995) Cardiac sodium channel mutations in patients with long QT syndrome, an inherited cardiac arrhythmia. Hum Molec Gen 4:1603–1607

Ward OC (1964) A new familial cardiac syndrome in children. J Irish Med Assoc 54:103–106

Wehrens XH, Abriel H, Cabo C et al (2000) Arrhythmogenic Mechanism of an LQT-3 Mutation of the Human Heart Na(+) Channel alpha-Subunit : A Computational Analysis. Circulation 102:584–590

Wei J, Yang IC, Tapper AR et al (1999) *KCNE1* polymorphism confers risk of drug-induced long QT syndrome by altering kinetic properties of IKs potassium channels. Circulation I-495 (abstract)

Weiss R, Barmada MM, Nguyen T et al (2002) Clinical and molecular heterogeneity in the Brugada syndrome: a novel gene locus on chromosome 3. Circulation 105:707–713

Wilde AA, Duren DR, VanLangen IM et al (1998) Auditory events as a trigger for arrhtyhmic events differentiate LQTS2 from LQTS1 patients. Circulation 98:776 (abstract)

Woosley RL, Chen Y, Freiman JP et al (1993) Mechanism of the cardiotoxic actions of terfenadine. JAMA 269:1532–1536

Yang P, Kanki H, Drolet B et al (2002) Allelic variants in Long QT disease genes in patients with drug-associated Torsades de Pointes. Circulation 105:1943–1948

Yang T, Snyders D, Roden DM (2001) Drug block of I(kr): model systems and relevance to human arrhythmias. J Cardiovasc Pharmacol 38:737–744

Zareba W, Moss AJ, Schwartz PJ et al (1998) Influence of the genotype on the clinical course of the long-QT syndrome. N Engl J Med 339:960–965

Zhang L, Timothy KW, Vincent GM et al (2000) Spectrum of ST-T-wave patterns and repolarization parameters in congenital long-QT syndrome : ECG findings identify genotypes. Circulation 102:2849–2855

Zhou ZF, Gong QM, Epstein ML et al (1998) HERG channel dysfunction in human long QT syndrome—intracellular transport and functional defects. J Biol Chem 273:21061–21066

Insulin Resistance and Cardiovascular Disease: New Insights from Genetics

A. Vidal-Puig[1] · E. D. Abel[2]

[1] Department of Clinical Biochemistry and Medicine, Addenbrooke's Hospital, University of Cambridge, Cambridge, UK

[2] Division of Endocrinology, Metabolism and Diabetes and Program in Human Molecular Biology and Genetics, School of Medicine, University of Utah, Building 533, Room 3410B, 15 North 2030 East, Salt Lake City, UT 84112, USA
e-mail: dale.abel@hmbg.utah.edu

1	**Insulin Resistance and the Metabolic Syndrome**	245
1.1	Definition of the Phenotype	245
1.2	Genetic Heterogeneity	246
1.3	Gene–Environment Interactions	246
2	**Insight into the Molecular Mechanisms of Insulin Resistance from Mouse Genetics**	247
2.1	Knock-Out Studies	247
2.2	Transgenic Mice	251
3	**Approaches to the Study of the Genetics of Insulin Resistance in Humans**	252
3.1	Positional Cloning	252
3.2	Candidate Gene Studies	252
3.3	New Developments That May Facilitate the Elucidation of the Genetic Basis of Insulin Resistance	253
4	**Genes Involved in Regulating Insulin Sensitivity in Humans Identified by the Positional Cloning Approach**	254
4.1	Calpain 10	254
4.2	Lamin A	254
5	**Genes Involved in Regulating Insulin Sensitivity in Humans Identified by the Candidate Gene Approach**	255
5.1	Insulin-Signaling Gene Candidates	256
5.1.1	Insulin Receptor Mutants	256
5.1.2	Insulin Receptor Substrate-1	257
5.1.3	Other Mutations in the Insulin-Signaling Cascade	257
5.2	Lipid Homeostasis Candidates	258
5.2.1	Peroxisome Proliferator-Activated Receptor Gamma	258
5.2.2	PPAR Alpha	259
5.2.3	CD36	260
5.3	Modulators of Insulin Sensitivity	260
5.3.1	ACRP 30	260
5.3.2	Resistin	261
5.3.3	PC1	261

6	Mechanisms for the Increased Risk of Cardiovascular Disease in Insulin Resistance	261
6.1	Genetic Studies in Rodents	261
6.1.1	Spontaneously Hypertensive Rat	262
6.1.2	Adiponectin/Acrp30	262
6.1.3	Nitric Oxide	263
6.1.4	Mouse Models with Dyslipidemia and/or Insulin Resistance	264
6.2	Human Genetic Studies	265
6.2.1	CD36	265
6.2.2	Lamin	265
6.2.3	PPARγ	266
6.2.4	Familial Hypercholesterolemia	266
6.2.5	Genetic Variation That May Modify or Increase the Association Between Insulin Resistance and Increased Risk for Cardiovascular Disease	267
7	Summary	269
	References	270

Abstract We are facing a growing worldwide epidemic of obesity and diabetes. The key underlying pathophysiology in these conditions is insulin resistance (IR), classically defined as an impaired ability of insulin to mediate its metabolic actions such as increasing glucose uptake into skeletal muscle, suppressing hepatic glucose production and suppressing lipolysis. However, the scope of insulin action is not restricted to the maintenance of glucose homeostasis. Rather it also has important roles in lipid and protein homeostasis and is probably the link that integrates many of the features of the metabolic syndrome: central obesity, hypertension, dyslipidemia (increased triglyceride, decreased HDL cholesterol and increase in small dense LDL), glucose intolerance or diabetes and hypercoagulability that together comprise the metabolic syndrome. As such, IR represents a major risk factor for cardiovascular disease. Monogenic forms of severe IR are recognized but these patients are rare. There is growing evidence that more common forms of IR also have an important genetic basis and it is possible that a few genes exerting a moderate effect over a general polygenic background may cause IR. Positional cloning and candidate gene studies have been employed to identify and pursue IR susceptibility genes in humans, while the genetic manipulation of mice has elucidated key molecules in the insulin-signaling pathway, suggesting new candidates for genetic studies in patients and drug targeting. The use of animal models and candidates gene studies in humans may also provide new insight into the mechanism underlying the increase in cardiovascular risk associated with IR.

Keywords Insulin resistance · Metabolic syndrome · Insulin signaling · Lipid homeostasis

1
Insulin Resistance and the Metabolic Syndrome

Insulin resistance (IR) is a state of impaired insulin action. In the presence of functional pancreatic beta cells, it is usually associated with a compensatory increase in insulin secretion to maintain normal glucose levels. When the compensatory pancreatic beta cell response fails, hyperglycemia and type 2 diabetes are the result. IR is generally defined in the context of insulin-mediated glucose lowering. It has become increasingly clear, however, that defective insulin signaling will broadly affect many other diverse important homeostatic functions such as vascular endothelial function (Steinberg and Baron 2002) or the regulation of appetite and reproduction (Bruning et al. 2000). There is growing evidence that IR represents a major risk factor for the development of cardiovascular disease. Indeed IR is commonly associated with other features such as central obesity, hypertension, dyslipidemia (increased triglyceride, decreased HDL cholesterol and increase in small dense LDL), glucose intolerance or diabetes and hypercoagulability that together comprise the metabolic syndrome (Zimmet et al. 2001). Many of these features are heritable and are enriched in certain populations. Thus, the likelihood of genetic predisposition to the metabolic syndrome is high, but it is also clear that environmental factors such as diet-induced obesity and inactivity, and low birth weight will also contribute to the development of IR (Zimmet et al. 2001).

Rare monogenic forms of severe IR due to genetic defects in the insulin receptor are recognized (Krook and O'Rahilly 1996). The genetic basis of more common forms of IR is less well understood due to a number of confounding factors (Fujisawa et al. 2002; Owen et al. 2002; Stern 2000; Vidal-Puig and Bjorbaek 1997).

1.1
Definition of the Phenotype

Several factors confound the definition of the IR phenotype in humans. First, there is no agreed level of insulinemia that constitutes IR, largely because the threshold for insulin action has not been clearly established. However, most would agree that a fasting insulin level greater than 150 pmol/l and/or insulin postglucose tolerance test over 1,500 pmol/l are criteria for marked IR. Second, IR has a relatively long gestation period. Indeed, the possibility that some genetic forms of IR will only become evident later in life may lead to the misclassification of subjects as controls in genetic studies. These confounding factors may seriously confound linkage analysis studies. Thus, an important challenge to increasing the power of IR genetic studies is to identify high-risk individuals and/or early stages of IR. In this regard, the possibility of using complete metabolic profiles or subphenotypes to better group homogeneous patterns of IR and metabolic syndrome has been suggested.

Third, the technology used to evaluate the degree of insulin sensitivity may also be a limiting/confounding factor. IR is evaluated directly or indirectly by measuring several parameters of carbohydrate metabolism. The complexity of these measurements is variable, ranging from simple determinations of plasma glucose and insulin levels, specific protocols of oral/intravenous glucose and insulin tolerance tests, derived indexes of insulin sensitivity such as HOMA or QUICKY (Katsuki et al. 2002) to the euglycemic hyperinsulinemic clamp technique, which provides the most accurate measurements of insulin sensitivity (Wallace and Matthews 2002). However, this last technique is very sophisticated, labor-intensive and clearly unsuitable for a routine clinical use. Thus, the use of these methods depends on the specific needs. For instance, while the use of clamps should be restricted to research protocols, the use of glucose and/or insulin tolerance tests may be more suitable for routine assessment in established clinical departments of endocrinology.

1.2
Genetic Heterogeneity

It is likely that IR may be associated with different genetic defects and metabolic alterations. If we consider that IR is a very common trait in type 2 diabetic patients, and that this disease is highly prevalent, it is conceivable that several genetic defects may run within the same family. Since IR is part of the metabolic syndrome, it is not unlikely that mutations in genes directly related to insulin sensitivity may act synergistically with other mutations in genes primarily affecting lipid homeostasis and/or blood pressure, with secondary effects on insulin sensitivity. Gene–gene interactions are expected, therefore, to play important roles in IR. Moreover, a relatively common observation is that genetic findings are not reproducible among different ethnic populations. This indicates a complex interaction between genetic background, specific mutations and sociocultural–environmental factors, implying that such polymorphisms may not be very important. In an attempt to exclude this factor, genetic studies will benefit from the use of relatively isolated pedigrees such as the Finns or Ashkenazi Jews.

1.3
Gene–Environment Interactions

The recent rise in the prevalence of obesity and diabetes is unlikely to be due to an "acute" genetic change in the population. It is more likely that environmental changes have triggered the epidemic, acting as sensitizing factors to the deleterious effects of predisposed genomes. The thrifty genotype theory proposes that genetic variants that optimize energy efficiency were selected in times of famine (Chukwuma and Tuomilehto 1998). These genetic variants do not pose any advantage for a new environment characterized by access to hypercaloric diet and lifestyle. The energy homeostatic system appears to be highly competent to solve situations where energy is in short supply, but is quite inefficient to deal with energy excess. Indeed, the excess of energy tends to be deposited in ectopic locations (outside its normal location in fat depots), causing IR. This form of lipotoxicity may be a common link between different pathogenic mechanisms leading to IR. Thus, the effect of the environment on insulin sensitivity may uncover but also confound the identification of primary genetic defects that predispose to IR.

2
Insight into the Molecular Mechanisms of Insulin Resistance from Mouse Genetics

In the past 2 decades, much progress has been made in understanding the molecular basis of insulin signaling (Fig. 1). It is evident that defects in any of these pathways could alter insulin signaling and result in IR. Many components of the insulin-signaling pathway have now been genetically modified in transgenic mice. The most informative data have come from gene knockout experiments. These studies are summarized in Tables 1 and 2. Global knockouts of the insulin receptor are uniformly lethal, with animals dying of ketoacidosis shortly after birth (Accili et al. 1996). However, many mice with deletions of downstream signaling pathways survived and produced important data.

2.1
Knock-Out Studies

Deletion of the insulin receptor substrate-1 (*IRS-1*) gene leads to IR and growth retardation, but glucose homeostasis is relatively preserved, because of compensatory islet cell hyperplasia (Araki et al. 1994; Tamemoto et al. 1994). In contrast, deletion of *IRS-2* results in IR and diabetes (Withers et al. 1998). Diabetes occurs because IRS-2 is an important regulator of β-cell survival via the regulation of β-cell specific transcription factors such as PDX1 (Kushner et al. 2002). PI3 kinase is believed to be a major regulatory node through which insulin signaling to metabolic pathways is mediated. PI3 kinase consists of a p85 regulatory subunit that interacts with IRS proteins via SH2 domains. p85 in turn interacts with the catalytic p110 subunit. Eight isoforms of the regulatory subunit of PI3 kinase have been described. They are alternatively spliced products of three genes (*p85α, p85β* and *P55^{PIK}*) that share some redundant and overlapping functions (Saltiel and Kahn 2001). Knockout strategies to analyze their function have been difficult to achieve. Whereas complete knockout of all products of the *p85α* gene result in early perinatal mortality, heterozygous knockouts exhibit an unexpected increase in insulin sensitivity (Fruman et al. 2000; Terauchi et al. 1999). These observations suggest that the stoichiometry of p85 to p110 also plays an important role in the regulation of signal transduction via PI3 kinase.

An important downstream regulator of the metabolic actions of insulin is Akt/PKB. Three isoforms of Akt exist and all have been deleted in mice. Akt1 is

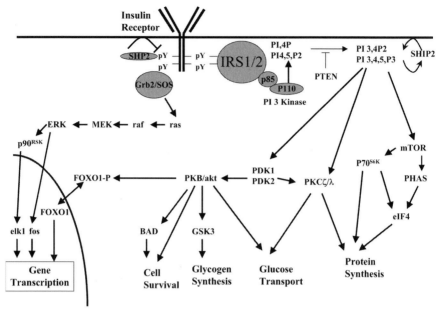

Fig. 1 Schematic representation of insulin signal transduction pathways. Activation of the insulin receptor (IR) increases the phosphorylation of the intracellular tyrosine kinase that in turn leads to increased association with and phosphorylation of insulin receptor substrates (*IRS1/2*). The signaling cascade diverges along pathways related to PI3 kinase or those related to mitogen-activated protein kinase (*ERK*). IRS proteins bind to the regulatory (*p85*) subunit of PI3 kinase which then interacts with the p110 catalytic subunit. Activation of PI3 kinase results in the generation of PI 3,4-diphosphate ($PI3,4P_2$) and PI 3,4,5-triphosphate ($PI3,4,5P_3$). These reactions are antagonized by PTEN. Insulin also activates the SH2 domain-containing inositol 5-phosphatase (*SHIP2*), which converts $PI3,4,5P_3$ to $PI3,4P_2$. $PI3,4P_2$ and $PI3,4,5P_3$ activate a variety of downstream kinases, including the mammalian target of rapamycin (*mTOR*), which regulates protein synthesis via PHAS/p70 S6 kinase ($p70^{S6k}$)/eukaryotic initiation factor 4 (*eIF4*). These lipid products also activate atypical protein kinase C isoforms ($PKC\zeta/\lambda$) and the phosphoinositide-dependent kinase (*PDK*) isoforms PDK1 and PDK2, which activate protein kinase B (*PKB/Akt*). In muscle and adipose tissue, activation of PKB and atypical protein kinases play an important role in glucose transporter translocation. PKB also regulates glycogen synthase kinase 3 (*GSK-3*), which may regulate glycogen synthesis and a variety of regulators of cell survival. PKB-mediated phosphorylation of the pro-apoptotic protein BAD inhibits apoptosis and phosphorylation of the forkhead transcription factors (*FOXO-1*) results in nuclear exclusion, thereby inhibiting its transcriptional activity. Binding of Grb2/son of sevenless (*Grb2/SOS*) to IRS mediates the activation of p21ras, thereby activating the ras/raf/mitogen-activated protein kinase (*ERK*) kinase (*MEK*)/ERK cascade. SHP-2 feeds back to inhibit IRS protein phosphorylation by directly dephosphorylating IRS1/2 and may also independently activate ERK. Activated ERK phosphorylates $p90^{RSK}$, which in turn phosphorylates c-fos, increasing its transcriptional activity. Similarly, ERK phosphorylates elk1 and increases its transcriptional activity

primarily involved in the regulation of growth and cell size, whereas Akt2 regulates insulin sensitivity in skeletal muscle and liver. Indeed, *Akt2*-null mice eventually develop diabetes, while *Akt 1*-null mice maintain normal glucose homeostasis (Cho et al. 2001a, b). Finally, insulin-mediated phosphorylation of forkhead transcription factors (foxo1) results in nuclear exclusion of the non-

Table 1 Phenotypes of mice with single-gene knockouts of insulin signaling pathway components

Gene	Phenotype	Reference
Insulin receptor (IR)	Normal at birth but die shortly thereafter from diabetic ketoacidosis	Accili et al. 1996
IRS-1	Insulin and IGF1 resistance, growth retardation, impaired glucose tolerance	Araki et al. 1994; Tamemoto et al. 1994
IRS-2	Insulin resistance, beta cell failure, type 2 diabetes	Withers et al. 1998
p85α	Increased insulin sensitivity and hypoglycemia in heterozygotes	Fruman et al. 2000; Terauchi et al. 1999
Akt1	Growth retardation	Cho et al. 2001b
Akt2	Type 2 diabetes, Insulin resistance in liver and muscle	Cho et al. 2001a
Foxo1	Haploinsufficiency rescues insulin resistance of heterozygous IR KO mice	Nakae et al. 2001, 2002
PTP1B	Increased insulin sensitivity and resistance to diet-induced obesity	Elchebly et al. 1999
SHIP2	Increased insulin sensitivity	Clement et al. 2001
GLUT4	Insulin resistance and cardiac hypertrophy, diabetes develops in heterozygotes but not in homozygotes	Katz et al. 1995; Stenbit et al. 1997
Syntaxin 4	Insulin resistance, impaired GLUT4 translocation in skeletal muscle	Yang et al. 2001a

Table 2 Tissue specific knockouts of components of the insulin signaling pathway

Gene	Tissue	Phenotype	Reference
Insulin receptor	Skeletal muscle	Normal glucose tolerance, hypertriglyceridemic, increased adiposity	Bruning et al. 1998
	Heart	Reduced heart size with decreased glucose and fatty acid oxidation rates	Belke et al. 2002
	β-Cell	Impaired glucose stimulated insulin release, glucose intolerance, impaired islet growth	Kulkarni et al. 1999
	Liver	Impaired glucose tolerance hyperinsulinemia	Michael et al. 2000
	Brain	Obesity, insulin resistance and hypothalamic hypogonadism	Bruning et al. 2000
GLUT4	Adipose tissue	Impaired glucose tolerance, secondary insulin resistance in liver and muscle	Abel et al. 2001
	Skeletal muscle	Severe insulin resistance. Glucose intolerance	Zisman et al. 2000
	Heart	Cardiac hypertrophy, impaired recovery from ischemia	Abel et al. 1999; Tian and Abel 2001
Glucokinase	Liver	Impaired glucose stimulated insulin release, impaired glycogen synthesis, hyperglycemia	Postic et al. 1999
	β-Cell	Early demise secondary to severe diabetes	Postic et al. 1999

phosphorylated receptor, which ultimately leads to de-repression of insulin-responsive genes. Thus, haploinsufficiency of *foxo1* in mice restores insulin sensitivity to heterozygous insulin-receptor-null mice and transgenic expression leads to IR (Nakae et al. 2001, 2002).

The insulin-signaling cascade can be modulated by the activity of phosphatases or by the serine phosphorylation of IRS proteins. Knockouts of the phosphatases PTP1B and SHIP 2 lead to enhanced insulin action (Clement et al. 2001; Elchebly et al. 1999) and overexpression of the LAR (leukocyte antigen-related) protein-tyrosine phosphatase in muscle causes IR (Zabolotny et al. 2001). Persistent activation of PI3 kinase in vivo or in vitro leads to serine phosphorylation of IRS-1, which retards insulin signaling by blocking the interactions between the insulin receptor, IRS-1 and the p85 subunit of PI3 kinase, which are mediated by SH2 domain interactions with phosphorylated tyrosines. Moreover, serine phosphorylated IRS-1 is subject to increased proteasomal degradation (White 2002). Other serine kinases such as PKCθ and c-Jun N terminal kinase (JNK) also catalyze the serine phosphorylation of IRS-1 (Aguirre et al. 2000). PKCθ-mediated phosphorylation of IRS-1 represents one potential link between increased fatty acid flux into muscle and the development of IR (Yu et al. 2002). JNK-mediated serine phosphorylation of IRS-1 represents a potential link between inflammation (signaling via cytokine receptors) and the development of IR (Hirosumi et al. 2002).

The ability of muscle to take up glucose is a universal defect in insulin resistant states. Deletions of *GLUT4* expression or function in mice have recapitulated IR. Mice with genetic ablation of the *GLUT4* gene exhibit surprisingly normal ambient blood glucose concentrations despite glucose intolerance and significant IR. They are growth retarded and exhibit a diminished life span. They develop cardiac hypertrophy, cardiac dysfunction and severely reduced adipose tissue (Katz et al. 1995). This model suggests that GLUT4 is essential for normal cardiac function. Mice that are heterozygous (+/−) for the *GLUT4* gene deletion express reduced levels of GLUT4. They are insulin resistant and 50%–60% of them become diabetic. Moreover, these mice develop increased blood pressure and cardiac changes that are reminiscent of diabetic cardiomyopathy (cardiac hypertrophy and patchy fibrosis). Thus, reduced levels of GLUT4 can contribute to the pathogenesis of IR (Stenbit et al. 1997).

Impaired docking of GLUT4 vesicles to the plasma membrane can also lead to IR, independent of changes in *GLUT4* expression. Mice with a heterozygous null mutation in the t-SNARE protein, syntaxin 4, develop impaired glucose tolerance and 50% reduction in skeletal muscle glucose uptake in vivo and in vitro, associated with reduced insulin-stimulated GLUT4 translocation. Intriguingly GLUT4 translocation is normal in adipocytes (Yang et al. 2001a). These data support the critical role for syntaxin 4 in GLUT4 vesicle docking in vivo, and suggest that tissue-specific differences exist in the mechanisms that regulate GLUT4 translocation in adipose tissue versus skeletal muscle.

The ability to engineer mice with gene deletions in selective cellular/tissue compartments has shed additional insight into the important cross talk between various organs in the pathogenesis of IR. Deletion of insulin receptors in muscle is associated with normal glucose tolerance and normal insulin concentrations, but these mice develop hypertriglyceridemia and increased concentrations of free fatty acids that are reminiscent of the metabolic syndrome (Bruning et al.

1998). In contrast, liver-specific knockouts of the insulin receptor develop significant hyperinsulinemia and glucose intolerance (Michael et al. 2000). Intriguingly, deletion of insulin receptors from β cells leads to islet cell dysfunction and glucose intolerance (Kulkarni et al. 1999). Deletion of GLUT4 glucose transporters from skeletal muscle results in IR, hyperinsulinemia and impaired glucose tolerance (Zisman et al. 2000). Surprisingly, deletion of GLUT4 selectively in adipose tissue resulted in a similar phenotype (Abel et al. 2001). Taken together, these results indicate specific contributions of various tissues/cell types to the pathogenesis of IR in vivo.

While these studies have shed important insight into the roles of various members of the insulin-signaling cascade in the pathogenesis of IR, it is unlikely that single gene deletions will account for IR syndromes in humans. However, combinations of mice with haploinsufficiency of various components of the insulin signaling pathway such as heterozygous deletions of the *insulin receptor* and *IRS-1*, combined heterozygous deletions of the *insulin receptor, IRS-1* and *IRS-2* or of *IRS-1* and *glucokinase* lead to the development of more profound IR and ultimately diabetes, despite the absence of diabetes in mice that are heterozygous for single genes (Bruning et al. 1997; Terauchi et al. 1997). These data suggest that downregulation of expression of multiple insulin signaling molecules could, in concert, contribute to the pathogenesis of IR and diabetes in humans.

2.2
Transgenic Mice

Transgenic studies have also shed important insights into the interaction of cellular overnutrition and the development of IR and revealed important interactions between lipid synthesis and partitioning between the adipocyte and the liver in the genesis of IR. Increased nutrient flux (glucose or fatty acids) ultimately leads to increased flux via the hexosamine biosynthetic pathway. Transgenic mice with overexpression of the rate-limiting enzymatic regulator of hexosamine biosynthesis (glutamine fructose amidotransferase, GFA) in muscle and adipose tissue (Hebert et al. 1996), β cells (Tang et al. 2000) or hepatocytes (Veerababu et al. 2000) ultimately develop IR and in the case of hepatocyte overexpression, they develop obesity and dyslipidemia. Accumulation of triglycerides and increased fatty acid flux into muscle and liver leads to IR. Thus, muscle- or liver-specific overexpression of lipoprotein lipase in transgenic mice leads to IR (Ferreira et al. 2001; Kim et al. 2001). Similarly, mouse models of lipodystrophy, in which white adipose tissue is genetically ablated, develop profound IR on the basis of increased lipid accumulation in muscle and liver, which can be reversed by transplantation of adipose tissue, or by treatment with adipose tissue-derived secretory proteins such as leptin and adiponectin (Colombo et al. 2002; Ebihara et al. 2001; Gavrilova et al. 2000; Yamauchi et al. 2001).

Sterol-regulatory element binding proteins (SREBP) are important transcriptional regulators of lipogenesis in liver and adipose tissue. Transgenic overexpression of SREBP-1c in adipose tissue produces a mouse model of lipodystrophy with severe IR (Shimomura et al. 1998). Interestingly, these mice develop fatty livers that are associated with increased hepatic expression of SREBP1-c. The hepatic changes are similar to those that develop in leptin-deficient ob/ob mice (Shimomura et al. 1999). Both of these models are hyperinsulinemic, and the hyperinsulinemia ultimately leads to downregulation of IRS-2 expression, thus exacerbating the IR. Despite hepatic IR from decreased IRS-2-mediated insulin signaling, alternative insulin signaling pathways continue to stimulate SREBP1-c expression (Shimomura et al. 2000). Similarly, increased expression of SREBP-1a expression was also observed in the livers of *IRS-2*-null mice that are insulin resistant and that develop hepatic steatosis (Tobe et al. 2001). Moreover, introduction of the *SREBP-1*-null allele into ob/ob mice reversed hepatic steatosis without restoring systemic insulin sensitivity (Yahagi et al. 2002). These models indicate potential mechanisms by which excessive nutrient fluxes or altered nutrient partitioning can interact to produce IR. These observations are of particular relevance to humans given the important role that diet-induced obesity plays in the pathogenesis of IR.

3
Approaches to the Study of the Genetics of Insulin Resistance in Humans

3.1
Positional Cloning

Positional cloning strategies use a genome-wide linkage analysis approach to identify regions of DNA linked with a specific disease, followed by identification of genes located in these regions that may cause the disease. This strategy relies heavily on technological support to navigate the whole genome efficiently using SNP (single nucleotide polymorphisms) markers. Linkage analysis, studying the segregation of IR with specific alleles of SNPs, suggests that the gene of interest is in linkage disequilibrium with a region close to this marker. This has the advantage of restricting the search to a smaller area of the genome, thereby decreasing the number of potential candidate genes. One of the advantages of this approach is that no physiological assumptions influence or bias the identification of the gene harboring the genetic variant.

3.2
Candidate Gene Studies

This approach takes advantage of, but also is limited by, the present understanding of the molecular determinants of insulin sensitivity (Elbein et al. 1995; Ikegami et al. 1996; Moller et al. 1996). These studies involve screening for variants followed by examination of their frequency in cases and controls. The list of potential candidates for IR is growing in parallel with our knowledge of the pathophysiology of this syndrome. An important criterion for a gene to be considered a candidate is that it is related to early changes associated with IR and not associated with other known phenotypes. But this approach has some limitations. For instance, the lamin A (*LMA*) gene which has been linked with a form of lipodystrophy-associated IR would not have fulfilled the second requirement, since some mutations in this gene are known to cause skeletal muscle dysfunction (Hegele et al. 2000; Lelliott et al. 2002; Shackleton et al. 2000).

The candidate gene approach has not been very successful because of problems related to age, gender or ethnicity of study populations and the increasing number of proposed candidates. Furthermore, the power of this approach to identify variants with modest effects is quite limited. In contrast to positional cloning approaches, most genetic screenings using the candidate approach have not considered regulatory regions that may be of great relevance. A few strategies have clearly improved the outcome of candidate gene studies. These include selecting extreme phenotypes (O'Rahilly 2002) combined with stricter criteria to select the candidates (such as a strong hypothesis or previous association studies) together with increasing genotyping capacity through incorporation of powerful high-throughput technology. The combination of these three strategies has allowed the identification of PPARγ mutants in insulin-resistant patients, lending further support to the involvement of this transcription factor in insulin sensitivity (Barroso et al. 1999). The focus on extreme phenotypes has been criticized on the grounds that the mutations identified by this strategy do not usually provide relevant information for the most common forms of IR. However, the identification of these mutants has validated the existence of a pathway involving PPARγ that controls insulin sensitivity and PPARγ as a target for pharmacological development. Positional cloning and candidate gene approaches are not exclusive. For instance, a candidate approach may be used to prioritize the study of genes located within the genomic location identified by positional cloning.

3.3
New Developments That May Facilitate the Elucidation of the Genetic Basis of Insulin Resistance

To solve such a complex problem, it is necessary to use powerful tools capable of integrating the analysis of multiple variables. Several developments have occurred that probably will facilitate tackling the problem of IR (Bennet et al. 2001). First, several collaborative groups have embarked on the assembly and characterization of large populations suitable for genetic studies of IR. However, still better-characterized family populations are necessary. Second, technology, mostly developed during the human genome project, now allows high-throughput genotyping strategies to screen these populations and access to powerful bioinformatic support can facilitate comprehensive data analysis at competitive cost. A major breakthrough in the last 10 years has been the development of automatic sequencing technology, which allows large population genotyping (Waterston et al. 2002). Also, the availability of the whole human genome sequence and a dense map of single nucleotide polymorphisms (SNPs) is expected to have a positive impact on the identification of biomedically important genes controlling insulin sensitivity (Sachidanandam et al. 2001). The availability of the human genome map has created unique opportunities to establish the genetic basis of complex traits such as IR in the context of the metabolic syndrome. However, whether or not the identification of relatively small risk variants will be useful for clinical screening and risk identification is still a concern (Willett 2002). In our opinion, it is more likely that the genomic information will need to be integrated with epidemiological studies of environmental factors.

4
Genes Involved in Regulating Insulin Sensitivity in Humans Identified by the Positional Cloning Approach

4.1
Calpain 10

Positional cloning has recently identified *calpain 10* as the first gene involved in IR (Baier et al. 2000; Permutt et al. 2000; Yang et al. 2001b). Linkage studies in Mexican-Americans had revealed a major locus for type 2 diabetes mellitus (*NIDDM1*) on chromosome 2 and positional cloning identified a common G→A polymorphism (UCSNP-43) within calpain 10. This sequence variant is associated with decreased calpain 10 gene expression and is associated with IR in Pima Indians. The *G/G* genotype is associated with decreased rate of insulin-mediated glucose turnover. However, these findings have not been clearly reproduced in the Caucasian population and more conclusive studies are awaited. One common outcome of positional cloning is that the newly identified gene is relatively unexpected. The role that calpain 10 may play in insulin sensitivity remains unclear (Elbein et al. 2002). Calpain 10 is a calcium-activated neutral cysteine protease whose substrates are not well characterized, and the most reasonable speculation is that proteins cleaved or activated by calpain 10 may have an important role in insulin sensitivity.

4.2
Lamin A

Lamin A is another important success of the positional cloning strategy. Mutations in the Lamin A gene are associated with the Duncan-type familial partial lipodystrophy-insulin resistance syndrome (FPLD). Since the Lamin A gene encodes a nuclear envelope structural protein it was not considered a priori a likely candidate for IR (Hegele et al. 2000; Shackleton et al. 2000). Positional cloning not only identified the cause of this disease but also revealed an unexpected role for the nuclear envelope in insulin sensitivity and adipogenesis. FPLD is an autosomal dominant disorder characterized by a marked loss of subcutaneous adipose tissue from the trunk, the gluteal region and extremities at puberty with a tendency to gain adipose tissue around the neck. Biochemical abnormalities associated with this lipodystrophy include IR, hyperlipidemia and altered leptin levels. Most of the mutations causing FPLD are located in a hot spot of the gene around amino acid 482 (e.g., R482W). Interestingly, mutations in other locations of this gene are thought to be responsible for three different disorders: conduction system disease (CMD1A; OMIM 115200), the autosomal dominant form of Emery-Dreifuss muscular dystrophy (EDMD2; OMIM 150330 and 181350), and limb girdle muscular dystrophy type 1B (LGMD1B; OMIM 119001). The reason for this tissue/depot specificity is unclear, but may be related to interactions with tissue-specific transcription factors as well as gene expression profile specificity. For instance, it has been shown that R482W interferes with the normal binding of SREBP1 (a transcription factor involved in fatty acid metabolism and insulin sensitivity).

5
Genes Involved in Regulating Insulin Sensitivity in Humans Identified by the Candidate Gene Approach

The most obvious candidates for a role in IR are those molecules involved in the transduction of the insulin signal (Moller et al. 1996) (Table 3). Patients with IR have functional defects in the insulin receptor, insulin receptor substrate proteins (IRSs) and PI-3-kinase signal transduction pathways. Insulin-resistant patients typically exhibit reduced phosphorylation of the IRS-1-PI-3-K complex, which can explain why these patients also have impaired glucose transport and glycogen synthesis. However, a reduction in glycogen synthase activity is also observed in prediabetic subjects, and importantly, these defects do not regress with improved metabolic control, suggesting that primary defects in glycogen synthesis may be a cardinal feature of IR. Thus, genetic defects in components of the insulin-signaling cascade are plausible mechanisms for IR. Many of the genes encoding these proteins such as the insulin receptor or the insulin-regulated glucose transporter 4 (GLUT4), have been studied extensively. The elucidation of new molecules in the insulin-signaling cascade has been followed by mutation fishing expeditions that in the best cases have identified mutations that account for a very small percentage of insulin-resistant or diabetic phenotypes.

In parallel with the increasing knowledge of the pathophysiological mechanisms of IR, the search for new candidates has been extended to molecules involved in lipid metabolism. Indeed, an emerging view suggests that IR may be primarily a lipid disorder that results in carbohydrate dysregulation. Carbohydrate and lipid metabolism are closely linked; for example, defects in glucose oxidation may lead to accumulation of malonyl CoA and intramuscular triglycerides that further impair insulin sensitivity.

As indicated above, the list of candidate genes for IR is increasing rapidly and a compilation of all the genetic studies is beyond the scope of this chapter. As a matter of exemplification, we have selected a few candidate genes based in our opinion on their importance and/or novelty.

Table 3 Insulin resistance candidate genes

Pathway	Candidates
Insulin signaling	Insulin receptor
	Insulin receptor substrates
	IRS proteins (1,2,3,4, Gab1, p62)
	Shc proteins (A,B,C)
	Grb10
	CCbl
	Other docking proteins
	SHIP, Fyn, CAP
	PI3Kinase family
	Catalytic subunit p110α, β
	Regulatory subunits p85α, p55α, p50α, p85α, p55γ
	Substrates of PI3 kinase
	Akt/PKB
	PKCζ/λ
	Glucose transport
	GLUTs 1, 2, 4
	Glycogen synthesis
	Glycogen synthase
	Protein phosphatase PP1
	Glycogen synthase kinase (GSK)
Carbohydrate homeostasis	Hexokinase 2
	Phosphoenolpyruvate carboxykinase
	Hepatic and muscle forms of pyruvate kinase
	Hepatic phosphofructokinase
Fatty acid homeostasis	β3 Adrenergic receptor
	SREBP1c
	CD36
	Carnitine palmitoyl transferase I
	Lipoprotein lipase
Modulators of insulin sensitivity	ACRP30, resistin
	TNF-α, TNF-α receptor
	PC1, 11 β HSD, caveolin-3

5.1
Insulin-Signaling Gene Candidates

5.1.1
Insulin Receptor Mutants

Insulin receptor mutants are typically associated with monogenic subtypes of IR (Krook and O'Rahilly 1996). The clinical phenotype of these patients tends to correlate with the residual function of the mutated insulin receptors, leading to a spectrum of diseases with progressively decreasing degrees of severity. These include the Donohue syndrome (DS), Rabson Mendenhall syndrome (RMS), Kahn's type A insulin resistance, and HAIR AN syndrome (hyperandrogenism, IR, acanthosis nigricans) (Moller and Vidal-Puig 1997; Vidal-Puig and Moller

1997). Typically DS is produced by homozygous/compound heterozygous mutations in the insulin receptor alpha subunit, which result in complete loss of the insulin receptor function. RMS is also produced by mutations in the alpha subunit of the insulin receptor causing partial loss of its function. Some of the patients with Type A IR or HAIR-AN syndrome also have mutations in the insulin receptor, but mainly located in the tyrosine kinase domain of the IR β subunit. These mutant insulin receptors still possess some residual function. Globally, mutations in the insulin receptor represent a small percentage (<0.01%) of the genetic defects producing IR, but the study of these natural mutations has been key to the understanding of the structure and function of the insulin receptor.

5.1.2
Insulin Receptor Substrate-1

Insulin receptor substrate-1 (IRS-1) was the first cloned member of a family of insulin receptor substrates that activate phosphatidylinositol (PI) 3-kinase and promote GLUT4 translocation. The *IRS-1* gene was considered, therefore, an important candidate to mediate IR. The most common variant of *IRS-1* is the *Gly→Arg972* change, which in in vitro assays was shown to impair insulin-stimulated IRS-1-associated PI3-kinase activity by interfering with the binding of IRS-1 to the p85 subunit of PI3-kinase (Federici et al. 2001; Hribal et al. 2000; Rosskopf et al. 2000). This mutant also decreased basal and insulin-stimulated glucose transport, Akt phosphorylation and glycogen synthesis. These data strongly suggested that the *Arg972* polymorphism of *IRS-1* may contribute to IR. The effect of this mutant on IR may be more prominent in obese individuals according to genotype/phenotype studies stratified by body mass index (BMI). However, the effects of the *IRS-1 Arg972* polymorphism are not limited to insulin sensitivity. This mutant also interferes with insulin secretion, suggesting that its prodiabetogenic effect may be the result of its combined effects on insulin secretion and sensitivity.

5.1.3
Other Mutations in the Insulin-Signaling Cascade

The mutation *Met→Ile* at codon 236 of the p85 alpha subunit of PI3-kinase significantly reduces insulin sensitivity and glucose tolerance in individuals homozygous for the variation. By itself, this mutant does not induce diabetes. Further downstream in the insulin-signaling cascade, an *Asp→Tyr* polymorphism was identified at codon 905 of the regulatory (PPP1R3) subunit of glycogen-associated protein phosphatase-1 (PP1G) (Xia et al. 1998). It has been suggested that this variant may be associated with IR, but further studies using adenoviral-mediated gene transfer do not support a role in impaired insulin-stimulated glycogen synthesis (Hansen et al. 2000; Wang et al. 2001). Nonetheless, defects in the PPP1R3 subunit may cause a syndrome of IR when associated with other minor but complementary defects in other pathways controlling insulin sensitivity. It has recently been reported that a heterozygous frameshift/ premature stop mutation in the PPP1R3A can, in association with mutations in PPARγ, cause IR (Savage et al. 2002) (see Sect. 5.2.1).

5.2
Lipid Homeostasis Candidates

5.2.1
Peroxisome Proliferator-Activated Receptor Gamma

Peroxisome proliferator-activated receptor gamma (PPARγ) is a key transcription factor in adipogenesis and the receptor for some thiazolidinediones, a new class of insulin-sensitizing drugs. These observations highlighted PPARγ as an important candidate in determining insulin sensitivity. A number of variations in the *PPARγ* gene have been identified. The most common and probably most relevant for the general population is the *Pro12Ala* polymorphism located at the amino terminus of *PPARγ*, in a region not involved in DNA or ligand binding. Initial genetic studies produced conflicting results. A recent meta-analysis has shown an association between the *Pro12Ala* variant, increased insulin sensitivity and better antilipolytic action of insulin (Altshuler et al. 2000). The marked controversy initially raised by this polymorphism is an example of the difficulties and limitations of genetic analyses of multifactorial diseases. In this case it may also suggest that the effect of *Pro12Ala* polymorphism is relatively minor. The experience illustrates the need for large population studies in order to have confidence in the results. A recent study showing the interaction between the *Pro12Ala PPAR γ* gene variant and specific nutrients (ratio saturated/unsaturated fatty acids) also illustrates how environmental factors can have a critical influence on the interpretation of IR genetic studies (Luan et al. 2001).

Another *PPARγ* variation has been found at position 115 within the amino terminus region. This one is located in the proximity of, and interferes with, the phosphorylation of Ser114. Individuals carrying this variant were obese, suggesting that mutations in a gene preferentially expressed in adipose tissue could lead to obesity (Ristow et al. 1998). Other rare mutations of *PPARγ* have been detected by searching in selected populations characterized by extreme phenotypes of IR. While this strategy tends to identify rare monogenic forms of the disease rather than variants that have more impact in the general population, it is extremely useful to confirm the involvement of specific genes in biological functions. Patients with the *Pro467Leu* and *Val290Met* variants of *PPARγ* exhibit not only insulin resistance but also hypertension and dyslipidemia, suggesting a broader role for PPARγ outside of regulating insulin sensitivity alone (Barroso et al. 1999). The characterization in vitro of the *Pro467Leu* and *Val290Met PPARγ* variants has provided the basis for two new concepts (Barroso et al. 1999). First, neither mutant can be activated by thiazolidinediones at low doses, and at higher doses the *Val290Met* mutant remains inactive. The correlation between the effects of thiazolidinediones in vitro and the response in vivo suggests that an individual-specific pharmacogenomic approach is feasible. Second, the PPARγ mutations may exert their effect through a dominant-negative mechanism (i.e., the mutant allele can interfere with the activity of the normal allele). It is possible that this dominant-negative effect is not confined to *PPARγ* but may also involve *PPARα* or *PPARδ*. This may be relevant in tissues where these receptors are normally co-expressed (e.g., vascular wall, heart, skeletal muscle) or abnormally co-expressed in pathological conditions (e.g., induced expression of PPARγ2 in fatty liver).

The study of extreme phenotypes also offers the possibility of identifying individuals who concentrate several mutations in their genomes (O'Rahilly 2002). As isolated genetic defects, some mutants may not be sufficient to produce a florid insulin resistant syndrome in vivo, but when associated with other minor mutations in a single genome, they may have devastating effects. A recent example of this has been the identification of the first di-genic form of IR in a family in which only members having double heterozygous mutations in *PPARγ* and in the muscle-specific regulatory subunit of PP1R3 had IR. The *PPARγ* mutant would be expected to exert its effects mainly in adipose tissue and fatty acid metabolism, while the effects of the *PP1R3* mutation should be restricted to the interference of glycogen synthesis in skeletal muscle. The deleterious effects of the association of both mutants are an illustration that crosstalk between tissues (adipose tissue/muscle) is of great relevance in the maintenance of energy homeostasis and that fatty acid and carbohydrate metabolism are closely regulated (Savage et al. 2002).

IR is a risk factor for atherosclerosis and coronary artery disease. There is pharmacological evidence that activation of PPARγ by thiazolidinediones improves insulin sensitivity and prevents the development of atherosclerosis. The beneficial effects of PPARγ activation on cardiovascular risk factors does not appear to be exclusively mediated by the improvement in insulin sensitivity. Indeed, PPARγ activation affects endothelial function, may modulate blood pressure (e.g., decreased levels of type 1 angiotensin II receptor gene), activates vascular smooth muscle cells' cholesterol efflux (ABC transporter) (Akiyama et al. 2002) and fibrinogen lysis (e.g., PAI-1), exerts antithrombotic effects (e.g., suppression of thromboxane receptor gene), and may limit postischemic cardiac injury (Reusch 2002). It is not yet known if mutations in PPARγ may account for pathologically relevant cardiovascular defects.

5.2.2
PPAR Alpha

This is a member of the PPAR family of nuclear receptors whose main function is to activate the genetic program of fatty acid oxidation. PPARα activation might improve insulin sensitivity by promoting fatty acid oxidation. PPARα also plays an important role in modulating inflammatory responses, which may be relevant to its cardiovascular actions. Genetic analysis of PPARα has revealed a *Leu162Val* polymorphism located in the DNA binding domain that is associated with increased numbers of apoB apolipoprotein-rich pro-atherogenic particles (Vohl et al. 2000). Paradoxically, in clinical studies, the *Leu162Val* polymorphism retards the progression of atherosclerosis, thereby diminishing the risk for heart disease (Flavell et al. 2000, 2002). This observation suggests that the protective effect of *Leu162Val* may be directly exerted at the level of the vascular wall. There are no data available about the effect of *PPARα* mutations on insulin sensitivity in humans, but mouse data suggests that PPARα may be necessary for the development of IR associated with high-fat diet. The influence of PPARα on the heart has emerged from the association of a G→C polymorphism in intron 7 with left ventricular growth in response to exercise.

5.2.3
CD36

The *Cd36* gene encodes a fatty acid transporter (CD36), which has been linked to the transmission of IR, defective fatty acid metabolism and increased blood pressure (Aitman 2001; Miyaoka et al. 2001; Petrie et al. 2001; Tanaka et al. 2001). CD36 is a receptor with several functions, including intracellular signaling and incorporation of long-chain fatty acids. Humans deficient in *Cd36* are glucose intolerant and more likely to be come diabetic. Glucose clamp studies on a small cohort of these patients revealed a significant defect in whole-body insulin-mediated glucose uptake (Miyaoka et al. 2001). Interestingly, CD36 is a target gene of PPARγ activation and recent studies in *Cd36*-deficient rodents suggest that it is the major mechanism by which currently available PPARγ agonists improve fatty acid transport and insulin sensitivity (Qi et al. 2002; Seda et al. 2003).

5.3
Modulators of Insulin Sensitivity

Adipose tissue is emerging as an important endocrine gland secreting hormones that regulate energy homeostasis. The most obvious example is leptin, which controls aspects such as food intake and energy expenditure. Recently other adipocyte-secreted hormones have been identified that may have important roles in insulin sensitivity. In fact, it has been claimed that alterations in ACRP30 and/or resistin may account for the IR associated with obesity. Thus, these molecules should be considered a priori candidates that mediate IR.

5.3.1
ACRP 30

Also known as adiponectin or adipo Q, ACRP 30 is an adipocyte-secreted protein whose absence has been associated with IR. ACRP30 plasma levels are decreased in subjects with IR. Furthermore, ACRP30 levels are up-regulated by thiazolidinediones (Combs et al. 2002; Yang et al. 2002), suggesting that ACRP30 may mediate some of the insulin-sensitizing effects of PPARγ activation. The

ACRP30 gene is located in chromosome *3q27*, a region previously identified as a diabetes-susceptibility locus using genome-wide scans. Several groups have searched the *ACRP30* gene for genetic variants linked to the development of IR. One group identified two SNPs, at positions 45 and 276 of the *ACRP30* gene, which may be associated with a higher IR index and increased BMI (Hara et al. 2002). Based on the same approach, Stumvoll et al. showed that the association of a common polymorphism in nucleotide 94 of exon 2 of the *ACRP30* gene with insulin sensitivity seems to be dependent on BMI (Stumvoll et al. 2002). In general, the effect of genetic variations in ACRP30 on insulin sensitivity is probably secondarily to changes in body fat mass.

5.3.2
Resistin

Resistin is another protein that antagonizes insulin action and seems to be down-regulated by PPARγ agonists in rodents. Based on the initial reports that resistin may be an important link between obesity and IR, several groups examined SNPs in noncoding sequences but concluded that genetic defects in resistin are unlikely to be a major cause of IR (Engert et al. 2002; Ma et al. 2002; Pizzuti et al. 2002; Sentinelli et al. 2002; Wang et al. 2002).

5.3.3
PC1

PC1 is a plasma cell differentiation antigen glycoprotein which inhibits insulin receptor signaling and is associated with IR. Several studies have identified a polymorphism in exon 4 of the PC-1 gene (*Lys121Glu*), which is strongly associated with IR (Frittitta et al. 2001). This polymorphism was correlated with insulin sensitivity independently of BMI, gender, age and waist circumference. However, this association has not been confirmed in all ethnic groups (Costanzo et al. 2001; Gu et al. 2000; Pizzuti et al. 1999; Rasmussen et al. 2000); for example, the *Lys121Glu* polymorphism is not associated with type 2 diabetes or IR among Danish Caucasians.

6
Mechanisms for the Increased Risk of Cardiovascular Disease in Insulin Resistance

6.1
Genetic Studies in Rodents

A variety of rodent models (naturally occurring mutations or genetically engineered mice) have provided interesting insights into potential links between IR and the development of cardiovascular disease.

6.1.1
Spontaneously Hypertensive Rat

The spontaneously hypertensive rat (SHR) is a rat model of the metabolic syndrome. These animals are hypertensive, insulin resistant, glucose intolerant and dyslipidemic. Using congenic chromosomal mapping and expression analysis, Aitman and colleagues mapped the quantitative trait loci (QTL) for defects in glucose and fatty acid metabolism, hypertriglyceridemia and hypertension to a single locus on chromosome 4 (Aitman et al. 1999). The defective gene was *Cd36*, or fatty acid translocase, encoding a fatty acid transport protein highly expressed in adipose tissue, skeletal and cardiac muscle (Brinkmann et al. 2002). Transgenic overexpression of *Cd36* lowered triglyceride and fatty acid levels and increased insulin sensitivity in mice, and transgenic rescue of SHR with *Cd36* ameliorated the IR and reduced serum levels of fatty acids in this strain (Ibrahimi et al. 1999; Pravenec et al. 2001b). Moreover, treatment of apolipoprotein E (Apo E) knockout mice (which develop IR and accelerated atherosclerosis on a high-fat diet) with a thiazolidinedione (troglitazone) resulted in up-regulation of *Cd36* expression in foam cells (Chen et al. 2001). In addition to amelioration of IR, the up-regulation of CD36, which presumably acts as a scavenger receptor for oxidized LDL, was associated with regression of atherosclerotic lesions. These data indicate that altered CD36 expression may represent a potentially unifying mechanism for the association of IR, dyslipidemia and atherosclerosis, which is also supported by studies in humans with CD36 deficiency (see Sect. 5.2.3).

It is unlikely, however, that CD36 deficiency can account for hypertension. This is based on the fact that transgenic rescue of *Cd36*-deficient rats did not correct hypertension (Pravenec et al. 2001b). Moreover, SHR rats from different sources that may have diverged from the founder lines have not been shown to have any alterations in *Cd36* gene expression and variable degrees of IR (Furukawa et al. 1998; Gotoda et al. 1999). A related strain, the stroke prone SHR, is also insulin resistant, but has normal expression of CD36 (Collison et al. 2000). Transgenic overexpression of the renin gene in rats leads to marked hypertension without accompanying IR (Vettor et al. 1994). Thus it is likely that other genetic defects contribute to hypertension in SHR, and that other genetic loci might also be involved in the insulin-resistant phenotype. This is supported by the recent discovery of a point mutation in the SREBP1c gene in SHR (Pravenec et al. 2001a).

6.1.2
Adiponectin/Acrp30

Adiponectin/Acrp30 has significant effects on insulin sensitivity in rodents in vivo (Berg et al. 2002). Many mouse models of IR and lipodystrophy are associated with reduced levels of Acrp30, which are reversed by exogenous administration of recombinant peptides (Yamauchi et al. 2001). The mechanisms for the insulin-sensitizing action of Acrp30 include enhanced hepatic insulin sensitivity and increased fatty acid oxidation in skeletal muscle (Combs et al. 2001; Tomas et al. 2002). In vitro studies suggested that Acrp30 is also antiatherogenic based on observations that it decreased monocyte adhesion to endothelial cells and reduced cytokine production, phagocytosis, lipid accumulation and LDL uptake in macrophages (Arita et al. 2002; Matsuda et al. 2002; Okamoto et al. 2002; Ouchi et al. 2000, 2001). The recent knockout of the Acrp30 gene in mice provided direct evidence that supports the hypothesis that Acrp30 may represent an important link between IR and atherosclerosis. Acrp30 knockout mice are not obese but develop IR (Kubota et al. 2002; Maeda et al. 2002) and a twofold increase in neointimal formation in response to external vascular cuff injury (Kubota et al. 2002; Matsuda et al. 2002). These data are also supported by data in humans, showing that Acrp30 levels were lowest in diabetics with coronary artery disease (Hotta et al. 2000).

6.1.3
Nitric Oxide

The generation of nitric oxide from endothelial cells plays an important role in cardiovascular homeostasis (see the chapter by Huang, this volume). The activity of endothelial cell nitric oxide synthase (eNOS/NOSIII) is acutely regulated by insulin (Zeng et al. 2000), which also chronically regulates eNOS gene expression (Kuboki et al. 2000). In insulin resistant states, there is a reduction in eNOS content in endothelial cells and the ability of insulin to increase eNOS activity and produce nitric oxide. This may represent one mechanism for endothelial dysfunction in IR and diabetes mellitus. Pharmacological blockade of NOS by intravenous administration of N^G-mono-methyl-L arginine (L-NMMA) leads to hypertension and IR in rats (Baron et al. 1995), and intracerebroventricular administration of L-NMMA induces peripheral IR and defective insulin secretion (Shankar et al. 1998). Two groups have demonstrated independently that mice with targeted disruption of eNOS develop IR in muscle and in liver (Duplain et al. 2001; Shankar et al. 2000). Moreover, eNOS knockout mice were hypertensive and dyslipidemic. That the syndrome was specific for NO deficiency was demonstrated by the observation that equally hypertensive 1-kidney/1-clip mice (a model of renovascular hypertension) were not insulin resistant. These data suggest that defective expression and function of eNOS could be a unifying mechanism that links vascular dysfunction and IR. Supportive evidence from human studies will be discussed below.

Two other genes (neuronal NOS/NOSI, and inducible NOS/NOSII) can also modulate nitric oxide availability. nNOS is expressed in the central nervous system and peripheral nervous plexi and is complexed with dystrophin in skeletal muscle. Deletion of nNOS resulted in a milder phenotype than eNOS knockout mice with IR developing only in skeletal muscle but not in the liver (Shankar et al. 2000). iNOS is expressed in macrophages and catalyzes NO production as part of the inflammatory response. Chronic inflammation has been implicated in the pathogenesis of diabetes and atherosclerosis. Furthermore, inflammatory cytokines lead to the development of IR in muscle, liver and adipose tissue. A role for iNOS in cytokine-induced IR was suggested by observations that exposure of rat skeletal muscle and cultured muscle and adipocytes to NO in the micromolar range (as might be expected by iNOS activation that generates much larger quantities of NO than eNOS) leads to IR (Kapur et al. 1997). Perreault and Marette recently showed that iNOS knockout mice did not become insulin resistant when placed on a high-fat diet, despite similar degrees of obesity as controls (Perreault and Marette 2001). These data were interpreted to support a role for iNOS induction in the pathogenesis of obesity related IR on the basis of increased production of inflammatory cytokines such as TNFα and interferon γ. Taken together these observations indicate that the NO pathway may be involved via multiple mechanisms in insulin-resistant states.

6.1.4
Mouse Models with Dyslipidemia and/or Insulin Resistance

The KK obese mouse is a naturally occurring mouse strain characterized by obesity, IR, glucose intolerance and dyslipidemia. A recessive trait in one strain (KK/San) is associated with abnormally low plasma lipid levels (Koishi et al. 2002). Using positional cloning strategies, an angiopoietin-like protein 3, encoded by *Angptl3*, was identified, which when overexpressed or injected intravenously increased plasma lipid levels (Koishi et al. 2002). Angptl3 is a naturally occurring inhibitor of lipoprotein lipase (Shimizugawa et al. 2002). Despite hypolipidemia, KK/San mice are insulin resistant and diabetic. These data suggest that the molecular basis for IR in this model does not occur on the basis of increased fatty acid delivery to skeletal muscle.

Mice with targeted deletion of the fatty acid-binding protein aP2 do not become insulin resistant when placed in a high-fat diet (Hotamisligil et al. 1996). aP2 is also expressed in macrophages, and the introduction of aP2-deficient macrophages to irradiated ApoE-null mice or introduction of the aP2-null allele into ApoE-null mice leads to regression of the atherosclerosis that characterizes ApoE knockout mice (Makowski et al. 2001). Thus, increased activity of aP2 might represent a link between obesity and atherosclerosis and may represent a novel drug target. Along similar lines, the introduction of a null allele for PPARα into ApoE KO mice reduced IR, blood pressure and the extent of atherosclerosis when these mice were placed on a high-fat diet (Tordjman et al. 2001). PPARα is an important transcriptional regulator of many genes involved in fatty acid metabolism. Whether or not the effects of reduced PPARα signaling on insulin sensitivity and atherosclerosis are mediated through aP2 has not been determined.

Transgenic overexpression if SREBP-1a in livers of mice results in massive overproduction of cholesterol and fatty acids in the liver (Shimano et al. 1996). The mice do not get overtly hyperlipidemic because of rapid clearance of VLDL by LDL receptors. However, when bred on to the LDL receptor-null background, these mice develop dramatic hypercholesterolemia and hypertriglyceridemia

(Horton et al. 1999). These studies provide an example of the way in which alterations in two gene products can interact to magnify a relevant phenotype.

Finally, some mice with deletions of components of the insulin-signaling pathway have been reported to exhibit phenotypes compatible with the metabolic syndrome. Thus, IRS-1 knockout mice, which are insulin resistant but not obese, have increased blood pressure, decreased endothelium-dependent vascular relaxation and hypertriglyceridemia as a consequence of reduced LPL activity (Abe et al. 1998). The overall message that can be gleaned from these studies is that there are multiple potential genetic mechanisms that may lead to phenotypes in rodents that mimic the metabolic syndrome in humans. These studies serve to identify potential pathways and mechanisms that can then be rigorously evaluated in human populations.

6.2
Human Genetic Studies

There are a number of single-gene defects in humans that recapitulate many features of the metabolic syndrome.

6.2.1
CD36

CD36 deficiency is not uncommon in Asian and African populations, both of which exhibit increased prevalence of IR compared to Caucasians. Humans with this deficiency have higher concentrations of triglycerides, lower concentrations of HDL cholesterol and increased blood pressure relative to age-, sex- and race-matched controls. Most patients with CD36 deficiency are nonobese. There is concordance between rodents and humans in the development of the metabolic syndrome on the basis of alterations in expression and function of CD36. Whether or not CD36 dysfunction is widely relevant to all cases of IR and whether the dyslipidemia is the primary anomaly in CD36-deficient patients such that IR follows as a secondary effect is unknown. Likewise it is unclear if IR itself alters CD36 activity and/or expression.

6.2.2
Lamin

Duncan-type familial partial lipodystrophy (FPLD) is a rare autosomal dominant form of IR. This monogenetic form of IR is due to missense mutations in the LMNA gene that encodes the nuclear envelope protein lamin A/C. Affected individuals are hyperinsulinemic, dyslipidemic (low HDL-C, high triglycerides) and commonly develop hypertension and type 2 diabetes (Hegele et al. 2000; Shackleton et al. 2000). Recent studies have now shown that these individuals develop premature coronary artery disease (Hegele 2001). These studies provide direct evidence that IR that develops on the basis of a single-gene mutation can ultimately recapitulate all of the clinical features of the metabolic syndrome, including accelerated atherosclerosis. The mechanisms by which a mutation in a nuclear envelope protein leads to lipodystrophic IR are unclear. Whether or not this mutation independently affects lipid homeostasis or the pathogenesis of atherosclerosis is also unknown. Studies in lipodystrophic rodents would suggest that lipodystrophy with repartitioning of lipids to muscle and liver is sufficient to induce IR and dyslipidemia. If this is true in patients with FPLD, then analysis of these individuals may provide insight into the association of hypertension and premature coronary artery disease with the metabolic syndrome.

6.2.3
PPARγ

PPARγ receptors are expressed predominantly in adipocytes and vascular cells (endothelium, vascular smooth muscle, monocytes and macrophages) and play a critical role in adipogenesis (Hsueh and Law 2001). Activation of PPARγ receptors by thiazolidinediones increase insulin sensitivity by multiple mechanisms, which include increased mobilization of lipid from the liver and muscle and storage in adipose tissue, and increased release of Acrp30/adiponectin (Berg et al. 2002; Hsueh and Law 2001). Thus, reduced expression of PPARγ might contribute to the pathogenesis of metabolic syndrome (particularly IR and dyslipidemia). Moreover, activation of PPARγ in rodent models of atherosclerosis leads to regression of atherosclerotic lesions (Collins et al. 2001). Rare mutations in PPARγ have been identified following analysis of families with severe IR. Two different dominant negative mutations were identified in three individuals with severe IR, diabetes and hypertension (discussed in more detail in Sect. 5.2.1). Although PPARγ is an attractive candidate gene, and despite the dramatic effects of pharmacological activation of PPARγ receptors, it remains to be established whether changes in this gene are responsible for the majority of cases of IR or the association of IR with accelerated vascular disease.

6.2.4
Familial Hypercholesterolemia

Familial hypercholesterolemia (FH) represents a single-gene defect (LDL receptor mutation) associated with an increased risk of coronary artery disease. Lean patients with FH are not insulin resistant when evaluated by glucose clamps. In a cross-sectional analysis of South Africans, there were no significant differences in insulin sensitivity between FH individuals with and without coronary artery disease (Raal et al. 1999). Moreover, although IR was correlated with hypertriglyceridemia and reduced HDL cholesterol concentrations in this population, there was no relationship between IR and its associated dyslipidemia and the presence or absence of coronary artery disease. In a larger case–control series of French Canadians in which patients with coronary artery disease with and without FH were compared, insulin and triglyceride concentrations were lower in individuals with FH. However, when both populations were stratified on the basis of abdominal girth and fasting insulin, the odds of developing coronary artery disease were highest in individuals with FH, IR and abdominal obesity (Gaudet et al. 1998). These data suggest that although hypercholesterolemia per se might not independently cause IR, the presence of IR will further increase the risk for developing coronary artery disease in individuals who are already genetically at high risk.

6.2.5
Genetic Variation That May Modify or Increase the Association Between Insulin Resistance and Increased Risk for Cardiovascular Disease

Variations in the expression of a number of genes may link IR and cardiovascular disease (see Table 4). No attempt will be made to exhaustively review all of these candidates, but there are a few that are worthy of further discussion.

As reviewed in Sect. 6.1.3, altered function of eNOS could represent an important link between IR and vascular dysfunction. It is widely accepted that diabetes and IR are associated with impaired endothelial function. Analysis of an eNOS polymorphism (4a/a) that is associated with premature coronary artery disease was not found to be associated with coronary artery disease in diabetics (Odawara et al. 1998). In another study that used flow-mediated endothelial-dependent vasodilatation (FMD) as an endpoint in a population of type 2 diabetics, the greatest degree of impairment in endothelial function was observed in individuals that carried at least a single *a* allele (~35% of the population) (Komatsu et al. 2002). The presence or absence of coronary artery disease was not determined in this study. More recently, circulating levels of the NOS inhibitor, asymmetric dimethyl arginine (AMDA), were found to be highest in insulin-resistant subjects and levels normalized after treatment with rosiglitazone (Stuhlinger et al. 2002). Alteration in activity and/or expression of enzymes that regulate the synthesis or degradation of AMDA might represent another link between IR and endothelial dysfunction.

Large randomized clinical trials have shown that long-term therapy with angiotensin-converting enzyme inhibitors (ACEIs) reduce the incidence of type 2 diabetes (Niskanen et al. 2001; Yusuf et al. 2000). Many studies in humans and animals with IR and/or hypertension have demonstrated that ACE inhibition is associated with enhanced whole body glucose uptake (Henriksen et al. 2001; Paolisso et al. 1992). Indeed, angiotensin II (ATII) had been shown to antagonize insulin signaling in cardiac muscle (Velloso et al. 1996). Moreover, there is evidence that ACE inhibition will increase glucose transport independently of increasing insulin signaling (Shiuchi et al. 2002). The effect of ACE or ATII receptor inhibition on glucose transport is seen only in insulin-resistant animals and not in animals with normal insulin sensitivity. These observations raise the possibility that increased activation of the renin–angiotensin system (RAS) could contribute to the pathogenesis of IR and its association with cardiovascular disease. Thus polymorphisms in the *ACE* gene that increase RAS activation

Table 4 Other potential genes of interest

Gene locus	Comment
IRS1	Might be more common in patients with NIDDM. Association with CAD in non-diabetics in one study (Baroni et al. 1999) but not with CAD in diabetics in another study (Ossei-Gerning et al. 1997)
Plasminogen activator inhibitor (*PAI-1*)	PAI levels are independently associated with insulin resistance independent of genotype (Juhan-Vague and Alessi 1997). PAI polymorphisms may amplify this association (Margaglione et al. 1998). Thus, PAI-1 levels are highest in insulin-resistant subjects with the susceptible genotype (Sartori et al. 2001). Likewise, stimulation of PAI-1 secretion by VLDL in vitro is greatest in umbilical vein endothelial cells from those with the susceptible genotype (Li et al. 1997)
Apolipoprotein genes	Various apolipoprotein genes have been examined for associations with insulin resistance and diabetes. Variations in these loci most likely may modify the dyslipidemia and modulate CVD risk but are unlikely to contribute directly to the pathogenesis of insulin resistance (Eichner et al. 2002)
Lipoprotein lipase (*LPL*)	The *Asn/Ser* genotype of LPL is associated with dyslipidemia in non-diabetics, and greater hypertriglyceridemia in the most insulin-resistant individuals. These associations were not seen in diabetics (Klannemark et al. 2000)
Insulin receptor (IR)	Increased frequency (75% vs 45%) of the C/C microsatellite polymorphism of the IR has been reported in Japanese hypertensives with hyperinsulinemia but not in hypertensives with normal insulin sensitivity (Fujioka et al. 1995)
Leptin	Leptin may contribute to obesity-induced hypertension. The *ll* polymorphism of the leptin gene was associated with hypertension independently of obesity and changes in insulin sensitivity (Shintani et al. 2002)
Red blood cell ion transport	Clustering of abnormalities in RBC Na/Li cotransport have been shown to be associated with insulin resistance and hypertension or a family history of hypertension (Romero et al. 2002a,b; Suchankova et al. 2002)
Results of genome-wide scans	Three studies in populations with high incidences of the insulin resistance syndrome and type 2 diabetes will be summarized. (1) In Mauritians of Indian descent, a linkage for premature CAD, type 2 diabetes and hypertension was found on chromosome *3q27* (*Acrp-30* gene). A similar association was previously shown in Caucasians. In addition, linkage for hypertension and type 2 diabetes was found on *8q23*, for premature CAD and dyslipidemia on *10q23* and hypertension and CAD on *16p13* (Francke et al. 2001). (2) In Hispanics from Los Angeles, a locus on chromosome 7 was linked to blood pressure, fasting insulin and leptin concentrations (Cheng et al. 2001; Xiang et al. 2001). (3) In Mexican-Americans in San Antonio a susceptibility locus influencing plasma triglycerides has been located on chromosome *15q* (Duggirala et al. 2000)

is an important candidate locus. The best-characterized polymorphism is the biallelic *ACE* polymorphism recognized by the absence (deletion, D) or the presence (insertion, I) of a 287-base pair ALU repeat inside intron 16. Serum ACE levels are highest in individuals homozygous for *D*, lowest in those with *II* alleles and intermediate in heterozygotes (Kennon et al. 1999). In nondiabetic as well as in diabetic populations, the presence of the *DD* allele increased the likelihood of developing ischemic heart disease (Staessen et al. 1997). In diabetics, the *D* allele confers increased susceptibility to all macrovascular disease and is associated with a greater risk for developing, and more rapid progression of, diabetic nephropathy. Studies examining the relationship between *ACE* polymorphisms and insulin sensitivity have yielded conflicting results (Kennon et al. 1999). Some studies have suggested that the prevalence of the *DD* polymorphism is greater among diabetics. Others have observed that the association only holds true in diabetics with hypertension, while others have seen either no association or the opposite effect (Bengtsson et al. 1999; Feng et al. 2002; Kennon et al. 1999; Thomas et al. 2001; Wong et al. 2001). Similarly, some studies have shown a greater prevalence of IR in diabetics with coronary artery disease, hypertension or nephropathy who harbor the *D* allele, but a few studies have suggested that *DD* homozygous subjects are more insulin sensitive (Kennon et al. 1999; Wong et al. 2001). The reason for such discrepant results is unclear and may relate to differences in study design and in the ethnicity of the study populations. In most of these studies, the levels and activity of ACE were not determined and so it is not known if the discrepant results are related to the potential effects of other modifiers on ACE activity.

Oxidative stress could contribute to IR. The association between levels of, and polymorphisms in, the serum enzyme paraoxonase-1 (PON1) has been investigated. PON1 protects lipoproteins from oxidation and decreased levels have been associated with increased cardiovascular disease risk (Imai et al. 2000). Polymorphisms of the PON1 promoter that reduce expression have been associated with increased levels of fasting blood glucose in diabetics. The *Leu55Met* polymorphism has been associated with impaired glucose tolerance and a greater likelihood of hyperinsulinemia and dyslipidemia (Barbieri et al. 2002; Deakin et al. 2002). All of these associations were observed in European Caucasians but not in other ethnic groups (Sanghera et al. 1998). In contrast, the Q192R polymorphism is associated with an increased cardiovascular disease risk in Chinese and Europeans with diabetes, and among diabetic Indian Asians (Osei-Hyiaman et al. 2001; Sanghera et al. 1997). Table 4 summarizes additional genetic studies of interest in human populations.

7
Summary

This review has summarized the current state of knowledge regarding potential genetic mechanisms that may predispose to the development of IR and that may account for the increased risk of cardiovascular disease in insulin-resistant individuals. It is clear that multiple genes are likely to interact in the pathogenesis of the metabolic syndrome and that the specific genetic mechanisms are likely to vary in different populations. Moreover, given the large number of potential genes that could be involved it will be difficult to ascertain the underlying genetic mechanisms in small study populations. Analysis will also be confounded by the interaction between environmental factors such as diet and exercise and genetic predisposition. Nevertheless, as more genetic insights become available or as the role of specific gene products is clarified, it is likely that this information will lead to new therapies for the metabolic syndrome and its sequelae and to improvement in our ability to screen for or identify those at greatest risk.

References

Abe H, Yamada, N Kamata et al (1998) Hypertension, hypertriglyceridemia, and impaired endothelium-dependent vascular relaxation in mice lacking insulin receptor substrate-1. J Clin Invest 101:1784–8

Abel ED, Kaulbach HC, Tian R et al (1999) Cardiac hypertrophy with preserved contractile function after selective deletion of GLUT4 from the heart J Clin Invest 104:1703–14

Abel ED, Peroni O, Kim JK et al (2001) Adipose-selective targeting of the GLUT4 gene impairs insulin action in muscle and liver. Nature 409:729–33

Accili D, Drago J, Lee EJ et al (1996) Early neonatal death in mice homozygous for a null allele of the insulin receptor gene Nat Genet 12:106–9

Aguirre V, Uchida T, Yenush L et al (2000) The c-Jun NH(2)-terminal kinase promotes insulin resistance during association with insulin receptor substrate-1 and phosphorylation of Ser(307). J Biol Chem 275:9047–54

Aitman TJ (2001) CD36, insulin resistance, and coronary heart disease. Lancet, 357, 651–2

Aitman TJ, Glazier AM, Wallace et al (1999) Identification of Cd36 (Fat) as an insulin-resistance gene causing defective fatty acid and glucose metabolism in hypertensive rats. Nat Genet 21:76–83

Akiyama TE, Sakai S, Lambert G et al (2002) Conditional disruption of the peroxisome proliferator-activated receptor gamma gene in mice results in lowered expression of ABCA1, ABCG1, and apoE in macrophages and reduced cholesterol efflux Mol Cell Biol 22:2607–19

Altshuler D, Hirschhorn JN, Klannemark M et al (2000) The common PPARgamma Pro12Ala polymorphism is associated with decreased risk of type 2 diabetes. Nat Genet 26:76–80

Araki E, Lipes MA, Patti ME et al (1994) Alternative pathway of insulin signalling in mice with targeted disruption of the IRS-1 gene. Nature 372:186–90

Arita Y, Kihara S, Ouchi N et al (2002) Adipocyte-derived plasma protein adiponectin acts as a platelet-derived growth factor-BB-binding protein and regulates growth factor-induced common postreceptor signal in vascular smooth muscle cell. Circulation 105:2893–8

Baier LJ, Permana PA, Yang X et al (2000) A calpain-10 gene polymorphism is associated with reduced muscle mRNA levels and insulin resistance. J Clin Invest 106:R69–73

Barbieri M, Bonafe M, Marfella R et al (2002) LL-paraoxonase genotype is associated with a more severe degree of homeostasis model assessment IR in healthy subjects. J Clin Endocrinol Metab 87:222–5

Baron AD, Zhu JS, Marshall S et al (1995) Insulin resistance after hypertension induced by the nitric oxide synthesis inhibitor L-NMMA in rats. Am J Physiol 269: E709–15

Baroni MG, D'Andrea MP, Montali A et al (1999) A common mutation of the insulin receptor substrate-1 gene is a risk factor for coronary artery disease. Arterioscler Thromb Vasc Biol 19:2975–80

Barroso I, Gurnell M, Crowley VE et al (1999) Dominant negative mutations in human PPARgamma associated with severe insulin resistance, diabetes mellitus and hypertension. Nature 402:880–3

Belke DD, Betuing S, Tuttle MJ et al (2002) Insulin signaling coordinately regulates cardiac size, metabolism, and contractile protein isoform expression. J Clin Invest 109:629–39

Bengtsson K, Orho-Melander M, Lindblad U et al (1999) Polymorphism in the angiotensin converting enzyme but not in the angiotensinogen gene is associated with hypertension and type 2 diabetes: the Skaraborg Hypertension and diabetes project. J Hypertens 17:1569-75

Bennet AM, Naslund TI, Morgenstern R et al (2001) Bioinformatic and experimental tools for identification of single- nucleotide polymorphisms in genes with a potential role for the development of the insulin resistance syndrome. J Intern Med 249:127-36

Berg AH, Combs TP, Scherer PE (2002) ACRP30/adiponectin: an adipokine regulating glucose and lipid metabolism, Trends Endocrinol Metab 13:84-9

Brinkmann JF, Abumrad NA, Ibrahimi A et al (2002) New insights into long-chain fatty acid uptake by heart muscle: a crucial role for fatty acid translocase/CD36. Biochem J 367:561-70

Bruning JC, Gautam D, Burks DJ et al (2000) Role of brain insulin receptor in control of body weight and reproduction. Science 289:2122-5

Bruning JC, Michael MD, Winnay JN et al (1998) A muscle-specific insulin receptor knockout exhibits features of the metabolic syndrome of NIDDM without altering glucose tolerance Mol Cell, 2, 559-69

Bruning JC, Winnay J, Bonner-Weir S et al (1997) Development of a novel polygenic model of NIDDM in mice heterozygous for IR and IRS-1 null alleles. Cell 88:561-72

Chen Z, Ishibashi S, Perrey S et al (2001) Troglitazone inhibits atherosclerosis in apolipoprotein E-knockout mice: pleiotropic effects on CD36 expression and HDL. Arterioscler Thromb Vasc Biol 21:372-7

Cheng LS, Davis RC, Raffel LJ et al (2001) Coincident linkage of fasting plasma insulin and blood pressure to chromosome 7q in hypertensive hispanic families. Circulation 104:1255-60

Cho H, Mu J, Kim JK et al (2001a) Insulin resistance and a diabetes mellitus-like syndrome in mice lacking the protein kinase Akt2 (PKB beta). Science 292:1728-31

Cho H, Thorvaldsen JL, Chu Q et al (2001b) Akt1/PKBalpha is required for normal growth but dispensable for maintenance of glucose homeostasis in mice. J Biol Chem,276:38349-52

Chukwuma C, Sr,Tuomilehto J (1998) The 'thrifty' hypotheses: clinical and epidemiological significance for non-insulin-dependent diabetes mellitus and cardiovascular disease risk factors. J Cardiovasc Risk 5:11-23

Clement S, Krause U, Desmedt F et al (2001) The lipid phosphatase SHIP2 controls insulin sensitivity. Nature 409:92-7

Collins AR, Meehan WP, Kintscher U et al (2001) Troglitazone inhibits formation of early atherosclerotic lesions in diabetic and nondiabetic low density lipoprotein receptor-deficient mice. Arterioscler Thromb Vasc Biol 21:365-71

Collison M, Glazier AM, Graham D et al (2000) Cd36 and molecular mechanisms of insulin resistance in the stroke-prone spontaneously hypertensive rat. Diabetes 49:2222-6

Colombo C, Cutson JJ, Yamauchi T et al (2002) Transplantation of adipose tissue lacking leptin is unable to reverse the metabolic abnormalities associated with lipoatrophy. Diabetes 51:2727-33

Combs TP, Berg AH, Obici S et al (2001) Endogenous glucose production is inhibited by the adipose-derived protein Acrp30. J Clin Invest 108:1875-81

Combs TP, Wagner JA, Berger J et al (2002) Induction of adipocyte complement-related protein of 30 kilodaltons by PPARgamma agonists: a potential mechanism of insulin sensitization. Endocrinology 143:998-1007

Costanzo BV, Trischitta V, Di Paola R et al (2001) The Q allele variant (GLN121) of membrane glycoprotein PC-1 interacts with the insulin receptor and inhibits insulin signaling more effectively than the common K allele variant (LYS121). Diabetes 50:831-6

Deakin S, Leviev I, Nicaud V et al (2002) Paraoxonase-1 L55 M polymorphism is associated with an abnormal oral glucose tolerance test and differentiates high risk coronary disease families. J Clin Endocrinol Metab 87:1268–73

Duggirala R, Blangero J, Almasy L et al (2000) A major susceptibility locus influencing plasma triglyceride concentrations is located on chromosome 15q in Mexican Americans. Am J Hum Genet 66:1237–45

Duplain H, Burcelin R, Sartori C et al (2001) Insulin resistance, hyperlipidemia, and hypertension in mice lacking endothelial nitric oxide synthase. Circulation 104:342–5

Ebihara K, Ogawa Y, Masuzaki H et al (2001) Transgenic overexpression of leptin rescues insulin resistance and diabetes in a mouse model of lipoatrophic diabetes. Diabetes 50:1440–8

Eichner JE, Dunn ST, Perveen G et al (2002) Apolipoprotein E polymorphism and cardiovascular disease: a HuGE review. Am J Epidemiol 155:487–95

Elbein SC, Chiu KC, Hoffman MD et al (1995) Linkage analysis of 19 candidate regions for insulin resistance in familial NIDDM. Diabetes 44:1259–65

Elbein SC, Chu W, Ren Q et al (2002) Role of calpain-10 gene variants in familial type 2 diabetes in Caucasians. J Clin Endocrinol Metab 87:650–4

Elchebly M, Payette P, Michaliszyn E et al (1999) Increased insulin sensitivity and obesity resistance in mice lacking the protein tyrosine phosphatase-1B gene. Science 283:1544–8

Engert JC, Vohl MC, Williams SM et al (2002) 5' flanking variants of resistin are associated with obesity Diabetes 51:1629–34

Federici M, Hribal ML, Ranalli M et al (2001) The common Arg972 polymorphism in insulin receptor substrate-1 causes apoptosis of human pancreatic islets. Faseb J 15:22–24

Feng Y, Niu T, XuX et al (2002) Insertion/deletion polymorphism of the ACE gene is associated with type 2 diabetes. Diabetes 51:1986–8

Ferreira LD, Pulawa LK, Jensen DR et al (2001) Overexpressing human lipoprotein lipase in mouse skeletal muscle is associated with insulin resistance. Diabetes 50:1064–8

Flavell DM, Jamshidi Y, Hawe E et al (2002) Peroxisome proliferator-activated receptor alpha gene variants influence progression of coronary atherosclerosis and risk of coronary artery disease. Circulation 105:1440–5

Flavell DM, Pineda Torra I, Jamshidi Y et al (2000) Variation in the PPARalpha gene is associated with altered function in vitro and plasma lipid concentrations in Type II diabetic subjects. Diabetologia 43:673–80

Francke S, Manraj M, Lacquemant C et al (2001) A genome-wide scan for coronary heart disease suggests in Indo- Mauritians a susceptibility locus on chromosome 16p13 and replicates linkage with the metabolic syndrome on 3q27. Hum Mol Genet 10:2751–65

Frittitta L, Baratta R, Spampinato D et al (2001) The Q121 PC-1 variant and obesity have additive and independent effects in causing insulin resistance. J Clin Endocrinol Metab 86:5888–91

Fruman DA, Mauvais-Jarvis F, Pollard DA et al (2000) Hypoglycaemia, liver necrosis and perinatal death in mice lacking all isoforms of phosphoinositide 3-kinase p85 alpha. Nat Genet 26:379–82

Fujioka Y, Takekawa K, Nakagawa Y et al (1995) Insulin receptor gene polymorphism and hyperinsulinemia in hypertensive patients. Hypertens Res 18:215–8

Fujisawa T, Ikegami H, Ogihara T (2002) [Clinical heterogeneity in elderly patients with diabetes mellitus]. Nippon Ronen Igakkai Zasshi 39:390–2

Furukawa LN, Kushiro T, Asagami T et al (1998) Variations in insulin sensitivity in spontaneously hypertensive rats from different sources, Metabolism 47:493–6

Gaudet D, Vohl MC, Perron P et al (1998) Relationships of abdominal obesity and hyperinsulinemia to angiographically assessed coronary artery disease in men with known mutations in the LDL receptor gene. Circulation 97:871–7

Gavrilova O, Marcus-Samuels B, Graham D et al (2000) Surgical implantation of adipose tissue reverses diabetes in lipoatrophic mice .J Clin Invest 105:271–8

Gotoda T, Iizuka Y, Kato N et al (1999) Absence of Cd36 mutation in the original spontaneously hypertensive rats with insulin resistance. Nat Genet 22:226–8

Gu HF, Almgren P, Lindholm E et al (2000) Association between the human glycoprotein PC-1 gene and elevated glucose and insulin levels in a paired-sibling analysis. Diabetes 49:1601–3

Hansen L, Reneland R, Berglund L et al (2000) Polymorphism in the glycogen-associated regulatory subunit of type 1 protein phosphatase (PPP1R3) gene and insulin sensitivity. Diabetes 49:298–301

Hara K, Boutin P, Mori Y et al (2002) Genetic variation in the gene encoding adiponectin is associated with an increased risk of type 2 diabetes in the Japanese population. Diabetes 51:536–40

Hebert LF, Jr, Daniels MC, Zhou J et al (1996) Overexpression of glutamine:fructose-6-phosphate amidotransferase in transgenic mice leads to insulin resistance. J Clin Invest 98:930–6

Hegele RA (2001) Premature atherosclerosis associated with monogenic insulin resistance. Circulation 103:2225–9

Hegele RA, Anderson CM, Cao, H (2000) Lamin A/C mutation in a woman and her two daughters with Dunnigan-type partial lipodystrophy and insulin resistance. Diabetes Care 23:258–9

Henriksen EJ, Jacob S, Kinnick TR et al (2001) Selective angiotensin II receptor receptor antagonism reduces insulin resistance in obese Zucker rats. Hypertension 38:884–90

Hirosumi J, Tuncman G, Chang L et al (2002) A central role for JNK in obesity and insulin resistance. Nature 420:333–6

Horton JD, Shimano H, Hamilton RL et al (1999) Disruption of LDL receptor gene in transgenic SREBP-1a mice unmasks hyperlipidemia resulting from production of lipid-rich VLDL. J Clin Invest 103:1067–76

Hotamisligil GS, Johnson RS, Distel RJ et al (1996) Uncoupling of obesity from insulin resistance through a targeted mutation in aP2, the adipocyte fatty acid binding protein. Science 274:1377–9

Hotta K, Funahashi T, Arita Y et al (2000) Plasma concentrations of a novel, adipose-specific protein, adiponectin, in type 2 diabetic patients. Arterioscler Thromb Vasc Biol 20:1595–9

Hribal ML, Federici M, Porzio O et al (2000) The Gly->Arg972 amino acid polymorphism in insulin receptor substrate- 1 affects glucose metabolism in skeletal muscle cells. J Clin Endocrinol Metab 85:2004–13

Hsueh WA, Law RE (2001) PPARgamma and atherosclerosis: effects on cell growth and movement Arterioscler Thromb Vasc Biol, 21, 1891–5

Ibrahimi A, Bonen A, Blinn WD et al (1999) Muscle-specific overexpression of FAT/CD36 enhances fatty acid oxidation by contracting muscle, reduces plasma triglycerides and fatty acids, and increases plasma glucose and insulin. J Biol Chem 274:26761–6

Ikegami H, Yamato E, Fujisawa T et al (1996) Analysis of candidate genes for insulin resistance in essential hypertension. Hypertens Res 19: Suppl 1, S31–4

Imai Y, Morita H, Kurihara H et al (2000) Evidence for association between paraoxonase gene polymorphisms and atherosclerotic diseases. Atherosclerosis 149:435–42

Juhan-Vague I, Alessi, MC (1997) PAI-1, obesity, insulin resistance and risk of cardiovascular events. Thromb Haemost 78:656–60

Kapur S, Bedard S, Marcotte B et al (1997) Expression of nitric oxide synthase in skeletal muscle: a novel role for nitric oxide as a modulator of insulin action. Diabetes 46:1691–700

Katsuki A, Sumida Y, Gabazza EC et al (2002) QUICKI is useful for following improvements in insulin sensitivity after therapy in patients with type 2 diabetes mellitus. J Clin Endocrinol Metab 87:2906–8

Katz EB, Stenbit AE, Hatton K et al (1995) Cardiac and adipose tissue abnormalities but not diabetes in mice deficient in GLUT4. Nature 377:151–5

Kennon B, Petrie JR, Small M et al (1999) Angiotensin-converting enzyme gene and diabetes mellitus Diabet Med 16:448–58

Kim JK, Fillmore JJ, Chen Y et al (2001) Tissue-specific overexpression of lipoprotein lipase causes tissue- specific insulin resistance. Proc Natl Acad Sci USA 98:7522–7

Klannemark M, Suurinkeroinen L, Orho-Melander M et al (2000) Interaction between the Asn291Ser variant of the LPL gene and insulin resistance on dyslipidaemia in high risk individuals for Type 2 diabetes mellitus. Diabet Med 17:599–605

Koishi R, Ando Y, Ono M et al (2002) Angptl3 regulates lipid metabolism in mice. Nat Genet 30:151–7

Komatsu M, Kawagishi T, Emoto M et al (2002) ecNOS gene polymorphism is associated with endothelium-dependent vasodilation in Type 2 diabetes. Am J Physiol Heart Circ Physiol 283:H557–61

Krook A, O'Rahilly S (1996) Mutant insulin receptors in syndromes of insulin resistance Baillieres. Clin Endocrinol Metab 10:97–122

Kuboki K, Jiang ZY, Takahara N et al (2000) Regulation of endothelial constitutive nitric oxide synthase gene expression in endothelial cells and in vivo : a specific vascular action of insulin. Circulation 101:676–81

Kubota N, Terauchi Y, Yamauchi T et al (2002) Disruption of adiponectin causes insulin resistance and neointimal formation. J Biol Chem 277:25863–6

Kulkarni RN, Bruning JC, Winnay JN et al (1999) Tissue-specific knockout of the insulin receptor in pancreatic beta cells creates an insulin secretory defect similar to that in type 2 diabetes. Cell 96:329–39

Kushner JA, Ye J, Schubert M et al (2002) Pdx1 restores beta cell function in Irs2 knockout mice J Clin Invest, 109, 1193–201

Lelliott CJ, Logie L, Sewter CP et al (2002) Lamin expression in human adipose cells in relation to anatomical site and differentiation state. J Clin Endocrinol Metab 87:728–34

Li XN, Grenett HE, Benza RL et al (1997) Genotype-specific transcriptional regulation of PAI-1 expression by hypertriglyceridemic VLDL and Lp(a) in cultured human endothelial cells. Arterioscler Thromb Vasc Biol 17:3215–23

Luan J, Browne PO, Harding AH et al (2001) Evidence for gene-nutrient interaction at the PPARgamma locus. Diabetes 50:686–9

Ma X, Warram JH, Trischitta V et al (2002) Genetic variants at the resistin locus and risk of type 2 diabetes in Caucasians. J Clin Endocrinol Metab 87:4407–10

Maeda N, Shimomura I, Kishida K et al (2002) Diet-induced insulin resistance in mice lacking adiponectin/ACRP30. Nat Med 8:731–7

Makowski L, Boord JB, Maeda K et al (2001) Lack of macrophage fatty-acid-binding protein aP2 protects mice deficient in apolipoprotein E against atherosclerosis. Nat Med 7:699–705

Margaglione M, Cappucci G, d'Addedda M et al (1998) PAI-1 plasma levels in a general population without clinical evidence of atherosclerosis: relation to environmental and genetic determinants. Arterioscler Thromb Vasc Biol 18:562–7

Matsuda M, Shimomura I, Sata M et al (2002) Role of adiponectin in preventing vascular stenosis The missing link of adipo-vascular axis. J Biol Chem 277:37487–91

Michael MD, Kulkarni RN, Postic C et al (2000) Loss of insulin signaling in hepatocytes leads to severe insulin resistance and progressive hepatic dysfunction. Mol Cell 6:87–97

Miyaoka K, Kuwasako T, Hirano K et al (2001) CD36 deficiency associated with insulin resistance. Lancet, 357:686–7

Moller D, Vidal-Puig A (1997) Genetics and Molecular Pathophysiology In Azziz, R, Nestler, J and Dewailly, D (eds), Androgen Excess Disorders in Women Lippincott-Raven, Philadelphia, pp 237–246

Moller DE, Bjorbaek C, Vidal-Puig A (1996) Candidate genes for insulin resistance. Diabetes Care 19:396–400

Nakae J, Biggs WH, 3rd, Kitamura T et al (2002) Regulation of insulin action and pancreatic beta-cell function by mutated alleles of the gene encoding forkhead transcription factor. Foxo1 Nat Genet 32:245–53

Nakae J, Kitamura T, Silver DL et al (2001) The forkhead transcription factor Foxo1 (Fkhr) confers insulin sensitivity onto glucose-6-phosphatase expression. J Clin Invest 108:1359–67

Niskanen L, Hedner T, Hansson L et al (2001) Reduced cardiovascular morbidity and mortality in hypertensive diabetic patients on first-line therapy with an ACE inhibitor compared with a diuretic/beta-blocker-based treatment regimen: a subanalysis of the Captopril Prevention Project. Diabetes Care 24:2091–6

O'Rahilly S (2002) Insights into obesity and insulin resistance from the study of extreme human phenotypes. E J E 147:435–441

Odawara M, Sasaki K, Tachi Y et al (1998) Endothelial nitric oxide synthase gene polymorphism and coronary heart disease in Japanese NIDDM. Diabetologia 41:365–6

Okamoto Y, Kihara S, Ouchi N et al (2002) Adiponectin reduces atherosclerosis in apolipoprotein E-deficient mice. Circulation 106:2767–70

Osei-Hyiaman D, Hou L, Mengbai F (2001) Coronary artery disease risk in Chinese type 2 diabetics: is there a role for paraxonase 1 gene (Q192R) polymorphism? Eur J Endocrinol 144:639–44

Ossei-Gerning N, Mansfield MW, Stickland MH et al (1997) Insulin receptor substrate-1 gene polymorphism and cardiovascular risk in non-insulin dependent diabetes mellitus and patients undergoing coronary angiography. Clin Lab Haematol 19:123–8

Ouchi N, Kihara S, Arita Y et al (2001) Adipocyte-derived plasma protein, adiponectin, suppresses lipid accumulation and class A scavenger receptor expression in human monocyte-derived macrophages. Circulation 103:1057–63

Ouchi N, Kihara S, Arita Y et al (2000) Adiponectin, an adipocyte-derived plasma protein, inhibits endothelial NF-kappaB signaling through a cAMP-dependent pathway. Circulation 102:1296–301

Owen KR, Shepherd M, Stride A et al (2002) Heterogeneity in young adult onset diabetes: aetiology alters clinical characteristics. Diabet Med 19:758–61

Paolisso G, Gambardella A, Verza M et al (1992) ACE inhibition improves insulin-sensitivity in aged insulin-resistant hypertensive patients. J Hum Hypertens 6:175–9

Permutt MA, Bernal-Mizrachi E, Inoue H (2000) Calpain 10: the first positional cloning of a gene for type 2 diabetes? J Clin Invest 106:819–21

Perreault M, Marette A (2001) Targeted disruption of inducible nitric oxide synthase protects against obesity-linked insulin resistance in muscle. Nat Med 7:1138–43

Petrie JR, Collison M, Connell JM et al (2001) CD36 deficiency and insulin resistance Lancet 358:242–3; discussion 244

Pizzuti A, Argiolas A, Di Paola R et al (2002) An ATG repeat in the 3'-untranslated region of the human resistin gene is associated with a decreased risk of insulin resistance. J Clin Endocrinol Metab 87:4403–6

Pizzuti A, Frittitta L, Argiolas A et al (1999) A polymorphism (K121Q) of the human glycoprotein PC-1 gene coding region is strongly associated with insulin resistance. Diabetes 48:1881–4

Postic C, Shiota M, Niswender KD et al (1999) Dual roles for glucokinase in glucose homeostasis as determined by liver and pancreatic beta cell-specific gene knock-outs using Cre recombinase. J Biol Chem 274:305–15

Pravenec M, Jansa P, Kostka V et al (2001a) Identification of a mutation in ADD1/SREBP-1 in the spontaneously hypertensive rat. Mamm Genome 12:295–8

Pravenec M, Landa V, Zidek V et al (2001b) Transgenic rescue of defective Cd36 ameliorates insulin resistance in spontaneously hypertensive rats. Nat Genet 27:156–8

Qi N, Kazdova L, Zidek V et al (2002) Pharmacogenetic Evidence That Cd36 Is a Key Determinant of the Metabolic Effects of Pioglitazone J Biol Chem 277:18501–18507

Raal FJ, Panz VR, Pilcher GJ et al (1999) Atherosclerosis seems not to be associated with hyperinsulinaemia in patients with familial hypercholesterolaemia. J Intern Med 246:75–80

Rasmussen SK, Urhammer SA, Pizzuti A et al (2000) The K121Q variant of the human PC-1 gene is not associated with insulin resistance or type 2 diabetes among Danish Caucasians. Diabetes 49:1608–11

Reusch JE (2002) Current concepts in insulin resistance, type 2 diabetes mellitus, and the metabolic syndrome. Am J Cardiol 90:19G–26G

Ristow M, Muller-Wieland D, Pfeiffer A et al (1998) Obesity associated with a mutation in a genetic regulator of adipocyte differentiation. N Engl J Med 339:953–9

Romero JR, Rivera A, Conlin PR (2002a) Red blood cell Na+/H+ exchange activity is insulin resistant in hypertensive patients. Clin Exp Hypertens 24:277–87

Romero JR, Rivera A, Monari A et al (2002b) Increased red cell sodium-lithium countertransport and lymphocyte cytosolic calcium are separate phenotypes in patients with essential hypertension. J Hum Hypertens 16:353–8

Rosskopf D, Frey U, Eckhardt S et al (2000) Interaction of the G protein beta 3 subunit T825 allele and the IRS-1 Arg972 variant in type 2 diabetes. Eur J Med Res 5:484–90

Sachidanandam R, Weissman D, Schmidt SC et al (2001) A map of human genome sequence variation containing 142 million single nucleotide polymorphisms. Nature 409:928–33

Saltiel AR, Kahn CR (2001) Insulin signalling and the regulation of glucose and lipid metabolism. Nature 414:799–806

Sanghera DK, Saha N, Aston CE et al (1997) Genetic polymorphism of paraoxonase and the risk of coronary heart disease. Arterioscler Thromb Vasc Biol 17:1067–73

Sanghera DK, Saha N, Kamboh MI (1998) The codon 55 polymorphism in the paraoxonase 1 gene is not associated with the risk of coronary heart disease in Asian Indians and Chinese. Atherosclerosis 136:217–23

Sartori MT, Vettor R, De Pergola G et al (2001) Role of the 4G/5G polymorphism of PaI-1 gene promoter on PaI-1 levels in obese patients: influence of fat distribution and insulin-resistance. Thromb Haemost 86:1161–9

Savage DB, Agostini M, Barroso I et al (2002) Digenic inheritance of severe insulin resistance in a human pedigree. Nat Genet 31:379–84

Seda O, Kazdova L, Krenova D et al (2003) Rosiglitazone fails to improve hypertriglyceridaemia and gluscoe tolerance in CD36-deficient BNSHR4 congenic rat strain. Physiol Genomics 12:000–000

Sentinelli F, Romeo S, Arca M et al (2002) Human resistin gene, obesity, and type 2 diabetes: mutation analysis and population study. Diabetes 51:860–2

Shackleton S, Lloyd DJ, Jackson SN et al (2000) LMNA, encoding lamin A/C, is mutated in partial lipodystrophy. Nat Genet 24:153–6

Shankar R, Zhu JS, Ladd B et al (1998) Central nervous system nitric oxide synthase activity regulates insulin secretion and insulin action. J Clin Invest 102:1403–12

Shankar RR, Wu Y, Shen HQ et al (2000) Mice with gene disruption of both endothelial and neuronal nitric oxide synthase exhibit insulin resistance. Diabetes 49:684–7

Shimano H, Horton JD, Hammer RE et al (1996) Overproduction of cholesterol and fatty acids causes massive liver enlargement in transgenic mice expressing truncated SREBP-1a. J Clin Invest 98:1575–84

Shimizugawa T, Ono M, Shimamura M et al (2002) ANGPTL3 decreases very low density lipoprotein triglyceride clearance by inhibition of lipoprotein lipase. J Biol Chem 277:33742–8

Shimomura I, Bashmakov Y, Horton JD (1999) Increased levels of nuclear SREBP-1c associated with fatty livers in two mouse models of diabetes mellitus. J Biol Chem 274:30028–32

Shimomura I, Hammer RE, Richardson JA et al (1998) Insulin resistance and diabetes mellitus in transgenic mice expressing nuclear SREBP-1c in adipose tissue: model for congenital generalized lipodystrophy. Genes Dev 12:3182–94

Shimomura I, Matsuda M Hammer RE et al (2000) Decreased IRS-2 and increased SREBP-1c lead to mixed insulin resistance and sensitivity in livers of lipodystrophic and ob/ob mice. Mol Cell 6:77–86

Shintani M, Ikegami H, Fujisawa T et al (2002) Leptin gene polymorphism is associated with hypertension independent of obesity. J Clin Endocrinol Metab 87:2909–12

Shiuchi T, Cui TX, Wu L et al (2002) ACE inhibitor improves insulin resistance in diabetic mouse via bradykinin and NO. Hypertension 40:329–34

Staessen JA, Wang JG, Ginocchio G et al (1997) The deletion/insertion polymorphism of the angiotensin converting enzyme gene and cardiovascular-renal risk. J Hypertens 15:1579–92

Steinberg HO, Baron AD (2002) Vascular function, insulin resistance and fatty acids. Diabetologia 45:623–34

Stenbit AE, Tsao, TS, Li J et al (1997) GLUT4 heterozygous knockout mice develop muscle insulin resistance and diabetes. Nat Med 3:1096–101

Stern MP (2000) Strategies and prospects for finding insulin resistance genes. J Clin Invest 106:323–7

Stuhlinger MC, Abbasi F, Chu JW et al (2002) Relationship between insulin resistance and an endogenous nitric oxide synthase inhibitor. Jama 287:1420–6

Stumvoll M, Tschritter O, Fritsche A et al (2002) Association of the T-G polymorphism in adiponectin (exon 2) with obesity and insulin sensitivity: interaction with family history of type 2 diabetes. Diabetes 51:37–41

Suchankova G, Vlasakova Z, Zicha J et al (2002) Erythrocyte membrane ion transport in offspring of hypertensive parents: effect of acute hyperinsulinemia and relation to insulin action. Ann N Y Acad Sci 967:352–62

Tamemoto H, Kadowaki T, Tobe K et al (1994) Insulin resistance and growth retardation in mice lacking insulin receptor substrate-1. Nature 372:182–6

Tanaka T, Nakata T, Oka T et al (2001) Defect in human myocardial long-chain fatty acid uptake is caused by FAT/CD36 mutations. J Lipid Res 42:751–9

Tang J, Neidigh JL, Cooksey RC et al (2000) Transgenic mice with increased hexosamine flux specifically targeted to beta-cells exhibit hyperinsulinemia and peripheral insulin resistance. Diabetes 49:1492–9

Terauchi Y, Iwamoto K, Tamemoto,H et al (1997) Development of non-insulin-dependent diabetes mellitus in the double knockout mice with disruption of insulin receptor substrate-1 and beta cell glucokinase genes Genetic reconstitution of diabetes as a polygenic disease. J Clin Invest 99:861–6

Terauchi Y, Tsuji Y, Satoh S et al (1999) Increased insulin sensitivity and hypoglycaemia in mice lacking the p85 alpha subunit of phosphoinositide 3-kinase. Nat Genet 21:230–5

Thomas GN, Tomlinson B, Chan JC et al (2001) Renin-angiotensin system gene polymorphisms, blood pressure, dyslipidemia, and diabetes in Hong Kong Chinese: a significant association of tne ACE insertion/deletion polymorphism with type 2 diabetes. Diabetes Care 24:356–61

Tian R, Abel, ED (2001) Responses of GLUT4-deficient hearts to ischemia underscore the importance of glycolysis. Circulation 103:2961–6

Tobe K, Suzuki R, Aoyama M et al (2001) Increased expression of the sterol regulatory element-binding protein-1 gene in insulin receptor substrate-2(-/-) mouse liver. J Biol Chem 276:38337–40

Tomas E, Tsao TS, Saha AK et al (2002) Enhanced muscle fat oxidation and glucose transport by ACRP30 globular domain: acetyl-CoA carboxylase inhibition and AMP-activated protein kinase activation Proc Natl Acad Sci USA 99:16309–13

Tordjman K, Bernal-Mizrachi C, Zemany L et al (2001) PPARalpha deficiency reduces insulin resistance and atherosclerosis in apoE-null mice. J Clin Invest 107:1025–34

Veerababu G, Tang J, Hoffman RT et al (2000) Overexpression of glutamine: fructose-6-phosphate amidotransferase in the liver of transgenic mice results in enhanced glycogen storage, hyperlipidemia, obesity, and impaired glucose tolerance. Diabetes 49:2070–8

Velloso LA, Folli F, Sun XJ et al (1996) Cross-talk between the insulin and angiotensin signaling systems Proc Natl Acad Sci USA 93:12490–5

Vettor R, Cusin I, Ganten D et al (1994) Insulin resistance and hypertension: studies in transgenic hypertensive TGR(mREN-2)27 rats Am J Physiol 267:R1503–9

Vidal-Puig A, Bjorbaek C (1997) [Molecular genetics of non insulin dependent diabetes mellitus] Med Clin (Barc) 109:107–14

Vidal-Puig A, Moller D (1997) Classification, Prevalence, Clinical Manifestations and Diagnosis In Azziz, R, Nestler, J and D, D (eds), Androgen Excess Disorders in Women, Philadelphia, pp 227–236

Vohl MC, Lepage P, Gaudet D et al (2000) Molecular scanning of the human PPARa gene: association of the L162v mutation with hyperapobetalipoproteinemia. J Lipid Res 41:945–52

Wallace TM, Matthews DR (2002) The assessment of insulin resistance in man. Diabet Med 19:527–34

Wang G, Qian R, Li Q et al (2001) The association between PPP1R3 gene polymorphisms and type 2 diabetes mellitus. Chin Med J (Engl) 114:1258–62

Wang H, Chu WS, Hemphill C et al (2002) Human resistin gene: molecular scanning and evaluation of association with insulin sensitivity and type 2 diabetes in Caucasians. J Clin Endocrinol Metab 87:2520–4

Waterston RH, Lander ES, Sulston, JE (2002) On the sequencing of the human genome. Proc Natl Acad Sci USA 99:3712–6

White MF (2002) IRS proteins and the common path to diabetes. Am J Physiol Endocrinol Metab 283:E413–22

Willett WC (2002) Balancing life-style and genomics research for disease prevention. Science 296:695–8

Withers DJ, Gutierrez JS, Towery H et al (1998) Disruption of IRS-2 causes type 2 diabetes in mice. Nature 391:900–4

Wong, TY, Szeto, CC, Chow, KM et al (2001) Contribution of gene polymorphisms in the renin-angiotensin system to macroangiopathy in patients with diabetic nephropathy. Am J Kidney Dis 38:9–17

Xia J, Scherer SW, Cohen PT et al (1998) A common variant in PPP1R3 associated with insulin resistance and type 2 diabetes. Diabetes 47:1519–24

Xiang AH, Azen SP, Raffel LJ et al (2001) Evidence for joint genetic control of insulin sensitivity and systolic blood pressure in hispanic families with a hypertensive proband. Circulation 103:78–83

Yahagi N, Shimano H, Hasty AH et al (2002) Absence of sterol regulatory element-binding protein-1 (SREBP-1) ameliorates fatty livers but not obesity or insulin resistance in Lep(ob)/Lep(ob) mice. J Biol Chem 277:19353–7

Yamauchi T, Kamon J, Waki H et al (2001) The fat-derived hormone adiponectin reverses insulin resistance associated with both lipoatrophy and obesity. Nat Med 7:941–6

Yang C, Coker KJ, Kim JK et al (2001a) Syntaxin 4 heterozygous knockout mice develop muscle insulin resistance. J Clin Invest 107:1311–8

Yang WS, Jeng CY, Wu TJ et al (2002) Synthetic peroxisome proliferator-activated receptor-gamma agonist, rosiglitazone, increases plasma levels of adiponectin in type 2 diabetic patients. Diabetes Care 25:376–80

Yang X, Pratley RE, Baier LJ et al (2001b) Reduced skeletal muscle calpain-10 transcript level is due to a cumulative decrease in major isoforms. Mol Genet Metab 73:111–3

Yu C, Chen Y, Cline GW et al (2002) Mechanism by which fatty acids inhibit insulin activation of IRS-1 associated phosphatidylinositol 3-kinase activity in muscle. J Biol Chem14:14

Yusuf S, Sleight P, Pogue J et al (2000) Effects of an angiotensin-converting-enzyme inhibitor, ramipril, on cardiovascular events in high-risk patients. The Heart Outcomes Prevention Evaluation Study Investigators. N Engl J Med 342:145–53

Zabolotny JM, Kim YB, Peroni OD et al (2001) Overexpression of the LAR (leukocyte antigen-related) protein-tyrosine phosphatase in muscle causes insulin resistance. Proc Natl Acad Sci USA 98:5187–92

Zeng G, Nystrom FH, Ravichandran LV et al (2000) Roles for insulin receptor, PI3-kinase, and Akt in insulin-signaling pathways related to production of nitric oxide in human vascular endothelial cells. Circulation 101:1539–45

Zimmet P, Alberti KG, Shaw J (2001) Global and societal implications of the diabetes epidemic. Nature 414:782–7

Zisman A, Peroni OD, Abel ED et al (2000) Targeted disruption of the glucose transporter 4 selectively in muscle causes insulin resistance and glucose intolerance. Nat Med 6:924–8

Genetic Disruption of Nitric Oxide Synthases and Cardiovascular Disease: Lessons from a Candidate Gene

P. L. Huang

Cardiovascular Research Center, Massachusetts General Hospital,
Harvard Medical School, Boston, MA 02114, USA
e-mail: huangp@helix.mgh.harvard.edu

1	The Nitric Oxide System	282
1.1	Nitric Oxide as an Important Biological Mediator	282
1.2	The Family of Nitric Oxide Synthase Enzymes	283
1.3	Downstream Effector Mechanisms	285
1.4	Roles of NOS Isoforms in the Cardiovascular System	285
1.4.1	nNOS	285
1.4.2	eNOS	286
1.4.3	iNOS	286
1.5	Regulation of NOS Enzyme Function	287
1.5.1	Dimerization	287
1.5.2	Subcellular Localization	287
1.5.3	Phosphorylation	288
2	Genetic Disruption of NOS Genes	289
2.1	General Considerations	289
2.2	Phenotypes of NOS Knockout Mice	290
2.2.1	nNOS Knockout Mice	290
2.2.2	eNOS Knockout Mice	290
2.2.3	iNOS Knockout Mice	291
2.3	Stroke and Cerebral Ischemia in NOS Knockout Mice	291
2.3.1	NO Levels in Ischemic Brain	291
2.3.2	Divergent Roles of NOS Isoforms	291
2.3.3	Response of NOS Knockout Mice to Cerebral Ischemia	292
2.3.4	Lessons Learned from NOS Knockout Mice	293
2.4	Atherosclerosis in NOS Knockout Mice	293
2.4.1	eNOS Knockout Mice as a Model for Endothelial Dysfunction	293
2.4.2	Vascular Injury in eNOS Knockout Mice	294
2.4.3	Diet-Induced Atherosclerosis in apoE/eNOS Double Knockout Mice	294
2.4.4	iNOS and Diet-Induced Atherosclerosis	294
3	NOS Gene Polymorphisms	295
3.1	General Considerations	295
3.2	Polymorphisms in eNOS	296
3.2.1	Coding Polymorphism: Glu298Asp	296
3.2.2	Promoter Polymorphisms	299
3.2.3	Intron Polymorphisms	300
3.3	nNOS and iNOS Polymorphisms	303
4	Conclusions	304
	References	304

Abstract The nitric oxide (NO) system is vitally important to the function of the cardiovascular system. Abnormalities in NO signaling have been linked to a wide variety of cardiovascular diseases, including atherosclerosis, hypertension, congestive heart failure, thrombosis, stroke, and diabetes mellitus. This chapter will review the roles that NO synthases play in normal vascular function and in the molecular mechanisms of disease processes, focusing on information gained from the study of mutant mice in which the NO synthase genes have been disrupted or modified. Specifically, we wil review the various genetic polymorphisms that have been described in the NO synthase genes, their association with disease processes and their functional effects on NO synthase function or expression levels. The cellular pathways that involve NO are complex and offer not only insights into how abnormalities in NO signaling can lead to disease but also opportunities and targets on which to intervene to prevent or treat disease.

Keywords Nitric oxide · Enzyme · Knockout · Endothelial · Neuronal · Inducible

1
The Nitric Oxide System

1.1
Nitric Oxide as an Important Biological Mediator

Soon after Alfred Nobel's discovery of how to synthesize large quantities of nitroglycerin in 1853 (Ringertz 2001), nitroglycerin was widely used to treat angina pectoris. However, the mechanism by which nitroglycerin dilated blood vessels was not well understood for many years. Indeed, the importance of nitric oxide (NO) as a biological mediator was not appreciated until relatively recently. In a landmark experiment, Furchgott and Zawadzki demonstrated that the vascular relaxation of isolated blood vessels in response to acetylcholine requires the presence of an intact endothelial layer (Furchgott and Zawadzki 1980). They proposed that acetylcholine does not act directly on vascular smooth muscle to relax it, but rather, that it acts on the endothelium. The endothelium then elaborates a factor, which they termed endothelium-derived relaxing factor (EDRF), which causes the vascular smooth muscle to relax. EDRF was noted to have unusual properties, including inactivation by heme, a short half-life in vitro, and the ability to diffuse freely across cell membranes. Ferid Murad found that nitroglycerin and related vasodilator compounds release the gas NO (Arnold et al. 1977; Katsuki et al. 1977). In 1986, both Furchgott and Ignarro independently proposed and provided experimental evidence that the gas NO was responsible for EDRF activity in blood vessels (Ignarro et al. 1987). The Nobel Prize for Physiology or Medicine in 1998 was awarded to Robert Furchgott, Louis Ignarro, and Ferid Murad for their pioneering work on NO.

1.2
The Family of Nitric Oxide Synthase Enzymes

NO is synthesized by the family of nitric oxide synthase (NOS) enzymes (Alderton et al. 2001). These enzymes oxidize the terminal guanidino nitrogen of the amino acid L-arginine to NO, producing citrulline from the remainder of the molecule. The reaction utilizes molecular oxygen, and FAD, FMN, NADPH, and tetrahydrobiopterin as cofactors, to catalyze a five-electron oxidation of the guanidino nitrogen. There are three major NOS isoforms, encoded by separate genes on separate chromosomes: neuronal NOS (nNOS, or type I NOS), inducible NOS (iNOS, or type 2 NOS), and endothelial NOS (eNOS, or type 3 NOS). Table 1 outlines the nomenclature of the NOS isoforms and lists some of their putative functions.

The NOS isoforms share common structural features, as shown in Fig. 1, including an oxygenase domain at the N-terminus and a reductase domain at the C-terminus. The C-terminus contains domains that bind the cofactors FAD,

Table 1 Nomenclature of NOS isoforms and potential function

Type	Isoform	Possible functions
I	Neuronal	Cell communication, learning, memory, retrograde neurotransmitter
		Excitatory amino acid (glutamate) neurotransmission
		Neurotransmission in NANC nerves
II	Inducible	Defense against pathogens
		Defense against tumors
		Inflammatory responses
III	Endothelial	Vascular relaxation (EDRF)
		Inhibition of smooth muscle proliferation
		Inhibition of platelet aggregation
		Inhibition of leukocyte adhesion

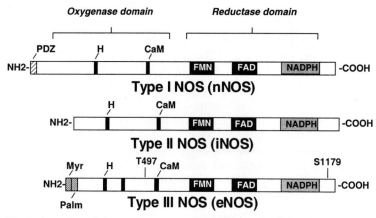

Fig. 1 Comparison of the protein structure of NOS isoforms showing co-factor binding domains

Table 2 Features of NOS isoforms

	Type I NOS	Type II NOS	Type III NOS
Common name	nNOS	iNOS	eNOS
Typical cell	Neurons	Macrophages	Endothelium
Other sites of expression	Smooth muscle	Endothelium	Smooth muscle
	Skeletal muscle	Smooth muscle	Platelets
	Lung epithelia	Liver	Hippocampal neurons
	Macula densa	Chondrocytes	
Chromosome	12	17	7
Expression pattern	Constitutive (inducible also)	Inducible	Constitutive (inducible also)
Regulation	Calcium dependent	Gene transcription	Calcium dependent Serine phosphorylation
Output	moderate (nM to µM)	High (µM)	Low (pM to nM)
Function	Signaling	Toxin	Signaling
Intracellular location	Soluble, sarcolemmal	Soluble	Caveolae (membrane-associated)
Means of localization	N-terminal PDZ domain	N/A	N-terminal myristoylation

FMN, and NADPH used in electron transfer reactions. The N-terminus contains regions involved in binding to heme, tetrahydrobiopterin, and calmodulin. The nNOS isoform has a unique PDZ domain at its N-terminus, which is involved in its membrane association and its localization to nerve terminals and neuromuscular junctions (Brenman et al. 1997). The eNOS isoform has unique sites near its N-terminus for myristoylation and palmitoylation (Janssens et al. 1992; Lamas et al. 1992; Marsden et al. 1992; Nishida et al. 1992; Sessa et al. 1992). The locations of two important phosphorylation sites in eNOS, Ser1179 and Thr497 (using a numbering convention that follows the bovine sequence) are shown as well.

Table 2 summarizes some key features of these isoforms and their genes. nNOS is expressed primarily in specific neurons in the brain and the peripheral nervous system. iNOS is expressed by macrophages, while eNOS is expressed by endothelial cells. Despite their common names, expression of these isoforms has been detected in a variety of cell types, and there is substantial overlap in their expression patterns. All three NOS isoforms play important roles in the cardiovascular system.

nNOS and eNOS share the property of dependence on intracellular calcium concentration for their activity. They are inactive at resting concentrations of calcium, but when there is a transient increase in intracellular calcium, calcium binds to calmodulin and activates the enzyme. The primary regulation of eNOS and nNOS enzyme activity is by intracellular calcium concentration. In contrast, iNOS has a tightly bound calmodulin moiety, and is not dependent on intracellular calcium transients for its activity. The enzymatic activity of iNOS is primarily regulated by transcriptional regulation of its expression, which can be induced by signals such as lipopolysaccharide or tumor necrosis factor. The output of the NOS enzyme in terms of NO generated also differs between isoforms, being the lowest for eNOS, higher for nNOS, and much higher for iNOS. This may reflect the roles of NO, which appear to be more related to signaling by nNOS and eNOS, and to toxicity or inflammation by iNOS.

1.3
Downstream Effector Mechanisms

In most cells, the physiological target of NO is soluble guanylate cyclase. NO activates guanylate cyclase by binding to its heme moiety. Activated guanylate cyclase produces cGMP, which mediates many of the biological effects of NO, including relaxation of vascular smooth muscle (vasodilation) and smooth muscle in the respiratory, gastrointestinal and genitourinary tracts (nonadrenergic, noncholinergic autonomic neurotransmission).

NO also binds to sulfhydryl groups to form nitrosothiol compounds on other targets (Stamler 1994), including hemoglobin, which may serve as a natural carrier for NO (Stamler et al. 1997). In addition, NO rapidly reacts with superoxide anion to form peroxynitrite anion, which itself is very reactive and may cause toxicity (Beckman et al. 1994; Beckman and Koppenol 1996). Large quantities of NO, made for instance by iNOS, directly inhibit mitochondrial respiratory chain enzymes (Brown 1995; Takehara et al. 1995; Okada et al. 1996) and overstimulate poly-ADP ribose synthase (Dawson et al. 1993; Dawson et al. 1994; Endres et al. 1998). Both of these effects result in depletion of cellular energy stores. Generally, these latter mechanisms underlie some of the toxicity of NO, while cGMP-dependent effects mediate many of the biological signaling roles of NOS.

1.4
Roles of NOS Isoforms in the Cardiovascular System

1.4.1
nNOS

In the brain, the nNOS isoform is expressed in specific populations of neurons, and appears to play two important roles. First, nNOS is important in the regulation of cerebral blood flow and in coupling cerebral metabolism with local perfusion. Second, nNOS generates NO as a retrograde messenger. Processes that involve activity-dependent refinement of neuronal networks such as learning and memory are thought to involve retrograde messengers. These messengers are released from a postsynaptic neuron and feedback to the presynaptic neurons that the signal has been received, allowing certain synaptic connections to be strengthened and others to be weakened. Because NO is a freely diffusible gas that can be produced by nNOS in the postsynaptic cell activated by receptor-mediated calcium influx, it has properties that allow it to serve as a retrograde messenger. Cerebral ischemia also causes marked stimulation of excitatory amino acid (glutamate) neurotransmitter receptors, so it is associated with marked activation of nNOS and increases the amount of NO generated by several orders of magnitude. Thus, overproduction of NO has important implications for cerebral ischemia and stroke.

In the peripheral nervous system, nNOS is expressed in nonadrenergic, noncholinergic autonomic nerves of the respiratory, gastrointestinal, and genitourinary tract. NO produced in these nitrergic nerves serves as a neurotransmitter, facilitating smooth muscle relaxation. nNOS is also found in the heart, where it may be involved in regulation of cardiac contractility.

1.4.2
eNOS

In the endothelium, NO produced by eNOS in endothelial cells diffuses to the underlying vascular smooth muscle, where it stimulates soluble guanylate cyclase. As an endogenous vasodilator, NO is important to the regulation of local blood flow and perfusion, as well as in regulation of blood pressure. In addition to regulating vascular smooth muscle relaxation, NO also suppresses smooth muscle cell proliferation (Mooradian et al. 1995), leukocyte–endothelial interactions (Bath 1993; Lefer et al. 1999), and platelet aggregation (Radomski et al. 1991; Freedman et al. 1999). These physiological effects of NO may normally inhibit the processes that underlie atherosclerosis.

Endothelial dysfunction, characterized by reduced or absent vasodilator responses to flow or biological stimuli, is a common feature of atherosclerosis, hypertension, diabetes mellitus, and hypercholesterolemia (Gimbrone 1989; Cai and Harrison 2000; Goligorsky et al. 2000). A characteristic of endothelial dysfunction is reduced production or bioavailability of NO in vascular tissues. There are multiple molecular mechanisms for endothelial dysfunction, including reduced eNOS protein levels, reduced eNOS enzymatic activity, uncoupling of eNOS activity leading to enhanced production of superoxide (Cai and Harrison 2000), abnormalities in eNOS trafficking to caveolae (Shaul 2002), and abnormalities in eNOS phosphorylation. These mechanisms are not mutually exclusive, and it is likely that multiple mechanisms operate simultaneously.

1.4.3
iNOS

iNOS expression is induced in macrophages as a response to infection, tissue damage, or inflammation. The iNOS isoform generates large amounts of NO, which may contribute to the excessive vasodilation and cardiac dysfunction seen in sepsis.

1.5
Regulation of NOS Enzyme Function

1.5.1
Dimerization

For enzymatic activity, NOS protein must bind cofactors and dimerize (Alderton et al. 2001). Dimerization is required for activity, because electron flow occurs *in trans*, between FAD, FMN, and NADPH moieties on different subunits. For nNOS and eNOS isoforms, monomer NOS proteins first bind to the cofactors FAD and FMN. The addition of the substrate L-arginine, the cofactor tetrahydrobiopterin and a heme group allows the NOS protein to dimerize, but the dimers are still inactive. Full activity results when intracellular calcium concentrations increase, resulting in calmodulin binding to the dimers. In contrast, the iNOS isoform binds FAD, FMN, and calcium/calmodulin even at low (resting intracellular) concentrations of calcium. The addition of L-arginine, tetrahydrobiopterin, and heme allows iNOS to dimerize and activates it.

1.5.2
Subcellular Localization

In some cells, nNOS is localized to neuronal synapses and neuromuscular junctions by an N-terminal PDZ domain (Brenman et al. 1996). In others, nNOS is a soluble protein. The iNOS isoform is soluble, and does not possess any known structural features that would dictate its subcellular localization.

The subcellular localization of eNOS is critical to its proper function. Two types of N-terminal fatty acid modifications—palmitoylation and myristoylation—are important for the proper intracellular localization of eNOS to specific plasmalemmal domains called caveolae. The eNOS gene encodes, at its very 5'-end, the consensus sequence MGNLKSV for myristoylation (Janssens et al. 1992; Lamas et al. 1992; Marsden et al. 1992; Nishida et al. 1992; Sessa et al. 1992). The methionine corresponding to the translation initiation codon is removed by a specific aminopeptidase, exposing glycine at the N-terminus. This glycine residue is the site for the addition of myristic acid, a process called N-myristoylation. Site-specific mutagenesis of the glycine residue results in a nonmyristoylated, soluble protein that shares identical in vitro enzymatic activity as wild-type eNOS, but with poor in vitro activity (Pollock et al. 1992; Busconi and Michel 1993; Liu and Sessa 1994). eNOS is also reversibly palmitoylated at Cys15 and Cys26. Mutations at these residues that cannot be palmitoylated also do not properly localize to caveolae (Garcia-Cardena et al. 1996; Shaul 2002).

In addition to myristoylation and palmitoylation, subcellular localization and enzyme activity are regulated by two proteins: hsp90 and caveolin. The heat shock protein hsp90, a protein involved in signal transduction and in protein-folding as a molecular chaperone, binds to eNOS, recruits it to caveolae and activates it (Garcia-Cardena et al. 1998). Caveolin, the major structural protein of caveolae, binds to eNOS and inhibits it; release of eNOS from caveolin is associated with increased NO production by eNOS (Garcia-Cardena et al. 1997; Michel et al. 1997; Feron et al. 1998).

1.5.3
Phosphorylation

eNOS activity is also regulated by phosphorylation of key serine and threonine residues. Dimmeler et al. and Fulton et al. showed that phosphorylation of Ser1179 by Akt kinase activates eNOS and renders the enzyme less dependent on intracellular calcium (Dimmeler et al. 1999; Fulton et al. 1999). Mutations at this residue have been useful to characterize the importance of its phosphorylation by Akt kinase. In the *S1179D* mutant, Ser1179 is replaced by aspartate, which mimics the negative charge of the phosphate group. S1179D is constitutively active and does not depend on intracellular calcium for its activity. In the *S1179A* mutant, Ser1179 is replaced by alanine, which cannot be phosphorylated. As a result, the S1179A enzyme can still be activated by calcium transients, but has far less activity than the wild-type eNOS enzyme.

In contrast to the Ser1179 site, phosphorylation at Thr497 by protein kinase C (PKC) inactivates eNOS (Harris et al. 2001). Furthermore, phosphorylation of Ser1179 and Thr497 shows a reciprocal pattern; dephosphorylation of Thr497 is coordinated with the phosphorylation of Ser1179 and vice-versa (Harris et al. 2001; Michell et al. 2001). Dephosphorylation is mediated by specific phosphatases PP1 and PP2A (Michell et al. 2001). Because Akt kinase is an important regulator of vascular cell survival, it is likely that Akt-mediated phosphorylation of eNOS plays a key role in modulation of eNOS activity. In addition to Akt kinase, eNOS is phosphorylated at Ser1179 by cAMP-dependent protein kinase and AMP-activated protein kinase (Chen et al. 1999). Recent work suggests that phosphorylation of eNOS at Ser1179 mediates the vascular responses to many diverse stimuli, including growth factors such as VEGF (Fulton et al. 1999), IGF-1 (Chen et al. 1999), and insulin (Montagnani et al. 2001), mechanical shear stress (Dimmeler et al. 1999), the rapid, nongenomic activation of eNOS by estrogens (Hisamoto et al. 2001), and the activation of eNOS by the HMG-CoA reductase inhibitors, or statins (Kureishi et al. 2000; Brouet et al. 2001). Furthermore, abnormalities in eNOS phosphorylation have been associated with abnormal endothelial function in diabetes (Du et al. 2001).

2
Genetic Disruption of NOS Genes

2.1
General Considerations

Many key studies on the biological roles of NO were accomplished with the use of pharmacological inhibitors of NOS enzymes, e.g., L-nitro-arginine and L-*N*-monomethyl-arginine. These substrate analogs bind to NOS, but cannot serve as substrates because of substitutions at the guandino nitrogen. Blockade of a physiological process by these agents, and its reversal with an excess of L-arginine was strong evidence of the involvement of NO in that process. In many cases, these effects could be complemented by the use of methylene blue (which inhibits guanylate cyclase), or phosphodiesterase inhibitors, to probe whether the NO effect is cGMP dependent. However, a limitation of the pharmacological approach is that many NOS inhibitors affect more than one NOS isoform, making it difficult to determine which effects were mediated by which isoforms. Further complicating this is that most tissues contain innervation, vasculature, and circulating cells, so all three isoforms could be present.

One powerful approach that has complemented the pharmacological approach is the genetic disruption of the NOS genes, resulting in the generation of knockout mice that lack each of the NOS genes. Fortunately, knockout mice for each of the three major NOS isoforms are viable and develop apparently normally. Because each NOS isoform is encoded by a separate gene, this approach offers a different, and sometimes greater specificity than do pharmacological NOS inhibitors. It also allows the study of how chronic absence of the NOS isoform affects physiology in intact animals.

Despite these important strengths, there are some important considerations in interpreting studies with gene knockout mice. First, with any gene knockout, there is always the possibility of a developmental abnormality. If one of the NOS isoforms plays a critical role in embryonic or postnatal development, its absence may lead to other secondary abnormalities that are difficult to predict. Second, there may be physiological compensation for the absence of individual NOS genes, either by changes in other remaining NOS isoforms, interacting pathways (other vasodilatory or vasoconstricting factors), or unrelated pathways. Indeed, the nNOS knockout mice show physiological compensation both by NO-independent and -dependent pathways (Crosby et al. 1995; Irikura et al. 1995; Ma et al. 1996). Third, genetic background may confound the results, particularly when the phenotype of the parental strains varies with the genetic background (Gerlai 2001). For this reason, it is important to backcross knockout animals to standard inbred strains (e.g., C57BL/6) to obtain congenic mice that differ from these standard strains solely by the disruption of the gene.

2.2
Phenotypes of NOS Knockout Mice

We will briefly describe each of the NOS knockout mice and their phenotypes. In the subsequent sections, we will focus on how interrogation of the knockout mice has led to new insights into the roles of NO in stroke and atherosclerosis.

2.2.1
nNOS Knockout Mice

Neuronal NOS knockout mice were generated by disrupting exon 2 of the mouse *nNOS* gene, which includes the initiation codon ATG (Huang et al. 1993). NO production is markedly diminished in the brain of nNOS knockout mice, as measured by a variety of methods, including NOS enzymatic assay, cGMP levels, and measurement of NO by spin trapping (Darius et al. 1995; Ichinose et al. 1995; Irikura et al. 1995; Zaharchuk et al. 1997). Detailed analysis indicates that these mutant mice do not express the nNOSα, the predominant splice form of nNOS, as detected by Western blot analysis or NADPH diaphorase staining. However, they do express the β and γ isoforms of nNOS, which lack the exon that was deleted. These isoforms account for less than 5% of the nNOS present in the brain. Since exon 2 encodes the PDZ domain, the remaining nNOS β and γ isoforms are soluble, unlike nNOSα, which is localized to postsynaptic regions and neuromuscular junctions by the PDZ domain. The most apparent phenotype in nNOS knockout mice is enlargement of the stomach, often to several times the normal size, demonstrating the importance of nNOS to smooth muscle relaxation of the pyloric sphincter. nNOS knockout mice are also resistant to focal and global cerebral ischemia, consistent with a role for nNOS-derived NO in cellular injury following ischemia (Huang et al. 1994; Dawson et al. 1996; Hara et al. 1996; Panahian et al. 1996; Zaharchuk et al. 1997), as will be described in detail in the following section.

More recently, knockout mice that lack exon 6 of the nNOS gene have been generated, resulting in absence of all detectable nNOS isoforms, including the β and γ isoforms. The exon 6 nNOS knockout mice have a more severe gastrointestinal phenotype, and require a liquid diet for survival past weaning. These mice also display reproductive abnormalities, some of which appear to be due to abnormalities in gonadotropin secretion (Gyurko et al. 2002).

2.2.2
eNOS Knockout Mice

eNOS knockout mice were generated by deleting the NADPH ribose and adenine binding sites, which are essential to the enzymatic activity (Huang et al. 1995). The mice appear to develop normally, and have no detectable anatomical abnormalities. In organ baths, isolated aortic rings from eNOS mutant mice do not relax in response to acetylcholine, although the vessels do relax to sodium nitroprusside and papaverine, providing genetic evidence that the *eNOS* gene is required for EDRF activity. The eNOS knockout mice are hypertensive, with mean arterial blood pressures that are 20–30 mmHg over values seen in wild-type animals. Using Millar catheters to measure left ventricular pressure, the contractile response of the eNOS knockout mice to the β-agonist isoproterenol is significantly increased compared to wild-type mice (Gyurko et al. 2000). These results suggest that eNOS normally serves to blunt the contractile response to β-agonists. As will be discussed in Sect. 11.3.4.1, eNOS knockout mice have been a useful animal model for endothelial dysfunction, because of their increased propensity to form neointima in response to vessel injury and diet-induced atherosclerosis.

2.2.3
iNOS Knockout Mice

Three separate groups independently disrupted the *iNOS* gene (Laubach et al. 1995; Macmicking et al. 1995; Wei et al. 1995). iNOS knockout mice are more sensitive to the intracellular pathogens *Listeria monocytogenes* and *Leishmania major*. iNOS knockout mice also have a blunted hypotensive response to septic shock induced by lipopolysaccharide.

2.3
Stroke and Cerebral Ischemia in NOS Knockout Mice

2.3.1
NO Levels in Ischemic Brain

Malinski et al. directly measured NO levels in the brain using a porphyrinic sensor and found that they increase by several orders of magnitude following cerebral ischemia (Malinski et al. 1993). The source of this dramatic rise in NO production appears to be the nNOS isoform, and is due in part to activation of pre-existing nNOS enzyme by increased intracellular calcium concentrations, and in part to up-regulation of nNOS expression. A second wave of NO production occurs later after the acute ischemic event, due to induction and activation of the iNOS isoform, principally within macrophages and glial cells in the infarct zone.

2.3.2
Divergent Roles of NOS Isoforms

The role of NO production following stroke has been studied using pharmacological inhibition of NOS. Depending on the type and dose of inhibitor used, the ischemia model tested, the route and timing of drug administration, and the species studied, the outcome of cerebral ischemia following NOS inhibition was highly variable (Iadecola et al. 1994). This was likely due to differential effects on more than one NOS isoform by the inhibitors. Knockout animals that lack the various NOS isoforms have been useful to clarify the different and sometimes opposing roles of nNOS, eNOS, and iNOS.

Neuronal overproduction of NO can contribute to cellular damage following ischemia by formation of peroxynitrite anion, activation of poly-ADP ribose polymerase (PARP), and inhibition of mitochondrial energy metabolism. In contrast, endothelial NOS is important for the maintenance of normal vascular tone, and during and after an ischemic event, may play important roles in attempting to restore or maintain blood flow to ischemic regions. Inducible NOS in astrocytes, macrophages and microglia also generates NO following ischemia.

2.3.3
Response of NOS Knockout Mice to Cerebral Ischemia

Neuronal NOS mutant mice are resistant to global and focal cerebral ischemia in vitro. In the middle cerebral artery occlusion (MCAO) model of focal ischemia, an intraluminal filament is used to occlude the middle cerebral artery. The MCAO model causes less morphological damage and smaller infarcts in nNOS knockout mice than in their wild-type counterparts (Huang et al. 1994). The nNOS knockout mice also have better functional neurological outcome. The neuroprotection is independent of changes in relative cerebral blood flow reductions as measured by laser Doppler flowmetry, indicating that the vascular effect of the primary ischemic insult is the same. These results are also seen with transient MCAO (Hara et al. 1996) and with global ischemia models (Panahian et al. 1996). These data confirm that nNOS-derived NO contributes to cellular damage following cerebral ischemia. Primary neuronal cultures grown from nNOS knockout mice are also resistant to both NMDA-induced neurotoxicity and oxygen-glucose deprivation (Dawson et al. 1996). On the other hand, these cells are as vulnerable to kainate toxicity as wild-type neurons. These data support the roles of NMDA stimulation and nNOS activation in excitotoxic neuronal injury.

In contrast, eNOS knockout mice develop larger infarcts and more severe neurological deficits from MCAO than wild-type animals (Huang et al. 1996). These infarcts correlate with more severe reductions in blood flow measured by laser Doppler flowmetry and by functional CT imaging (Lo et al. 1996). Inhibition of the remaining eNOS isoform in nNOS knockout mice also results in worsening of the outcome. These results confirm that the eNOS isoform serves important protective roles in maintaining blood flow following ischemia. Since many pharmacological NOS inhibitors target eNOS as well as nNOS, these results may explain the aggravation of outcome found with nonspecific NOS inhibition.

Like nNOS knockout mice, iNOS knockout mice develop smaller infarcts following MCAO when observed at a later time point (Iadecola et al. 1995, 1997). Other markers for severity of ischemic injury such as rCBF reduction, neutrophil accumulation, and astrocytic proliferation were comparable between the knockouts and their wild-type littermates, indicating that the ischemic resistance of the mice is due to lack of iNOS and that iNOS contributes to late neurotoxicity following ischemia.

2.3.4
Lessons Learned from NOS Knockout Mice

These results indicate that nNOS and iNOS are important mediators of early and late toxicity following ischemia, while eNOS serves important vascular protective roles. Thus, selective blockade of nNOS and iNOS and, potentially, the augmentation of eNOS might be beneficial in the treatment of stroke. The benefit of eNOS augmentation may extend beyond the domain of stroke treatment to stroke prophylaxis, ideally in patients who are at high-risk for an ischemic insult. In several large clinical studies, the use of HMG-CoA reductase inhibitors, a class of drugs commonly known as statins, was associated with reductions in the incidence of stroke, independent of cholesterol-lowering effects. In experimental animals, simvastatin and lovastatin protect against cerebral infarction and neurological dysfunction after MCAO (Endres et al. 1998). One of the mechanisms for these effects may involve up-regulation of eNOS activity by stabilization of eNOS mRNA (Laufs et al. 2000) or by Akt phosphorylation of eNOS (Kureishi et al. 2000). The beneficial effect of the statins is absent in eNOS mutant mice, indicating that the protection is dependent on eNOS expression (Endres et al. 1998).

2.4
Atherosclerosis in NOS Knockout Mice

2.4.1
eNOS Knockout Mice as a Model for Endothelial Dysfunction

A common feature of atherosclerosis, hypertension, diabetes and hypercholesterolemia is endothelial dysfunction, in which the vasculature fails to respond to stimuli that normally elicit vasodilation. Since endothelial NO production is responsible for vasodilation, reduced or altered eNOS activity would lead to endothelial dysfunction. It is not clear whether endothelial dysfunction is causally related to the mechanisms of atherosclerosis, or whether it is merely associated with them. In this regard, eNOS knockout mice are a useful tool. They represent one extreme model of endothelial dysfunction, in which there is no endothelial NO production. While many studies have used homozygous eNOS knockout mice, studies in heterozygous animals also support the contention that *eNOS* gene dosage, and consequently the level of eNOS expression, are important to normal vascular function (Shesely et al. 1996; Kojda et al. 2001). These results suggest that minor changes in the level of eNOS expression or activity, for example those caused by polymorphisms, may have important consequences in terms of disease susceptibility.

2.4.2
Vascular Injury in eNOS Knockout Mice

An early step in the development of atherosclerosis is injury to the endothelium, followed by a vascular response to injury (Ross 1993). Normally, this response consists of proliferation of cells in the medial smooth muscle layer, followed by migration of the cells to the subendothelial space, forming the neointima. eNOS knockout mice subjected to vascular injury models show an exaggerated injury response, with increased proliferation of medial smooth muscle cells, and increased neointimal thickness (Moroi et al. 1998; Zhang et al. 1999; Yogo et al. 2000).

2.4.3
Diet-Induced Atherosclerosis in apoE/eNOS Double Knockout Mice

To test the effect of eNOS deficiency on diet-induced atherogenesis, double knockout mice were bred using eNOS knockout mice and apoE knockout mice. ApoE knockout mice develop predictable atherosclerotic lesions in their aortas, a process than can be accelerated by feeding a high-fat, Western-type diet, in which fat accounts for 40% of the total caloric content. apoE/eNOS double knockout mice place eNOS deficiency on top of the apoE knockout mouse model of diet-induced atherosclerosis.

apoE/eNOS double knockout mice develop nearly twice the atherosclerotic burden that apoE knockout mice do in the absence of eNOS gene deletion (Kuhlencordt et al. 2001). eNOS deficiency markedly accelerates and worsens the disease pattern of atherosclerosis in Western-diet-fed apoE knockout animals. In addition, apoE/eNOS double knockout mice display coronary artery disease, ischemic heart disease, left ventricular dysfunction, and vascular complications of aortic aneurysm and dissection (Kuhlencordt et al. 2001). Hypertension alone does not account for the development of aortic aneurysm and dissection, in that pharmacological control of blood pressure does not reduce the incidence of these complications (Chen et al. 2001). The phenotype of apoE/eNOS double knockout mice closely resembles the spectrum of cardiovascular complications seen in human atherosclerosis. It is also the first murine model to demonstrate spontaneous distal coronary arteriosclerosis associated with left ventricular dysfunction. These findings support the concept that restoration of eNOS function in patients with atherosclerosis is an important therapeutic goal.

2.4.4
iNOS and Diet-Induced Atherosclerosis

In contrast to eNOS, which serves important protective roles, iNOS may contribute to lesion formation by increasing oxidative stress and oxidizing LDL. apoE/iNOS double knockout mice fed a Western-type diet showed aortic lesion areas that were significantly reduced, by about 20%, compared to apoE knockout animals. There was no effect on lipoprotein profile, although plasma levels of lipoperoxides were significantly reduced, suggesting that reduction in iNOS-mediated oxidative stress may explain the protection from lesion formation in double knockout animals. These findings suggest that iNOS may be proatherogenic. Furthermore, one should be cautious about nonspecific supplementation with L-arginine, in an effort to increase eNOS-derived NOS, since iNOS may also be affected (Chen et al. 2003; Loscalzo 2003).

3
NOS Gene Polymorphisms

3.1
General Considerations

Genetic polymorphisms in the *NOS* genes have been reported to be associated with susceptibility to cardiovascular diseases, including hypertension, atherosclerotic disease of the coronary arteries and carotid arteries, renal disease, diabetic complications, and a several rare diseases. These studies were performed in different populations, and the results vary in the association of specific polymorphisms with cardiovascular disease susceptibility, with some studies reporting strong associations, and others reporting none. Indeed, when taken together, these studies show a great deal of inconsistency and contradictory results (Wang and Wang 2000; Wattanapitayakul et al. 2001; Hingorani 2003).

An important consideration is the study design and methodology used. There are several methods for detecting association between polymorphisms and disease. One method is the case–control method, in which two groups of subjects, one with the disease, and one without (normal controls) are compared for the frequency of the genotypes at the polymorphism studied. In case–control studies, it is important to match the important characteristics of the two groups so that the control group serves as a true control. Case–control studies may lead to spurious associations due to population stratification, or the presence of genetically different groups in the population. A second method is the cohort method, to follow individuals from the general population for development of a phenotype, for example hypertension, or for response to a given stimulus, for example, vascular response to acetylcholine. All the subjects would be genotyped to determine whether the polymorphisms are associated with the phenotype or response. In a variation of the cohort method, the sample could be drawn, not from the general population, but from patients affected by a given disease. An example of this type of study is following patients with polycystic kidney disease for development of renal failure, or following patients with coronary artery stents for development of restenosis. Finally, there are family-based methods to test association of specific polymorphism genotypes with inheritance of the disease phenotype, most commonly using the transmission disequilibrium test.

In addition to study methodology, there are other potential reasons for the wide discrepancies in the many published studies. First, the relationship between polymorphisms and cardiovascular phenotype may depend on the population studied. For example, each of the three major types of eNOS gene polymorphisms varies significantly between Caucasian, African-American, and Asian populations (Tanus-Santos et al. 2001). Second, some studies may be underpowered (i.e., not have enough subjects to show a clear association). Third, studies may give rise to spurious results because control groups and study groups may differ (population stratification). Fourth, given the emerging understanding of the haplotype block structure of the human (Dawson et al. 2002) and mouse genomes (Wade et al. 2002), even finding a particular association is not a guarantee that the polymorphism itself is responsible for (causally related to) the phenotype, as opposed to being closely associated with it. For example, if a haplotype block, which is generally the minimum unit for recombination, contains an *eNOS* polymorphism as well as a polymorphism in a different gene, the *eNOS* polymorphism will appear linked to phenotypes that are in actuality caused by the other gene. Furthermore, if the association between the *eNOS* polymorphism and the polymorphism in the other gene differs between populations, then analysis of some populations will detect the linkage with the eNOS gene, while others will not.

The effect of genetic polymorphisms on disease phenotype is more easily understood in cases where the polymorphism affects protein structure (e.g., a coding mutation), or expression (e.g., mutation in a promoter or intron affecting expression levels). It is less clear in the case of a polymorphism that does not cause a detectable change in protein function or expression levels. Thus, determining the functional significance of a given polymorphism by its expression either in vitro or in vitro is a key step in validating whether a given association with disease susceptibility is real.

3.2
Polymorphisms in eNOS

The polymorphisms in the *eNOS* gene can be divided into those that affect the coding regions, the promoter region, and the introns. The positions of these polymorphisms are shown in Fig. 2.

3.2.1
Coding Polymorphism: *Glu298Asp*

The only known polymorphism to affect the coding region of the *eNOS* gene is *Glu298Asp*. First identified by several groups in Japan, this polymorphism occurs in exon 7, and is the result of a G→T conversion at nucleotide position 894 in the eNOS cDNA (thus it is also known as the *G894T* polymorphism). The codon GAG, which encodes glutamate, is replaced by GAT, which encodes aspartate. The substitution of glutamate with aspartate is technically a missense mutation, albeit a conservative one, in that both amino acids are acidic.

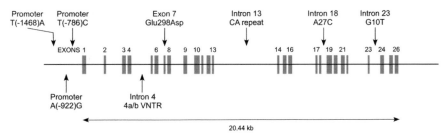

Fig. 2 Distribution of polymorphisms in eNOS

The *Glu298Asp* mutation was first positively associated with essential hypertension (Miyamoto et al. 1998), myocardial infarction (Hibi et al. 1998; Shimasaki et al. 1998), and coronary vasospasm (Yoshimura et al. 1998) in populations in Kyoto, Kumamoto, and Yokohama in Japan. Subsequently, it was reported and studied in other populations. To date, there have been a total of 43 reports on this polymorphism and its relationship to disease, but of these, 22 suggest a positive association, and 21 do not. Of these, 27 studies were case–control studies, where the genotypes of subjects in an affected group and an unaffected control group are compared. Six were cohort studies testing association between the polymorphism and specific responses in the general population, and nine were cohort studies testing for specific responses or phenotypes in patients with specific diseases. One was a family-based study using transmission disequilibrium testing to look for association of the polymorphism with the inheritance of coronary artery disease. Table 3 summarizes the references, specific disease or response phenotypes examined, population studied, number of subjects and controls, and whether a statistical association was found.

From the three-dimensional structure of eNOS, the Glu298 residue is predicted to be exposed on the exterior of the dimer surface, and would not be expected to be close to the catalytic site or cofactor-binding domains, which are located in the cleft region (Jachymova et al. 2001). Several studies have examined the effect of the *Glu298Asp* mutation on the function or expression of eNOS protein. In human subjects, measurements of forearm blood flow responses to acetylcholine (an endothelium-dependent response) or nitroprusside (an endothelium-independent response) were the same across *eNOS* genotypes at amino acid position 298, as was NO production, suggesting that this polymorphism does not result in a detectable change in vascular function in vitro (Schneider et al. 2000). Similarly, in human saphenous vein grafts obtained from patients with coronary artery disease, the *eNOS* genotype at amino acid position 298 was not associated with differences in vascular responses to acetylcholine, bradykinin, calcium ionophore, or nitroprusside (Guzik et al. 2001). However, another study found that although there were no differences in the vascular response of the brachial artery at baseline, the Asp298 phenotype was associated with lower flow-mediated dilation in combination with smoking, raising the possibility of interactions between genotype and environmental factors such as cigarette

Table 3 Studies on association between Glu298Asp polymorphism and cardiovascular diseases

Case–control studies				
Reference	Disease	Population	Subjects (patient/normal)	Association
Yasujima et al. 1998	Hypertension	Japan	166/174	Yes
Miyamoto et al. 1998	Hypertension	Japan	405/463	Yes
Lacolley et al. 1998	Hypertension	France	309/123	No
Kato et al. 1999	Hypertension	Japan	549/513	No
Shoji et al. 2000	Hypertension	Japan	183/193	Yes
Jachymova et al. 2001	Hypertension	Czech Republic	119/85	Yes
Karvonen et al. 2002	Hypertension	Finland	600/600	No
Hingorani et al. 1999	Coronary artery disease	UK	547/321	Yes
Pulkkinen et al. 2000	Coronary artery disease	Finland	308/110	No
Granath et al. 2001	Coronary artery disease	Australia	573/624	No
Wang et al. 2001	Coronary artery disease	Taiwan	218/218	No
Colombo et al. 2002	Coronary artery disease	Italy	201/114	Yes
Yoshimura et al. 2000	Coronary artery spasm	Japan	201/345	Yes
Shimasaki et al. 1998	Myocardial infarction	Japan	285/607	Yes
Hibi et al. 1998	Myocardial infarction	Japan	225/357	Yes
Ukkola et al. 2001	Diabetic atherosclerosis	Finland	239/245	No
Markus et al. 1998	Stroke, cerebrovascular disease	UK	361/236	No
Karvonen et al. 2002	Carotid artery stenosis	Finland	600/600	No
Ghilardi et al. 2002	Carotid artery stenosis	Italy	88/133	No
Noiri et al. 2002	End-stage renal disease	Japan	185/304	Yes
De Prado et al. 2002	End-stage renal disease	Spain	84/93	No
Ohtoshi et al. 2002	Diabetes mellitus	Japan	233/301	No
Yoshimura et al. 2000	Pre-eclampsia	Japan	152/170	Yes
Hefler et al. 2002	Recurrent miscarriage	Austria	130/67	No
Salvarani et al. 2002	Behçet's disease	Italy	73/135	Yes
Fatini et al. 2002	Systemic sclerosis	Italy	73/112	Yes
Heltianu et al. 2002	Fabry's disease	Romania	19/39	Yes
Droma et al. 2002	High altitude pulmonary edema	Japan	41/51	Yes
Cohort studies in general population				
Reference	Response studied	Population	Subjects	Association
Benjafield and Morris 2000	Hypertension	Australia	112	No
Tsujita et al. 2001	Hypertension	Japan	4,055	No
Chen et al. 2001	Hypertension	USA (African descent)	1,021	Yes
Lembo et al. 2001	Carotid atherosclerosis	Italy	375	Yes
Naber et al. 2001	Reduced coronary blood flow	Germany	97	Yes
Grossmann et al. 2001	Vascular response to Ach	Germany	37	No
Cohort studies in disease populations				
Reference	Response studied	Population	Subjects	Association
Philip et al. 1999	Response to phenylephrine in patients undergoing cardiac surgery	France	68	Yes

Table 3 (continued)

Cohort studies in disease populations				
Reference	Response studied	Population	Subjects	Association
Persu et al. 2002	Renal failure in patients with PCKD1	Belgium	173	Yes
Walker et al. 2003	HTN and renal disease in patients with PCKD 1	USA	215	No
French et al. 2003	Flow-limiting stenoses in patients with myocardial infarction	USA	395	No
Guzik et al. 2001	Vascular responses to Ach in patients with CAD	UK	104	No
Gomma et al. 2002	Restenosis in CAD patients with stents	UK	205	Yes
Ohtoshi et al. 2002	Insulin resistance in patients with diabetes mellitus	Japan	233	No
Viklicky et al. 2002	Graft survival in renal transplant patients	Czech Republic	?	No
Jachymova et al. 2001	Resistance to treatment in patients with hypertension	Czech Republic	119	Yes
Family-based studies				
Reference	Disease and response studied	Population	Subjects	Association
Via et al. 2003	Coronary artery disease	Spain	101	No

smoking (Leeson et al. 2002). In vitro, there is evidence that the *Glu298Asp* polymorphism correlates with increased platelet aggregation (Tanus-Santos et al. 2002).

Another approach to test the functional effects of the polymorphism has been in vitro or in vitro expression of the variant, to test whether it affects eNOS enzymatic function. Chinese hamster ovary (CHO) cells stably transfected with WT eNOS cDNA consistently produced greater amounts of NO than did CHO cells transfected with eNOS cDNA carrying the *Asp298* variant, suggesting that the polymorphism affects NO output. Furthermore, in transfected cells, primary human endothelial cells, and human hearts, eNOS carrying Asp, but not Glu, at the 298 position was preferentially cleaved, resulting in the generation of N-terminal 35-kD and 100-kD C-terminal products (Tesauro et al. 2000). Thus, this coding polymorphism may generate protein products with differing susceptibility to proteolytic cleavage.

3.2.2
Promoter Polymorphisms

Three polymorphisms have been identified in the 5' flanking promoter regions of the eNOS gene. These are the *T(-1468)A*, *T(-786)C*, and *A(-922)G* polymorphisms. The designation indicates the nucleotide normally at the position, the location of the polymorphism with respect to the transcription start site (defined as position 1), and the nucleotide present in people who carry the polymorphism. Thus, *T(-786)C* is located 786 bases upstream from the transcription start site of the eNOS gene, and T is the nucleotide normally found at this position, while C is the nucleotide present in people who carry the polymorphism.

The *T(-1468)A*, *T(-786)C*, and *A(-922)G* polymorphisms are linked (Nakayama et al. 1999). Pooling any of these polymorphisms as a single factor for multiple logistical regression analysis, Nakayama et al. found that possession of any of these three *eNOS* promoter polymorphisms was an independent risk factor for coronary artery spasm. Other studies have focused on the *T(-786)C* polymorphism, and examined its association with hypertension, coronary artery disease, diabetes, stent restenosis, and renal failure. As with the coding polymorphisms, the methodologies used include case–control studies, cohort studies, and family-based approaches. Table 4 summarizes the references, disease or response studied, population studied, sample size, and major findings.

Because these polymorphisms are found in the 5′ flanking region of the *eNOS* gene, they cannot affect the sequence of the encoded protein. However, they could potentially affect levels of expression by affecting promoter function and/or transcript stability or function. Nakayama expressed luciferase reporter genes under the control of the eNOS promoter with and without the *T(-786)C* polymorphism in human endothelial cells (Nakayama et al. 1999). The wild-type or normal eNOS promoter with T at the −786 position showed higher activity than did the promoter carrying the polymorphism, with C at the same position. Furthermore, hypoxia-induced expression was also reduced with the *T(-786)C* polymorphism. However, using similar methodology, Wang et al. found that the transcription efficiency of the eNOS promoter with T at the −786 position is lower than the promoter with C at the position (Wang et al. 2002). The reasons for this discrepancy are not clear, as similar constructs, reporter genes, and human endothelial cells were used. Wang reported that cigarette smoking extracts significantly increase the transcription efficiency of the normal promoter, while the efficiency of the promoter with C at the position decreased. Thus, the *T(-786)C* polymorphism may have functional effects on eNOS promoter function, both for basal transcriptional activity and for responses to other stimuli such as hypoxia or cigarette smoking.

3.2.3
Intron Polymorphisms

Several polymorphisms have been found in the introns of the *eNOS* gene. These include single nucleotide variations found in intron 18 (*A27C*) and intron 23 (*G10T*), where the designation shows the normal nucleotide, the location within the intron, and variant nucleotide found in the polymorphism. These polymorphisms have been studied for association with hypertension in case–control studies, and no association was found for either one (Bonnardeaux et al. 1995; Lacolley et al. 1998; Miyamoto et al. 1998).

Table 4 Studies on association between T(-786)C promoter polymorphisms and cardiovascular diseases

Case control studies				
Reference	Disease	Population	Subjects (patient/normal)	Association
Kajiyama et al. 2000	Hypertension	Japan	401/456	No
Sim et al. 1998	Coronary artery disease	Australia	633/160	No
Granath et al. 2001	Coronary artery disease	Australia	573/624	No
Nakayama et al. 1999	Coronary artery spasm	Japan	174/161	Yes
Yoshimura et al. 2000	Coronary artery spasm	Japan	201/345	Yes
Ghilardi et al. 2002	Carotid artery stenosis	Italy	88/133	Yes
Ohtoshi et al. 2002	Diabetes mellitus	Japan	233/301	No
Fatini et al. 2002	Systemic sclerosis	Italy	73/112	No
Cohort studies in general population				
Reference	Response studied	Population	Subjects	Association
Tsujita et al. 2001	Hypertension	Japan	4,055	No
Cohort studies in disease patients				
Reference	Response studied	Population	Subjects	Association
Gomma et al. 2002	Restenosis in CAD patients with stents	UK	205	Yes
French et al. 2003	Flow-limiting stenoses in patients with myocardial infarction	USA	395	No
Ohtoshi et al. 2002	Insulin resistance in patients with diabetes mellitus	Japan	233	Yes
Zanchi et al. 2000	Renal failure in patients with diabetes mellitus	USA	347	Yes
Family-based studies				
Reference	Disease	Population	Subjects	Association
Via et al. 2003	Coronary artery disease	Spain	101	No
Zanchi et al. 2000	Diabetic renal failure	USA	132	Yes

Another type of intron polymorphism involves the variable number of tandem repeats (VNTR) found in the intron. Intron 13 contains a variable number of repeats of the dinucleotide sequence CA. Some studies have found associations between a high number of CA repeats and coronary artery disease (Stangl et al. 2000) or hypertension (Nakayama et al. 1997), although other studies have not (Bonnardeaux et al. 1995).

Intron 4 contains a variable number of repeats of a 27-base sequence. Most individuals have five copies of the 27-base repeat, which is designated the intron *4b* VNTR, while the variant found in the polymorphism contains only four copies of the repeat, designated intron *4a* VNTR. Most studies do not show an association between the intron *4a/b* VNTR polymorphism and hypertension, cor-

Table 5 Studies on association between intron polymorphisms and cardiovascular diseases

Case control studies					
Reference	Polymorphism	Disease	Population	Subjects (patient/normal)	Association
Miyamoto et al. 1998	Intron 18, 23	Hypertension	Japan	405/463	No
Lacolley et al. 1998	Intron 23	Hypertension	France	309/123	No
Stangl et al. 2000	Intron 13	Coronary artery disease	Germany	1,000/1,000	Yes
Miyamoto et al. 1998	4a/b VNTR	Hypertension	Japan	405/463	No
Yasujima et al. 1998	4a/b VNTR	Hypertension	Japan	166/174	No
Shoji et al. 2000	4a/b VNTR	Hypertension	Japan	183/193	No
Granath et al. 2001	4a/b VNTR	Coronary artery disease	Australia	573/624	No
Pulkkinen et al. 2000	4a/b VNTR	Coronary artery disease	Finland	308/110	No
Hwang et al. 2002	4a/b VNTR	Coronary artery disease	Taiwan	219/?	No
Yoshimura et al. 2000	4a/b VNTR	Coronary artery spasm	Japan	201/345	Yes
Hibi et al. 1998	4a/b VNTR	Myocardial infarction	Japan	225/357	No
Droma et al. 2002	4a/b VNTR	High-altitude pulmonary edema	Japan	41/51	Yes
Heltianu et al. 2002	4a/b VNTR	Fabry's disease	Romania	19/39	Yes
Salvarani et al. 2002	4a/b VNTR	Behçet's disease	Italy	73/135	No
Cohort studies in disease patients					
Reference	Polymorphism	Response studied	Population	Subjects	Association
French et al. 2003	4a/b VNTR	Flow-limiting stenoses in patients with myocardial infarction	USA	395	No
Pulkkinen et al. 2000	4a/b VNTR	Hypertension in patients with diabetes and CAD	Finland	251	Yes
Zanchi et al. 2000	4a/b VNTR	Renal failure in patients with diabetes mellitus	United States	347	Yes
Family-based studies					
Reference	Polymorphism	Disease	Population	Subjects	Association
Bonnardeaux et al. 1995	Intron 13	Hypertension	France	346	No
Via et al. 2003	4a/b VNTR	Coronary artery disease	Spain	101	No
Zanchi et al. 2000	4a/b VNTR	Diabetic renal failure	United States	347	Yes

onary artery disease, or myocardial infarction. Table 5 summarizes the results of studies on eNOS intron polymorphisms.

Functionally, intron polymorphisms do not affect the structure of the encoded protein, as the intron sequences are spliced out during mRNA processing. However, alterations in intron sequences may potentially affect the rate of eNOS gene transcription or mRNA precursor stability, so they may affect eNOS expression levels. Tsukada et al. measured NO byproducts in the human plasma of subjects with both *4a* and *4b* VNTRs, and found a 10%–20% decrease in subjects carrying the *4a* allele (Tsukada et al. 1998). Wang et al. tested the effect of both *4a* and *4b* VNTRs on transcription as an enhancer element and found differences (Wang et al. 2002), although the functional significance of these differences is not known.

Yoshimura et al. showed that the intron *4a* allele and *T(-786)C* polymorphisms are in linkage disequilibrium (Yoshimura et al. 2000). They reported that by multiple logistic regression analysis, the *T(-786)C* polymorphism is more predictive for coronary spasm than the intron *4a* VNTR. Thus, they suggest that the linkage between the two polymorphisms may underlie weak associations seen with the intron 4 polymorphism that are in fact due to the *T(-786)C* polymorphism.

3.3
nNOS and iNOS Polymorphisms

In contrast to the many studies on *eNOS* gene polymorphisms, there have been relatively few studies on polymorphisms in the *nNOS* or *iNOS* genes. The *nNOS* gene contains known polymorphisms in its promoter region, introns, and the 3′ untranslated region. The promoter region contains a variable number of dinucleotide repeats, a polymorphism that has been reported to be associated with Parkinson's disease (Lo et al. 2002). Intron 20 of the *nNOS* gene contains a variable number of repeats of the trinucleotide AAT. The number of repeats appears to be inversely correlated with the level of exhaled NO and associated with acute chest syndrome in subjects with sickle cell anemia (Sullivan et al. 2001). The 3′ untranslated region, located in exon 29 of the nNOS gene, contains a variable number of CA repeats, which results in heterogeneity in the size of the nNOS mRNA. There does not appear to be any association between this polymorphism and Alzheimer's disease (Liou et al. 2002).

The *iNOS* gene contains two known polymorphisms in its promoter region. One consists of repeats of AAAT/AAAAT, located between 756 and 716 bases upstream from the transcription start site. This does not appear to be associated with migraine headaches (Lea et al. 2001). Another polymorphism consists of repeats of CCCTT. One study reported it to be associated with dementia (Xu et al. 2000), but another did not (Singleton et al. 2001).

Few studies have examined the role of these or other *nNOS* or *iNOS* polymorphisms in cardiovascular disease. Furthermore, there have been few studies on the functional effects of these polymorphisms on the structure or expression levels of either nNOS or iNOS protein.

4
Conclusions

The NO system is complex, with the individual NOS isoforms playing distinct and sometime opposing roles in physiology and disease pathogenesis. Studies in knockout mice have demonstrated the protective roles of eNOS in maintaining blood flow following cerebral ischemia, and in normal vascular function, suppressing the development of atherosclerosis. In contrast, nNOS-derived NO (early) and iNOS-derived NO (late) contributes to tissue damage and toxicity following cerebral ischemia. Furthermore, iNOS participates in atherogenesis by contributing to lipid oxidation.

The extreme phenotypes of eNOS knockout mice indicate the importance of normal endothelial NO production to vascular function. They further demonstrate that endothelial dysfunction is indeed causally related to the molecular mechanisms of atherogenesis. Results in eNOS knockout mice suggest that minor effects on eNOS structure, function, or level of expression, such as those caused by polymorphisms, may indeed affect disease susceptibility.

Many studies on genetic association of *eNOS* gene polymorphisms and cardiovascular disease have led to contradictory results. The reasons for such divergent results include the different frequencies of the polymorphisms in ethnic populations, potentially insufficient sample size and power, spurious results from population stratification, and linkage disequilibrium between polymorphisms causally related to phenotype and those that are not. Indeed, only in few instances have polymorphisms been demonstrated to have a clear-cut functional effect on eNOS structure, function, or level of expression

Despite these unresolved issues, the NO system is rich with molecular targets for intervention. These include substrate availability, cofactor availability, subcellular localization and its determinants (myristoylation, palmitoylation, interactions with caveolin and hsp90), regulation at key phosphorylation sites by Akt kinase and other kinases or phosphatases, and downstream effector molecules, including guanylate cyclase and superoxide. A better understanding of how these targets interact in disease pathogenesis may lead to new approaches to disease prevention and treatment.

References

Alderton WK, Cooper CE, Knowles RG (2001) Nitric oxide synthases: structure, function and inhibition. Biochem J 357:593–615

Arnold WP, Mittal CK, Katsuki S et al (1977) Nitric oxide activates guanylate cyclase and increases guanosine 3':5'-cyclic monophosphate levels in various tissue preparations. Proc Natl Acad Sci USA 74:3203–7

Bath PM (1993). The effect of nitric oxide-donating vasodilators on monocyte chemotaxis and intracellular cGMP concentrations in vitro. Eur J Clin Pharmacol 45:53–8

Beckman JS, Chen J, Ischiropoulos H et al (1994) Oxidative chemistry of peroxynitrite. Methods Enzymol 233:229–240

Beckman JS, Koppenol WH (1996) Nitric oxide, superoxide, and peroxynitrite: the good, the bad, and ugly. Am J Physiol 271: C1424–37

Bonnardeaux A, Nadaud S, Charru A et al (1995) Lack of evidence for linkage of the endothelial cell nitric oxide synthase gene to essential hypertension. Circulation 91:96–102

Brenman JE, Christopherson KS, Craven SE et al (1996) Cloning and characterization of postsynaptic density 93, a nitric oxide synthase interacting protein. J Neurosci 16:7407–15

Brenman JE, Xia H, Chao DS et al (1997) Regulation of neuronal nitric oxide synthase through alternative transcripts. Dev Neurosci 19:224–31

Brouet A, Sonveaux P, Dessy C et al (2001) Hsp90 and caveolin are key targets for the proangiogenic nitric oxide-mediated effects of statins. Circ Res 89:866–73

Brown GC (1995) Nitric oxide regulates mitochondrial respiration and cell functions by inhibiting cytochrome oxidase. FEBS Lett 369:136–9

Busconi L, Michel T (1993) Endothelial nitric oxide synthase. N-terminal myristoylation determines subcellular localization. J Biol Chem 268:8410–3

Cai H, Harrison DG (2000) Endothelial dysfunction in cardiovascular diseases: the role of oxidant stress. Circ Res 87:840–4

Chen J, Kuhlencordt P, Urano F et al (2003) Effects of chronic treatment with L-arginine on atherosclerosis in apoE knockout and apoE/inducible NO synthase double-knockout mice. Arterioscler Thromb Vasc Biol 23:97–103

Chen J, Kuhlencordt PJ, Astern J et al (2001) Hypertension does not account for the accelerated atherosclerosis and development of aneurysms in male apolipoprotein e/endothelial nitric oxide synthase double knockout mice. Circulation 104:2391–4

Chen ZP, Mitchelhill KI, Michell BJ et al (1999) AMP-activated protein kinase phosphorylation of endothelial NO synthase. FEBS Lett 443:285–9

Crosby G, Marota JJ, Huang PL (1995) Intact nociception-induced neuroplasticity in transgenic mice deficient in neuronal nitric oxide synthase. Neuroscience 69:1013–1017

Darius S, Wolf G, Huang PL et al (1995) Localization of NADPH-diaphorase/nitric oxide synthase in the rat retina: an electron microscopic study. Brain Res 690:231–5

Dawson E, Abecasis GR, Bumpstead S et al (2002) A first-generation linkage disequilibrium map of human chromosome 22. Nature 418:544–8

Dawson TM, Dawson VL, Snyder SH (1993) Nitric oxide as a mediator of neurotoxicity. NIDA Res Monogr 136:258–71

Dawson TM, Dawson VL, Snyder SH (1994) Molecular mechanisms of nitric oxide actions in the brain. Ann N Y Acad Sci 738:76–85

Dawson VL, Kizushi VM, Huang PL et al (1996) Resistance to neurotoxicity in cortical cultures from neuronal nitric oxide synthase-deficient mice. J Neurosci 16:2479–87

Dimmeler S, Fleming I, Fisslthaler B et al (1999) Activation of nitric oxide synthase in endothelial cells by Akt-dependent phosphorylation. Nature 399:601–5

Du XL, Edelstein D, Dimmeler S et al (2001) Hyperglycemia inhibits endothelial nitric oxide synthase activity by posttranslational modification at the Akt site. J Clin Invest 108:1341–8

Endres M, Laufs U, Huang Z et al (1998) Stroke protection by 3-hydroxy-3-methylglutaryl (HMG)-CoA reductase inhibitors mediated by endothelial nitric oxide synthase. Proc Natl Acad Sci USA 95:8880–5

Endres M, Scott G, Namura S et al (1998) Role of peroxynitrite and neuronal nitric oxide synthase in the activation of poly(ADP-ribose) synthetase in a murine model of cerebral ischemia-reperfusion. Neurosci Lett 248:41–4

Feron O, Michel JB, Sase K et al (1998) Dynamic regulation of endothelial nitric oxide synthase: complementary roles of dual acylation and caveolin interactions. Biochemistry 37:193–200

Feron O, Saldana F, Michel JB et al (1998) The endothelial nitric-oxide synthase-caveolin regulatory cycle. J Biol Chem 273:3125–8

Freedman JE, Sauter R, Battinelli EM et al (1999) Deficient platelet-derived nitric oxide and enhanced hemostasis in mice lacking the NOSIII gene. Circ Res 84:1416–21

Fulton D, Gratton JP, McCabe TJ et al (1999) Regulation of endothelium-derived nitric oxide production by the protein kinase Akt. Nature 399:597–601

Furchgott RF, Zawadzki JV (1980) The obligatory role of endothelial cells in the relaxation of arterial smooth muscle by acetylcholine. Nature 288:373–6

Garcia-Cardena G, Fan R, Shah V et al (1998) Dynamic activation of endothelial nitric oxide synthase by Hsp90. Nature 392:821–4

Garcia-Cardena G, Martasek P, Masters BS et al (1997) Dissecting the interaction between nitric oxide synthase (NOS) and caveolin. Functional significance of the nos caveolin binding domain in vivo. J Biol Chem 272:25437–40

Garcia-Cardena G, Oh P, Liu J et al (1996) Targeting of nitric oxide synthase to endothelial cell caveolae via palmitoylation: implications for nitric oxide signaling. Proc Natl Acad Sci USA 93:6448–53

Gerlai R (2001) Gene targeting: technical confounds and potential solutions in behavioral brain research. Behav Brain Res 125:13–21

Gimbrone MA Jr. (1989) Endothelial dysfunction and atherosclerosis. J Card Surg 4:180–3

Goligorsky MS, Noiri E, Tsukahara H et al (2000) A pivotal role of nitric oxide in endothelial cell dysfunction. Acta Physiol Scand 168:33–40

Guzik TJ, Black E, West NE et al (2001) Relationship between the G894T polymorphism (Glu298Asp variant) in endothelial nitric oxide synthase and nitric oxide-mediated endothelial function in human atherosclerosis. Am J Med Genet 100:130–7

Gyurko R, Kuhlencordt P, Fishman MC et al (2000) Modulation of mouse cardiac function in vivo by eNOS and ANP. Am J Physiol Heart Circ Physiol 278: H971–81

Gyurko R, Leupen S, Huang PL (2002) Deletion of exon 6 of the neuronal nitric oxide synthase gene in mice results in hypogonadism and infertility. Endocrinology 143:2767–74

Hara H, Huang PL, Panahian N et al (1996) Reduced brain edema and infarction volume in mice lacking the neuronal isoform of nitric oxide synthase after transient MCA occlusion. J Cereb Blood Flow Metab 16:605–11

Harris MB, Ju H, Venema VJ et al (2001) Reciprocal phosphorylation and regulation of endothelial nitric-oxide synthase in response to bradykinin stimulation. J Biol Chem 276:16587–91

Hibi K, Ishigami T, Tamura K et al (1998) Endothelial nitric oxide synthase gene polymorphism and acute myocardial infarction. Hypertension 32:521–6

Hingorani AD (2003) Endothelial nitric oxide synthase polymorphisms and hypertension. Curr Hypertens Rep 5:19–25

Hisamoto K, Ohmichi M, Kurachi H et al (2001) Estrogen induces the Akt-dependent activation of endothelial nitric-oxide synthase in vascular endothelial cells. J Biol Chem 276:3459–67

Huang PL, Dawson TM, Bredt DS et al (1993) Targeted disruption of the neuronal nitric oxide synthase gene. Cell 75:1273–86

Huang PL, Huang Z, Mashimo H et al (1995) Hypertension in mice lacking the gene for endothelial nitric oxide synthase. Nature 377:239–42

Huang Z, Huang PL, Ma J et al (1996) Enlarged infarcts in endothelial nitric oxide synthase knockout mice are attenuated by nitro-L-arginine. J Cereb Blood Flow Metab 16:981–987

Huang Z, Huang PL, Panahian N et al (1994) Effects of cerebral ischemia in mice deficient in neuronal nitric oxide synthase. Science 265:1883–5

Iadecola C, Pelligrino DA, Moskowitz MA et al (1994) Nitric oxide synthase inhibition and cerebrovascular regulation. J Cereb Blood Flow Metab 14:175–92

Iadecola C, Zhang F, Casey R et al (1997) Delayed reduction of ischemic brain injury and neurological deficits in mice lacking the inducible nitric oxide synthase gene. J Neurosci 17:9157–9164

Iadecola C, Zhang F, Xu X (1995) Inhibition of inducible nitric oxide synthase ameliorates cerebral ischemic damage. Am J Physiol 268: R286–92

Ichinose F, Huang PL, Zapol WM (1995) Effects of targeted neuronal nitric oxide synthase gene disruption and nitroG-L-arginine methylester on the threshold for isoflurane anesthesia. Anesthesiology 83:101–8

Ignarro LJ, Buga GM, Wood KS et al (1987) Endothelium-derived relaxing factor produced and released from artery and vein is nitric oxide. Proc Natl Acad Sci USA 84:9265–9

Irikura K, Huang PL, Ma J et al (1995) Cerebrovascular alterations in mice lacking neuronal nitric oxide synthase gene expression. Proc Natl Acad Sci USA 92:6823–7

Jachymova M, Horky K, Bultas J et al (2001) Association of the Glu298Asp polymorphism in the endothelial nitric oxide synthase gene with essential hypertension resistant to conventional therapy. Biochem Biophys Res Commun 284:426–30

Janssens SP, Shimouchi A, Quertermous T et al (1992) Cloning and expression of a cDNA encoding human endothelium-derived relaxing factor/nitric oxide synthase. J Biol Chem 267:14519–22

Katsuki S, Arnold W, Mittal C et al (1977) Stimulation of guanylate cyclase by sodium nitroprusside, nitroglycerin and nitric oxide in various tissue preparations and comparison to the effects of sodium azide and hydroxylamine. J Cyclic Nucleotide Res 3:23–35

Kojda G, Cheng YC, Burchfield J et al (2001) Dysfunctional regulation of endothelial nitric oxide synthase (eNOS) expression in response to exercise in mice lacking one eNOS gene. Circulation 103:2839–44

Kuhlencordt PJ, Gyurko R, Han F et al (2001) Accelerated atherosclerosis, aortic aneurysm formation, and ischemic heart disease in apolipoprotein E/endothelial nitric oxide synthase double-knockout mice. Circulation 104:448–54

Kureishi Y, Luo Z, Shiojima I et al (2000) The HMG-CoA reductase inhibitor simvastatin activates the protein kinase Akt and promotes angiogenesis in normocholesterolemic animals. Nat Med 6:1004–10

Lacolley P, Gautier S, Poirier O et al (1998) Nitric oxide synthase gene polymorphisms, blood pressure and aortic stiffness in normotensive and hypertensive subjects. J Hypertens 16:31–5

Lamas S, Marsden PA, Li GK et al (1992) Endothelial nitric oxide synthase: molecular cloning and characterization of a distinct constitutive enzyme isoform. Proc Natl Acad Sci USA 89:6348–52

Laubach VE, Shesely EG, Smithies O et al (1995) Mice lacking inducible nitric oxide synthase are not resistant to lipopolysaccharide-induced death. Proc Natl Acad Sci USA 92:10688–10692

Laufs U, Endres M, Stagliano N et al (2000) Neuroprotection mediated by changes in the endothelial actin cytoskeleton. J Clin Invest 106:15–24

Lea RA, Curtain RP, Shepherd AG et al (2001) No evidence for involvement of the human inducible nitric oxide synthase (iNOS) gene in susceptibility to typical migraine. Am J Med Genet 105:110–3

Leeson CP, Hingorani AD, Mullen MJ et al (2002) Glu298Asp endothelial nitric oxide synthase gene polymorphism interacts with environmental and dietary factors to influence endothelial function. Circ Res 90:1153–8

Lefer DJ, Jones SP, Girod WG et al (1999) Leukocyte-endothelial cell interactions in nitric oxide synthase-deficient mice. Am J Physiol 276: H1943–50

Liou YJ, Hong CJ, Liu HC et al (2002) No association between the neuronal nitric oxide synthase gene polymorphism and Alzheimer Disease. Am J Med Genet 114:687–8

Liu J, Sessa WC (1994) Identification of covalently bound amino-terminal myristic acid in endothelial nitric oxide synthase. J Biol Chem 269:11691-4

Lo E, Hara H, Rogowska J et al (1996) Temporal correlation mapping analysis of the hemodynamic penumbra in mutant mice deficient in endothelial nitric oxide synthase gene expression. Stroke 27:1381-1385

Lo HS, Hogan EL, Soong BW (2002) 5'-flanking region polymorphism of the neuronal nitric oxide synthase gene with Parkinson's disease in Taiwan. J Neurol Sci 194:11-3

Loscalzo J (2003) Adverse effects of supplemental L-arginine in atherosclerosis: consequences of methylation stress in a complex catabolism? Arterioscler Thromb Vasc Biol 23:3-5

Ma J, Ayata C, Huang PL et al (1996) Regional cerebral blood flow response to vibrissal stimulation in mice lacking type I NOS gene expression. Am J Physiol 270: H1085-90

MacMicking JD, Nathan C, Hom G et al (1995) Altered responses to bacterial infection and endotoxic shock in mice lacking inducible nitric oxide synthase. Cell 81:641-50

Malinski T, Bailey F, Zhang ZG et al (1993) Nitric oxide measured by a porphyrinic microsensor in rat brain after transient middle cerebral artery occlusion. J Cereb Blood Flow Metab 13:355-8

Marsden PA, Schappert KT, Chen HS et al (1992) Molecular cloning and characterization of human endothelial nitric oxide synthase. FEBS Lett 307:287-93

Michel JB, Feron O, Sacks D et al (1997) Reciprocal regulation of endothelial nitric-oxide synthase by Ca2+-calmodulin and caveolin. J Biol Chem 272:15583-6

Michel JB, Feron O, Sase K et al (1997) Caveolin versus calmodulin. Counterbalancing allosteric modulators of endothelial nitric oxide synthase. J Biol Chem 272:25907-12

Michell BJ, Chen Z, Tiganis T et al (2001) Coordinated control of endothelial nitric-oxide synthase phosphorylation by protein kinase C and the cAMP-dependent protein kinase. J Biol Chem 276:17625-8

Miyamoto Y, Saito Y, Kajiyama N et al (1998) Endothelial nitric oxide synthase gene is positively associated with essential hypertension. Hypertension 32:3-8

Montagnani M, Chen H, Barr VA et al (2001) Insulin-stimulated activation of eNOS is independent of Ca2+ but requires phosphorylation by Akt at Ser(1179). J Biol Chem 276:30392-8

Mooradian DL, Hutsell TC, Keefer LK (1995) Nitric oxide (NO) donor molecules: effect of NO release rate on vascular smooth muscle cell proliferation in vitro. J Cardiovasc Pharmacol 25:674-8

Moroi M, Zhang L, Yasuda T et al (1998) Interaction of genetic deficiency of endothelial nitric oxide, gender, and pregnancy in vascular response to injury in mice. J Clin Invest 101:1225-32

Nakayama M, Yasue H, Yoshimura M et al (1999) T-786->C mutation in the 5'-flanking region of the endothelial nitric oxide synthase gene is associated with coronary spasm. Circulation 99:2864-70

Nakayama T, Soma M, Takahashi Y et al (1997) Association analysis of CA repeat polymorphism of the endothelial nitric oxide synthase gene with essential hypertension in Japanese. Clin Genet 51:26-30

Nishida K, Harrison DG, Navas JP et al (1992) Molecular cloning and characterization of the constitutive bovine aortic endothelial cell nitric oxide synthase. J Clin Invest 90:2092-6

Okada S, Takehara Y, Yabuki M et al (1996) Nitric oxide, a physiological modulator of mitochondrial function. Physiol Chem Phys Med NMR 28:69-82

Panahian N, Yoshida T, Huang PL et al (1996) Attenuated hippocampal damage after global cerebral ischemia in mice mutant in neuronal nitric oxide synthase. Neuroscience 72:343-354

Pollock JS, Klinghofer V, Forstermann U et al (1992) Endothelial nitric oxide synthase is myristylated. FEBS Lett 309:402-4

Radomski MW, Palmer RM, Moncada S (1991) Modulation of platelet aggregation by an L-arginine-nitric oxide pathway. Trends Pharmacol Sci 12:87–88

Ringertz N (2001) Alfred Nobel–his life and work. Nat Rev Mol Cell Biol 2:925–8

Ross R (1993) The pathogenesis of atherosclerosis: a perspective for the 1990 s. Nature 362:801–9

Schneider MP, Erdmann J, Delles C et al (2000) Functional gene testing of the Glu298Asp polymorphism of the endothelial NO synthase. J Hypertens 18:1767–73

Sessa WC, Harrison JK, Barber CM et al (1992) Molecular cloning and expression of a cDNA encoding endothelial cell nitric oxide synthase. J Biol Chem 267:15274–6

Shaul PW (2002) Regulation of endothelial nitric oxide synthase: Location, location, location. Annu Rev Physiol 64:749–774

Shesely EG, Maeda N, Kim HS et al (1996) Elevated blood pressures in mice lacking endothelial nitric oxide synthase. Proc Natl Acad Sci USA 93:13176–81

Shimasaki Y, Yasue H, Yoshimura M et al (1998) Association of the missense Glu298Asp variant of the endothelial nitric oxide synthase gene with myocardial infarction. J Am Coll Cardiol 31:1506–10

Singleton AB, Gibson AM, McKeith IG et al (2001) Nitric oxide synthase gene polymorphisms in Alzheimer's disease and dementia with Lewy bodies. Neurosci Lett 303:33–6

Stamler JS (1994) Redox signaling: nitrosylation and related target interactions of nitric oxide. Cell 78:931–6

Stamler JS, Jia L, Eu JP et al (1997) Blood flow regulation by S-nitrosohemoglobin in the physiological oxygen gradient. Science 276:2034–7

Stangl K, Cascorbi I, Laule M et al (2000) High CA repeat numbers in intron 13 of the endothelial nitric oxide synthase gene and increased risk of coronary artery disease. Pharmacogenetics 10:133–40

Sullivan KJ, Kissoon N, Duckworth LJ et al (2001) Low exhaled nitric oxide and a polymorphism in the NOS I gene is associated with acute chest syndrome. Am J Respir Crit Care Med 164:2186–90

Takehara Y, Kanno T, Yoshioka T et al (1995) Oxygen-dependent regulation of mitochondrial energy metabolism by nitric oxide. Arch Biochem Biophys 323:27–32

Tanus-Santos JE, Desai M, Deak LR et al (2002) Effects of endothelial nitric oxide synthase gene polymorphisms on platelet function, nitric oxide release, and interactions with estradiol. Pharmacogenetics 12:407–13

Tanus-Santos JE, Desai M, Flockhart DA (2001) Effects of ethnicity on the distribution of clinically relevant endothelial nitric oxide variants. Pharmacogenetics 11:719–25

Tesauro M, Thompson WC, Rogliani P et al (2000) Intracellular processing of endothelial nitric oxide synthase isoforms associated with differences in severity of cardiopulmonary diseases: cleavage of proteins with aspartate vs. glutamate at position 298. Proc Natl Acad Sci USA 97:2832–5

Tsukada T, Yokoyama K, Arai T et al (1998) Evidence of association of the ecNOS gene polymorphism with plasma NO metabolite levels in humans. Biochem Biophys Res Commun 245:190–3

Wade CM, Kulbokas EJ, 3rd, Kirby AW et al (2002) The mosaic structure of variation in the laboratory mouse genome. Nature 420:574–8

Wang J, Dudley D, Wang XL (2002) Haplotype-specific effects of endothelial NO synthase promoter efficiency: modifiable by cigarette smoking. Arterioscler Thromb Vasc Biol 22: e1–4

Wang XL, Wang J (2000) Endothelial nitric oxide synthase gene sequence variations and vascular disease. Mol Genet Metab 70:241–51

Wattanapitayakul SK, Mihm MJ, Young AP et al (2001) Therapeutic implications of human endothelial nitric oxide synthase gene polymorphism. Trends Pharmacol Sci 22:361–8

Wei XQ, Charles IG, Smith A et al (1995) Altered immune responses in mice lacking inducible nitric oxide synthase. Nature 375:408–11

Xu W, Liu L, Emson P et al (2000) The CCTTT polymorphism in the NOS2A gene is associated with dementia with Lewy bodies. Neuroreport 11:297–9

Yogo K, Shimokawa H, Funakoshi H et al (2000) Different vasculoprotective roles of NO synthase isoforms in vascular lesion formation in mice. Arterioscler Thromb Vasc Biol 20: E96-E100

Yoshimura M, Yasue H, Nakayama M (1998) A missense Glu298Asp variant in the endothelilal nitric oxide sythase gene is associated with coronary spasm in the Japanese. Hum Genet 103:65–69

Yoshimura M, Yasue H, Nakayama M et al (2000) Genetic risk factors for coronary artery spasm: significance of endothelial nitric oxide synthase gene T-786->C and missense Glu298Asp variants. J Investig Med 48:367–74

Zaharchuk G, Hara H, Huang PL et al (1997) Neuronal nitric oxide synthase mutant mice show smaller infarcts and attenuated apparent diffusion coefficient changes in the peri-infarct zone during focal cerebral ischemia. Magn Reson Med 37:170–5

Zhang L, Fishman MC, Huang PL (1999) Estrogen mediates the protective effects of pregnancy and chorionic gonadotropin in a mouse model of vascular injury. Arterioscler Thromb Vasc Biol 19:2059–65

Association of Thrombotic Disease with Genetic Polymorphism of Haemostatic Genes: Relevance to Pharmacogenetics

E. G. D. Tuddenham

Haemostasis Research Group, Clinical Sciences Centre, Hammersmith Hospital,
Du Cane Road, London, W12 ONN, UK
e-mail: edward.tuddenham@ic.ac.uk

1	Introduction	312
2	Polymorphisms Associated with Venous Thrombosis	312
2.1	Factor V	312
2.2	Prothrombin	314
2.3	Factor XIIIA	314
2.4	Factor VIII	314
3	Polymorphisms Associated with Arterial Thrombosis	315
3.1	Fibrinogen	315
3.2	Factor VII	315
3.3	Factor XIII	316
3.4	PAI-1	316
3.5	Platelet Membrane Glycoprotein IIb/IIIa	317
3.6	Platelet Membrane Glycoprotein Ib-IX-V	317
3.7	Platelet Membrane Glycoprotein Ia/IIa	319
3.8	Connexin 37	319
3.9	Stromelysin-1	319
4	Conclusions and Relevance to Pharmacogenetics	319
	References	321

Abstract Venous thrombosis (VT) has been strongly linked to polymorphic variation in genes for blood coagulation and its regulation. The variants causing higher protein expression levels or encoding protein structure alteration affecting function have mechanistically clear effects on thrombin generation or fibrinolysis, which correlate with increased risk of VT. About half the incidence of VT can be attributed to these genetic variants. The risk of VT conferred by possessing more than one of the prothrombotic variants is at least additive. Predictably, anticoagulants and antiplatelet agents are effective in reducing this risk. Arterial thrombosis, by contrast, is much less firmly associated with polymorphism in genes related to haemostasis or vascular biology. However recent studies with large, carefully selected patient and control groups have supported strong associations between genetic variants of a few genes relevant to haemostasis or vascular integrity. Some of these polymorphisms show differential response to aspirin, which may explain conflicting results from earlier studies. Hence future studies should address therapeutic response as well as risk prediction.

Keywords Venous thrombosis · Arterial thrombosis · Platelet

1
Introduction

An association between genetic variation in haemostasis and risk of thrombotic disease has been inherently plausible since 1862 when Virchow enunciated his famous triad of the aetiopathology of thrombosis, namely changes in the vessel wall, altered blood flow or hypercoagulable blood. The first firmly established genetic cause of thrombosis was familial antithrombin deficiency, obviously an example of hypercoagulable blood (Egeberg 1965). Affected members of families segregating autosomal dominant defects in the antithrombin gene suffer from early onset venous thromboembolism, a clinical tendency for which Egeberg coined the term thrombophilia. Although a large number of essentially private mutations causing antithrombin deficiency have been described, it remains a very rare cause of thrombosis, which almost exclusively affects the large veins. A series of other rare causes of familial venous thrombosis were discovered during the past 20 years, including mutations causing defects of protein C, protein S or fibrinogen. These will not be discussed further here as the frequency of any individual mutation in the general population is well below 1% and therefore non-polymorphic. This chapter will first survey genetic polymorphism associated with VT and then the genetic background to arterial thrombosis will be presented. Finally implications for pharmacogenetics and directions for future research will be discussed.

2
Polymorphisms Associated with Venous Thrombosis

Polymorphisms associated with venous thrombosis include factor V, prothrombin, factor XIIIA and factor VIII (Fig. 1).

2.1
Factor V

By studying families segregating a thrombophilic phenotype, but in whom no previously described defect was present, Dahlback in 1993 discovered the phenomenon of activated protein C resistance (APCR) (Dahlback et al. 1993).This was soon shown to be due to a variant of factor V with a point mutation at one of the activated protein C cleavage/inactivation sites, factor V R506Q, designat-

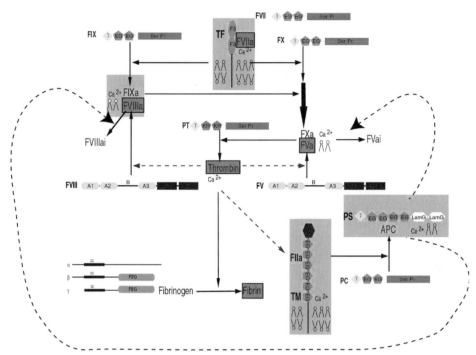

Fig. 1 Coagulation network. Highlighted in orange boxes are gene products for which polymorphic variation has been linked to venous and/or arterial thrombosis. Individual coagulation factors are represented as protein module symbols, emphasizing the highly homologous structure of the principle plasma factors. Positive and negative feedback reactions initiated by thrombin are shown as *red and blue broken lines*, respectively. Reactions requiring or occurring on phospholipid surfaces are contained within *grey boxes*. In the case of the IXa/VIIIa and Xa/Va complexes, the relevant surface is an activated platelet, hence linking thrombin generation to platelet activation. Not shown is the fact that thrombin itself is a powerful platelet activator via protease-activated receptor-1. *TF*, tissue factor; *TM*, thrombomodulin; *PS*, protein S; *PC*, protein C; *γ*, gamma carboxyglutamic acid; *EG*, epidermal growth factor; *KR*, kringle; *CLECT*, c lectin; *LamG*, laminin type G; *Ser Pr*, serine protease; *F3*, fibronectin type 3. (Image created by J. McVey)

ed factor V Leiden by its discoverers (Bertina et al. 1994). Factor V Leiden occurs in about 5% of Europeans. In heterozygotes, it confers fourfold increased lifetime risk of developing venous thrombosis and this risk is enhanced by coincident genetic or acquired risk factors such protein C deficiency, oral contraception, pregnancy or immobilization (Martinelli et al. 1998). Homozygotes have a much further increased risk of thrombosis and can be considered as having a form of thrombophilia. The mutation causing factor V Leiden occurs on a single haplotype background and is thought to have spread through European populations due to heterozygote advantage, namely decreased risk of fatal postpartum haemorrhage (Zivelin B et al. 1997). There is some evidence that another factor V haplotype, designated *HR2*, increases the risk of VT if combined with the Leiden haplotype in double heterozygosity (Faioni et al. 1999). Activated protein

C resistance as a phenotype in the absence of factor V Leiden confers enhanced thrombotic risk (de Visser et al. 1999). This was partly due to elevated factor VIII, which increases APCR, but the risk remained after factoring out the factor VIII level. Hence there remain other unidentified causes of increased APCR with associated thrombotic risk that may be inherited or acquired.

2.2
Prothrombin

The discovery of factor V Leiden prompted a search for other polymorphic variants in clotting factors linked to increased incidence of VT. So far only one such variant has been firmly established, the substitution of G by A at nucleotide 20210 in the 3' UTR of the prothrombin gene. The prevalence of the *A* allele is about 2% in Caucasians and confers a 2.8-fold relative risk for VT compared to controls with the commoner allele *G*(Poort et al. 1996). The mechanistic basis of the enhanced risk appears to be that 20210A is associated with higher plasma prothrombin concentration. Although prothrombin 20210A occurs on a single haplotype background (Zivelin et al. 1998), the reason for its prevalence has not been elucidated.

2.3
Factor XIIIA

Factor XIII subunit A is a transglutaminase essential for normal haemostasis, which it promotes by cross-linking fibrin strands and thus stabilizing the clot mechanically and delaying fibrinolysis. A common polymorphism in the protein-coding region of FXIIIA is Val34Leu. The codon affected is three amino acids N-terminal of the thrombin activation site at Arg37-Gly38. Factor XIIIA34Leu is activated more rapidly by thrombin than FXIIIA34Val (Kangsadalampai and Board 1998) but paradoxically seems to protect against VT (Franco and Reitsma 1999) and myocardial infarction (see Sect. 2.4). A possible explanation for this may lie in the response to aspirin. It has recently been shown that aspirin inhibits the activation of XIIIA34Leu much more than the commoner *Val* allele(Undas et al. 2003).

2.4
Factor VIII

The Leiden Thrombophilia Study(Koster et al. 1995), in a case control analysis, first established that elevated factor VIII plasma levels are a risk factor for VT. Factor VIII activity greater than 150 IU/dl was associated with a 4.8-fold increased relative risk of VT (CI, 2.3–10). Several subsequent studies have confirmed this association and have shown that elevated factor VIII in VT patients is not due to activation or chronic inflammation(O'Donnell et al. 1997). It turns out to be one of the commonest findings in unselected patients referred for thrombophilia work-up (M. Laffan, personal communication). No polymorphic variation has been found in or near the factor VIII gene that can account for raised levels in such patients(Mansvelt et al. 1998), although the known effect of non-O blood group is evident and accounts for an excess of blood groups A and B in sufferers from VT. Further work on other quantitative trait loci with an effect on factor VIII levels may reveal the polymorphisms that control the wide normal range of this factor, which does show familial clustering (Schambeck et al. 2001) and can therefore be assumed to be under genetic control.

3
Polymorphisms Associated with Arterial Thrombosis

3.1
Fibrinogen

There is consistent and strong evidence for an association between elevated plasma levels of fibrinogen and arterial disease (Meade et al. 1986). Smoking elevates fibrinogen levels (Meade TW et al. 1987), which links smoking to haemostasis, but otherwise there is no mechanistic or statistical correlation between variation at the fibrinogen gene locus and arterial disease.

The fibrinogen gene locus consists of three linked genes in the order γ, α, β. There are multiple polymorphic sites in all three genes but most studies have focussed on the β gene. In the ECTIM study(Behague I et al. 1996), $-455G/A$ was strongly associated with plasma levels ($p<0.0003$), but the association was only found in smokers. The rarer allele of $-455G/A$ was associated with more severe coronary artery disease but not with myocardial infarction. Numerous studies on fibrinogen and arterial disease are well summarized in an excellent review(Lane and Grant 2000), where the authors conclude that "The most striking point about these association studies is their inconsistency in relating genotype, fibrinogen level and arterial disease". Another extensive review (Simmonds et al. 2001) reached the same conclusion. It is also telling that a very large recently published study of 112 candidate gene polymorphisms from Japan(Yamada et al. 2002) failed to find any association between β fibrinogen polymorphism and myocardial infarction.

3.2
Factor VII

Studies on factor VII and cardiovascular risk have been plagued by similar problems to those noted above for fibrinogen, namely, that no clear connection has been established between genotype and disease, despite there being a clear association of certain polymorphisms and plasma levels of the factor. An early study(Meade et al. 1986) found that elevated plasma levels were predictive of death but subsequent studies have not been consistent in regard to this relationship. Also, there are technical difficulties in the plasma assays, which make different studies hard to compare. Investigators have therefore sought an association between polymorphisms at the factor VII locus and arterial disease. Of many studies (summarized in reviews by Lane and Grant 2000; Simmonds et al. 2001), only one has found such an association (Iacovello et al. 1998), where *Arg353* plus intron7 *HVR4* was predictive of myocardial infarction. Again it is noteworthy that the large Japanese study of myocardial infarction(Yamada et al. 2002) found no such association, but this discrepancy may be due to the different populations under study (Italian vs Japanese).

3.3
Factor XIII

As noted above, fibrin-stabilizing factor is a transglutaminase essential for normal haemostasis. Cross-linked fibrin produced by its action is more resistant to fibrinolysis; hence lower levels of factor XIII activity might be expected to be protective against thrombosis.

A common polymorphism in the A chain of factor XIII (the chain containing the active enzyme) Val34Leu affects the enzyme activity. Factor XIII34Leu is more rapidly activated by thrombin, but the specific transglutaminase activity of the product is the same as that of factor XIII34Val. However, the porosity of the clot produced in 34Leu plasma is lower. Paradoxically, factor XIII34Leu has been found to be protective against myocardial infarction (Kohler et al. 1998). A resolution to this paradox may be found in the recent observation that aspirin causes a more pronounced inhibition of activation of factor XIII in 34Leu carriers than 34Val carriers (Undas et al. 2003).

3.4
PAI-1

PAI-1 is the main inhibitor of tissue plasminogen activator (tPA) in plasma, where it exists in molar excess over tPA. Congenital deficiency of PAI-1 is associated with a haemorrhagic tendency. Increased levels of PAI-1 are consistently associated with risk of myocardial infarction (Thogersen et al. 1998). The polymorphism that accounts for this association is now clearly identified as −675 4G/5G, with the *4G* allele being linked to higher plasma levels and higher risk. The recent study from Japan (Yamada el al. 2002) found an association of *4G-675* with MI significant at $p<0.001$ in women but not in men, conferring a risk ratio of 1.6 (1.2–2.1). Although small, this risk is certainly convincing in Japanese women. The only study of similar power in Europeans(Gardemann et al. 1999) showed a similar elevated risk ratio of 1.31 (1.04–1.65) for coronary artery disease in subjects with the *4G* allele, which also interacted with smoking and hypertension to elevate the severity of atheroma in the European study. It has also been noted that almost 50% of the variance in PAI-1 levels may be attributed to the insulin-resistant state in men, making that more important than polymorphic gene variants(Henry et al. 1998).

3.5
Platelet Membrane Glycoprotein IIb/IIIa

When activated, this receptor binds fibrinogen, von Willebrand factor and vitronectin. Platelet-to-platelet cross-linking via GPIIb/IIIa receptors and fibrinogen is the principle mechanism of platelet aggregation. The first report of a polymorphism in GPIIb/IIIa linked to myocardial infarction or unstable angina was that of Weiss et al.(1996), who found that Pro33 (PLA2) was over-represented in cases compared to normal controls (39.4% vs 19.1%). This small study (71 cases and 68 controls) led to many other studies of varying size with inconsistent results (reviewed in Lane and Grant 2000 and Simmonds et al. 2001). The large Japanese study(Yamada et al. 2002) found no association. Again an explanation of this paradox may lie in the recent observation that the presence of the *PlA2* allele is associated with enhanced thrombin formation and an impaired antithrombotic action of aspirin, which might favour coronary thrombosis in the *PlA2* carriers (Undas et al. 2001) (Fig. 2). Hence future studies need to take into account not only the *PLA* genotype but also the history of aspirin usage in subjects and controls.

3.6
Platelet Membrane Glycoprotein Ib-IX-V

Although four gene products comprise the GPIb-IX-V complex, which is essential for platelet adhesion to surfaces under high shear, only polymorphisms in the *GPIbα* gene have been studied in relation to arterial disease. The so-called macroglycopeptide region of the *GPIbα* gene contains a 39-bp variable number tandem repeat coding for 13 amino acids, which may be present in 1, 2, 3 or 4 copies. No influence on platelet function has been reported but an association with ischaemic heart disease was found in one study(Murata et al. 1997). Other studies found no such association (reviewed in Simmonds et al. 2001).

Another polymorphism in the *GPIb α* gene is located within the Kozak sequence immediately upstream of the initiator codon at –5T/C. This polymorphism was reported to have has a major effect on receptor density(Afshar-Khargan et al. 1999). A subsequent study denied such an effect(Corral et al. 2000). Although the jury is out in regard to the phenotypes produced by the Kozak variants, a recent study(Douglas et al. 2002) found a strong association between the *TT* genotype and myocardial infarction ($p<0.001$). Paradoxically, this is the genotype expected from in vitro studies to confer lower expression of the receptor. Unfortunately, the Kozak polymorphism was not included in the Japanese study of 112 polymorphisms but the VNTR was and showed no significant association with myocardial infarction (Yamada et al. 2002).

Hence it may be concluded that at this point further studies are needed on the phenotypic effect of these polymorphisms, their association with arterial disease and the effect of anti-platelet agents before firm conclusions can be reached.

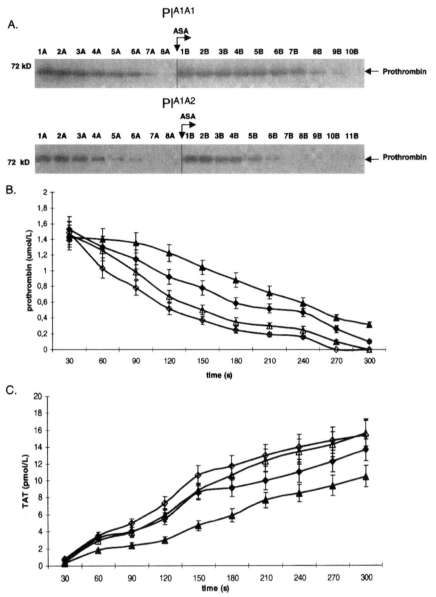

Fig. 2 A Immunoblots of prothrombin in PlA1A1 (*top*) and PlA1A2 (*bottom*) subjects. **A** *lines* show bleeding-time blood samples taken before low-dose aspirin (*ASA*) ingestion; **B** *lines* show samples after aspirin ingestion. **B** Quantitative analysis of prothrombin consumption showing concentrations of prothrombin in 12 *PlA1A1* subjects (*triangles*) and in 12 *PlA2* carriers (*diamonds*) before (*open symbols*) and after aspirin treatment (*closed symbols*). **C** Time courses for thrombin–antithrombin III complex (*TAT*) formation showing concentrations of thrombin–antithrombin III complexes in 12 *PlA1A1* subjects (*triangles*) and in 12 *PlA2* carriers (*diamonds*) before (*open symbols*) and after aspirin treatment (*closed symbols*). Values are plotted as means ± SEM. (Reproduced from Undas et al. 2001, with permission)

3.7
Platelet Membrane Glycoprotein Ia/IIa

This receptor is responsible for platelet activation caused by contact with collagen. It is present at a wide range of surface densities in platelets from normal individuals. A polymorphism in the *Ia* gene, *807C/T* (a silent exonic dimorphism), is strongly associated with receptor density and with myocardial infarction and stroke in the young (Santoso et al. 1999). There also seem to be gene–environment interactions for this polymorphism (reviewed in Simmonds et al. 2001). Nevertheless, no robust association with myocardial infarction was found in Japanese subjects for this polymorphism(Yamada et al. 2002). It still remains for fully joined up studies to be performed in which phenotype, genotype, drug effect and environmental interaction are all taken into account.

3.8
Connexin 37

This protein is intimately involved in gap-junctional communication between endothelial cells. A polymorphism at *1019C/T* has been the focus of several studies. The *C* allele was over represented in men with atherosclerotic plaques in a small Swedish study (Boerma et al. 1999). However, in Japanese men, the *T* allele of connexin 37 was highly significantly associated with myocardial infarction ($p<0.001$) and was the only such association detected out of 112 genetic polymorphisms in different genes that were screened (Yamada et al. 2002).

3.9
Stromelysin-1

This protein (otherwise known as matrix metalloproteinase 3) is important in vascular matrix metabolism (see the chapter by Henney, this volume). A polymorphism at *–1171 5A/6A* emerged as one of only two highly significant associations for myocardial infarction ($p<0.001$) in Japanese women(Yamada et al. 2002), the other being PAI-1. Furthermore, the *6A/6A* genotype vs the *5A/5A* gave a risk ratio of 4.9, the highest in the whole study and one of the highest claimed for a polymorphism in arterial disease. This association has been strongly confirmed in a prospective study by Humphries et al. (2002), who also found enhanced risk associated with the *5A/6A* and *6A/6A* genotypes. However, whereas the risk was modestly increased in the latter by smoking, it was greatly increased in the *5A/5A* group by smoking.

4
Conclusions and Relevance to Pharmacogenetics

In an ideal world of rational therapeutics, decisions as to prophylaxis and treatment would take into account genetic predisposition, drug interaction, environmental factors and their several interactions, before prescribing lifestyle advice and/or treatment. In regard to venous thromboembolism, this ideal situation is plausibly approaching if not actually already here. Thrombophilia screening now identifies clear polymorphic risk factors such as factor V Leiden in more than half of patients so screened. The genotype is directly correlated with a mechanistically clear prothrombotic phenotype for two of the polymorphisms, and rational therapy to correct the phenotype is available. Advice as to type and length of anticoagulant and/or antiplatelet treatment can be based on established recurrence risks. However, the treatment currently available is not yet ideal in that warfarin requires regular monitoring and lowers natural anticoagulant levels as well procoagulant factors. This is likely to change soon with the introduction of orally active direct thrombin inhibitors such as ximelagatran. The genetic basis of elevated factor VIII levels has not yet been established, but the phenotypic assays are straightforward and the risk of recurrence has been quantified(Kyrle et al. 2000). No drug that directly inhibits factor VIII is currently available but the finding that this is a very common risk factor should stimulate the search for such agents.

Arterial disease has proved to be an elusive target for identification of robust genotypic and phenotypic correlations. No doubt this is due to the highly polygenic nature of arterial degenerative disease and the strong effect of environmental factors, which by their interaction with varying genotypes can conceal genetic predispositions. Most studies have been underpowered and/or did not attempt to join up genotype, phenotype, environmental variables and concurrent drug treatment. This situation is now changing and several strong, mechanistically plausible correlations have been reported. The most promising of these are PAI-1 *-675 4G/5G*, GPIIb/IIIa *33Leu/Pro*, GPIbα Kozak *-5T/C*, connexin 37 *1019C/T* and stromelysin-1 *-1171 5A/6A* polymorphisms. The first two and stromelysin-1 show strong environmental interaction with smoking, underlining the importance of lifestyle modification. In relation to GPIIb/IIIa, another confounding factor is the response to aspirin, also noted for factor XIII *34Val/Leu*. This observation may be taken as a stimulus to develop or identify antiplatelet agents that are effective in the presence of the aspirin-resistant alleles.

The importance of the PAI-1 polymorphism highlights the fact that we do not currently have effective fibrinolytic drugs that are orally active and can be safely taken for prolonged periods, which should make this a target for future drug development. In regards to the vascular wall polymorphisms, connexin 37 and stromelysin-1, the mechanism of association and therefore appropriate treatment are presently unclear, but the strong risk found must be a stimulus to further targeted pharmaceutical research. The finding of these latter risk associations confirms the first item of Virchow's triad: changes in the vessel wall. It is certainly in line with the most progressive expectations of the genome project (Chakravarti and Little 2003), that large scale risk association studies be undertaken, using high-throughput genetic methods to simultaneously ascertain hundreds of candidate gene polymorphisms in population groups and appropriate controls, who have also provided samples for phenotype assays and whose current drug treatment, diet and lifestyle are well monitored. Undoubtedly these are expensive and demanding studies to organize, but the yield in terms of rational therapeutics must surely justify them in terms of effective prevention and treatment strategies for the future.

References

Afshar-Khargan V, Li CCQ, Khoshnevis-Asl M et al (1999) Kozak sequence polymorphism of the glycoprotein (GP) Ibα gene is a major determinant of the plasma membrane levels of the platelet GPIb-IX-V complex. Blood 94:186–191
Behague I, Poirier O, Nicaud V et al (1996) β fibrinogen gene polymorphisms are associated with plasma fibrinogen and coronary artery disease in patients with myocardial infarction. The ECTIM Study Circulation 93:440–449
Bertina RM, Koelman BP, Koster T et al (1994) Mutation in Blood coagulation factor V associated with resistance to activated protein C. Nature 369:64–67
Boerma M, Forsberg L, van Zeijl L et al (1999) A generic polymorphism in Connexin 37 as a prognostic marker for the atherosclerotic plaque development. J Intern Med 246:211–218
Chakravarti A, Little P (2003) Nature, nurture and human disease. Nature 421:412–414.
Corral J, Lozano ML, Gonzalez-Conejero R et al (2000) A common polymorphism flanking the initiator codon of GPIbα does not affect expression and is not a major risk factor for arterial thrombosis. Thromb Haemost 83:23–28
Dahlback B, Carlsson M, Svensson PJ (1993) Familial thrombophilia due to a previously unrecognized mechanism characterized by poor anticoagulant response to activated protein C: prediction of a cofactor to activated protein C. Proc Natl Acad Sci USA 90:1004–1008
de Vissser MCH, Rosendaal FR, Bertina RM (1999) A reduced sensitivity for activated protein C in the absence of factor V Leiden increases the risk or venous thrombosis. Blood 93:1271–1276
Douglas H, Michaelides K, Gorog DA et al (2002) Platelet membrane glycoprotein Ibα gene –5T/C Kozak sequence polymorphism as an independent risk factor for the occurrence of coronary thrombosis. Heart 87:70–74
Egeberg O (1965) Inherited antithrombin III deficiency causing thrombophilia. Thromb Diath Haemorrh 13:516–530
Faioni EM, Franchi F, Bucciarelli P et al. (1999) Coinheritance of the HR2 haplotype in the factor V gene confers an increased risk of venous thrombosis to carriers of factor V R506Q (factor V Leiden) Blood 94:3062–3066
Franco RF, Reitsma PH, Lourenco D et al (1999) Factor XIII Cal34Leu is a genetic factor involved in the aetiology of venous thrombosis. Thromb Haemost 81:676–679
Gardemann A, Lohe J, Katz N et al (1998) Association of the platelet glycoprotein IIIa PLA1/2 gene polymorphism to coronary artery disease but not to nonfatal myocardial infarction in low risk patients. Thromb Haemost 80:214–217
Henry M, Tregouet DA, Alessi MC et al (1998) Metabolic determinants are much more important than genetic polymorphisms in determining the PAI-1 activity and antigen plasma concentrations: a family study with part of the Stanislas Cohort. Arterioscler Thromb Vasc Biol 18:84–91
Humphries SE, Martin S, Cooper J et al (2002) Interaction between smoking and the stromelysin-1 (MMP3) gene 5A/6A promoter polymorphism and risk of coronary heart disease in healthy men. Ann Hum Genet 66:343–52
Iacovello L, di Castelnuovo A, de Knijff P et al (1998) Polymorphisms in the coagulation factor VII gene and the risk of myocardial infarction. N Engl J Med 338:79–85

Kangsadalampai S, Board PG (1998) The Val34Leu polymorphism in the A subunit of the coagulation Factor XIII contributes to the large normal range in activity and demonstrates that the activation peptide plays a role in catalytic activity. Blood 92:2766–2770

Kohler HP, Stickland MH, Ossei-Gerning N et al (1998) Association of a common polymorphism in the factor XIII gene with myocardial infarction. Thromb Haemost 79:8–13

Koster T, Blann AD, Briet E et al (1995) Role of clotting factor VIII in effect of von Willebrand factor on occurrence of deep vein thrombosis. Lancet 345:152–155

Kyrle PA, Minar E, Hirschl M et al (2000) High plasma levels of factor VIII and the risk of recurrent venous thromboembolism. N Engl J Med 343:457–62

Mansvelt EPG, Laffan M, McVey JH et al (1998) Analysis of the F8 gene in individuals with high plasma factor VIII:C levels and associated venous thrombosis. Thromb Haemost 80:561–565

Martinelli I, Mannucci PM, de Stefano V et al (1998) Different risks of thrombosis in four coagulation defects associated with inherited thrombophilia: a study of 150 families. Blood 92:2353–2358

Meade TW, Mellows S, Brozovic M et al (1986) Haemostatic function and ischaemic heart disease: principal results of the Northwick Park Heart Study. Lancet 2:533–537

Meade TW, Imeson J, Stirling Y (1987) Effects of changes in smoking and other characteristics on clotting factors and the risk of ischaemic heart disease. Lancet ii:986–988

Murata M, Matsubara Y, Kawano K et al (1997) Coronary artery disease and polymorphisms in a receptor mediating shear stress-dependent platelet activation. Circulation 96:3281–3286

O'Donnell J, Tuddenham EGD, Manning R et al (1997) High prevalence of elevated factor VIII levels in patients referred for thrombophilia screening: role of increased synthesis and relationship to acute phase reaction. Thromb Haemost 77:825–828

Poort SR, Rosendaal FR, Reitsma PH et al (1996) A common genetic variation in the 3' untranslated region of the prothrombin gene is associated with elevated plasma levels and an increase in venous thrombosis. Blood 88:3698–3703

Santoso S, Kunicki TJ, Kroll H et al (1999) Association of the platelet glycoprotein Ia C807T gene polymorphism with nonfatal myocardial infarction in younger patients. Blood 93:2449–2453

Schambeck CM, Hinney K, Haubitz I et al (2001) Familial clustering of high factor VIII levels in patients with venous thromboembolism. Arterioscler Thromb Vasc Biol 21:289–292

Simmonds RE, Hermida J, Rezzenda SM et al (2001) Haemostatic genetic risk factors in arterial thrombosis. Thromb Haemost 86:374–385

Thogersen AM, Jansson JH, Boman K et al (1998) High plasminogen activator inhibitor and tissue plasminogen activator levels in plasma precede a first acute myocardial infarction in both men and women: evidence for the fibrinolytic system as an independent primary risk factor. Circulation 98:2241–2247

Undas A, Brummel K, Musial J et al (2001) PlA2 Polymorphism of β3 integrins is associated with enhanced thrombin generation and impaired
antithrombotic action of aspirin at the site of microvascular injury. Circulation 104:2666–2672

Undas A, Sydor WJ, Brummel K et al (2003) Aspirin alters the cardioprotective effects of the factor XIII Val34Leu polymorphism. Circulation 107:17–20

Weiss EJ, Bray PF, Tayback M et al (1996) A polymorphism of a platelet glycoprotein receptor as an inherited risk factor for coronary thrombosis N Engl J Med 334:1090–1094

Yamada Y, Izawa H, Ichihara S et al (2002) Prediction of the risk of myocardial infarction from the polymorphism in candidate genes. N Engl J Med 347:1916–1923

Zivelin A, Griffin JH, Xu X, et al (1997) A single genetic origin for a common Caucasian risk factor for venous thrombosis. Blood 89:397–402

Zivelin A, Rosenberg N, Faier S et al (1998) A single genetic origin for the common prothrombotic G20210A polymorphism in the prothrombin gene. Blood 92:1119–1124

The Influence of Genetic Factors on Leukocyte and Endothelial Cell Adhesion Molecules

R. M. Rao · A. I. Russell · T. Vyse · D. O. Haskard

BHF Cardiovascular Medicine Unit, ICSM, National Heart and Lung Institute, Hammersmith Hospital, 2602 J Block, Du Cane Road, London, W12 ONN, UK
e-mail: d.haskard@ic.ac.uk

1	Introduction	324
2	Selectins	325
2.1	P-Selectin	326
2.2	E-Selectin	326
2.3	L-Selectin	328
3	Selectin Ligands	328
3.1	P-Selectin Glycoprotein Ligand-1	328
3.2	Leukocyte Adhesion Deficiency Type II	329
4	Immunoglobulin Superfamily	329
4.1	ICAM-1	329
4.2	VCAM-1	331
4.3	PECAM-1	331
5	Integrins	332
5.1	β_2 Integrins	332
6	The Therapeutic Impact of Leukocyte-Endothelial Adhesion Molecules	333
7	Conclusion	333
	References	334

Abstract Atherosclerosis and its complications arise largely as a result of inflammatory processes. A key component of this response is the emigration of mononuclear cells into the blood vessel wall, which is orchestrated by adhesion molecules expressed both on leukocytes and on endothelium. Over the last few years, these molecules have been extensively studied, leading to a well-described paradigm of leukocyte recruitment, termed the adhesion cascade. Adhesion molecule families are classified according to structure and function and, broadly speaking, are genetically well conserved across species. However, it has become apparent that in human genetics variations exist within a number of these molecules. In some cases, these variations have been directly associated with the development or worsening of inflammatory disease and in others a functional molecular significance has been elucidated. Only recently has pharmacological inhibition of adhesion molecules been successful in human disease, although a number of experimental models exist utilizing both monoclonal antibodies and small molecule inhibitors. As yet, examination of the effects of genetic factors on such treatments has not been reported. As both the appreciation of the importance of genetic variations of adhesion molecules increases and the development of specific therapies further expands into the clinical arena, it seems possible that both will be of importance in the diagnosis and management of atherosclerotic disease.

Keywords Adhesion molecule · Polymorphism · Selectin · Integrin · Immunoglobulin superfamily · Leukocyte · Endothelium

Doctors Rao and Russell contributed equally to this article

1
Introduction

Over the last few years, it has become clear that atherosclerosis and its complications are inflammatory processes, characterized by dysfunctional leukocyte-endothelial cell (EC) interactions and mononuclear infiltration into arterial tissues (Ross 1999). Leukocyte extravasation into tissues occurs through a step-by-step process, involving initial tethering from free flow, rolling on the endothelial surface (mediated in large part by selectins and their ligands), stable adhesion (mediated in the main by integrins and their ligands), and transendothelial migration. The adhesion molecules that mediate these interactions can be classified into families based on molecular structure (Springer 1995). Differences in structure influence function, as established from observations of leukocyte attachment in vitro using flow chambers or in vivo using intra-vital microscopy. The sequence of molecular interactions leading to leukocyte emigration has been termed the adhesion cascade (Fig. 1).

The multi-step paradigm of leukocyte–EC interactions helps explain the striking heterogeneity of leukocyte infiltrates seen at sites of inflammation. Broadly speaking, acute inflammatory stimuli give rise to predominantly granulocytic infiltrates (neutrophilic in response to bacteria and eosinophilic in response to allergens and protozoa). On the other hand, sub-acute and chronic responses are extremely variable, giving rise to mixed populations of myeloid and lymphoid cells in tissues that organize over time into recognized special distributions. Similarly, lymphoid organs have a complex morphology, through which distinct sub-populations of circulating lymphocytes continuously traffic. The potential of a given leukocyte to extravasate is determined by the sequential and/or parallel interaction of several adhesion molecule pairs, the participation of which is determined by the local microvascular environment.

Genetic influences could impact on the expression and/or function of any of a number of adhesion molecules and thus affect the number or type of leuko-

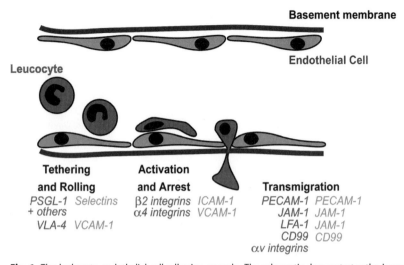

Fig. 1 The leukocyte-endothelial cell adhesion cascade. The schematic demonstrates the important adhesion molecules involved in the extravasation of leukocytes. Their precise functions are outlined in more detail in the text. Flow is from left to right and migration through the endothelium and basement membrane is denoted *downwards*

cytes recruited to inflammatory foci. The main focus of this review is on the impact of genetic variations in adhesion molecules on atherosclerosis-related cardiovascular disease. In some instances, we have highlighted potential associations with other diseases, particularly where polymorphisms have been found to influence function. A detailed description of the biological role of each protein is beyond the scope of the review.

2 Selectins

Selectins mediate early adhesion events between leukocytes and activated endothelium (Kansas 1996). There are three selectins (L-selectin on leukocytes, E-selectin on activated ECs and P-selectin on activated ECs and platelets), encoded by genes (*SELL*, *SELE*, and *SELP*, respectively) that lie in a 150-kb-length segment of the long arm of human chromosome 1, *1q24.2* (www.emsembl.org; Watson et al. 1990). In each of the selectin genes, a lectin domain is coded within a single exon, followed by the epidermal growth factor (EGF)-like domain, a number of short consensus repeats (each contained in a single exon), a transmembrane domain and a cytoplasmic domain. The chromosomal location of selectin genes places them within overlapping susceptibility intervals linked to autoimmune diseases, systemic lupus erythematosus (SLE), multiple sclerosis (MS), type I diabetes mellitus, and inflammatory bowel disease. Polymorphisms have been characterized in each of the three selectin genes and in some instances have been associated with cardiovascular disease. Since there is strong linkage disequilibrium across the whole selectin gene cluster, it is difficult to confidently relate disease to individual polymorphisms (Takei et al. 2002; A.I. Russell, T.J. Vyse, unpublished data).

2.1
P-Selectin

Vascular endothelial expression of P-selectin is increased in atherosclerotic plaques (Johnson-Tidey et al. 1994). Furthermore, P-selectin knock-out mice show protection from atherosclerosis, due both to EC and platelet P-selectin deficiency (Johnson et al. 1997; Burger and Wagner 2003). Following expression on the cell surface, a proportion of P-selectin molecules are shed into the circulation, where they may have a procoagulant effect (Andre et al. 2000).

The *SELP* gene consists of 17 exons extending over 50 kb. Hermman et al. (1998) sought genomic variation in the *SELP* coding sequence and approximately 2 kb of the 5′-flanking region (likely to harbor regulatory elements). In 40 chromosomes from survivors of myocardial infarction (MCI), 13 single nucleotide polymorphisms (SNPs) were identified, including four non-synonymous changes. Strong patterns of linkage disequilibrium and conserved haplotypes were evident across *SELP*. The SNPs were genotyped in 647 French and Northern Irish male survivors of MCI and 758 age-matched controls within the Etude Cas-Témoins de L'Infarctus du Myocarde (ECTIM) study. The minor *Pro715* allele of the *Thr715Pro* polymorphism, which lies in the ninth consensus repeat domain, was significantly under-represented in the patient group compared to that of controls in both populations ($P_c<0.02$), independent of traditional risk factors for coronary artery disease (CAD). Interestingly, this allele was significantly less common in the French subjects.

The protective effect of *Pro715* was confirmed in the ECTIM extension study, which notably also contained a large number of female cases (Kee et al. 2000). However, the retrospective nature of this study limits interpretation, since under-representation of *Pro715* in the survivors examined might have been due to increased mortality from MCI. No association between this allele (or two 5′ flanking SNPs) and stable or unstable angina was evident in a large cohort of Germans (Barbaux et al. 2001).

2.2
E-Selectin

The E-selectin gene (*SELE*) consists of 14 exons spanning 13 kb of DNA. The most extensively characterized polymorphism is an $A\rightarrow C$ transversion at nucleotide 561 in exon 4, resulting in a non-conservative serine to arginine substitution at amino acid 128 (Wenzel et al.1994). The minor *128R* allele has been associated with early-onset and angiographically severe atherosclerosis in independent cross-sectional studies in Caucasian populations (Wenzel et al. 1994; Ye et al. 1999). This allele has also been found to be over-represented in two cohorts of patients that developed post-angioplasty restenosis (Rauchhaus et al. 2002), and to be positively associated with coronary artery calcification in women below 50 years of age (Ellsworth et al. 2001). Although a recent study from Japan failed to find the *128R* allele associated with myocardial infarction (Yamada et al. 2002), the *128R* allele may be less common in Japanese subjects (Takei et al. 2002).

The *S128R* polymorphism has also been associated with systemic lupus erythematosus (SLE) (El Magadmi et al. 2001), a disease in which coronary artery disease and stroke are increasingly important causes of late morbidity and mortality (Urowitz et al. 2000). This raises the intriguing possibility of shared pathogenic pathways in these vascular diseases. Interestingly, Amoli et al. (2002) failed to find an association between *S128R* E-selectin and large or small vessel vasculitis.

Functionally, the E-selectin *S128R* polymorphism has been the subject of a number of reports. While the crystal structures of E-selectin (Graves et al. 1994) and E-selectin bound to its carbohydrate ligand sialyl Lewis^x (sLe^x) (Somers et al. 2001) have shown that sLe^x binds the lectin domain at its apex, two lines of evidence suggest that this polymorphism is functionally relevant:

1. The EGF domain is necessary for optimal E-selectin adhesive function (Pigott et al. 1991) and, by extrapolation from studies on P- and L-selectin, may critically influence ligand-binding specificity and/or affinity (Kansas et al. 1994; Dwir et al., 2000)
2. An E-selectin 128R fusion protein demonstrated altered specificity of in vitro binding to myeloid cells lines under static conditions when compared to E-selectin 128S (Revelle et al. 1996). Thus, the *128R* mutation led to binding of E-selectin to K562 cells, which do not express fucosyl-transferase IV or VII, do not express sLe^x, and are not able to bind E-selectin 128S (Revelle et al. 1996).

We have recently shown that enhanced interactions between E-selectin 128R and myeloid cell lines also occur under physiological flow conditions, and that neutrophil adhesion is similarly affected (Rao et al. 2002). Moreover, in our studies E-selectin 128R was found to bind a broader range of lymphocytes than that which binds E-selectin 128S (Rao et al. 2002). Normally, E-selectin binding to T lymphocytes is restricted to a subpopulation of memory cells, which express the cutaneous lymphocyte antigen (CLA) and which home to skin. In contrast, E-selectin 128R bound an additional population of CLA-negative memory lymphocytes, raising the possibility that this polymorphism might affect the control of immune responses by causing lymphocytes to be recruited inappropriately into inflamed tissues. A more recent study has reiterated these findings for neutrophil and mononuclear cell attachment to ECs following transduction with adenovirus encoding 128S or 128R E-selectin (Yoshida et al. 2003). Interestingly, the authors also observed a greater degree of constitutive phosphorylation of extracellular signal-related kinase (ERK)-1 and -2 and p38 mitogen-activated protein kinase (MAPK) in ECs transduced with 128R E-selectin compared to 128S, suggesting the polymorphism may influence E-selectin-mediated intracellular signaling.

The 5′ untranslated region of *SELE* contains a polymorphic variant, *98G>T*, which has also been found to be associated with early-onset CAD (Zheng et al. 2001). Non-conservative leucine for phenylalanine exchange at nucleotide 554 (*L554F*) results from 1,839C>T in exon 11. As this region encodes the transmembrane domain, this SNP could disturb the anchoring or orientation of the protein within the cell membrane. The minor *T* allele has been associated with early atherosclerosis (Wenzel et al.1996), whilst on the other hand the major *C* allele was over-represented in French patients with atherosclerosis (Sass et al. 2000).

2.3
L-Selectin

There are a number of polymorphisms in the L-selectin gene, but so far none have been directly associated with cardiovascular disease. There are reports of associations of L-selectin polymorphisms with IgA nephropathy and with type I diabetes mellitus (Kretowski and Kinalska 2000; Takei et al. 2002).

3
Selectin Ligands

Selectins are cell surface lectins, and bind glycoproteins (and possibly glycolipids). The capacity of specific glycoproteins to bind selectins is determined both by their protein backbone as well as by their glycosylation.

3.1
P-Selectin Glycoprotein Ligand-1

P-selectin glycoprotein ligand-1 (PSGL-1) is the principle leukocyte ligand for P-selectin and also binds E-selectin and L-selectin (McEver and Cummings 1997). It is a homodimer of approximately 250 kDa, made up of two mucin-like subunits. The protein core of each subunit consists of a series of decameric repeat sequences, similar to those observed in the structurally similar platelet molecule, glycoprotein GpIb-alpha. Recently, three *PSGL-1* alleles have been described, encoding a variable number of tandem repeat sequences (*VNTR*) (Afshar-Khargham et al. 2001). The *A* allele is the most common (16 repeats), the *B* allele has 15 and the *C* allele has 14 repeats. In small cohorts of Spanish patients, smaller *VNTR* alleles were significantly associated with reduced risk of cerebrovascular disease, whereas no associations were detected with CAD or deep vein thrombosis (Lozano et al. 2001). These interesting preliminary results require replication before firm conclusions may be drawn, and no studies have as yet included combined analysis of both selectin and ligand polymorphisms.

It is possible that the number of VNTRs may influence PSGL-1 function, as binding of activated platelets to neutrophils from individuals with the *A/C* allele was reduced compared to neutrophils from individuals with either *A/A* or *A/B* alleles present (Lozano et al. 2001).

3.2
Leukocyte Adhesion Deficiency Type II

In the rare leukocyte adhesion deficiency type II (LAD-II), neutrophil selectin ligands (e.g., PSGL-1) are inappropriately post-translationally modified, resulting in defective leukocyte rolling on endothelium. Affected individuals suffer from recurrent infections despite a marked leukocytosis, abnormalities in growth and psychomotor retardation, hypotonia, seizures, strabismus as well red blood cells lacking the H antigen (the Bombay phenotype) (Etzioni and Tonetti 2000). This condition has been recently classified as part of a wider spectrum of disorders as congenital disorder of glycosylation type II-c. The molecular abnormality is a defect in GDP-fucose biosynthesis and more specifically a defect in a GDP-fucose transporter (FUCT1), responsible for transport of fucose into the Golgi (Karsan et al. 1998). Missense mutations have been identified in the fourth (C439T) and ninth transmembrane domains (C923G) (Lübke et al. 2001; Lühn et al. 2001). LAD-II is a childhood disease, and it is not known whether it has any influence on the development of cardiovascular disease.

4
Immunoglobulin Superfamily

The immunoglobulin supergene family consists of a large number of molecules that are structurally related to antibodies (Williams and Barclay 1988; Wang and Springer 1998). Within this family falls a sub-group of molecules (ICAM-1, -2, -3, VCAM-1, MadCAM-1, PECAM-1), which act as adhesion ligands for integrins.

4.1
ICAM-1

Intercellular adhesion molecule-1 (ICAM-1) plays a major role in normal immune function, acting as the ligand for the lymphocyte function-associated antigen 1 (LFA-1, CD11a/CD18) and macrophage antigen 1 (Mac-1, CD11b/CD18). Besides contributing to leukocyte extravasation, ICAM-1 also supports leukocyte effector function and lymphocyte proliferation (van de Stope and van der Saag 2001). ICAM-1 is the major human rhinovirus receptor (Greve et al. 1989) and also binds *Plasmodium falciparum*-infected erythrocytes (Berendt et al. 1989; Ockenhouse et al.1992). The encoding gene for this protein (*ICAM1*) is a seven-exon structure, which maps to 19p13.3–13.2, within a cluster of genes for other immunoglobulin superfamily members.

Two coding SNPs in *ICAM1* immunoglobulin domains are known in Caucasian populations: *G241R*, a very rare substitution in the Mac-1 binding site (exon 4) and the more frequent *K469E* in the fifth immunoglobulin-like domain (exon 6) (Vora et al. 1994). These polymorphisms have been examined in numerous small case–control association studies in different clinical conditions, yet the functional implications of these variants are unknown.

The genetic epidemiology of ICAM1 polymorphisms and atherosclerosis is inconsistent. Jiang et al. (2002) report a significant increase in the *E469* allele in patients with both stable CAD and MCI ($p<0.001$), and homozygosity for *E469* may be a risk factor in peripheral arterial occlusive disease (Gaetani et al. 2002).

Transplant-associated coronary disease (TxCAD) is a leading cause of late graft loss. A small study of TxCAD in UK heart transplants found a borderline significant protective effect of the *E469* allele of the cadaveric donor and a trend to reduced frequency of this allele in a sub-group defined by multiple rejection episodes (Borozdenkova et al. 2001). In a study of the analogous process in renal transplants, chronic renal allograft failure (CRAF), the rare *R241* allele was increased in CRAF patients compared to long-term survivors and healthy controls (McLaren et al. 1999). The distribution of *E469* was uniform and although this allele was associated with rapid graft loss, this finding is of questionable significance given the small number of patients involved. Clearly prediction of those at increased risk of graft rejection would be of great clinical value, but it is premature to advocate *ICAM1* genotyping as part of the pre-transplantation assessment. Large-scale collaborative studies will be necessary in order to recruit sufficient cases to adequately address this issue.

Positive associations with the G241R and/or K469E polymorphisms have also been observed in a number of inflammatory diseases, including Behçet's syndrome (Verity et al. 2000), inflammatory bowel disease (Braun et al. 2001) and rheumatoid arthritis (Macchioni et al. 2000). Conflicting findings have been reported in type 1 diabetes mellitus (Kristiansen et al. 2000; Nejentsev et al. 2000; Nishimura et al. 2000), multiple sclerosis (Marrosu et al. 2000; Mycko et al. 1998) and polymyalgia rheumatica/giant cell arteritis (Salvarani et al. 2000; Amoli et al. 2001).

Binding of erythrocytes infected with *Plasmodium falciparum* to small vessel endothelium may contribute to the virulence of this organism. This process may be mediated by a number of receptors, including adhesion molecules such as E- and P-selectin, VCAM-1 and ICAM-1. Fernandez-Reyes et al. (1997) defined genomic variation in the N-terminal immunoglobulin domain of the *ICAM1* gene in 24 asymptomatic children from Kilifi, Kenya, an area where malaria is endemic. A single mutation was observed at nucleotide *179A>T*; this specifies a lysine to methionine exchange (K29M), which the authors named ICAM1^{Kilifi}. This variant was observed with an allele frequency of ~0.3 in Kenya and the Gambia, but was absent in Europeans. Results from a case–control association study suggested that the ICAM1^{Kilifi} is a dose-dependent risk factor for cerebral malaria. In contrast, no association was found between ICAM1^{Kilifi} and severity of malaria in the Gambia (Bellamy et al. 1998). Amino-acid 29 lies at the N-terminal of ICAM-1, and M29 ICAM-1 shows reduced binding of T lymphocytes, fibrinogen and some strains of *Plasmodium falciparum*-infected erythrocytes (Craig et al. 2000). The selective advantage of the M29 ICAM-1 is not clear, and it will be interesting to determine whether it protects from atherosclerosis and auto-immune diseases.

4.2
VCAM-1

Mutation screening of the *VCAM1* gene in Caucasians suggests that the VCAM-1 protein is non-polymorphic in this population (Wenzel et al. 1996). In contrast, Taylor et al. (2002) have recently described 33 biallellic SNPs in African-Americans, and have proposed that the non-synonymous mutation *G1238C* may be associated with protection from stroke in sickle cell disease. Although it is not known whether *G1238C* influences function, the association between a VCAM-1 polymorphism and vascular complications of sickle cell disease is potentially relevant as sickle cell erythrocytes express the VCAM-1 ligand $\alpha_4\beta_1$ integrin and may bind VCAM-1 expressed by cytokine-activated EC (Gee and Platt 1995).

4.3
PECAM-1

PECAM-1 (CD31) expression is widely distributed on ECs, circulating platelets, myeloid cells, and some T lymphocytes. PECAM-1 probably therefore contributes to disparate pathological processes including thrombosis, hemostasis, immunity, and the inflammatory response.

The gene encoding PECAM-1 (*PECAM1*) is located at *17q23* and is highly polymorphic, containing a *CA* repeat in intron 6 and a number of coding SNPs (cSNPs). By direct sequencing of *PECAM,* cDNA derived from 21 normal individuals, Behar et al. identified a high-frequency non-synonymous C→G polymorphism, which results in a conservative leucine-to-valine exchange at residue 125 within the first extracellular domain (Behar et al. 1996). When 46 bone marrow transplant (BMT) recipients and their HLA-matched sibling donors were genotyped for this SNP, 71% of those with acute graft versus host disease (GVHD) were discordant at this locus. Although this association was confirmed in a cohort of 118 BMT patients (Grumet et al. 2001), two further equally powered studies yielded null results (Nichols et al. 1996; Maruya et al. 1998). However, Maruya et al. (1998) identified two novel linked cSNPs, *Asn563Ser* and *Gly670Arg*, which were associated with graft versus host disease, a result independently confirmed in at least two cohorts (Balduini et al. 2001; Grumet et al. 2001). The relative importance of non-identity at the *PECAM1* locus in this important complication of BMT is therefore unclear and further work is needed. The strong linkage disequilibrium (LD) between these SNPs clearly complicates association studies and haplotype analyses would aid discrimination of the etiological alleles.

Studies examining *PECAM1* genotype in atherosclerosis have proved similarly inconclusive. Wenzel et al. (1999) reported a significant increase in *Val125/Asn563* homozygosity in German patients with early atherosclerosis compared to controls, and minor effects have also been reported in a large Caucasian cohort (Gardemann et al. 2000). Conversely, when these SNPs were examined in a group of 136 Japanese patients with MCI and 253 controls, homozygosity for the *563Ser/670Arg* haplotype was increased in the myocardial infarction patient group compared to controls (Sasaoka et al. 2001). These contrasting results could result from racial differences in the pathogenesis of atherosclerosis, or could reflect LD patterns between the *PECAM1* locus and a true etiological polymorphism. On the other hand, it is possible that this contradiction is due to the inadequate power of small, cross-sectional studies.

5
Integrins

The integrins constitute a large family of non-covalently linked $\alpha\beta$ heterodimeric glycoproteins (subunits of 95–200 kDa) that are so named because of their integral function in connecting the intracellular cytoskeleton with the extracellular environment. One or more integrins is found on most animal cell types, with the characteristics of the integrin(s) affecting the other cells and matrix components with which the cell can interact. Integrins play an essential role in the migration and function of leukocytes, with α_4 integrins (e.g., VLA-4) and β_2 integrins (e.g., LFA-1, Mac-1) particularly important for leukocyte-EC interactions (Harris et al. 2000). As yet no functionally significant genetic variations of VLA-4 have been described.

5.1
β_2 Integrins

Absent or severely reduced (<10%) β_2 integrin expression results in the rare clinical syndrome of leukocyte adhesion deficiency type 1 (LAD I) (Arnaout 1990). All patients described to date have mutations in the gene encoding the $\beta2$ subunit (CD18), present in single copy on 21q22.3, and inheritance is autosomal recessive. Patients show defects in leukocyte motility, adherence and phagocytosis, leading to recurrent periodontitis, skin infections and delayed wound healing. Those with less than 1% $\beta2$ subunit expression suffer from life-threatening infection and require bone marrow transplantation. No vascular abnormalities have thus far been described in these patients. Heterozygote relatives express 40%–60% normal levels of β_2-integrins and are clinically normal.

More recently, patients have been described with LAD I variant syndromes in which leukocytes express adequate β_2 integrin expression but have evidence of β_2 integrin dysfunction (Hogg et al. 1999). One patient had a mutation leading to proline for serine substitution at position 138 and arginine for glycine at position 273 in the β_2 integrin sub-unit. Further studies revealed that the *S138P* mutation can support the formation of non-functioning CD11/CD18 heterodimers whereas the *G273R* mutation cannot form heterodimers at all. The mutations both lie in the important I domain of the β_2 sub-unit, with the S138 representing a second serine in a conserved DLSYS motif within the metal ion-dependent adhesion site (MIDAS).

The *CD18* gene has been examined for coding sequence variants that might influence inflammatory disease, perhaps providing a *forme fruste* of LAD I. However, at least in Caucasians, CD18 appears to be uniform at the amino acid level (Meller et al. 2001). An *Ava*II restriction fragment length polymorphism within exon 11 (1,323C>T) is implicated by association with systemic vasculitis, specifically with antibodies to proteinase 3 (Gencik et al. 2000). Since this SNP is silent its effect may be at the level of transcription/translation regulation, or by virtue of LD with other SNPs in 5' or 3' flanking sequence.

There is an increasing literature suggesting a functional role for β_2 integrins in lumen re-narrowing after coronary arterial procedures. Koch et al. (2001) prospectively investigated the silent 1,323C>T polymorphism in the *CD18* gene in 1207 consecutive patients undergoing coronary stent placement. These investigators identified a gene-dose dependent protective effect of the major *T* allele in restenosis. There are no published data linking this gene with de novo atherosclerosis, and null results have been reported in stroke (Zee et al. 2002).

6
The Therapeutic Impact of Leukocyte-Endothelial Adhesion Molecules

Following the successful use in acute coronary syndromes of anti-platelet integrin (GpIIb/IIIa) inhibitors, either in the form of monoclonal antibodies or synthetic compounds, the potential for similar inhibition of leukocyte adhesion molecules is currently being investigated (Marshall and Haskard 2002). The majority of these compounds are still in development and full review of all the compounds in development is beyond the scope of this chapter. Promising results in clinical trials have been obtained with anti-LFA-1 in psoriasis (Gottlieb et al. 2002) and with anti-VLA-4 (Natalizumab) in multiple sclerosis (Miller et al. 2003) and Crohn's disease (Ghosh et al. 2003). As yet there are no published data indicating the importance of an adhesion molecule polymorphism in response to treatment.

7
Conclusion

This survey of adhesion molecule polymorphisms and their disease associations illustrates important issues that broadly apply to genetic epidemiology as a whole. The disease states described above are in the main complex traits, in which a genetic contribution interacts with environmental and stochastic factors to result in manifestations of disease. In complex traits, epidemiological and genome-wide linkage studies in humans and animal models suggest that the genetic susceptibility is the result of the interaction between a number of genes, each with modest effects. The implication, therefore, is that large, adequately powered studies are required in order to confidently identify, or *exclude* susceptibility genes. In many disease states, including cardiovascular disease, this is further complicated by heterogeneity in disease phenotype.

Whilst some strong associations are evident, such as those relating the selectin genes with atherosclerosis, a significant majority of studies remain inconclusive. The literature cited here is littered with small-scale retrospective case–control association studies, in which power considerations have often been inadequately addressed. Frequently, weakly positive findings have been reported with failure to correct for the multiple hypotheses testing results from examination of a number of polymorphisms within a gene or family of genes, presumably compounded by publication bias. It is therefore vital that the disease associations reviewed here are more rigorously addressed.

It is striking to note that only very few of the clinically associated genetic abnormalities have been functionally addressed. This may reflect acquisition and publication bias, although notable exceptions exist. Establishing functional significance for the genes described above remains a critical step in the implication of these genes and their products in the pathogenesis and etiology of cardiovascular disease. Thus far, the most extensive data concern the E-selectin *S128R* polymorphism, demonstrating both quantitative and qualitative effects on interactions with myeloid and lymphoid cells under static and, more importantly, flow conditions.

In summary, putative associations between adhesion molecule polymorphisms and inflammatory or autoimmune disease mostly require further consideration. A clear definition of the genetic basis of complex disease will *ipso facto* enhance our understanding of the disease process and therefore ultimately lead to improved patient care, both through rationally designed therapeutics and identification of those individuals genetically predisposed to disease.

References

Afshar-Khargham V, Diz-Kucukkaya R, Ludwig EH, Marian AJ, Lopez JA (2001) Human polymorphism of P-selectin glycoprotein ligand-1 attributable to variable numbers of tandem decameric repeats in the mucin-like region. Blood 97:3306–3307

Amoli MM, Shelley E, Mattey DL, Garcia-Porrua C, Thomson W, xx AH, Ollier WE, Gonzalez-Gay MA (2001) Lack of association between intercellular adhesion molecule-1 polymorphisms and giant cell arteritis. J Rheumatol 28:1600–1604

Amoli MM, Alansari A, El Magadmi M, Thomson W, Hajeer AH, Calvino MC, Garcia-Porrua C, Ollier WE, Gonzalez-Gay MA (2002) Lack of association between A561C E-selectin polymorphism and large and small-sized blood vessel vasculitides. Clin Exp.Rheumatol. 20:575–576

Andre P, Hartwell D, Hrachovinova I, Saffaripour S, Wagner DD (2000) Pro-coagulant state resulting from high levels of soluble P-selectin in blood. Proc Natl Acad Sci USA 97:13835–13840

Arnaout MA (1990) Leukocyte adhesion molecules deficiency: its structural basis, pathophysiology and implications for modulating the inflammatory response. Immunological Reviews 114:145–179

Balduini CL, Frassoni F, Noris P, Klersy C, Iannone AM, Bacigalupo A, Giorgiani G, Di Pumpo M, Locatelli F (2001) Donor-recipient incompatibility at CD31-codon 563 is a major risk factor for acute graft-versus-host disease after allogeneic bone marrow transplantation from a human leucocyte antigen-matched donor. Br J Haematol 114:951–953

Barbaux SC, Blankenberg S, Rupprecht HJ, Francomme C, Bickel C, Hafner G, Nicaud V, Meyer J, Cambien F, Tiret L (2001) Association between P-selectin gene polymorphisms and soluble P- selectin levels and their relation to coronary artery disease. Arterioscler Thromb Vasc Biol 21:1668–1673

Behar E, Chaoi NJ, Hiraki DD, Krishnaswamy S, Brown BW, Zehnder JL, Grumet FC (1996) Polymorphism of adhesion molecule CD31 and it's role in acute graft-versus-host disease. N Engl J Med 334:286–291

Bellamy R, Kwiatkowski D, Hill AV (1998) Absence of an association between intercellular adhesion molecule 1, complement receptor 1 and interleukin 1 receptor antagonist gene polymorphisms and severe malaria in a West African population. Trans R Soc Trop Med Hyg 92:312–316

Berendt AR, Simmons D, Tansey J, Newbold CI, Marsh K (1989) Intercellular adhesion molecule-1 is an endothelial cell adhesion receptor for Plasmodium falciparum. Nature 341:57–59

Borozdenkova S, Smith J, Marshall S, Yacoub M, Rose M (2001) Identification of ICAM-1 polymorphism that is associated with protection from transplant associated vasculopathy after cardiac transplantation. Human Immunol 62:247–255

Braun C, Zahn R, Martin K, Albert E, Folwaczny C (2001) Polymorphisms of the ICAM-1 gene are associated with inflammatory bowel disease, regardless of the p-ANCA status. Clin Immunol 101:357–360

Burger, P.C., Wagner DD (2003) Platelet P-selectin facilitates atherosclerotic lesion development. Blood 101:2661–2666.

Craig A, Fernandez-Reyes D, Mesri M, McDowell A, Altieri DC, Hogg N, Newbold CI (2000) A functional analysis of a natural variant of intercellular adhesion molecule-1 (ICAM-1Kilifi). Hum Mol Genet 9:525–530

Dwir O, Kansas GS, Alon R (2000) An activated L-selectin mutant with conserved equilibrium binding properties but enhanced ligand recognition under shear flow. J Biol Chem 275:18682–18691

El Magadmi M, Alansari A, Teh LS, Ordi J, Gul A, Inanc M, Bruce I, Hajeer A (2001) Association of the A561C E-selectin polymorphism with systemic lupus erythematosus in 2 independent populations. J Rheumatol 28:2650–2652

Ellsworth DL, Bielak LF, Turner ST, Sheedy II PF, Boerwinkle E, Peyser PA (2001) Gender- and age-dependent relationships between E-selectin S128R polymorphism and coronary artery calcification. J Mol Med 79:390–398

Etzioni A, Tonetti M (2000) Leukocyte adhesion deficiency II- from A to almost Z. Immunology Reviews 178:138–147

Fernandez-Reyes D, Craig AG, Kyes SA, Peshum N., Snow RW, Berendt A, Marsh K, Newbold CI (1997) A high frequency African coding polymorphism in N-terminal domain of ICAM-1 predisposing to cerebral malaria in Kenya. Human Molecular Genetics 6:1357–1360

Gaetani E, Flex A, Pola R, Papaleo P, De Martini D, Pola E, Aloi F, Flore R, Serricchio M, Gasbarrini A, Pola P (2002) The K469E polymorphism of the ICAM-1 gene is a risk factor for peripheral arterial occlusive disease. Blood Coagul Fibrinolysis 13:483–488

Gardemann A, Knapp A, Katz N, Tillmanns H, Haberbosch W (2000) No evidence for the CD31 C/G gene polymorphism as an independent risk factor of coronary heart disease. Thromb Haemost 83:629

Gee BE, Platt OS (1995) Sickle reticulocytes adhere to VCAM-1. Blood 85:268–274

Gencik M, Meller S, Borgmann S, Sitter T, Menezes Saecker AM, Epplen JT (2000) The association of CD18 alleles with anti-myeloperoxidase subsets of ANCA-associated systemic vasculitides. Clin Immunol 94:9–12

Ghosh, S., E.Goldin, F.H.Gordon, H.A.Malchow, J.Rask-Madsen, P.Rutgeerts, P.Vyhnalek, Z.Zadorova, T.Palmer, S.Donoghue, and the Natalizumab Pan-European Study Group. 2003. Natalizumab for Active Crohn's Disease. *N Engl J Med* 348:24–32.

Gottlieb, A.B., J.G.Krueger, K.Wittkowski, R.Dedrick, P.A.Walicke, and M.Garovoy. 2002. Psoriasis as a model for T-cell-mediated disease: immunobiologic and clinical effects of treatment with multiple doses of efalizumab, an anti-CD11a antibody. *Arch. Dermatol.* 138:591–600

Graves BJ, Crowther RL, Chandran C, Rumberger JM, Li S, Huang K-S, Presky DH, Familletti JM, Wolitzky BA, Burns DK (1994) Insight from E-selectin/ligand interaction from the crystal structure and mutagenesis of the lec/EGF domains. Nature 367:532–538

Greve JM, Davis G, Meyer AM, Forte CP, Yost SC, Marlor CW, Kamarck ME, McClelland A (1989) The major human rhinovirus receptor is ICAM-1. Cell 56:839–847

Grumet FC, Hiraki DD, Brown BWM, Zehnder JL, Zacks ES, Draksharapu A, Parnes J, Negrin RS (2001) CD31 mismatching affects marrow transplantation outcome. Biol Blood Marrow Transplant 7:503–512.

Harris ES, McIntryre TM, Prescott SM, Zimmerman GA (2000) The leukocyte integrins. J Biol Chem 275:23409–23412

Herrmann SM, Ricard S, Nicaud V, Mallet C, Evans A, Ruidavets JB, Arveiler D, Luc G, Cambien F (1998) The P-selectin gene is highly polymorphic: reduced frequency of the Pro715 allele carriers in patients with myocardial infarction. Human Molecular Genetics 97:1277–1284

Hogg N, Stewart MP, Scarth SL, Newton R, Shaw JM, Law SK, Klein N (1999) A novel leukocyte adhesion deficiency caused by expressed but nonfunctional beta2 integrins Mac-1 and LFA-1. J Clin Invest 103:97–106

Jiang H, Klein RM, Niederacher D, Du M, Marx R, Horlitz M, Boerrigter G, Lapp H, Scheffold T, Krakau I, Gulker H (2002) C/T polymorphism of the intercellular adhesion molecule-1 gene (exon 6, codon 469). A risk factor for coronary heart disease and myocardial infarction. Int J Cardiol 84:171–177

Johnson RC, Chapman SM, Dong ZM, Ordovas JM, Mayadas TN, Herz J, Hynes RO, Schaefer EJ, Wagner DD (1997) Absence of P-selectin delays fatty streak formation in mice. J Clin Invest 99:1037–1043

Johnson-Tidey RR, McGregor JL, Taylor PR, Poston RN (1994) Increase in the adhesion molecule P-selectin in endothelium overlying atherosclerotic plaques: coexpression with intercellular adhesion molecule-1. Am J Pathol 144:952–961

Kansas GS (1996) Selectins and their ligands: current concepts and controversies. Blood 88:3259–3287

Kansas GS, Saunders KB, Ley K, Zakrzewicz A, Gibson RM, Furie BC, Furie B, Tedder TF (1994) A role for the epidermal growth factor-like domain of P-selectin in ligand recognition and cell adhesion. J Cell Biol 124:609–618

Karsan A, Cornejo CJ, Winn RK, Schwartz BR, Way W, Lannir N, GershoniBaruch R, Etzioni A, Ochs HD, Harlan JM (1998) Leukocyte adhesion deficiency type II is a generalized defect of de novo gdp-fucose biosynthesis—endothelial cell fucosylation is not required for neutrophil rolling on human nonlymphoid endothelium. J Clin Invest 101:2438–2445

Kee F, Morrison C, Evans AE, McCrum E, McMaster D, Dallongeville J, Nicaud V, Poirier O, Cambien F (2000) Polymorphisms of the P-selectin gene and risk of myocardial infraction in men and women in the ECTIM extension study. Heart 84:548–552

Koch W, Bottiger C, Mehilli J, von Beckerath N, Neumann FJ, Schomig A, Kastrati A (2001) Association of a CD18 gene polymorphism with a reduced risk of restenosis after coronary stenting. Am.J Cardiol 88:1120–1124

Kretowski A, Kinalska I (2000) L-selectin gene T668C mutation in type 1 diabetes patients and their first degree relatives. Immunol Lett. 74:225–228

Kristiansen OP, Nosøe RL, Holst H, Reker S, Larsen ZM, Johannsen J, Nerup J, Pociot F, Mandrup-Poulsen T, The Danish IDDM Epidemiology and Genetics Group, The Danish Study Group of IDDM in children (2000) The intercellular adhesion molecule-1 K469E polymorphism in type 1 diabetes. Immunogenetics 52:107–111

Lozano ML, Gonzalez-Conejero R, Corral J, Rivera J, Iniesta JA, Martinez C, Vicente V (2001) Polymorphisms of P-selectin glycoprotein ligand-1 are associated with neutrophil-platelet adhesion and with ischaemic cerebrovascular disease. Br J Haematol 115:969–976

Lübke T, Marquardt T, Etzioni A, Hartmann E, von Figura K, Körner C (2001) Complementation cloning identifies CDG-IIc, a new type of congenital disorders of glycosylation, as a GDP-fucose transporter deficiency. Nat Genet 28:73–76

Lühn K, Wild MK, Eckhardt M, Gerardy-Schahn R, Vestweber D (2001) The gene defective in leukocyte adhesion deficiency II encodes a putative GDP-fucose transporter. Nat Genet 28:69–72

Macchioni P, Boiardi L, Casali B, Nicoli D, Farnetti E, Salvarani C (2000) Intercellular adhesion molecule 1 (ICAM-1) gene polymorphisms in Italian patients with rheumatoid arthritis. Clin Exp Rheumatol 18:553–558

Marrosu MG, Schirru L, Fadda E, Mancosu C, Lai M, Cocco E, Cucca F (2000) ICAM-1 gene is not associated with multiple sclerosis in sardinian patients. J Neurol 247:677–680

Marshall D, Haskard DO (2002) Clinical overview - where are we now? Semin Immunol 14:133–140.

Maruya E, Saji H, Seki S, Fujii Y, Kato K, Kai S, Hiraoka A, Kawa K, Hoshi Y, Ito K, Yokoyama S, Juji T (1998) Evidence that CD31, CD49b, and CD62L are immunodominant minor histocompatibility antigens in HLA identical sibling bone marrow transplants. Blood 92:2169–2176

McEver RP, Cummings RD (1997) Role of PSGL-1 binding to selectins in leukocyte recruitment. J Clin Invest 100:485–492

McLaren AJ, Marshall SE, Halder NA, Mullighan CG, Fuggle SV, Morris PJ, Welsh KI (1999) Adhesion molecule polymorphisms in chronic renal failure. Kidney Int 55:1977–1982.

Meller S, Jagiello P, Borgmann S, Fricke H, Epplen JT, Gencik M (2001) Novel SNPs in the CD18 gene validate the association with MPO-ANCA+ vasculitis. Genes Immun. 2:269–272

Miller DH, Khan OA, Sheremata WA, Blumhardt LD, Rice GPA, Libonati MA, Willmer-Hulme AJ, Dalton CM, Miszkiel KA, O'Connor PW, and the International Natalizumab Multiple Sclerosis Trial Group. 2003. A Controlled Trial of Natalizumab for Relapsing Multiple Sclerosis. *N Engl J Med* 348:15–23

Mycko MP, Kwinkowski M, Tronczynska E, et al (1998) Multiple sclerosis: the increased frequency of the ICAM-1 exon 6 gene point mutation genetic type K469. Ann Neurol 44:70–76.

Nejentsev S, Laine A-P, Simell O, Ilonen J (2000) Intercellular adhesion molecule-1 (ICAM-1) K469E polymorphism: no association with type I diabetes among Finns. Tissue Antigens 55:568–570

Nichols WC, Antin JH, Lunetta KL, Terry VH, Hertel CE, Wheatley MA, Arnold ND, Siemieniak DR, Boehnke M, Ginsburg D (1996) Polymorphism of adhesion molecule CD31 is not a significant risk factor for graft-versus-host disease. Blood 88:4429–4434

Nishimura M, Obayashi H, Maruya E, Ohta M, Tegoshi H, Fukui M, Hasegawa G, Shigeta H, Kitagawa Y, Nakano K, Saji H, Nakamura N (2000) Association between type 1 diabetes age-at-onset and intercellular adhesion molecule-1 (ICAM-1) gene polymorphism. Human Immunol. 61:507–510

Ockenhouse CF, Betageri R, Springer TA, Staunton DE (1992) Plasmodium falciparum-infected erythrocytes bind ICAM-1 at a site distinct from LFA-1, Mac-1, and human rhinovirus. Cell 68:63–69

Pigott R, Needham LA, Edwards RM, Walker C, Power C (1991) Structural and functional studies of the endothelial activation antigen endothelial leucocyte adhesion molecule-1 using a panel of monoclonal antibodies. J Immunol 147:130–135

Rao RM, Haskard D.O., Landis RC (2002) Enhanced recruitment of Th2 and CLA negative T lymphocytes by the S128R polymorphism of E-selectin. J Immunol 169:5860–5865

Rao RM, Clarke JL, Ortlepp S, Robinson MK, Landis RC, Haskard D.O. (2002) The S128R polymorphism of E-selectin mediates neuraminidase-resistant tethering of myeloid cells under shear flow. Eur J Immunol 32:251–260

Rauchhaus M, Gross M, Schulz S, Francis DP, Greiser P, Norwig A, Weidhase L, Coats AJ, Dietz R, Anker SD, Glaser C (2002) The E-selectin SER128ARG gene polymorphism and restenosis after successful coronary angioplasty. Int J Cardiol 83:249–257

Revelle BM, Scott D, Beck PJ (1996) Single amino acid residues in the E- and P-selectin epidermal growth factor domains can determine carbohydrate binding specificity. J Biol Chem 271:16160–16170

Ross R (1999) Atherosclerosis–an inflammatory disease. N.Engl.J Med. 340:115–126

Salvarani C, Casali B, Boiardi L, Ranzi A, Macchioni P, Nicoli D, F x, Brini M, Portioli I (2000) Interercellular adhesion molecule 1 gene polymorphisms in polymyalgia rheumatica/giant cell arteritis: association with risk and severity. J Rheumatol 27:1221

Sasaoka T, Kimura A, Hohta SA, Fukuda N, Kurosawa T, Izumi T (2001) Polymorphisms in the platelet-endothelial cell adhesion molecule-1 (PECAM-1) gene, Asn563Ser and Gly670Arg, associated with myocardial infarction in the Japanese. Ann.N.Y.Acad.Sci. 947:259–269

Sass C, Pallaud C, Zannad F, Visvikis S (2000) Relationship between E-selectin L/F554 polymorphism and blood pressure in the Stanislas cohort. Hum Genet 107:58–61

Somers WS, Tang J, Shaw GD, Camphausen RT (2001) Insights into the molecular basis of leukocyte tethering and rolling revealed by structures of P- and E-selectin bound to SLe^x and PSGL-1. Cell 103:467–479

Springer TA (1995) Traffic signals on endothelium for lymphocyte recirculation and leukocyte emigration. Annu Rev Physiol 57:827–872

Takei T, Iida A, Nitta K, Tanaka T, Ohnishi Y, Yamada R, Maeda S, Tsunoda T, Takeoka S, Ito K, Honda K, Uchida K, Tsuchiya K, Suzuki Y, Fujioka T, Ujiie T, Nagane Y, Miyano S, Narita I, Gejyo F, Nihei H, Nakamura Y (2002) Association between single-nucleotide polymorphisms in selectin genes and immunoglobulin A nephropathy. Am J Hum Genet 70:781–786

Taylor JG, Tang DC, Savage SA, Leitman SF, Heller SI, Serjeant GR, Rodgers GP, Chanock SJ (2002) Variants in the VCAM1 gene and risk for symptomatic stroke in sickle cell disease. Blood 100:4303–4309

Urowitz M, Gladman D, Bruce I (2000) Atherosclerosis and systemic lupus erythematosus. Curr Rheumatol Rep 2:19–23

van de Stope A, van der Saag PT (2001) Intercellular adhesion molecule-1. J Mol Med 74:13–33

Verity DH, Vaughan RW, Kondeatis E, Madanat W, Zureikat H, Fayyad F, Marr JE, Kanawati CA, Wallace GR, Stanford MR, The Anglo-Jordanian Behçet's disease research group (2000) Intercellular adhesion molecule-1 gene polymorphisms in Behçet's disease. Eur J Immunogenet 27:73–76

Vora DK, Rosenbloom CL, Beaudet AL, Cottingham RW (1994) Polymorphisms and linkage analysis for ICAM-1 and the selectin gene cluster. Genomics 21:473–477

Wang J, Springer TA (1998) Structural specializations of immunoglobulin superfamily members for adhesion to integrins and viruses. Immunol Rev 163:197–215

Watson ML, Kingsmore SF, Johnston GI, Siegelman MH, Le Beau MM, Lemons RS, Bora NS, Howard TA, Weissman IL, McEver RP, Seldin MF (1990) Genomic organization of the selectin family of leukocyte adhesion molecules on human and mouse chromosome 1. J Exp Med 172:263–272

Wenzel K, Hanke R, Speer A (1994) A polymorphism in the human E-selectin gene detected by PCR-SSCP. Hum Genet 94:319–330

Wenzel K, Felix S, Kleber FX, Brachold R, Menke T, Schattke S, Schulte KL, Glaser C, Rohde K, Baumann G, et al (1994) E-selectin polymorphism and atherosclerosis: an association study. Hum Mol Genet 3:1935–1937

Wenzel K, Ernst M, Rohde K, Baumann G, Sper A (1996) DNA polymorphisms in adhesion molecule genes: a new risk factor for early atherosclerosis. Hum Genet 97:15–20

Wenzel K, Baumann G, Felix SB (1999) The homozygous combination of Leu125Val and Ser563Asn polymorphisms in the PECAM1 (CD31) gene is associated with early severe coronary heart disease. Hum.Mutat 14:545

Williams AF, Barclay AN (1988) The immunoglobulin superfamily-domains for cell surface recognition. Ann Rev Immunol 6:381–405

Yamada Y, Izawa H, Ichihara S, Takatsu F, Ishihara H, Hirayama H, Sone T, Tanaka M, Yokota M (2002) Prediction of the risk of myocardial infarction from polymorphisms in candidate genes. N Engl J Med 347:1916–1923

Ye SQ, Usher D, Virgil D, Zhang LQ, Yochim SE, Gupta R (1999) A PstI polymorphism detects the mutation of serine128 to arginine in CD 62E gene—a risk factor for coronary artery disease. J Biomed Sci 6:18–21

Yoshida M, Takano Y, Sassaoka T, Izumi T, Kimura A (2003) E-selectin polymorphism associated with myocardial infarction causes enhanced leukocyte-endothelial cell interactions under flow conditions. Areteriooscler Thromb Vasc Biol 23:xx-xx.

Zee RY, Bates D, Ridker PM (2002) A prospective evaluation of the CD14 and CD18 gene polymorphisms and risk of stroke. Stroke 33:892–895

Zheng F, Chevalier JA, Zhang LQ, Vurgil D, Kwiterovich PO (2001) An HphI polymorphism in the E-selectin gene is associated with premature coronary artery disease. Clinical Genetics 59:58–64

Genetic Regulation of Metalloproteinase Activity: Pathogenic and Therapeutic Implications

A. M. Henney

Enabling Science and Technology (Biology), AstraZeneca Pharmaceuticals, Mereside, Alderley Park, Macclesfield, Cheshire, SK10 4TG, UK
e-mail: adriano.henney@astrazeneca.com

1	Introduction to Matrix Metalloproteinases	342
2	Regulation of Matrix Metalloproteinase Activity	344
3	Genetic Variation as a Modulator of Activity	345
4	MMP Genetics and Disease Pathogenesis	348
5	Therapeutic Implications	352
6	Conclusions	353
	References	354

Abstract The remodelling of connective tissue matrices is fundamentally important for normal growth and development, as well as for the repair and maintenance of the functional integrity of tissues. These remodelling processes are tightly controlled to maintain a fine balance between the extracellular matrix-degrading enzymes and their inhibitors and thus regulate tissue function and development. Matrix degradation is mediated by a variety of proteases, but one family of enzymes in particular, the matrix metalloproteinases (MMPs), have connective tissue macromolecules as their primary targets. Wherever and whenever there is a need for connective tissues to be remodelled, a failure to control the process appropriately will present the possibility of its contribution to the advancement of disease. A number of MMPs have been implicated in the pathogenesis of various chronic diseases such as atherosclerosis, aneurysms, arthritis and cancer. These are complex traits where the disease is influenced by a combination of common variation in a constellation of genes and the effect of a wide range of environmental variables. Thus, the mechanisms underlying the development of the disease will be modulated by genetic diversity and the effect this has on an individual's response to environmental challenges such as smoking, diet and exercise. This review focuses on the evidence supporting a role for MMPs in the development and progression of cardiovascular disease, for the potential for MMP gene polymorphisms to modulate disease and the impact they may have on therapy.

Keywords Matrix metalloproteinases · Polymorphisms · Atherosclerosis · Genetics

1
Introduction to Matrix Metalloproteinases

The Matrix metalloproteinases (MMPs, matrixins) constitute a family of closely related enzymes whose primary function is to mediate the degradation and turnover of extracellular connective tissue matrices. Induction and regulation of these processes at appropriate times is essential for the resorption and remodelling of tissues during reproduction, the development of embryos and morphogenesis. The members of the MMP family are key components of the cascade of events that contributes to the routine repair and replacement of worn or damaged matrices in organs and tissues, which is necessary for the maintenance of integrity and function throughout life (Murphy and Reynolds 1993; Birkedal-Hansen et al. 1993; Nagase and Woessner 1999). But their function is now known to extend beyond that of matrix remodelling to include an indirect effect on the control of cellular differentiation and function (Streuli 1999; Sternlicht and Werb 2001). It is increasingly accepted that disruption of the tight control of these essential physiological processes contributes to the pathogenesis of a wide variety of diseases, ranging from arthritis through periodontal disease to cancer and cardiovascular disease. The central role of MMPs in these processes has prompted significant interest in attempts to modulate their activity pharmacologically.

The vertebrate MMP family is composed of more than twenty zinc-dependent endopeptidases, which are generally expressed at low levels in normal adult tissue but are up-regulated during normal and pathological remodelling processes (Nagase 1996; Fini et al. 1998). Table 1 summarizes the mammalian members that have been most widely studied to date, but there are also related MMPs identified in non-vertebrate species, and newer additions to this family which are discussed in the review by Sternlicht and Werb (2001).

All MMPs share common features that allow their classification as a family:

1. Synthesized as latent zymogens requiring activation: this is true for the majority of the members listed in the table, although some can be activated intracellularly by furin.
2. Degrade extracellular matrix components. As Table 1 illustrates, virtually all members are capable of using a wide variety of matrix components as substrates. There are differences in preference between the four major classes, for example the *collagenases* are relatively specific for the interstitial collagens, whereas the *gelatinases* have a preference for denatured and basement membrane collagens and elastin. The *stromelysins*, in addition to a generally broader set of substrate preferences, have the interesting additional capability of activating a variety of latent MMPs. As their name implies, the *membrane-type* MMPs differ from the other family members in that they

Table 1 The family of mammalian matrix metalloproteinases (from Henney et al. 2000, with permission)

Subgroup	Name (alternative)	Number	Chromosomal location	Known substrates
Collagenases	Interstitial collagenase (fibroblast-type collagenase)	MMP-1	11q22.3	Collagen I, II, III, VII, VIII, X, gelatin, aggrecan, versican, PLP, casein, α_1PI, α_2M, ovostatin, nidogen, myelin base protein, pro-TNFα, l-selectin, MMP-2, MMP-9
	Neutrophil collagenase (PMN-type collagenase)	MMP-8	11q21	Collagen I, II, III, V, VII, VIII, X, gelatin, aggrecan, α_1PI, α_2AP, fibronectin
	Collagenase-3	MMP-13	11q22.3	Collagen I, II, III, IV, gelatin, aggrecan, perlecan, tenascin, PAI-2
Gelatinases	Gelatinase A (72-kD type IV collagenase)	MMP-2	16q21	Gelatin, collagen I, IV, V, VII, X, XI, XIV, elastin, fibronectin, aggrecan, versican, PLP, myelin base protein, pro-TNFα, α_1PI, MMP-9, MMP-13, β amyloid
	Gelatinase B (92-kD type IV collagenase)	MMP-9	20q11.2–13.1	Gelatin, collagen IV, V, VII, X, XIV, elastin, aggrecan, versican, PLP, fibronectin, nidogen, α_1PI, myelin base protein, pro-TNFα
Stromelysins	Stromelysin-1	MMP-3	11q22.3	Collagen III, IV, IX and X, PLP, fibronectin, laminin, elastin, gelatin, aggrecan, perlecan, versican, casein, ovostatin, pro-TNFα, α_1PI, α_2M, myelin base protein, MMP-1, MMP-7, MMP-8, MMP-9, MMP-13
	Stromelysin-2	MMP-10	11q22.3	Collagen III, IV, V, gelatin, aggrecan, elastin, casein, fibronectin, PLP, MMP-1, MMP-8
	Stromelysin-3	MMP-11	22q11.2	α_1PI
	No trivial name allocated	MMP-19	12q14	
Membrane-type	MT-MMP-1	MMP-14	14q11–q12	Collagen I, II, III, gelatin, elastin, casein, fibronectin, aggrecan, vitronectin, MMP-2, MMP-13, pro-TNFα, laminin B chain, dermatan sulphate proteoglycan
	MT-MMP-2	MMP-15	16q13–q21	MMP-2, gelatin, tenascin, laminin, fibronectin, nidogen
	MT-MMP-3	MMP-16	8q21–q22.1	MMP-2
	MT-MMP-4	MMP-17	12q24.33	
Not classified	Matrilysin (PUMP-1)	MMP-7	11q21–q22	Collagen IV, X, gelatin, elastin, aggrecan, PLP, fibronectin, laminin, casein, transferrin, pro-TNFα, α_1PI, MMP-1, MMP-2, MMP-9, myelin base protein, entactin
	Macrophage metalloelastase	MMP-12	11q22.2–22.3	Collagen IV, gelatin, elastin1 a$_1$PI, fibronectin, vitronectin, laminin, pro-TNFα, myelin base protein

PLP, proteoglycan link protein; α2M, α2macroglobulin; α1PI, α1proteinase inhibitor.

are tethered to the cell membrane and appear at the cell surface, having been activated intracellularly.
3. Function at neutral pH.
4. Contain a conserved sequence in their catalytic domains (HEXXHXX-GXXH), which is responsible for binding Zinc.
5. Are inhibited by specific tissue inhibitors of metalloproteinases (TIMPs), of which four have been identified.

The reader is referred to specific reviews for more detailed information on the biochemistry of MMPs and related proteases (Werb 1997; Barrett, Rawlings and Woessner 1998; Woessner 1998; Nagase and Woessner 1999; Sternlicht and Werb 2001).

2
Regulation of Matrix Metalloproteinase Activity

Metalloproteinase activity is tightly regulated, firstly at the level of gene expression, then at the point of activation of the latent enzyme and, finally, through inhibition of the active enzymes by the action of specific tissue inhibitors of metalloproteinases (TIMPs) and non-specific inhibitors (e.g. α_2-macroglobulin).

In healthy adult tissues, connective tissue turnover is normally very low and the levels of MMP expression will consequently also be low. However, expression of both MMPs and their inhibitors is inducible in response to a variety of stimuli triggering matrix remodelling (Matrisian 1990; Fini et al. 1998). Analysis of the promoter regions of MMP genes has identified *cis*-acting elements known to regulate expression, for example AP-1 and PEA-3, which interact, respectively, with the Fos/Jun and Ets families of transcription factors. Matrix metalloproteinase expression has been seen to increase in response to cytokines and some growth factors (e.g. IL-1, IL-6, PDGF, FGF), and is down-regulated by hormones and other growth factors (e.g. retinoids, glucocorticoids, TGFβ), all operating at the transcriptional level. Chemical stimuli are not the only transcriptional triggers. Cell–cell and cell–matrix interactions will also modulate MMP expression. For example, the attachment of macrophages to fibronectin via the $\alpha_5\beta_1$ integrin will induce MMP-9 expression (Xie et al. 1998), whilst the interaction of T cells with endothelial cells via VLA-4 and VCAM-1 induces expression of MMP-2 (Romanic and Madri 1994). The transcriptional regulation of MMPs has been reviewed by Fini et al. (1998). These are the primary cellular response mechanisms to signals triggering growth and repair, as well as those associated with complex inflammatory and other disease-associated pathways.

Most of the MMPs are secreted as inactive zymogens, which need to be activated in the pericellular and extracellular environment before they can degrade their substrates (Nagase 1997; Werb 1997). In the case of the membrane-type MMPs, evidence suggests that these are activated inside the cell by furin before migration to the cell surface (Pei and Weiss 1996), as is also the non-membrane type MMP-11 (Pei and Weiss 1995). The precise mechanism for physiological activation of the remainder of the MMP family has been the subject of debate for a number of years, but elegant experiments using gene knockout technology targeting the plasminogen activation system has demonstrated that plasmin is an important activator in vivo (Carmeliet et al. 1997). It appears that the activation mechanism is mediated by urokinase bound to its receptor, rather than tissue plasminogen activator, effectively focusing the activation step close to the cell surface and thus, possibly, helping to limit the extent to which active enzymes are present in the wider matrix. The involvement of plasmin in MMP activation offers an additional level of regulation, in that the cleavage of plasminogen to produce plasmin itself is the subject of a complex cascade.

One of the secreted members of the family, MMP-2, is the exception to the plasmin activation rule. In this case, MMP-2 appears to be activated at the cell surface through the assembly of a ternary complex consisting of the zymogen, the inhibitor TIMP-2 and MMP-14 (Strongin et al. 1995). It is unclear whether other MMPs may also be activated at the surface of the cell by similar, MT-MMP-mediated mechanisms.

Once activated, the potent destructive potential of these enzymes is held in check by specific natural inhibitors, TIMPs, as well as by other non-specific protease inhibitors such as α_2-macroglobulin. While TIMPs do not need any post-translational processing to exhibit their activity, they appear in many cases to be co-regulated with MMPs at the transcriptional level, thus maintaining an appropriate metabolic balance during remodelling events (Gomez et al. 1997; Das et al. 1997). To date, four TIMPs have been identified, encoded by four separate genes: TIMPs 1, 2 and 4 are secreted in a soluble form, whilst TIMP-3 appears to be bound to the extracellular matrix. With the exception of TIMP-4, which is more selective, the remainder of the TIMPs act effectively against all MMPs by forming non-covalent, 1:1 stoichiometric complexes with MMPs and blocking access to their substrate. It has been suggested that MMP-2 and MMP-9 can form complexes with TIMPs in their zymogen forms as well as active forms, whereas collagenase-type and stromelysin-type MMPs only form complexes after exposure of the active site. The evolution, structure and function of TIMPs is reviewed by Brew et al. (2000).

3
Genetic Variation as a Modulator of Activity

The effect of these proteases on the extracellular matrix will depend on the balance that is struck between the degradation of existing matrix components and their replacement by the synthesis of new ones. Given that such tight control exists over the activity of MMPs, it is reasonable to hypothesize that aberrations in these control mechanisms would likely contribute to a disturbance in the natural balance required for remodelling processes, creating conditions which favour excess deposition, or conversely excess removal of matrix proteins. Such imbalances could arise through the effects of naturally occurring, common genetic polymorphisms in regulatory sequences or coding regions, having a direct

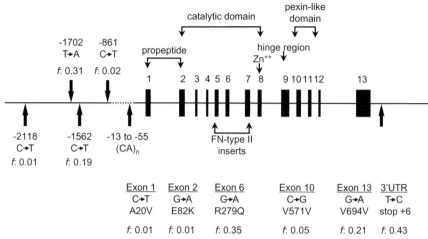

Fig. 1 Schematic representation of the human MMP-9 gene. The location of promoter polymorphisms is indicated by base position 5′ of the start of transcription, together with the nature of the base change and its estimated population frequency (*f*). The coding region variants are shown by exon and amino acid number, and identify the nature of the base change and any consequent amino acid change. *FN*, fibronectin; Zn^{++}, location of zinc-binding domain. (From Henney et al. 2000, with permission)

impact on levels of expression or protein function, and resulting in variations in the dynamics of local tissue remodelling. The combination of such inter-individual genetic variation with the effects of non-genetic or environmental factors contributes to the risk of disease progression, including the potential to suffer acute sequelae. This variability gene concept has been well described in the context of a number of diseases over the last 10 years (Sing and Moll 1990; Humphries et al. 1992; 1998a).

Of the various regulatory steps affecting the action of MMPs, the critical step in the majority of cases is transcription, given that expression of the vast majority of MMPs will only occur when active tissue remodelling is necessary. Studies on the impact of genetic polymorphism on MMP activity were prompted by the discovery that MMP-3 was apparently over-expressed in inflammatory atherosclerotic lesions (Henney et al. 1991). Initial studies focusing on the promoter identified a single base insertion-deletion polymorphism (5A/6A) 1612 bases 5′ of the start of transcription, estimated to have a population frequency of 0.49 (Ye et al. 1995). Using in vitro assays, it was subsequently shown that this stretch of the promoter bound nuclear proteins differentially, depending on the allelic sequence, and that this resulted in differences in expression driven by the two alleles (Ye et al. 1996)

Subsequent studies on MMP-9 (St. Jean et al. 1995; Zhang et al. 1999a) discovered a number of variants in the promoter and coding region, and these are summarized in Fig. 1. All of the polymorphisms described across the 9 kb of the *MMP-9* gene analysed are in tight linkage disequilibrium. Of those variants

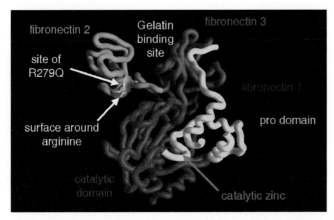

Fig. 2 Crystal structure of the human MMP-9 protein showing the catalytic domain, with the attached propeptide and zinc bound in the catalytic site. The three fibronectin domains are identified emerging from central core of the enzyme, creating a channel which may facilitate the binding of the substrate. Graphical interpretation of the model used the program GRASP (Nicholls et al. 1991, and data taken from Elkins et al. 2002)

identified in the promoter, two appeared to interfere with transcription: the SNP at position –1562 demonstrates allelic modulation of *MMP-9* expression when constructs are expressed in in vitro assays using macrophages (Zhang et al. 1999b); the $(CA)_n$ repeat is a multi-allelic microsatellite polymorphism, the allele frequencies of which are distributed bi-modally (St. Jean et al. 1995). This microsatellite was originally used to map *MMP-9* to *20q11-q13*, but more recently promoter function assays have shown that the length of the repeat affects the strength of nuclear protein binding (Peters et al. 1999; Shimajiri et al. 1999). Further, when expressed in fibroblasts, promoters containing shorter repeat alleles resulted in 60% lower transcriptional activity than those with longer repeat elements (Peters et al. 1999).

Of the variants identified elsewhere in the gene, two were neutral, two were very rare and in the propeptide region, and one appeared in the 3′ UTR. The remaining G→A transition found in exon 6 caused a potentially significant amino acid change, with the substitution of the positively charged arginine for the uncharged glutamine. The location of this change is close to the border of one of the fibronectin-type II domains, structural elements in the MMPs that are unique to the two gelatinases, MMP-2 and MMP-9, and that have been shown to mediate the binding of these enzymes to their substrates. In vitro studies expressing the two variants have not shown any allelic differences in overall enzyme activity, using conventional biochemical assays (unpublished data). However, the potential effect on the dynamics of the interaction between enzyme and substrate can be explored through modelling. Using protein structure information for MMP-9 (Elkins et al. 2002), we have mapped the position of the amino acid variant onto a computer model of the protein (Fig. 2). The interpretation of this model is that the gelatin substrate would likely bind along the length of the channel between the fibronectin domains, potentially bringing it into contact with the site of the amino acid substitution. The change in charge at this site might influence the interaction though; for example, arginine may recognize more readily than glutamine any adjacent glutamic acid or aspartic acid residues in the substrate (unpublished data). This information could not have been derived from looking solely at the gene or protein sequence, and offers support for designing not only more sensitive assays to measure the potential effect of the substitution on the strength of binding between enzyme and substrate, but also small molecules to bind to the active site (Rowsell et al. 2002).

A similar study of polymorphisms across the gene encoding the other gelatinase, MMP-2, has recently been reported (Price et al. 2001). Again, variation in the promoter appeared to have potential functional relevance: the $C \rightarrow T$ transition at $-1,306$, in particular, affected the binding of the transcription factor Sp-1and showed very significant allelic modulation of transcription. In the coding region, many of the variants were neutral, but one caused a non-conservative substitution of glycine for serine (G456S) in the hinge region of the protein. Once again this region is thought to be important in the interaction between enzyme and substrate, and the authors report that the glycine is conserved across species, suggesting that it has an important function.

The analysis of the $5'$ sequences of other MMP genes has identified similar promoter polymorphisms to those described above, again with the apparent potential to affect transcription in vitro. An insertion/deletion (*1G/2G*) SNP in *MMP-1* at position -1607 was identified (Rutter et al. 1998). This variation affected the core sequence for binding the nuclear protein Ets-1, and was shown to modulate transcriptional activity (Rutter et al. 1998). In the *MMP-7* promoter, two SNPs were found, at positions $-181(A/G)$ and -153 (*C/T*), both modulating nuclear protein binding and transcriptional activity in vitro (Jormsjö et al. 2001). The SNP found in the *MMP-12* promoter at position -82 (*A/G*) influenced the binding of the transcription factor AP-1 (Jormsjö et al. 2000), an important regulator of expression of many of the MMP genes. In vitro data not only confirmed allelic differences in transcription of the gene, but also a differential response to insulin and Phorbol 12-myristate 13-acetate, both of which are known to activate transcription through AP-1 (Angel et al. 1987). Analysis of the coding regions and $3'$ UTR in these three MMPs has so far been limited to *MMP-12* and the report of the substitution of an asparagine for serine at position 357 (Joos et al. 2002). The effect on protein function of this non-conservative amino acid substitution is unknown

4
MMP Genetics and Disease Pathogenesis

Wherever and whenever there is a need for connective tissues to be remodelled, a failure to control the remodelling process appropriately will present the possibility of its contribution to the advancement of disease. Given that these mechanisms are fundamental to normal growth, development and repair, there is wide scope for MMPs to affect disease pathogenesis. Many studies have described patterns of expression of the family of MMPs in a wide variety of diseases, suggesting a role for these enzymes in the aberrant tissue remodelling contributing to the pathology. These include their involvement in cartilage destruction in arthritis, the development of aneurysms, chronic lung disease, cancer cell invasion and metastasis, as well as the growth and rupture of atherosclerotic plaques (Henney et al. 1991; Murphy and Reynolds 1993; Birkedal-Hansen et al. 1993, Nagase and Woessner 1999; Segura-Valdez et al. 2000; Imai et al. 2001; Galis and Khatri 2002). The function of MMPs in the context of cardiovascular disease has been comprehensively reviewed in a recent series of thematic articles in *Circulation Research* covering myocardial infarction (Creemers et al. 2001) and heart failure (Spinale 2002), as well as atherogenesis and vascular remodelling (Galis and Khatri 2002).

In general terms, much of the evidence supporting a role for MMPs in these diseases has been accumulated indirectly from the study of isolated diseased tissues, using various techniques to measure levels of expression of the various proteins. But, in recent years, additional in vivo evidence of MMP involvement has been acquired from the detection of elevated levels of these proteins in plasma or serum samples from patients with a variety of diseases, the presumption being that MMPs leech out of the affected tissues and into the circulation (Zucker et al. 1999). Such analyses have detected raised MMP levels in patients with cancer (Jung et al. 1997; Torii, et al. 1997; Lein et al. 2000), rheumatoid and osteoarthritis (Zucker et al. 1994; Naito et al. 1999), multiple sclerosis (Lichtinghagen et al. 1999), acute myocardial infarction (Kai et al. 1998) and abdominal aortic aneurysm (McMillan and Pearce 1999). In the case of chronic lung disease, correlations between disease endpoints and enzyme/inhibitor levels in sputum samples have been observed (Bosse et al. 1999; Cataldo et al. 2000).

Despite this gathering weight of circumstantial evidence supporting a role for MMPs in these diseases, there has been very little to indicate a direct causal relationship. Here, genetics offers an additional tool with which to assess more directly the role of the various members of the MMP family, as well as their activators and inhibitors, in the development and progression of disease. In contrast to the matrix proteins such as the collagens, which MMPs degrade during the remodelling process, there are very few single gene defects causing severe mendelian disorders like osteogenesis imperfecta that unequivocally would ascribe a causal relationship to a specific disease. A notable exception is the *TIMP-3* gene, one of the natural inhibitors of MMP activity, where mutations have been described in patients with the ocular condition Sorsby's fundus dystrophy (Weber et al. 1994). This was the first evidence linking a rare mutation in a gene involved in regulating MMP activity to a disease caused by defective connective tissue remodelling. Before this report, we had speculated for some time that common genetic variants, which disrupt or modulate the cascade controlling MMP activity, may contribute to the pathogenesis of multifactorial diseases in which deficiencies in connective tissue remodelling are a feature. The demonstration of a link between mutations in *TIMP-3* and disease provided a proof of

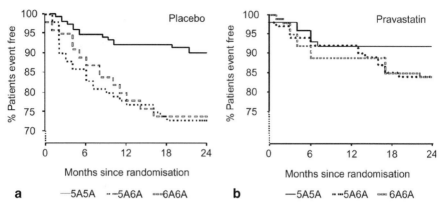

Fig. 3a, b Clinical event-free survival curves for each MMP-3 genotype in the placebo (**a**) and pravastatin (**b**) treatment groups, suggesting that carriers of the *6A* allele had a greater chance of improved event-free survival on pravastatin. (From de Maat et al. 1999, with permission)

principle, albeit in a rare disorder rather than a common complex trait, that correct maintenance of the balance between active enzyme and inhibitor is critical to preserving the function of the extracellular matrix.

The MMP polymorphisms have all been applied to genetic association studies designed to test the hypothesis that allelic variation contributes to the disease phenotype, through the effect of small differences on protein function as described above. In the MMP family, the first study to adopt this approach considered the role of *MMP-3* in the progression of coronary atherosclerosis and myocardial infarction (Ye et al. 1995). Genetic analysis in three clinical trials showed that a *MMP-3* promoter variant was associated with lesion progression as assessed by repeated quantitative coronary angiography (Ye et al. 1995; Humphries et al. 1998b; de Maat et al. 1999). In the REGRESS trial, which included a treatment arm given pravastatin, there was a significant 2.3-fold increased risk of development of angiographically visible new lesions in those homozygous for the *6A* allele relative to other genotypes (de Maat et al. 1999). The association between the *6A* allele and the risk of more rapid disease progression was confirmed in a second independent study (Humphries et al. 1998b). This genotype was associated with lower promoter activity, suggesting a tendency to greater matrix deposition locally in the vessel wall, which would add to the fibrotic mass of vascular lesions, a key element of lesion growth. This view is supported by recent independent evidence that the *6A* homozygous genotype is associated with increased carotid artery wall thickness (Gnasso et al. 2000). Possibly of greater interest and potential significance was the unexpected discovery in the REGRESS study that there was a significant interaction between genotype and pravastatin medication on clinical event-free survival (Fig. 3). This result suggested a potential pharmacogenetic impact of the *MMP-3* promoter variant in patients treated with this statin, which would merit further investigation.

In contrast, a Japanese association study of *MMP-3* and myocardial infarction observed that carrying the *5A* allele, responsible for higher levels of transcription, predisposed individuals to a greater than twofold higher relative risk of myocardial infarction (Terashima et al. 1999), presumably through excess proteolysis resulting in less stable caps, the commonest cause of myocardial infarction (Davies 1995). This potential influence of excess enzyme activity in the vessel wall has also been invoked in a recent study to explain the observation that coronary aneurysms too are associated with the more active *5A* allele (Lamblin et al. 2002). The *5A* allelic effect has been further reinforced in a very recent logistic regression study, which showed a significantly elevated risk of myocardial infarction (odds ratio, 4.7) in 1294 women homozygous for the *5A* allele (Yamada et al., 2002). The authors went on to propose that there might be grounds for using the *MMP-3 5A-1171/6A* polymorphism as a genetic marker of myocardial infarction risk.

A number of case–control studies of atherosclerosis have also been used to explore the potential role of SNPs in *MMP-9* (Zhang et al. 1999b), *MMP-12* (Jormsjö et al. 2000) and most recently *MMP-7* (Jormsjö et al. 2001). The *C-1562T* promoter variant in *MMP-9* has been shown to be significantly associated with severity of atherosclerosis. The *T* allele is linked with higher promoter activity in vitro (Zhang et al. 1999b), and in preliminary studies, with increased plasma levels of the enzyme (unpublished data). In clinical practice, stenoses of 50% or greater in three coronary arteries were found in 26% of the *T* allele carriers compared with only 15% of the *C* homozygotes. This finding is supported by an independent post-mortem study, which analysed coronary lesions in detail. This showed that the *T* allele was a significant risk factor for the formation of complex lesions in all main coronary arteries, and also increased significantly the risk of myocardial infarction (Pöllänen et al. 2001).

In the case of *MMP-12*, the *A-82G* SNP, which is located adjacent to an AP-1 binding site, has been shown to be associated with narrowing of coronary arteries in diabetics with coronary artery disease, an interesting observation given the in vitro data showing differential transcriptional responses to an insulin stimulus (Jormsjö et al. 2000). The promoter SNPs identified in *MMP-7* also influence lumenal diameter in hypercholesterolaemic patients undergoing PTCA using stents (Jormsjö et al. 2001). This observation parallels that seen for *MMP-3* in the REGRESS trial, where allele-specific effects were seen in the mildly hypercholeseterolaemic placebo group, and suggests a possible interaction between lipids and transcription of these MMPs (de Maat et al. 1999).

Preliminary studies have also been carried out on the potential effects of these functional SNPs on aneurysm growth in a cohort of patients with small abdominal aortic aneurysms. From this work it appears that, of the three MMPs studied, the rate of small aneurysm growth over nearly 3 years is influenced by variation in the *MMP-2* promoter (P. Eriksson, personal communication). Polymorphisms in the *MMP-9* gene have also been studied in relation to abdominal aortic aneurysms (St. Jean et al. 1995; Yoon et al. 1999) and intracranial aneurysms (Peters et al. 1999; Yoon et al. 1999) with a positive association only being found in the latter study (Peters et al. 1999).

Finally, some examples of studies in non-cardiovascular disorders offer further support for MMPs contributing to disease pathogenesis. An analysis of polymorphisms in *MMP-1*, *-9* and *-12* to investigate their impact on chronic obstructive lung disease, showed that variants in *MMP-1* (*G-1607/GG*) and *MMP-12* (*N357S*), but not *MMP-9* (*C-1562T*), are associated with the rate of decline in lung function in groups of smokers. The authors concluded that either the MMP polymorphisms themselves are the causative factors in smoking-related lung injury, or they are closely linked to others that are responsible for the damage (Joos et al. 2002), but they did not speculate on the cellular mechanisms that might be affected. Three studies have reported an association between the *MMP-1* promoter SNP and cancer invasiveness. Here, the insertion (*2G*) allele driving increased expression is associated with a significantly increased risk of deeply invasive cutaneous malignant melanoma (Ye et al. 2001), colorectal cancer invasiveness (Ghilardi et al. 2001) and risk of early development of lung cancer (Zhu et al. 2001). The role of this enzyme in facilitating cell movement through connective tissue matrices is as potentially relevant to local vascular remodelling phenomena as it is to its potential function in tumour invasion and metastasis. In that sense, these studies serve to highlight an association between this mechanism and MMP-1 which has not been noted in cardiovascular studies thus far, but which may merit consideration.

The combined weight of all these reports suggests that common genetic variants in this family, which modify rather than drastically disrupt the function of MMPs, may be important in determining the extent to which connective tissue is remodelled to establish and progress complex cardiovascular disease traits. In this way, subtle differences in the control of activity between individuals could contribute to variation not only in disease progression but also to the response to environmental challenge in its broadest sense.

5
Therapeutic Implications

The MMPs have been regarded as potentially tempting pharmacological targets for some time, and much effort has been devoted to the pursuit of small molecule inhibitors for use, in particular, in the treatment of cancer and arthritis. A comprehensive review summarizes the medicinal chemistry information currently available on the various classes of MMP inhibitors that have been tested, and also covers questions related to the feasibility of selective inhibition of specific members of the MMP family (Whittaker et al. 1999).

Evidence does exist in the literature to suggest that blocking MMP activity will have a beneficial effect in cardiovascular disease. For example, vascular over-expression of TIMP-1 (Allaire et al. 1998), dosing with doxycyline (Curci et al. 1998) or various hydroxamate-based inhibitors such as British Biotech's BB94 (Bigatel 1999; Moore et al. 1999) reduced or abolished the progression of aneurysms in experimental models in rodents. Similar studies in ApoE-deficient mouse models of atherosclerosis also showed reduced lesion progression (Rouis et al. 1999), whilst others measuring vascular remodelling in rat carotid arteries (Cowan et al. 2000) further reinforce the message that MMP inhibition is beneficial. George (2000) reviews the therapeutic potential of MMP inhibition in treating atherosclerosis in detail. Most recently there has been a mounting interest in the involvement of the MMP family in ventricular remodelling after myocardial infarction. The therapeutic potential in this area for MMP inhibitors has been fuelled by the discovery that when they are administered acutely in mouse models of infarction, these compounds lead to a reduction in left ventricular dilatation and, hence, a reduced risk of heart failure (Creemers et al. 2001).

All of these experiments serve to reinforce the role of MMPs in disease mechanisms within these models and suggest that the development of inhibitors might be a potential therapeutic option. But, in addition to the normal considerations undertaken during the development of any drug, there are a number of questions of particular relevance in mounting a campaign in this area:

1. Given that these enzymes are critical to a range of normal physiological processes, what side effects might arise from the administration of a systemic inhibitor, and how many of these are likely to cause problems if the drug is given for prolonged periods in different age groups?
2. Accepting that the design of a small molecule inhibitor for use in treating cardiovascular disease is feasible, what hurdles would we encounter during development? So far all the evidence we have of the potential beneficial effect is in animal models; how would we go about designing a proof of concept study to evaluate the real potential in man and minimize the risk of hypothesis failure in later development?
3. If a proof-of-concept study in a particular disease group proved encouraging, how would we design a dose-ranging trial, and what endpoints we could measure and over how long to see an effect? Many of the diseases discussed here are chronic, developing over many years and with few, if any, easily measurable surrogate endpoints.

As with all clinical trials, this information would be needed to assess the feasibility and cost of drug development, and this would need to be considered alongside the commercial estimates of the market potential. These questions are not intended to deter or diminish interest, but more to act a stimulus for discussion in considering the best way to overcome the hurdles in pursuit of MMP inhibition as a therapeutic option.

6
Conclusions

Substantial evidence is available implicating MMPs as major players in the remodelling of connective tissue in a variety of cardiovascular diseases. The precise contribution made by individual members of the family during the development of the diseases remains unclear, but emerging technologies, in particular in genetics, may offer the possibility of unravelling the complexities of these processes and point the way to the best therapeutic targets. Human genetic studies have begun to suggest that some MMP gene markers may have potential as determinants of risk, or as markers of pharmacogenetic interactions, which could be used in the future to segment patient groups, but this will require a lot more investigation.

The available evidence suggests that inhibition of MMP activity may indeed be an option for the treatment of some cardiovascular disorders. However, although this view is influenced partly through the successful action of small molecules, the evidence is based mainly on animal models, which may or may not be appropriate or accurate parallels to the human disease. The only real way to evaluate the potential of MMP inhibitors will be to take a safe and potent inhibitor into an appropriate human proof of concept study, the design of which may be difficult, but perhaps not impossible.

Acknowledgements. The author gratefully acknowledges the generous support of the British Heart Foundation for work on this subject in his former laboratory. Thanks are also due to Drs. Shu Ye and William McPheat for critical comments, and to Tony Slater for the analysis and interpretation of MMP-2 and MMP-9 structural studies.

References

Allaire E, Forough R, Clowes M et al (1998) Local over-expression of TIMP-1 prevents aortic aneurysm degeneration and rupture in a rat model J Clin Invest 102:1413–1420

Angel P, Imagawa M, Chiu R et al (1987) Phorbol ester-inducible genes contain a common cis element recognised by a TPA-modulated trans-acting factor. Cell 49:729–739

Barrett, A J, Rawlings, N D, and Woessner Jr, J F (1998) Handbook of Proteolytic Enzymes, Academic Press, London

Bigatel D A, Elmore J R, Carey D J et al (1999) The matrix metalloproteinase inhibitor BB-94 limits expansion of experimental abdominal aortic aneurysms. J Vasc Surg 29:130–138

Birkedal-Hansen H, Moore WGI, Bodden MK et al (1993) Matrix Metalloproteinases: a review. Critical Reviews in Oral Biology and Medicine, 4:197–250

Bosse M, Chakir J, Rouabhia M et al (1999) Serum matrix metalloproteinase-9: tissue inhibitor of metalloproteinase-1 ratio correlates with steroid responsiveness in moderate to severe asthma. Am J Respir Crit Care Med 159:596–602

Brew K, Dinakarpandian D, Nagase H (2000) Tissue inhibitors of metalloproteinases: evolution, structure and function. Biochim Biophys Acta 1477:267–283

Carmeliet P, Moons L, Lijnen R et al (1997) Urokinase-generated plasmin activates matrix metalloproteinases during aneurysm formation. Nat Genet 17:439–444

Cataldo D, Munaut C, Noel A et al (2000) MMP-2 and MMP-9-linked gelatinolytic activity in the sputum from patients with asthma and chronic obstructive pulmonary disease. Int Arch Allergy Immunol 123:259267

Cowan K N, Jones P L, Rabinovitch, M (2000) Elastase and matrix metalloproteinase inhibitors induce regression, and tenascin-C antisense prevents progression, of vascular disease. J Clin Invest 105:21–34

Creemers E E J M, Cleutjens J P M, Smits J F M et al (2001) Matrix metalloproteinase inhibition after myocardial infarction: A new approach to prevent heart failure? Circ Res 89:201–210

Curci J A, Petrinec D, Liao S, et al (1998) Pharmacologic suppression of experimental abdominal aortic aneurysms: a comparison of doxycycline and four chemically modified tetracyclines. J Vasc Surg 28:1082–1083

Das SK, Yano S, Wang J et al (1997) Expression of matrix metalloproteinases and tissue inhibitors of metalloproteinases in the mouse uterus during the peri-implantation period. Dev Genet 21:44–54

Davies, MJ (1995) Acute coronary thrombosis- the role of plaque disruption and its initiation and prevention. Eur Heart J 16 (suppl L):3–7

de Maat MP, Jukema JW, Ye S et al (1999) Effect of the stromelysin-1 promoter on efficacy of pravastatin in coronary atherosclerosis and restenosis. Am J Cardiol 83:852–856

Elkins PA, Ho YS, Smith WW et al (2002) Structure of the C-terminally truncated human ProMMP9, a gelatin binding matrix metalloproteinase. Acta Crystallogr Sect D 58:1182–1192

Fini ME, Cook JR, Mohan R et al, (1998) Regulation of matrix metalloproteinase gene expression. In: Parks WC, Mecham RP (eds) Matrix Metalloproteinases Academic Press, California, pp 300–356

Galis ZS, Khatri JJ (2002) Matrix metalloproteinases in vascular remodelling and atherogenesis: the good, the bad and the ugly. Circ Res 90:251–262

George SJ (2000) Therapeutic potential of matrix metalloproteinase inhibitors in atherosclerosis. Exp Opin Invest Drugs 9:993–1007

Ghilardi G, Biondi ML, Mangoni J et al (2001) Matrix metalloproteinase-1 promoter polymorphism 1G/2G is correlated with colorectal cancer invasiveness. Clin Cancer Res 7:2344–2346

Gnasso A, Motti C, Irace C et al (2000) Genetic variation in human stromelysin gene promoter and common carotid artery geometry in healthy male subjects. Arterioscler Thromb Vasc Biol 20:1600–1605

Gomez DE, Alonso DF, Yoshiji H et al (1997) Tissue inhibitors of metalloproteinases: structure, regulation and biological functions. Eur J Cell Biol 74:111–122

Henney AM, Wakeley PR, Davies M et al (1991) Localization of stromelysin gene expression in atherosclerotic plaques by in situ hybridization. Proc Nat Acad Sci USA 88:8154–8158

Henney AM Ye S, Zhang B et al (2000) Genetic diversity in the matrix metalloproteinase family. Effects on function and disease progression Proceedings of the Vth Saratoga Conference on Atherosclerosis, Ann NY Acad Sci 902:27–39

Humphries SE, Green FR, Henney AM et al (1992) DNA polymorphisms: the variability gene concept and the risk of coronary artery disease In: Bearn, AG (ed) Genetics of Coronary Heart Disease, Institute of Medical Genetics, University of Oslo, Oslo, pp 123–142

Humphries SE, Fisher RM, Miller G et al (1998a) Predisposing genes, high-risk environments and coronary artery disease: LPL and MMP-3 as examples In Jacotot B, Mathe D, Fruchart JC (eds) Proceedings of the XIth International Symposium on Atherosclerosis Elsevier Science, Singapore: pp 1055–1059

Humphries SE, Luong LA, Talmud PJ et al (1998b) The 5A/6A polymorphism in the promoter of the stromelysin-1 (MMP-3) gene predicts progression of angiographically determined coronary artery disease in men in the LOCAT gemfibrozil study, Lopid Coronary Angiography Trial Atherosclerosis 139:49–56

Imai K, Dalal SS, Chen ES et al (2001) Human collagenase (matrix metalloproteinase-1) expression in the lungs of patients with emphysema. Am J Respir Crit Care Med 163:786–791

Joos L, He, J-Q Shepherdson M et al (2002) The role of matrix metalloproteinase polymorphisms in the rate of decline in lung function. Hum Mol Genet 11:569–576

Jormsjö S, Ye S, Moritz J et al (2000) Allele-specific regulation of matrix metalloproteinase-12 gene activity is associated with coronary artery luminal dimensions in diabetic patients with manifest coronary artery disease. Circ Res 86:998–1003

Jormsjö S, Whatling C, Walter D H et al (2001) Allele-specific regulation of matrix metalloproteinase-7 promoter activity is associated with coronary artery luminal dimensions among hypercholesterolemic patients. Arterioscler Thromb Vasc Biol 21;1834–1839

Jung K, Nowak L, Lein M et al (1997) Matrix metalloproteinases 1 and 3, tissue inhibitor of metalloproteinase-1 and the complex of metalloproteinase-1/tissue inhibitor in plasma of patients with prostate cancer. Int J Cancer 74:220–223

Kai H, Ikeda H, Yasukawa H et al (1998) Peripheral blood levels of matrix metalloproteinases-2 and -9 are elevated in patients with acute coronary syndromes. J Am Coll Cardiol 32:368–372

Lamblin N, Bauters C, Hermant X et al (2002) Polymorphisms in the promoter regions of MMP-2, MMP-3, MMP-9 and MMP-12 genes as determinants of aneurysmal coronary artery disease. JACC 40:43–48

Lein M, Jung K, Laube C et al (2000) Matrix-metalloproteinases and their inhibitors in plasma and tumor tissue of patients with renal carcinoma. In J Cancer 85:801–804

Lichtinghagen R, Seifert T, Kracke A et al (1999) Expression of matrix metalloproteinase-9 and its inhibitors in mononuclear blood cells of patients with multiple sclerosis. J Neuroimmunol 99:19–26

Matrisian, LM (1990) Metalloproteinases and their inhibitors in matrix remodelling. Trends Genet 6:121–125

McMillan WD, Pearce WH (1999) Increased plasma levels of metalloproteinase-9 are associated with abdominal aortic aneurysms. J Vasc Surg 29:122–127

Moore G, Liao S, Curci JA et al (1999) Suppression of experimental abdominal aortic aneurysms by systemic treatment with a hydroxamate-based matrix metalloproteinase inhibitor (RS 132908) J Vasc Surg 29:522–532

Murphy G, Reynolds J (1993) Extracellular matrix degradation In: Royce, PM & Steinman, B (eds) Connective tissue and its heritable disorders Wiley-Liss, Inc, New York, pp 287–316

Nagase H (1996) In: Hooper N (ed) Zinc metalloproteases in health and disease Taylor and Francis, London, pp 153–204

Nagase, H (1997) Activation mechanisms of matrix metalloproteinases Biol Chem 378:151–160

Nagase H, Woessner Jr JF (1999) Matrix Metalloproteinases. J Biol Chem 274:21491–21494

Naito K, Takahashi M, Kushida K et al (1999) Measurement of matrix metalloproteinases (MMPs) and tissue inhibitor of metalloproteinases-1 (TIMP-1) in patients with knee osteoarthritis: comparison with generalized osteoarthritis Rheumatology (Oxford) 38:510–515

Nicholls A, Sharp KA, Honig B (1991) Protein folding and association: insights from the interfacial and thermodynamic properties of hydrocarbons. Proteins 11:281–296

Pei D, Weiss SJ (1995) Furin-dependent intracellular activation of the human stromelysin-3 zymogen. Nature 375:244–247

Pei D, Weiss SJ (1996) Transmembrane-deletion mutants of the membrane-type matrix metalloproteinase-1 process progelatinase A and express intrinsic matrix-degrading activity. J Biol Chem 271:9135–9140

Peters DG, Kassam A, St Jean PL et al (1999) Functional polymorphism in the matrix metalloproteinase-9 promoter as a potential risk factor for intracranial aneurysm. Stroke 30:2612–2616

Pöllänen PJ, Karhunen PJ, Mikkelsson J et al (2001) Coronary artery complicated lesion area is related to functional polymorphism of matrix metalloproteinase 9 gene: An autopsy study. Arterioscler Thromb Vasc Biol 21:1446–1450

Price SJ, Greaves DR, Watkins, H (2001) Identification of novel, functional genetic variants in the human matrix metalloproteinase-2 gene. J Biol Chem 276:7549–7558

Romanic AM, Madri JA (1994) The induction of 72-kD gelatinase in T cells upon adhesion to endothelial cells is VCAM-1 dependent. J Cell Biol 125(5):1165–1178

Rouis M, Adamy C, Duverger N et al (1999) Adenovirus-mediated over-expression of tissue inhibitor of metalloproteinase-1 reduces atherosclerotic lesions in apolipoprotein E-deficient mice. Circulation 100:533–540

Rowsell S, Hawtin P, Minshull CA et al (2002) Crystal structure of human MMP-9 in complex with a reverse hydroxamate inhibitor. J Mol Biol 319:173–181

Rutter JL, Mitchell TI, Buttice G et al (1998) A single nucleotide polymorphism in the matrix metalloproteinase-1 promoter creates an Ets binding site and augments transcription Cancer Res 58:5321–5325

St. Jean PL, Zhang XC, Hart BK et al (1995) Characterisation of a dinucleotide repeat in the 92 kDa type IV collagenase gene (CLG4B), localisation of CLG4B to chromosome 20 and the role of CLG4B in aortic aneurysmal disease. Ann Hum Genet 59:17–24

Segura-Valdez L, Pardo A, Gaxiola M et al (2000) Upregulation of gelatinases A and B, collagenases 1 and 2 and increased parenchymal cell death in COPD Chest 117:684–694

Shimajiri S, Arima N, Tanimoto A et al (1999) Shortened microsatellite d(CA)21 sequence down-regulates promoter activity of matrix metalloproteinase 9 gene. FEBS Lett 455:70–74

Sing CF, Moll PP (1990) Genetics of atherosclerosis. Ann Rev Genet 24:171–187

Spinale FG (2002) Matrix metalloproteinases: Regulation and dystregulation in the failing heart 90:520–530

Sternlicht MD, Werb, Z (2001) How matrix metalloproteinases regulate cell behaviour. Ann Rev Cell Dev Biol 17:463–516

Streuli, C (1999) Extracellular matrix remodelling and cellular differentiation. Curr Op Cell Biol 11:634–640

Strongin AY, Collier I, Bannikov G et al (1995) Mechanism of cell surface activation of 72-kDa type-IV collagenase- isolation of the activated form of the membrane metalloproteinase. J Biol Chem 270:5331–5339

Terashima M, Akita H, Kanazawa K et al (1999) Stromelysin promoter 5A/6A polymorphism is associated with acute myocardial infarction. Circulation 99:2717–2719

Torii A, Kodera Y, Uesaka K et al (1997) Plasma concentration of matrix metalloproteinase 9 in gastric cancer. Br J Surg 84:133–136

Weber BH, Vogt G, Pruett RC et al (1994) Mutation in the tissue inhibitor of metalloproteinases-3 (TIMP3) in patients with Sorsby's fundus dystrophy. Nat Genet 8:352–356

Werb Z (1997) ECM and cell surface proteolysis: regulating cellular ecology. Cell 91:439–442

Whittaker M, Floyd CD, Brown P, et al (1999) Design and therapeutic application of matrix metalloproteinase inhibitors, Chem Rev 99:2735–2776

Woessner Jr J F (1998) The matrix metalloproteinase family In: Parks, W C, Mecham, R P (eds) Matrix Metalloproteinases Academic Press, California, pp 1–14

Xie B, Laouar A, Huberman E (1998) Fibronectin-mediated cell adhesion is required for induction of 92-kDa Type IV collagenase/gelatinase (MMP-9) gene expression during macrophage differentiation The signaling role of protein kinase C-β. J Biol Chem 273:11576–11582

Yamada Y, Izawa H, Ichihara S et al (2002) Prediction of the risk of myocardial infarction from polymorphisms in candidate genes N Eng J Med 347:1916–1923

Ye S, Watts GF, Mandalia S, et al (1995) Genetic variation in the human stromelysin promoter is associated with progression of coronary atherosclerosis. British Heart Journal, 73:209–215

Ye S, Eriksson P, Hamsten A, et al (1996) Progression of coronary atherosclerosis is associated with a common genetic variant of the human stromelysin-1 promoter which results in reduced gene expression. J Biol Chem 271:13055-13060

Ye S, Dhillon S, Turner SJ et al (2001) Invasiveness of cutaneous malignant melanoma is influenced by matrix metalloproteinase 1 gene polymorphism Cancer Res 61:1296-1298

Yoon S, Tromp G, Vongpunsawad S et al (1999) Genetic analysis of MMP3, MMP9, and PAI-1 in Finnish patients with abdominal aortic or intracranial aneurysms, Biochem Biophys Res Commun 265:563-568

Zhang BP, Henney AM, Eriksson P et al (1999a) Genetic variation at the matrix metalloproteinase-9 locus on chromosome 20q122-131. Human Genetics 105:418-423

Zhang BP, Ye S, Hermann SM et al (1999b) Functional polymorphism in the regulatory region of the gelatinase B gene in relation to severity of coronary atherosclerosis. Circulation 99:1788-1794

Zhu Y, Spitz M, Lei L et al (2001) A single nucleotide polymorphism inthe promoter of matrix metalloproteinase-1 enhances lung cancer susceptibility. Cancer Res 61:7825-7829

Zucker S, Lysik RM, Zarrabi MH et al (1994) Elevated plasma stromelysin levels in arthritis. J Rheumatol 21:2329-2333

Zucker S, Hymowitz M, Conner C et al (1999) Measurement of matrix metalloproteinases and tissue inhibitors of metalloproteinases in blood and tissues Clinical and experimental applications. Ann N Y Acad Sci 878:212-227

Current Perspectives on Gene and Cell-Based Therapies for Myocardial Protection, Rescue and Repair

L. G. Melo[1,2] · A. S. Pachori[1] · D. Kong[1] · V. J. Dzau[1]

[1] Department of Medicine, Brigham and Women's Hospital and Harvard Medical School, 75 Francis Street, Boston, MA 02115, USA
[2] Department of Physiology, Queen's University, 20 Stuart Street, Kingston, Ontario, K7L 3N6, Canada
e-mail:luis.melo@usask.ca

1	Introduction	360
2	**Strategies and Tools for Genetic Manipulation of the Myocardium**	361
2.1	Strategies for Genetic Manipulation of the Myocardium	361
2.2	Selection of Therapeutic Target, Vector and Delivery Strategy.	362
2.3	Nonviral Vectors	363
2.4	Viral Vectors	365
2.5	Cell-Mediated Gene Delivery	366
3	**Gene Therapy for Myocardial Protection**	367
3.1	Targets for Gene Therapy	367
3.2	Gene Therapy for Protection from Ischemia and Reperfusion Injury	367
3.3	Gene Therapy for Hypertension, Atherosclerosis and Thromboresistance . . .	371
3.3.1	Gene Therapy for Hypertension	372
3.3.2	Gene Therapy for Atherosclerosis, Thromboresistance and Plaque Stabilization	373
3.4	Gene Therapy for Restenosis and Vascular Proliferative Disease	375
4	**Gene Therapy for Myocardial Rescue**	377
4.1	Gene Therapy for Myocardial Ischemia	377
4.2	Cell-Based Therapy for Myocardial Ischemia	380
4.3	Gene Therapy for Rescue of Contractile Function	381
4.4	Gene Therapy for Myocardial Hypertrophy and Remodeling	383
4.5	Gene Therapy for Inherited and Congenital Heart Disease.	384
4.6	Cell-Based Therapy for Myocardial Regeneration.	386
5	**Perspectives and Future Directions**	389
References		390

Abstract Despite significant therapeutic advances, heart disease remains the prevalent cause of premature death across all age and racial groups, accounting for a significant proportion of all hospital admissions and putting enormous financial strain on health delivery systems. Recent developments in the understanding of the molecular mechanisms of myocardial disease have led to the identification of novel therapeutic targets, and the availability of efficient cardiotropic vectors offers the opportunity for the design of gene therapies for both protection and rescue of the myocardium. Genetic therapies have been devised to treat complex diseases such as myocardial ischemia, hypertension, atherosclerosis, restenosis and inherited myopathies in various animal models. Some of these experimental therapies have made a successful transition to clinical trial and are now being considered for use in human patients. The recent isolation of regeneration-competent endothelial and cardiomyocyte precursor cells from adult bone marrow provides the opportunity for repair of the damaged heart using autologous cell transplantation. Cell-based therapies may have potential application in neovascularization and regeneration of ischemic and infarcted myocardium, in blood vessel reconstruction and in bioengineering of artificial organs and vascular prostheses. With advances in the field occurring at a rapid pace, we can expect the development of vectors and delivery methods with enhanced safety, efficacy and specificity. The advent of genomic screening technology will allow not only the identification of novel therapeutic targets, but will also facilitate the detection of disease-causing polymorphisms and permit the design of individualized gene and cell-based therapies.

Keywords Gene therapy · Myocardial rescue · Stem cells

1
Introduction

Despite significant advances in the clinical management of cardiovascular disease, acute myocardial infarction (MI) and heart failure (HF) due to coronary artery disease (CAD), cardiomyopathy and systemic vascular disease remain the prevalent causes of premature death across all age and racial groups (Kannel and Belanger 1991). The complexity of the pathological processes leading to heart disease and the lack of specific predictive markers has been a major impediment to the development of effective preventive therapies, despite the identification of various risk factors and sensitive risk assessment technologies (Stein 2002; Wilson et al. 1998; D'Agostino et al. 2000). Consequently, the focus has been on the design of so-called rescue treatments for overt symptoms of the disease, such as hyperlipidemia, myocardial ischemia, left ventricular pump failure and hemodynamic overload (McMurray and Pfeffer 2002). Although these therapies have, undeniably, improved the clinical outlook for patients afflicted by MI and HF, morbidity and mortality associated with these diseases remain high, indicating the need for more effective treatments.

The current availability of efficient cardiotropic vector systems such as adeno-associated virus (AAV) (Robbins and Ghivizzani 1998; Monahan and Samulski 2000) and the recent identification of several gene targets associated with heart disease (Colucci 1997; Givertz and Colucci 1998) offer opportunities for the design of gene therapies for myocardial protection and rescue. The ability of AAV to confer long-term and stable protein expression with a single administration of the therapeutic gene (Svensson et al. 1999; Kaplitt et al. 2000; Kimura et al. 2001) renders them ideally suited for delivery of therapeutic genes such as antioxidant, proangiogenic or contractility-enhancing genes to patients afflicted or at risk of developing CAD and HF. In addition, the repair and vascularization of injured myocardium using autologous cell transplantation may be possible with the recent identification and isolation of cardiomyocyte and endothelial progenitor cells from adult bone marrow and peripheral blood (Asahara et al. 1997; Kocher et al. 2001, Makino et al. 1999; Jackson et al. 2001). Such experimental gene and cell-based strategies for myocardial protection, rescue and regeneration are being intensely pursued by various groups and several small-scale trials support the feasibility of these approaches.

In this chapter, we review the major advances in gene and cell-based therapies for heart disease, with emphasis on strategies for protection and rescue of the ischemic, their clinical feasibility and a perspective on future developments in the field. We will highlight the breakthroughs, the challenges in making the transition from preclinical evaluation to clinical application and the opportunities lying ahead in this exciting and growing field.

2
Strategies and Tools for Genetic Manipulation of the Myocardium

2.1
Strategies for Genetic Manipulation of the Myocardium

A wide selection of therapeutic strategies, vectors and delivery methods are available for genetic manipulation of the myocardium with variable degrees of efficiency (Li et al. 2000; Akhtar et al. 2000; Morishita et al. 1998; Robbins and Ghivizzani 1998). The most common gene therapy strategy for the myocardium involves the exogenous delivery and expression of genes whose endogenous activity may either be defective or attenuated due to a mutation or a pathological process. Such gain-of-function gene transfer strategies have been widely used with a variety of therapeutic genes, including proangiogenic and survival factors (Losordo et al. 1998; Matsui et al. 2001) antioxidant enzymes (Melo et al. 2002) and anti-inflammatory cytokines (Brauner et al. 1997). Gene blockade strategies have also been devised for the inhibition of genes that may be involved in the development of heart disease. Acute inhibition of transcription and translation can be achieved by treatment with short single-stranded antisense oligodeoxynucleotides, ribozymes and more recently, using RNA interference technology. (Akhtar et al. 2000; Kimura et al. 2001; Simons et al. 1992; Mann et al. 1997; Hannon 2002). Inhibition of transcription factor DNA binding using double-stranded decoy oligonucleotides containing DNA consensus binding sequences for several transcriptional factors has also been employed to inhibit the transactivating activity of target transcription factors (Morishita et al. 1995, 1997, 1999). In many instances, short-term inhibition (loss of function) of a pathogenic gene is sufficient to prevent the development of disease. For example, the inhibition of cell-cycle regulatory proteins using decoy oligonucleotides was shown to prevent neointimal hyperplasia and subsequent restenosis following balloon angioplasty or bypass grafting (Morishita et al. 1995).

2.2
Selection of Therapeutic Target, Vector and Delivery Strategy

The choice of therapeutic target, vector and delivery strategy is, to a large extent, governed by the pathological features of the disease, the putative role of the target gene(s) in the pathophysiological process, and the timing of intervention (Isner 2002). The efficiency of gene transfer to the myocardium is highly dependent on the type of vector, the route and the dosage and volume of delivery of the genetic material (Alexander et al. 1999; Wright et al. 2001). The permissiveness and cycling status of the target cells plays a central role in the selection of vector. For example, for terminally differentiated cells such as adult cardiomyocytes, only vectors capable of transferring genetic material to quiescent cells should be used. Other vector characteristics such as the capacity to accommodate the transgene, ability to integrate into the host genome and immunogenicity should also be taken into consideration. The pathophysiology of the disease dictates whether transient or long-term transgene expression is warranted. Thus, vector systems capable of integrating into the genome of the host and providing sustained expression should be used for gene transfer in conditions requiring prolonged transgene expression, whereas non-integrating vectors should be used for conditions requiring transient expression.

With regard to the route of administration, intracoronary delivery of the therapeutic material is favored for global myocardial diseases such as heart failure and cardiomyopathy, as well as for regional ischemia via a patent artery. The selectivity of coronary endothelium and the barrier imposed by the basement membrane may restrict the diffusion of some vectors, thus limiting distribution and uptake of the therapeutic transgene. On the other hand, localized delivery of the therapeutic material by intramyocardial injection may be an acceptable method for regional delivery to areas of myocardial disease with ischemia or infarction. This approach has been used for the delivery of angiogenic and cytoprotective genes to ischemic myocardium (Losordo et al. 1999; Isner 2002; Melo et al. 2002). The major shortcoming of direct injection is the fact that transgene expression is restricted to the area surrounding the site of injection. In some cases, this may require multiple injections to adequately cover the affected area.

A variety of catheter types have been developed for both intracoronary and intramyocardial delivery with the assistance of *trans*-esophageal echocardiographic guiding and mapping techniques (Sylven et al. 2002; Herttuala and Martin 2000; Isner 2002). Other methods, such as pericardial injection and retroperfusion have been employed in myocardial gene transfer with limited success, and consequently are not widely used (Fromes et al. 1999; Boekstegers et al. 2000; Herttuala and Martin 2000).

The overall safety and specificity of gene transfer protocols may be enhanced by incorporating regulatory elements that can direct tissue-specific expression as well as regulated expression of the transgene in response to underlying pathophysiological cues such as hypoxia, oxidative stress or inflammation (Prentice et al. 1997; Shibata et al. 2000; Nicklin et al. 2001). This degree of physiological control of transgene expression would allow the therapeutic protein to be produced only where and when needed, and could avert potential biological and ethical problems associated with nonregulated constitutive transgene expression, such as cytotoxicity and germ cell line transmission (Lee et al. 2000).

2.3
Nonviral Vectors

Gene transfer vectors may be classified under three broad categories as nonviral, viral and cell-based (Table 1). Nonviral vectors include naked plasmids, cationic liposome and hybrid formulations, synthetic peptides and several physical methods (Wright et al. 2001; Song et al. 1997; Labhasetwar et al. 1998; Cartier and Reszka 2002, Harrison et al. 1998; Mann et al. 1999). Myocardial gene transfer efficiency using nonviral vectors is low due to rapid degradation of the vector, resulting in transient transgene expression. Nevertheless, there have been reports where naked plasmid-mediated gene transfer into the myocardium led to a sustained therapeutic effect (Losordo et al. 1998; Shyu et al. 2002). The efficiency of plasmid gene transfer can be increased by encapsulating the plasmid in neutral liposomes fused to the viral coat of the Sendai virus (hemagglutaning virus of Japan, HVJ) (Dzau et al. 1996), but transgene expression with this vector system is transient, rendering it unsuitable for use in chronic heart disease.

Electroporation has been used for transfer of naked DNA into embryonic chick hearts ex vivo with moderate efficiency (Harrison et al. 1998), but this protocol is impractical for myocardial gene transfer in humans. Application of nondistending pressure in an enclosed environment has been used to deliver oligonucleotides ex vivo to the heart (Mann et al. 1999) and vein grafts (Poston et al. 1998), highlighting a potential application of this technique for genetic engineering of blood vessels and other organs in preparation for transplantation. Recently, several groups have reported that application of ultrasound at the time of gene delivery enhances transgene uptake significantly (Shohet et al. 2000: Beeri et al. 2002; Schratzberger et al. 2001), suggesting that this could be used as an adjunctive in myocardial gene transfer protocols. Other nonviral methods of gene transfer such as cationic liposomes, calcium phosphate and particle bombardment have shown limited efficacy in myocardial gene therapy (Li and Huang 2000). A promising new delivery technology uses synthetic peptide carriers containing a nuclear localization signal to facilitate nuclear uptake of the target cDNA (Cartier and Reszka 2002). These peptide-DNA heteroplexes are recognized by intracellular receptor proteins and imported into the nucleus, where the target cDNA is transcribed.

Table 1 Vectors used for transfer and manipulation of genetic material in cardiovascular tissues

Vector	Chromosomal integration	Transfer efficiency in vivo	Onset of transgene expression	Sustainability of therapeutic effect	Level of expression	Target cells	Host immune response	Potential risks
Nonviral								
Cationic liposomes	No	+	Rapid	Short	+	Quiescent and dividing	+	Cytotoxicity
HVJ-liposomes	No	+++	Rapid	Short	++	Quiescent and dividing	+	Cytotoxicity
Naked plasmid	No	+	Moderate	Short	+	Quiescent and dividing	+	Cytotoxicity
Viral								
Retrovirus	Yes	++	Rapid	Life-long	++	Dividing	+	Cytotoxicity oncogenesis
Lentivirus	Yes	+++	Rapid	Life-long	+++	Quiescent and dividing	+	Cytotoxicity viral mutation
Adenovirus	No	+++++	Rapid	Moderate	+++++	Quiescent and dividing	++++	Cytotoxicity viral mutation
Adeno-associated virus	Yes	+++	Slow	Life-long	+++	Quiescent and dividing	+	Oncogenesis viral mutation
Herpes simplex virus	No	+++	Moderate	Long	+++	Quiescent and dividing	+++	Cytotoxicity viral mutation
Alphavirus	No	++	Very rapid	Short	+++	Quiescent and dividing	+++	Cytotoxicity viral mutation

2.4
Viral Vectors

Recombinant viruses have become the preferred vectors for myocardial gene transfer because they can deliver genetic material into cells with higher efficiency than nonviral vectors (Robbins and Ghivizzani 1998; Mah et al. 2002), and some are capable of sustaining expression of the therapeutic gene for longer periods of time. Unfortunately, a robust immune reaction may be triggered by the host in response to the viral proteins synthesized by the vector, which may reduce the efficiency of gene transfer and the sustainability of transgene expression (Krasnyhk et al. 2000). Furthermore, although the viral vectors used in gene therapy are replication-deficient, there is the possibility, albeit remote, that these vectors may revert to replication proficiency, thus raising safety concerns about biological hazards such as oncogenesis and insertional mutagenesis (Mah et al. 2002).

Adenoviruses are the most widely used viral vectors (for review see Krasnykh et al. 2000). These viruses can transduce a wide variety of myocardial cell types and can accommodate large DNA inserts. The vector infects both dividing and terminally differentiated cells. However, the cytotoxicity associated with induction of the immune response and the episomal localization of the viral genomes results in rapid loss of transgene expression even in the absence of cell division (Krasnyckh et al. 2000). A new generation of gutted adenoviral vectors has been developed in which the host inflammatory response is highly attenuated by removing all of the adenoviral coding sequences (Hartigan-O'Connor et al. 1999). These adenoviral vectors can accommodate very large DNA fragments and may be useful for delivering multiple genes.

Adeno-associated virus (AAV) has emerged as the vector of choice for myocardial gene transfer because of its high myocardial tropism and ability to stably transduce terminally differentiated myocytes with high efficiency (Kaplitt et al. 2000; Svensson et al. 1999). Intramyocardial delivery of AAV is more efficient than intracoronary delivery, but the efficiency of the latter method can be improved by transient permeabilization of the endothelium with histamine. The vector is poorly immunogenic (Chirmule et al. 1999), minimizing inflammatory damage. The major limitation of the vector is its inability to accommodate large DNA inserts (transgene size is restricted to 4 kb or less) (Monahan and Samulski 2000). Trans-splicing between two separate AAV vectors has recently been used as a strategy for delivery of genes greater than 4 kb (Yan et al. 2000).

RNA-based retroviral and lentiviral vectors have not found widespread application in myocardial gene transfer protocols for several biological and technical reasons (Hu and Pathak 2000; Daly and Chernajovski 2000). These vectors integrate into the host genome leading to the possibility of long-term transgene expression (Hu and Pathak 2000). However, retroviral integration requires cell division, rendering these vectors inefficient in transduction of adult cardiomyocytes. Furthermore, retrovirally-delivered transgenes are prone to transcription silencing, which may significantly shorten the duration of transgene expression. Production of high-titer retrovirus preparations is difficult, but recent improvements in packaging systems, such as the use of pseudotyped viral coats incorporating the vesicular stomatitis virus G-protein (VSV-G), have greatly improved the stability of the viral particles and have allowed transduction of a wider spectrum of cell types with relatively high efficiency (Daly and Chernajovski 2000). Lentiviruses are relative newcomers in cardiovascular gene therapy (Trono 2000). In contrast to the oncoretroviruses, human immunodeficiency virus (HIV-1)-related retroviruses can infect both dividing and quiescent cells. Moderate transgene expression was recently seen in the heart following transduction with a pseudotyped lentivirus (Sakoda et al. 1999; Zhao et al. 2002).

Other viral vector systems currently used for gene transfer such as herpes simplex viruses (HSV) and alphaviruses have had limited application in myocardial gene transfer. The ability of HSV-based vectors to accommodate very large DNA fragments provides an advantage for the transfer of very large genes such as dystrophin or sarcoglycans for treatment of inherited cardiomyopathies (Coffin et al. 1996). Alphaviruses are positive strand RNA viruses based on the Semliki Forest virus (SFV) and Sendibis virus (Schlesinger 2001). These viruses have recently been used for very rapid and efficient transduction of several cells and tissues in vitro (Datwyler et al. 1999). These viruses are capable of expressing transgenes within 24 h of transduction in the heart with minimal cytotoxicity, suggesting their potential application for gene manipulation in acute myocardial disease such as myocardial infarction.

2.5
Cell-Mediated Gene Delivery

A number of cell types have also been used as vectors for delivery of genetic material to tissues. The recent identification and isolation of endothelial and cardiomyocyte precursor stem cells from adult bone marrow and peripheral blood (Makino et al. 1999; Asahara et al. 1997) provides a nondepleting, self-renewing autologous cell source that can simultaneously be used as substrate for regeneration and reconstruction of injured myocardium and blood vessels and as vehicles for delivery of therapeutic genes. For example, the cells could be engineered ex vivo to express cytoprotective and/or proangiogenic genes that would promote survival of the grafted cells and neovascularization of the infarcted myocardium (Iwaguro et al. 2002). Macrophages, erythrocytes and vascular endothelial cells have also been successfully transduced ex vivo with retroviral vectors and used as shuttles for efficient delivery of therapeutic genes into tissues (Griffiths et al. 2001; Magnani et al. 2002). Macrophages genetically engineered to express protective genes under endogenous regulation by hypoxia may have potential application for targeted delivery of genes in myocardial ischemia.

3
Gene Therapy for Myocardial Protection

3.1
Targets for Gene Therapy

Several genes have emerged as potential targets for gene therapy for myocardial disease (Table 2). In the setting of myocardial protection, the overexpression of cytoprotective and survival genes, such as antioxidant enzymes (Woo et al. 1998; Okudo et al. 2001; Melo et al. 2002; Li et al. 2001), antiapoptotic proteins (Brocheriou et al. 2000), protein kinase B/Akt (Miao et al. 2000; Matsui et al. 2001) and/or the inhibition of pro-inflammatory cytokines (Brauner et al. 1997), pro-apoptotic (Holly et al. 1999) and pro-oxidant (Fukui et al. 2001) genes, have emerged as potential therapeutic targets for cardioprotection from studies in various animal and cellular models of myocardial ischemic injury.

Gene manipulations yielding overexpression of vasodilator substances (Lin et al. 1995, 1997) and thrombolytic proteins (Nishida et al. 1999; Waugh et al. 1999) or inhibition of vasoconstrictor pathways (Makino et al. 1999; Wang et al. 1999) have also shown protective effects against hypertension and atherosclerosis-induced myocardial injury. Gene therapy strategies for plaque stabilization and inhibition of platelet adhesion may also be of benefit in reducing the occurrence of thrombotic events and myocardial infarction. Potential therapeutic targets include the inhibition of proinflammatory mediator CD40/CD40L signaling (Lee et al. 1999) and the glycoprotein IIb/IIIa receptor (Kereiakes 1998; Kingma et al. 2000).

Other potential strategies in the postinfarction period include inhibition of genes involved in regulation of ventricular remodeling and chamber dilatation such as the matrix metalloproteinases (MMPs) that participate in extracellular matrix degradation (Spinale 2002). Strategies aimed at modulating the activity of proliferation-regulating genes in the vessel wall have shown efficacy in limiting neointimal hyperplasia (Morishita et al. 1995; Mann et al. 1997; Kibbe et al. 2000), suggesting that these approaches may yield potential as adjunct therapy for prevention of restenosis and graft atherosclerosis in cases where surgical revascularization or percutaneous transluminal angioplasty are indicated for treatment for myocardial ischemia. Inhibition of proinflammatory cytokines and adhesion molecules may find application as immunosuppressive therapy in acute myocardial infarction and in transplantation (Stepkowski 2000; Feeley et al. 2000; Poston et al. 1999; Brauner et al. 1997).

3.2
Gene Therapy for Protection from Ischemia and Reperfusion Injury

The continuum of myocardial injury that is initiated by a coronary ischemic event and perpetuated by reperfusion (I/R injury) may be clinically manifested in patients undergoing thrombolytic therapy following an acute coronary epi-

Table 2 Targets for gene-based therapy for acquired, inherited and congenital heart disease

Strategy	Therapeutic target	Genetic manipulation	Vector	Application
Protection/prevention				
Antioxidant enzymes	HO-1, SOD, catalase, GPx	Overexpression	AAV, LV	CAD, MI
Heat shock proteins	HSP70, HSP90, HSP27	Overexpression	AAV, LV	CAD, MI
Anti-inflammatory	I-CAM, V-CAM, NF-κB, TNF-α	Inhibition	AS-ODN, Decoy ODN, AAV-AS-ODN, RV-AS-ODN	Graft atherosclerosis transplantation
Survival genes	Bcl-2, Akt	Overexpression	AAV, LV	CAD, MI, HF
Pro-apoptotic genes	Bad, p53, Fas ligand	Inhibition	AS-ODN, Decoy ODN, AAV-AS-ODN	MI, HF
Coronary vessel tone	eNOS, adenosine (P1, P3) receptors	Overexpression	RV, AAV(?)	CAD, HF
Rescue				
Pro-angiogenic genes	VEGF, FGF, HGF	Overexpression	AAV	CAD, MI, HF
Contractility	β-Adrenergic receptors, SERCA 2A, V1 receptor	Overexpression	AAV	HF
	BARK, Phospholamban	Inhibition	AAV	HF
Plaque stabilization	CD40	Overexpression	RV, AAV(?)	CAD
Thromboprotection	PAI-1, plasminogen activator Tissue factor	Inhibition	AS-ODN	CAD, MI
	TPA, hirudin, urokinase Thrombomodulin, COX-1, PGI$_2$ synthase	Overexpression	AAV	CAD, MI
Blood pressure	Kallikrein, eNOS, ANP	Overexpression	AAV, RV	Hypertension, HF
	ACE, AGT, AT$_1$	Inhibition	AAV-AS-ODN	
Vascular cell proliferation	NOS, Ras dominant negative	Overexpression	AD, RV, AAV(?)	Graft atherosclerosis Restenosis
	E2F, c-myb, c-myc, PCNA	Inhibition	AS-ODN, Decoy-ODN	
Inherited heart disease				
Channelopathies	SCN5A, I$_k$	Overexpression/inhibition	α-MHC-AAV	Arrhythmia

Table 2 (continued)

Strategy	Therapeutic target	Genetic manipulation	Vector	Application
Cardiomyopathy	Sarcomeric proteins, sarcoglycans	Overexpression	α-MHC-AAV	DCM
Congenital heart disease				
Heart and vessel defects	Endoglin, NKx2.5, TBX5, TFAP2B	Overexpression	α-MHC-AAV	Septal defects, patent ductus arteriosus, arteriovenous malformations, looping conductance defects

AAV, adeno-associated virus; AS-ODN, antisense oligodeoxynucleotide; CAD, coronary artery disease; DCM, dilated cardiomyopathy; HF, heart failure; LV, lentivirus; MI, myocardial infarction; α-MHC, alpha myosin heavy chain; RV, retrovirus; HO-1, heme oxygenase-1; SOD, superoxide dismutase; GPx, glutathione peroxidase; HSP70, 70-kDa heat shock protein; HSP90, 90-kDa heat shock protein; I-CAM, intracellular adhesion molecule; V-CAM, vascular adhesion molecule; NF-κB, nuclear factor kappa B; TNF-α, tumor necrosis factor alpha; eNOS, endothelial nitric oxide synthase; VEGF, vascular endothelial growth factor; FGF, fibroblast growth factor; HGF, hematopoietic growth factor; SERCA2A, sarcoplasmic/endoplasmic reticulum Ca^{2+} ATPase; V1, vasopressin-1 receptor; βARK, beta adrenergic receptor kinase; PAI-1, plasminogen activator inhibitor-1; TPA, tissue plasminogen activator; COX-1, cyclooxygenase-1; PGI_2 synthase, prostacyclin synthase; ANP, atrial natriuretic peptide; ACE, angiotensin-converting enzyme; AGT, angiotensinogen; AT_1, angiotensin II-type 1 receptor; NOS, nitric oxide synthase; PCNA, proliferating cell nuclear antigen; SCN5A, cardiac sodium channel gene 5A.

sode. The increase in reactive oxygen species (ROS) formation during reperfusion of the ischemic myocardium may eventually deplete the buffering capabilities of endogenous antioxidant systems, thereby exacerbating the cytotoxic effects of these reactive molecules (Park and Lucchesi 1999). The development of gene therapies for acute myocardial infarction has been difficult because the time required for transcription and translation of therapeutic genes with the current generation of vectors exceeds the time window for successful intervention. An alternative gene therapy for myocardial protection is to prevent I/R injury by the transfer of cytoprotective genes into the myocardium of high-risk patients prior to ischemia using a gene delivery method that could confer long-term therapeutic gene expression. This novel concept of so-called preventive gene therapy would protect the heart from future I/R injury, thereby minimizing the need for acute intervention (Melo et al. 2002). Given the prominent role of oxidative stress in I/R injury, a therapeutic approach aimed at increasing endogenous antioxidant reserves should, in principle, be a useful strategy for prevention and protection in patients at risk of acute myocardial infarction. This strategy would potentiate the native protective response of the myocardium, rendering it resistant to future ischemic insults.

We have evaluated the feasibility of antioxidant enzyme gene transfer as a long-term first line of defense against I/R-induced oxidative injury, using an rAAV vector for intramyocardial delivery of heme oxygenase-1 (*HO-1*) gene in a rat model of myocardial I/R injury ((Melo et al. 2002). Our findings show that *HO-1* gene delivery to the left ventricular risk area several weeks in advance of myocardial infarction results in approximately 80% reduction in infarct size. The reduction in myocardial injury in the treated animals is accompanied by decreases in oxidative stress, inflammation and interstitial fibrosis. Consistent with the histopathology, echocardiographic assessment showed postinfarction recovery of left ventricular function in the *HO-1*-treated animals, whereas the untreated control animals presented evidence of ventricular enlargement and significantly depressed fractional shortening and ejection fraction. Thus, these findings suggest that AAV-mediated delivery of H*O*-1 may be a viable therapeutic option for long-term myocardial protection from I/R injury in patients with CAD.

Comparable findings were found with extracellular superoxide dismutase (ecSOD) gene transfer (Li et al. 2001; Chen et al. 1998). This secreted metalloenzyme plays an essential role in maintenance of redox homeostasis by dismutating the oxygen free radical superoxide. Our findings showed that long-term survival after acute myocardial infarction is improved in the ecSOD-treated animals relative to the animals treated with the control vector, in parallel with smaller infarcts and decreased myocardial inflammation (Agrawal et al. 2001). Efficient protection from I/R injury has also been achieved by overexpression of other major antioxidant enzyme systems, such as Cu/Zn SOD (Woo et al. 1998) catalase (Zhu et al. 2000) and glutathione peroxidase (Yoshida et al. 1996), stress-induced heat shock proteins such as HSP 70 (Suzuki et al. 2002) and HSP 27 (Vander Heide 2002), survival genes (*Bcl-2, Akt*) (Chatterjee et al. 2002; Matsui et al. 2001), as well as immunosuppressive cytokines (Brauner et al. 1997), adenosine A_1 and A_3 receptors (Yang et al. 2002), kallikrein (Agata et al. 2002), caspase inhibitor (Holly et al. 1999) and hepatocyte growth factor (Ueda et al 1999).

The inhibition of proinflammatory genes involved in the pathogenesis of I/R injury offers another option for cardioprotection. Morishita et al. (1997) showed that pretreatment with a decoy oligonucleotide capable of inhibiting the *trans*-activating activity of the proinflammatory transcription factor NF-κB reduces myocardial infarct after coronary artery ligation in rats. Similarly, intravenous administration of antisense oligonucleotide against angiotensin-converting enzyme mRNA (Chen et al. 2001) or angiotensin AT_1 receptor (Yang et al. 2002) significantly reduces myocardial dysfunction and injury following ischemia and reperfusion. Although the rapid in vivo degradation of oligonucleotides would preclude their use in long-term myocardial protection, they may be useful in treatment of acute myocardial ischemia and cardiac transplantation (Stepkowski 2000) by providing a tool for inhibiting of pro-oxidant, proinflammatory and immunomodulatory genes activated by ischemia and reperfusion. For example, treatment with antisense oligonucleotide directed against intercellular adhesion molecule-1 (ICAM-1) was shown to prolong cardiac allograft tolerance and long-term survival when administered ex vivo prior to transplantation into the host (Poston et al. 1999). Such an approach could be beneficial in the preparation of donor hearts for transplantation. Thus oligonucleotide-mediated inhibition of anti-inflammatory genes and adhesion molecules in donor organs in advance of transplantation could be used to suppress the acute inflammatory response that ensues upon reperfusion of the transplanted organ in the recipient.

The suitability of these experimental therapies for myocardial protection in humans remains to be established. Further work is required to elucidate the mechanism by which exogenous gene delivery of antioxidant enzymes confers myocardial protection from ischemic injury. Conceivably, the increase in basal pro-oxidant scavenging activity imparted by constitutive overexpression of antioxidant enzymes may confer cytoprotection by preconditioning the myocardium to future I/R episodes. Nevertheless, these preclinical studies provide compelling evidence that antioxidant gene therapy may be a viable strategy for protection from ischemic myocardial injury.

3.3
Gene Therapy for Hypertension, Atherosclerosis and Thromboresistance

Gene therapies for treatment of systemic hypertension and dyslipidemia may be beneficial for myocardial protection because these diseases are primary risk factors for coronary artery disease and myocardial infarction (Stein 2002). The complexity of both diseases, however, poses some difficulties in the selection of appropriate targets and delivery strategies. Although the preponderance of cases is multigenic, resulting from complex interactions between genes, environment and lifestyle (Stein 2002), various drug therapies have been efficacious in treating both diseases. For example, drugs targeting the renin–angiotensin system, adrenergic signaling and calcium channel activity have been used successfully in the management of hypertension (Hall 1999), whereas the statin class of 3-hydroxy-3-methylglutaryl-coenzyme A (HMG-CoA) inhibitors are effective in reducing plasma cholesterol and atherogenesis (Knopp 1999). This indicates that despite the complexity of primary hypertension and atherosclerosis, select pharmacological targets have a predominating effect on disease progression, thus rendering them amenable to genetic manipulation.

On the other hand, it may be difficult to justify the development of genetic therapies for these diseases, given the efficacy of current drug therapies. Nevertheless, several promising experimental gene therapies for hypertension, atherosclerosis and thromboresistance have evolved. The prohibitive cost of the drug therapies currently used for the clinical management of these diseases, together with the need for continuous treatment, the occurrence of undesirable side effects and the related problem of non-compliance provides a rationale for gene therapy as an alternative to the current drug therapies.

3.3.1
Gene Therapy for Hypertension

Two gene therapy strategies for hypertension have been tested in animal models. One strategy involves the inhibition of pressor pathways using antisense oligonucleotides against components of the renin—angiotensin system (RAS) (Tang et al. 1999; Kimura et al. 2001; Makino et al. 1999) or the β-adrenergic signaling pathway (Zhang et al. 2000). Using AAV for intravenous delivery of angiotensinogen antisense cDNA, Tang et al. (1999) showed a dose-dependent decrease in arterial blood pressure in adult spontaneously hypertensive rats (SHR) in association with reduced angiotensinogen levels. Using a similar strategy, Kimura et al. (2001) showed that a single intracardiac injection of angiotensinogen antisense cDNA to newborn SHR rats delayed the onset and severity of hypertension up to 6 months in these animals, leading to decreased ventricular hypertrophy and remodeling. Comparable results have been reported with other components of the RAS signaling system, including antisense inhibition of ACE (Wang et al. 1999) and AT_1 receptor (Katovich et al. 1999; Martens et al. 1998). Effective reduction in blood pressure has also been achieved by antisense inhibition of β_1-adrenergic receptor (Zhang et al. 2000), suggesting that this strategy could be used as an alternative to pharmacological β-blockade.

The other gene therapy strategy for hypertension is based on the overexpression of genes encoding vasodilatory moieties, such as nitric oxide synthase (NOS), atrial peptides and kinins. The efficacy of vasodilatory peptide overexpression in reducing blood pressure in hypertensive animals has been documented in several studies (Lin et al. 1995, 1997, 1998; Yoshida et al. 2000; Dobrzynski et al. 2000; Chao and Chao 1997). Intravenous delivery of a plasmid encoding human endothelial NOS under the CMV promoter led to a sustained hypotensive effect in SHR rats in parallel with increased urinary cGMP and nitrite/nitrate levels (Lin et al. 1997). Comparable findings were reported with transfer of *HO-1* and *ecSOD* genes. Retrovirally mediated intracardiac delivery of *HO-1* to 5-day-old SHR attenuated the development of hypertension, in association with decreased vascular reactivity (Sabaawy et al. 2001), whereas intravenous delivery of adenovirus encoding *ecSOD* significantly reduced arterial pressure in 20-week old SHR rats (Chu et al. 2003). Others have shown that systemic delivery of atrial natriuretic factor (Lin et al. 1995, 1998), kallikrein (Yoshida et al. 2000) or adrenomedullin (Dobrzynski et al. 2000) genes with a constitutively active adenoviral vector decreases blood pressure and attenuates renal and myocardial damage in salt-fed Dahl salt-sensitive and DOCA-salt rats.

To date, the use of antisense gene therapy in the treatment of hypertension has not been tested in human trials despite its simplicity and compelling preclinical evidence about its safety and efficacy. Enthusiasm for these novel approaches is tempered by the efficacy of current drug therapies. Nevertheless, the prospect of achieving long-term control of blood pressure in hypertensive patients by gene therapy with minimal side effects is an appealing attribute that is likely to facilitate acceptance of gene therapy as a viable alternative to pharmacological therapies.

3.3.2
Gene Therapy for Atherosclerosis, Thromboresistance and Plaque Stabilization

Plaque rupture and subsequent coronary thrombosis and occlusion are the major causes of acute coronary episodes that result in myocardial infarction and sudden cardiac death (Rentrop 2000). Gene therapy aimed at reducing the cholesterol level and/or at increasing thromboresistance and tensile strength within the plaque may offer a novel and potentially effective alternative option to achieve long-term plaque stabilization and prevent the occurrence of acute coronary events (Feldman and Isner 1995).

The effect of lipid-lowering gene therapy has been evaluated mainly in inherited disorders of lipid metabolism, such as familial hypercholesterolemia (FH) and apoE deficiency, because of their monogenic etiology and refractoriness to drug treatment. Initial attempts at correcting FH involved transplantation of autologous hepatocytes stably transduced ex vivo with a retroviral vector constitutively expressing the LDL receptor in heritable hyperlipidemic Watanabe rabbits (Chowdhury et al. 1991). This initial study showed a 30%–50% decrease in plasma cholesterol levels for up to 6 months. The success of this animal study led to a small clinical trial, but the outcome was less impressive, showing a reduction of 6%–23% in plasma LDL levels in three out of five treated patients (Grossman et al. 1995), with a relatively short duration, possibly attributable to retroviral gene silencing.

Other potential targets for correction of genetic hyperlipidemia include replacement of lipoprotein lipase and hepatic lipase genes (Zsigmond et al. 1997; Applebaum-Bowden et al. 1996), Apo-E (Rinaldi et al. 2000), VLDL receptor (Oka et al. 2001) and scavenger receptor B-1 (SR-B1) (Laukkanen et al. 2000). In most cases, the improvement in serum lipid profiles was transient, probably due to the immune response to the vector used. The low immunogenicity of AAV provides an advantage in this regard, and recently Chen et al (2000) and Harris et al. (2002) showed that AAV-mediated delivery of VLDL receptor or apolipoprotein-E led to sustained reduction of serum lipid levels and inhibition of aortic atherosclerosis into LDLR- or Apo-E-deficient mice, respectively.

Novel lipid-lowering and plaque-stabilizing strategies are emerging (for review see Rader and Tietge 1999). For example, the overexpression of apoprotein ApoA-1 in mice by intravenous adenoviral gene delivery increases serum HDL levels (Tangirala et al. 1999). Blockade of monocyte infiltration and activation in the arterial wall by inhibition of monocyte chemoattractant protein-1 (MCP-1) receptor activation was shown to retard the onset of atheroma and to limit progression and destabilization of established atherosclerotic lesions in ApoE mice (Inoue et al. 2002). Overexpression of antithrombotic genes at sites in the vessel wall at risk of thrombosis may be a feasible protective strategy for vulnerable plaque and prevention of acute coronary events, and delivery of anticoagulant, antifibrinolytic and antiplatelet genes such as thrombomodulin (Waugh et al. 1996), tissue-specific plasminogen activator (t-PA) (Dichek et al. 1996), tissue factor pathway inhibitor (Zoldhelyi et al. 2000; Golino et al. 2001), prostacyclin synthase (Numaguchi et al. 1999) and cyclooxygenase I (Zoldhelyi et al. 1996) to the injured vessel wall has been reported to reduce the incidence of thrombosis.

Gene transfer of cytoprotective genes such as *HO-1* and nitric oxide synthase (NOS) has also shown to exert vasculoprotective effects. Adenovirus-mediated delivery of *HO-1* significantly reduced the development of aortic lesions in ApoE-deficient mice, in parallel with a decrease in iron deposition (Juan et al. 2001). The vasculoprotective effect of *HO-1* is likely due to its anti-inflammatory and antioxidant properties (Morse and Choi 2000). Another important target for vascular protection is NOS (Channon et al. 2000). *NOS* gene transfer provides a mechanism to increase NO bioactivity and enhance the antiatherogenic properties of the vessel wall. Indeed, delivery of inducible nitric oxide synthase (iNOS) (Shears et al. 1997) and neuronal nitric oxide synthase (nNOS) (Qian et al. 2000) by adenovirus has been reported to abrogate aortic allograft atherosclerosis in rats and to significantly reduce inflammatory cell infiltration and lipid deposition in carotid arteries of cholesterol-fed rabbits, respectively. Various strategies have been developed for the transfer of therapeutic genes into atherosclerotic vessels. Local expression of genes in the arterial wall has been achieved using catheters for delivery of plasmid and viral vectors in vivo and ex vivo (Kullo et al. 1999; Rekhter and Simari 1998). Gene transfer to coronary arteries is technically challenging and refinements to the delivery catheters currently in use are necessary to improve efficiency. Nevertheless, genes have successfully been delivered into the coronary arteries of dogs and pigs using perfusion balloon catheters (Kullo et al. 1999). As primary thrombolytic therapy for acute myocardial infarction, gene transfer of anticoagulant genes is not feasible, at least with the current generation of vectors, because the time required for production of the therapeutic protein falls outside the time window for successful intervention following coronary thrombosis. Antithrombotic gene therapy may have a role as an adjuvant to primary thrombolytic therapy to prevent the recurrence of thrombosis and reocclusion of the affected vessel.

3.4
Gene Therapy for Restenosis and Vascular Proliferative Disease

Surgical revascularization procedures using percutaneous transluminal angioplasty (PTCA), stenting or coronary artery bypass grafting (CABG) is a common treatment option for CAD. However, despite significant improvements in pharmacological therapies and the introduction of biocompatible and drug-coated stents, these procedures are still accompanied by significant failure rates due to restenosis and graft atherosclerosis. The ability to deliver antiproliferative and antithrombotic genes and to inhibit pro-proliferative genes in the vessel wall allows genetic engineering of native vessels or grafts to render them resistant to atherosclerosis and neointimal hyperplasia.

Genetic strategies to inhibit neointimal smooth muscle proliferation have major implications for treatment of vascular proliferative diseases. Using adenovirus to deliver thrombomodulin to jugular vein segments ex vivo prior to interpositional grafting in rabbits, Kim et al. (2002) reported that genetic engineering of the graft led to thromboresistance and graft survival. Zoldhelyi et al. (2000) showed that delivery of tissue factor pathway inhibitor to balloon-injured atherosclerotic carotid arteries of Watanabe rabbits reduced neointima proliferation and inhibited thrombus formation. Adenoviral delivery of the suicide gene, thymidine kinase, into carotid arteries of Watanabe rabbits inhibits neointimal proliferation after balloon angioplasty (Steg et al. 1996), demonstrating the potential of cytotoxic gene therapy for inhibition of restenosis.

Cytostatic gene therapy has also yielded promising results in the treatment of vasculoproliferative disease. This strategy involves the inhibition of key proteins regulating cell cycle progression (Braun-Dullaeus et al. 1998). Treatment of jugular veins in vivo with HVJ-liposome complexes containing antisense oligonucleotide against cell cycle regulators PCNA and cdc2 kinase inhibited atherosclerosis and neointimal hyperplasia after carotid artery interpositional grafting in rabbits maintained on a high cholesterol diet (Morishita et al. 1993). We have shown that ex vivo genetic engineering of vein grafts with a decoy deoxyoligonucleotide consisting of the consensus binding sequence of *E2F-1*, a transcriptional factor involved in cell-cycle progression, resulted in prolonged resistance to neointimal hyperplasia and improved graft patency (Morishita et al. 1995). These findings led to a large-scale phase I prospective, randomized, double-blind trial of human saphenous vein graft treatment with *E2F* decoy (PREVENT-1) (Mann et al. 1999). Using non-distending pressure to deliver the *E2F* decoy oligonucleotide ex vivo prior to arterial interpositional grafting, the authors reported that *E2F* decoy treatment was safe and prevented graft atherosclerosis concomitant with inhibition of cell cycle progression. These results have recently been confirmed in a phase II trial designed to evaluate the effect of *E2F* decoy treatment on CABG failure (Grube et al., American Heart Association Meeting, Nov. 2001, see commentary by McCarthy 2001). Interestingly, we reported recently that the *E2F* decoy selectively targets vascular smooth muscle cell proliferation without affecting the endothelial cell proliferative burst that is essential for healing after vein grafting (Ehsan et al. 2002). We believe that this sparing effect on endothelium contributes to the enhanced endothelial function that we previously reported in vein grafts treated with cell cycle regulatory proteins (Mann et al. 1997).

Other cytostatic strategies have yielded variable degrees of success in experimental models of restenosis. Treatment with antisense against cell-cycle regulatory genes cdk2 kinase and proliferating cell nuclear antigen (Morishita et al. 1994), p21 (Chang et al. 1995a) and p27 (Chen et al. 1997) cyclin-dependent kinase inhibitors, non-phosphorylatable retinoblastoma gene product (Chang et al. 1995b) p53 (Yonemitsu et al. 1998) and the proto-oncogenes c-*myb* (Gunn et al. 1997) and c-*myc* (Shi et al. 1994), have all been reported to inhibit neointimal hyperplasia in animal models of arterial injury. Similar results have been reported for the inhibition of intracellular signaling mediators of mitogen-dependent kinases, NF-κB, Bcl-x_L and growth factors, or overexpression of Fas ligand, gax and GATA-6 transcription factors and cytokines such as β-interferon and VEGF (for review see Morishita et al. 1998; Kibbe et al. 2000). The application of VEGF gene transfer may be particularly useful in re-establishing vascular wall homeostasis after injury because of the ability of this endothelium-specific cytokine to promote re-endothelializaton of the denuded arterial wall (Van Belle et al. 1997).

Local delivery of angiotensin-converting enzyme antisense oligonucleotide was shown to reduce neointima formation in a rat carotid injury model (Morishita et al. 2000), suggesting that locally derived angiotensin may play a role in vascular injury. In vivo delivery of endothelial and iNOS genes is quite efficacious in reducing neointimal thickening in balloon-injured vessels (von der Leyen et al. 1995; Tzeng et al. 1996; for review see von der Leyen and Dzau 2001). This has led to at least one phase I clinical trial (REGENT-I) to evaluate the efficacy of catheter-based iNOS gene delivery to prevent restenosis of coronary arteries treated by PTCA. Local delivery of antioxidant enzymes such as HO-1 (Tulis et al. 2001) and ecSOD (Laukkanen et al. 2002) by adenovirus has also been reported to inhibit neointima hyperplasia in various animal models of restenosis, possibly due to reduction in inflammation and oxidative stress during the early phase of vascular injury and the subsequent inhibition of vascular smooth muscle proliferation.

Despite these promising preclinical data, the use of gene therapy as a therapeutic modality for restenosis and vasculoproliferative disease still has to overcome various feasibility, safety and efficacy issues. Improvements in vector and delivery technologies are warranted. The complexity of the pathological processes leading to restenosis suggests that genetic manipulation of multiple targets may be more appropriate than strategies directed at a single therapeutic target. Vascular cell types, such as endothelial cells, may be genetically modified ex vivo to express cytoprotective or antiproliferative genes and used for repair of damaged vessels and vascular prostheses and stents bioengineered to render them thromboresistant and less susceptible to restenosis. Preclinical studies have already demonstrated proof of concept for some of these strategies, and future clinical trials should determine their feasibility and safety for use in humans.

4
Gene Therapy for Myocardial Rescue

Gene therapy strategies for rescuing failing myocardium may be attainable in certain situations (Table 2). Therapeutic angiogenesis by delivery of genes coding proangiogenic growth factors, such as VEGF, fibroblast growth factor (FGF) and hepatocyte growth factor (HGF), has been shown to promote neovascularization and functional recovery of ischemic myocardium in several animal models and in humans with coronary artery disease (Mack et al. 1998; Giordano et al. 1996; Ueda et al. 1999; Losordo et al. 1998). Other potential strategies for rescuing contractile function in the failing myocardium include overexpression of the sarcoplasmic reticulum calcium ATPase (SERCA2a) (Myamoto et al. 2000), β-adrenergic receptor (Maurice et al. 1999) and adenylate cyclase (Roth et al. 1999) (Table 2). An exciting new field is emerging with the recent identification and isolation of endothelial and cardiomyocyte precursor stem cells from adult bone marrow (Makino et al. 1999; Asahara et al. 1997). The ability to expand and genetically modify these cells ex vivo offers the opportunity to use them as an autologous cellular substrate for the generation of new blood vessels (therapeutic vasculogenesis), repairing infarcted myocardium and in tissue engineering.

4.1
Gene Therapy for Myocardial Ischemia

The vascular endothelium usually remains in a quiescent, nonproliferative state, and with the exception of the female reproductive tract and neoplastic disease, postnatal neovascularization is rare (Carmeliet 2000a). Wounding, inflammation and oxidative stress activates the endothelium, resulting in cell proliferation, migration and formation of new vascular networks by angiogenesis (Carmeliet 2000a). In patients and animal models with ischemic heart disease, the progressive occlusion of the coronary artery leads to a chronic imbalance in myocardial oxygen supply and demand, which stimulates the development of collateral vessels aimed at maintaining tissue perfusion and oxygenation (Ware and Simons 1997). This native adaptive response of the myocardium, however, does not provide adequate compensation in face of severe ischemia and depression of cardiac function ensues, which in time leads to heart failure.

Evidence of enhanced neovascularization and functional recovery of ischemic myocardium has been reported in several animal and human studies after exogenous supplementation of proangiogenic factors by gene transfer (Tio et al. 1999; Mack et al. 1998; Giordano et al. 1996; Ueno et al. 1997; Ueda et al. 1999; Rosengart et al. 1999; Symes et al. 1999; Hammond et al. 2001). This novel strategy, commonly known as therapeutic angiogenesis, offers a potentially efficacious method for the treatment of coronary artery disease where percutaneous angioplasty or surgical revascularization has been excluded. Proof of principle has been demonstrated in several animal models of hindlimb and myocardial ischemia by gene transfer of *VEGF* (Rosengart et al. 1999; Symes et al. 1999; Lee et al. 2000), *FGF* (Giordano et al. 1996; Ueno et al. 1997; Tabata et al. 1997) and hepatocyte growth factor (*HGF*) (Ueda et al. 1999; Taniyama et al. 2002; Aoki et al. 2000). In all cases, improvement in tissue perfusion was accompanied by morphological and angiographic evidence of new vessel formation, thereby establishing a relationship between improved tissue viability and neovascularization. For example, Mack et al. (1998) showed that intramyocardial delivery of $VEGF_{121}$ by adenovirus led to an improvement in regional myocardial perfusion and left ventricular function in response to stress in an ameroid constrictor model of chronic myocardial ischemia in pigs. Using intracoronary injection of an adenovirus vector encoding human *FGF-5*, Giordano et al. (1996) also showed a significant improvement in blood flow and a reduction in stress-induced functional abnormalities (in association with an increase in capillary-to-fiber ratios) as early as 2 weeks after ameroid placement around the proximal left circumflex coronary artery in pigs.

Transmyocardial laser revascularization has been reported to provide relief of angina in patients with ischemic heart disease by forming channels that may improve collateral blood flow (Yamamoto et al. 2000). Transmyocardial laser therapy in combination with pro-angiogenic gene transfer has been tested as a potential synergistic approach to maximally stimulate myocardial angiogenesis. Sayeed-Shah and colleagues (1998) demonstrated that intramyocardial delivery of plasmid-encoding *VEGF* in the region treated by transmyocardial laser revascularization yielded superior recovery of ventricular function than either therapy alone, providing evidence for an added benefit of this combinatorial approach. To our knowledge, this strategy has not been tested in human patients.

Several phase I and II clinical trials of angiogenic gene therapy have been carried out with patients suffering from myocardial and limb ischemia (Rosengart et al. 1999; Symes et al. 1999; Vale et al. 2001; Grines et al. 2002; Losordo et al. 1999; for review see Bashir et al. 2002). These safety trials, although consisting of small nonrandomized patient samples, demonstrate the potential of angiogenic gene therapy for treatment of ischemic heart disease.

Losordo et al. (1998) carried out a phase I study in five male patients 53–71 years of age with angiographic evidence of coronary artery disease that did not respond to conventional anti-anginal therapy. The authors reported that direct intramyocardial delivery of naked plasmid encoding $VEGF_{165}$ into the ischemic myocardium resulted in significant reduction of anginal symptoms and modest improvement in left ventricular function concomitant with reduced ischemia and improved Rentrop score. Using adenovirus for intramyocardial delivery of $VEGF_{121}$ into an area of reversible ischemia in the left ventricle as sole or adjunct therapy in patients undergoing conventional coronary artery bypass grafting, Rosengart et al. (1999) showed improvements in regional ventricular function and wall motion in the region of vector administration in both groups of patients. Vale and colleagues (2001) carried out a randomized, single-blinded placebo-controlled phase I trial in patients with chronic myocardial ischemia using catheter-based delivery of naked $VEGF_{165}$ assisted by electromechanical NOGA mapping of the left ventricle. The results of this study indicated significant reductions in weekly anginal attacks for as long as 1 year after gene delivery in the treated patients, in contrast to the patients receiving placebo. The reduction in anginal episodes was accompanied by improved myocardial perfusion as evidenced by SPECT-sestamibi perfusion scanning and electromechanical mapping. Recently Grines and colleagues (2002) completed the Angiogenic GENe Therapy (AGENT) double-blinded, randomized, placebo-controlled trial using dose-escalating adenovirus-mediated intracoronary delivery of *FGF-4* in patients with angina, in order to evaluate the safety and efficacy of this protocol in reducing ischemic symptoms. The authors reported increased exercise tolerance and improved stress echocardiograms at 4 and 12 weeks after gene transfer in the patients that received *FGF-4* gene therapy compared to the patients receiving placebo. Unfortunately, the long-term outcome beyond 12 weeks has not been reported.

The success of these initial small-scale phase I and phase II trials warrant larger and more adequately controlled later phase trials. Several issues relating to feasibility, safety and sustainability require further investigation before therapeutic angiogenesis may be envisaged as a viable therapeutic option for treatment of ischemic heart disease. The broad issue of safety of the approach requires systematic evaluation. This is particularly relevant in light of recent evidence that transplantation of myoblasts constitutively expressing *VEGF* under a retroviral promoter into mouse hearts led to intramural angiomas followed by heart failure and death (Lee et al. 2000; see commentary by Carmeliet 2000). This observation underscores the necessity for regulated expression of pro-angiogenic factors.

Such a strategy may require the incorporation of promoter sequences, for example hypoxia-sensitive responsive elements, capable of rendering expression of the therapeutic transgene subservient to the pathophysiological changes in myocardial oxygen tension. This concept has recently been validated by Su et al. (2002), who demonstrated that hypoxia-induced *VEGF* expression in ischemic myocardium from an AAV vector encoding *VEGF* under transcriptional control by the erythropoietin hypoxia responsive element (HRE). Another approach to achieve regulated therapeutic angiogenesis uses engineered transcription factors capable of activating endogenous *VEGF* expression as a strategy to induce *VEGF* expression in pathophysiological conditions (Vincent et al. 2000; Rebar et al. 2002). These novel strategies may allow endogenous regulation of angiogenesis so that the magnitude of neovascularization is graded to the severity of the ischemic insult.

Further work is also necessary to determine the safest and most efficacious route and method of gene delivery to avert potentially hazardous side effects, such as neovascularization of occult neoplasms or peripheral vascular effects that may result in edema and hypotension. In this context, the optimal strategy may require targeted tissue delivery by incorporation of cell-specific promoters for expression of the transgene exclusively at the target sites. It also needs to be established whether the desired long-term therapeutic effect can be achieved with a single administration of the therapeutic gene or whether multiple treatments may be required.

4.2
Cell-Based Therapy for Myocardial Ischemia

An alternative strategy for therapeutic angiogenesis involves the use of endothelial precursor cells as angiogenic substrate. Several reports have documented the existence of blood-borne endothelial progenitor cells (EPC) originating from a common hemangioblast precursor in adult bone marrow (Asahara et al. 1997; Shi et al. 1998; Asahara et al. 1999). These endothelial lineage cells have the properties of an endothelial progenitor ($CD34^+$, $Flk-1^+$) and are recruited to foci of neovascularization such as ischemic muscle (Shintani et al. 2001) and the myocardium (Kawamoto et al. 2001), where they differentiate into functional endothelial cells, indicating that they may play a role in postembryonic vasculogenesis in ischemic tissues.

The therapeutic potential of these cells as vehicles for tissue salvage and/or regeneration from ischemia has been demonstrated. Local implantation of autologous bone marrow-derived cells in rat (Ikenaga et al. 2001) and mouse (Murohara et al. 2000; Kalka et al. 2000) ischemic hindlimbs induces angiogenesis and partially restores blood flow and exercise capacity in the ischemic limb. Similarly, transplantation of ex vivo-expanded human EPCs into nude rats (Kawamoto et al. 2001; Kocher et al. 2001) and pigs (Fuchs et al. 2001) with myocardial ischemia leads to increased capillary density and improved ventricular function. More recently it was reported that the number of circulating EPCs increases in patients with acute myocardial infarction (Shintani et al. 2001) and is lower in patients with coronary artery disease (Vasa et al. 2001a), indicating that these cells may play an essential role in neovascularization of the myocardium in response to ischemia.

The ability to culture and genetically engineer EPCs ex vivo with vectors expressing therapeutic genes suggests that these cells may be ideally suited as a substrate for cell-based gene therapy for neovascularization of ischemic tissues. In this scheme, EPCs genetically modified to express angiogenic growth factors could serve as a cell substrate for new vessel growth by vasculogenesis, driven by local proliferation and differentiation of the transplanted cells, and as a source of pro-angiogenic growth factors for growth of pre-existing vessels by sprouting.

This concept was recently validated by Iwaguro et al. (2002). Using athymic mice with hindlimb ischemia, this group showed that the transplantation of murine EPC transduced ex vivo with an adenoviral vector expressing *VEGF* resulted in more efficient neovascularization and blood flow recovery that treatment with untransduced EPCs. The improved neovascularization in the animals treated with VEGF-transduced EPCs appears to be, at least in part, due to enhanced EPC proliferation and adhesion. Thus, *VEGF* gene transfer exerts phenotypic modulation of the EPCs, thereby potentiating biological properties that favor the angiogenic response. A potential noninvasive approach for angiogenesis of ischemic myocardium in CAD may involve the mobilization of EPCs to the ischemic region using conventional pharmacological therapeutic agents used in treatment of CAD such as statins. Recently, several groups showed that statin therapy increases the number of EPCs in patients with stable CAD (Vasa et al. 2001b; Dimmeler et al. 2001), suggesting that the mobilization of EPCs and subsequent neovascularization of ischemic myocardium may contribute to the therapeutic benefit of these drugs. Walter et al. (2002) showed that statin therapy accelerates re-endothelization of balloon-injured arterial segments in rats, leading to reduction in neointimal thickening.

4.3
Gene Therapy for Rescue of Contractile Function

Rescue of contractile function in the failing myocardium is another major goal of myocardial gene therapy. The failing myocardium is characterized by alterations in calcium handling, decreased myofilament sensitivity, excessive catecholamine release and adrenergic receptor down-regulation and desensitization (Towbin and Bowles 2002), resulting in decreased contractility. β-Adrenergic receptors (β-AR) are G-protein-coupled receptors that play an essential role in regulation of myocardial contractility and inotropic state in response to neurohumoral stimulation (Rockman et al. 2002). Activation of β-AR (predominantly β_1) by norepinephrine leads to the phosphorylation of several proteins involved in regulation of excitation–contraction coupling, including the sarcolemmal L-type Ca^{2+} channels, ryanodine receptors, sarcoplasmic reticulum calcium ATPase (SERCA2) inhibitor phospholamban, troponin I and myosin binding protein C (Bers 2002) via stimulation of the adenylate cyclase-cAMP-PKA signaling cascade.

The β-AR-signaling and calcium-regulating pathways have been used as targets for treatment of heart failure (Towbin and Bowles 2002; Mann 1999). The use of β-blockers and calcium channel antagonists have lead to remarkable improvements in the long-term survival and quality of life of patients with advanced heart failure. Recent preclinical studies suggest that genetic manipulation of these therapeutic targets may be a viable and potentially effective alternative to the pharmacotherapies currently in use in the management of heart failure. Adenovirus-mediated intracoronary delivery of the β_2-AR gene led to improvements in basal and isoproterenol-stimulated LV contractility and hemodynamic function in rabbits (Maurice et al. 1999; Shah et al. 2000), and rescued β-AR signaling in ventricular myocytes from failing hearts. Similarly, adenovirus delivery of the β-ARKct peptide inhibitor improved postinfarction LV function significantly in rabbits after myocardial infarction, in parallel with increased β-AR-stimulated adenylate cyclase activity and cAMP generation (Shah et al. 2001). Recently, Roth et al. (2002) demonstrated that cardiac-specific overexpression of adenylate cyclase type VI (AC_{VI}) improved ventricular function, restored β-AR-stimulated cAMP generation and increased long-term survival in mice rendered cardiomyopathic by overexpression of Gq protein. These findings suggest that gene transfer protocols aimed at normalizing β-AR signaling may have application as a strategy for functional rescue of the failing heart. Exogenous overexpression of β-AR receptors and signaling proteins by gene transfer may compensate for the decrease in endogenous β-AR density and sensitivity resulting from chronic sympathetic activation in heart failure, thereby normalizing left ventricular function.

Gene therapy strategies for normalization of myocardial cytosolic calcium transients have also yielded promising results in experimental models of heart failure (Miyamoto et al. 2000; del Monte et al. 2001; Hajjar et al. 2000). The ratio of phospholamban to SERCA2a is increased in heart failure, resulting in decreased Ca^{2+} ATPase activity and reduced calcium uptake by the SR (Schmidt et al. 1997; Towbin and Bowles 2002). Adenovirus-mediated overexpression of SERCA2a in neonatal cardiac myocytes enhanced contraction by increasing peak $[Ca^{2+}]i$ release and a decrease in resting $[Ca^{2+}]i$ (Hajjar et al. 1997). In a rat model of heart failure induced by aortic banding, intracoronary SERCA2a gene delivery by adenovirus at approximately the time of transition from compensated hypertrophy to heart failure restored systolic and diastolic function concomitant with an increase in basal Ca^{2+}-ATPase activity (Myamoto et al. 2000) and improved phosphocreatine/ATP ratio and long-term survival (del Monte et al. 2001). Furthermore, SERCA2a gene transfer normalized cytosolic transients and restored contractile function in ventricular myocytes isolated from patients with end-stage heart failure (del Monte et al. 1999) and improved diastolic function in aged rats (Schmidt et al. 2000). Presumably, overexpression of SERCA2a restores the normal stoichiometry between phospholamban and the Ca^{2+}-ATPase, preventing cytosolic calcium overload and left ventricular dysfunction. Conversely, antisense inhibition of phospholamban was shown to improve contractility in cultured rat neonatal myocytes (Eizema et al. 2000) and in ventricular myocytes of end-stage heart failure patients (del Monte et al. 2002), in association with improved calcium sensitivity of SERCA and reduced time for recovery of the Ca^{2+} transient.

Despite these promising findings, the available data have not established the long-term efficacy and safety of adenoviral-mediated myocardial expression of adrenergic and calcium-regulating proteins. Adenoviral vectors exhibit significant myocardial cytotoxicity at high concentration and induce a robust inflammatory response that may cause damage to the infected cells and lead to loss of transgene expression. Sustained expression of the therapeutic transgene may be essential for rescue of the failing heart, necessitating the use of a vector type such as AAV. Secondly, the physiological consequences of long-term β-AR and *SERCA2a* gene transfer needs to be established. Although transgenic mice with cardiac-specific overexpression of β_2-AR or SERCA2a do not show any morphological evidence of myocardial pathology (Baker et al. 1998; Milano et al. 1994), it is not known whether viral-mediated expression of these proteins has any secondary effects besides calcium regulation. Concerns have recently been raised that the increase in SERCA2a expression by gene transfer in the failing heart may impose extra demands on myocardial energy expenditure due to increased inotropic state, and may cause adverse electrophysiological events such as arrhythmias. Such potential adverse effects could accelerate myocardial cell death and precipitate the progression of heart failure and will have to be addressed before inotropic gene therapy could make the transition from the preclinical stage to clinical trial.

4.4
Gene Therapy for Myocardial Hypertrophy and Remodeling

Inhibition of ventricular remodeling is a prime target in the treatment of heart failure, and the long-term survival benefits of therapies such as ACE inhibition and β-blockade in patients suffering from MI or heart failure are attributed in part to a decrease in LV remodeling. Pharmacological inhibition of these pathways attenuates the hypertrophic and remodeling process and delays the progression of disease (McMurray and Pfeffer 2002). More recently, treatment with MMP inhibitors was shown to attenuate post-infarction LV dilation effectively (Asakura et al. 2002), suggesting that this could be a therapeutic strategy for heart failure.

Genetic manipulation of these targets may prove an alternative to current pharmacological approaches for treatment of heart failure. Gene therapies aimed at inhibiting hypertrophic and profibrotic pathways should be useful in limiting the extent of remodeling. For example, inhibition of AT_1-R signaling by antisense reduces cardiac hypertrophy in a renin-overexpressing transgenic rat, independent of systemic effects (Pachori et al. 2002), suggesting a role of local ANG II in inducing the hypertrophic phenotype. A similar approach could be used for inhibition of cardiotrophic factors such as calcineurin and protein kinases (Taigen et al. 2000). Antisense inhibition of myocardial TGF-β1 factor signaling and metalloproteinase activity could be employed as strategies to reduce fibrosis and remodeling. Conversely, myocardial overexpression of antihypertrophic factors may be used as a strategy to reverse hypertrophy in failing hearts. Li et al. (1997) demonstrated that cardiac-specific overexpression of insulin-like growth factor-1 (IGF-1) in mice prevented myocyte death in the viable myocardium and attenuated ventricular dilation and hypertrophy after MI. Similarly, cardiac overexpression of glycogen synthase-3β, an endogenous antagonist of calcineurin action, was reported to inhibit hypertrophy in response to chronic β-adrenergic stimulation and pressure overload (Antos et al. 2002). Overexpression of cyclin-dependent kinase inhibitor p16 has also been shown to reduce cardiac hypertrophy in response to ET-1 (Nozato et al. 2001), in agreement with findings that cyclin-dependent kinase inhibitors play an essential role in inhibition of pressure-induced hypertrophy (Tamamori 1998). Systemic overexpression of vasodilatory genes such as NOS (Lin et al. 1997), ANP (Lin et al. 1998) and kallikrein (Yoshida et al. 2000) were effective in reducing cardiac hypertrophy and fibrosis in hypertensive rat models.

Whether such genetic approaches yield therapeutic potential in the treatment of human heart failure remains to be investigated. The molecular complexity of heart failure in humans suggests that combinatorial therapeutic strategies aimed at modifying the activity of multiple targets involved in cardiac hypertrophy and remodeling may be more effective than selective therapeutic approaches focusing on a single target. For example, the optimal gene therapy for heart failure may combine antiremodeling strategies, such as antisense inhibition of MMP activity, with contractility enhancing strategies, such as overexpression of SERCA2a or β1-AR.

4.5
Gene Therapy for Inherited and Congenital Heart Disease

In principle, myocardial disease resulting from single-gene mutations could be corrected by exogenous delivery of the normal gene (gain of function approach). The major hurdle in this regard has been the unavailability of a suitable vector capable of sustained transgene expression in the myocardium. The introduction of AAV and lentivirus vector systems, which have the capacity to transduce and integrate into the genome of terminally-differentiated cells such as cardiomyocytes, may partially overcome this problem

Although gene therapies for inherited cardiomyopathies and channelopathies have not yet been tested in humans, preclinical data suggest the feasibility of these therapies. Several animal models have been engineered to express the genetic mutations found in humans with cardiomyopathy and LQT syndromes. In all cases, the histopathological and physiological abnormalities characteristic of human cardiomyopathy and channelopathies were observed (Maass and Leinwand 2000; Ikeda and Ross 2000; Balser 2002), rendering these animals ideal for testing and evaluating specific gene therapies. The feasibility of gene therapy for inherited cardiomyopathy has been demonstrated by Kawada et al. (2002), who showed that intramyocardial delivery of δ-sarcoglycan to 5-week-old TO-2 Syrian hamsters using an AAV vector completely rescued the progression of cardiomyopathy and led to a drastic increase in life expectancy. The transgene was expressed throughout life and led to improved sarcolemmal integrity, reduced calcification and normalization of myocardial contractility and hemodynamics, in association with re-expression of α-, β-, γ- and δ-sarcoglycan and reconstitution of the dystrophin-associated glycoprotein complex. Similarly, Ikeda et al. (2002)

showed that coronary retroinfusion of adenovirus vector coding for δ-sarcoglycan in 8- to 12-week old BIO 14.6 hamsters resulted in restoration of δ-, α- and β-sarcoglycan to the sarcolemma and improvement in ventricular function compared to age-matched untreated CM hamsters. Using an Epstein-Barr virus based plasmid vector, Tomaiyasu et al. (2000) showed that intramyocardial delivery of β_2-AR to aged BIO 14.6 hamsters led to improved basal and agonist-stimulated left ventricular contractility and hemodynamics.

Myocardial delivery of genes encoding defective channel proteins or regulatory G proteins may provide a strategy for correction of the genetic defects associated with inherited and acquired LQT syndromes. Donahue et al. (2000) were able to reduce heart rate following atrial fibrillation in pigs by local delivery of the $G\alpha_{i2}$ gene to the atrioventricular node by adenovirus, suggesting that this approach may have application in the treatment of atrial arrhythmias. The *HERG* gene encodes the K^+ channels mediating the faster component of the delayed rectifier potassium current (I_{Kr}) that is critical for myocardial repolarization. Nuss et al. (1999) showed that adenoviral transfer of the human *HERG* gene to adult rabbit ventricular myocytes maintained in primary culture led to abbreviated action potentials and drastically reduced the incidence of early after depolarizations after a train of action potentials. This was found to be associated with increased duration of the refractory period.

Despite the identification of several single-gene mutations associated with heart and vessel abnormalities, the development of effective genetic therapies for congenital heart disease has been problematic for various biological and technical reasons. The precision, both in time and mechanism, by which these developmentally regulated genes exert their effects on heart morphogenesis and development dictates that any external corrective measure such as replacement of defective genes needs to be performed before the developmental programs affected by the mutated genes are activated because the anatomical and functional defects emanating from these mutations may be irreversible. The ability to intervene and reprogram a defective gene within the crucial developmental time window requires the availability of diagnostic tools that would permit detection of such mutations before the onset of disease and access to an effective system for in utero gene delivery. Although the current technology enables genetic screening for detection of many disease-causing mutations and polymorphisms, the prohibitive cost of this technology restricts its use to cases with a strong familial history, leaving many undiagnosed cases. Secondly, the dependency of normal heart and vessel development and maturation on precise stage-specific regulation of these morphogenetic genes mandates that corrective strategies be amenable to regulation by the endogenous mechanisms responsible for normal development. Furthermore, the heterogeneity of congenital heart disease due to polymorphisms and differential responsiveness to environmental and other secondary factors suggests that remedial genetic therapies would need to be individualized.

4.6
Cell-Based Therapy for Myocardial Regeneration

Despite evidence of myocyte replication in the heart (Beltrami et al. 2001), the vast majority of adult cardiomyocytes are terminally differentiated and unable to divide (Soonpaa and Field 1998). Consequently, the regenerative capacity of the infarcted myocardium is limited (Li et al. 1996). Hypertrophy and, possibly, hyperplasia of the surviving myocytes may provide initial structural and functional compensation. However, in time these processes lead to maladaptive remodeling of the ventricle and heart failure (St. John Suton and Sharpe 2000).

Cell transplantation (cellular cardiomyoplasty) may offer a potential alternative for reconstitution of infarcted myocardium and recuperation of cardiac function (Reinlib and Field 2000). This approach is based on the premise that repopulation of the necrotic myocardium with replication-competent cells will rescue contractile function and re-establish the structural integrity that is disrupted by myocardial infarction. Several cell-based regenerative strategies have evolved using a variety of substrates such as skeletal muscle myoblasts (Taylor et al. 1998), fetal (Li et al. 1997) and embryonic cardiomyocytes (Min et al. 2002) and autologous marrow-derived mesenchymal cardiomyocyte progenitors (Tomita et al. 1999; Orlic et al. 2001a; Jackson et al. 2002; Toma et al. 2002). However, the therapeutic efficacy of cellular cardiomyoplasty has been inconsistent, and several technical and safety issues remain unresolved. For example, the optimal time for grafting after injury, the source and availability of cellular substrate, the delivery method and the immune tolerance of the host to the grafted cells are important technical and safety considerations.

The use of an adult self-regenerating autologous source of progenitor cells with the potential for differentiating into cardiomyocytes would appear ideal for various reasons. First, the technical problems associated with immunohistocompatibility would be eliminated, removing the need for adjunctive immunosuppressive therapy after transplantation, which could potentially translate into improved graft survival. Secondly, the use of an autologous source of cells for transplantation would circumvent many of the legal, ethical and moral hurdles implicit in the use of embryonic and fetal tissue. More significantly, the self-renewing capability of progenitor cells would provide a readily available and sustainable substrate pool for autologous cell transplantation protocols, simply requiring consent from the patient. Mesenchymal cells from the bone marrow stroma of long bones may offer a viable option for cellular cardiomyoplasty using autologous cells. These cells exhibit a high degree of plasticity (Krause 2002; Jiang et al. 2002) and can differentiate into functional cardiomyocytes under specific culture conditions (Hakuno et al. 2002; Jackson et al. 2001; Jiang et al. 2002). Mesenchymal cells can be induced to differentiate into synchronously-beating cardiomyocytes in vitro after treatment of primary cultures of mouse bone marrow with the cytosine analog 5-azacytidine (Makino et al. 1999; Tomita et al. 1999). The differentiated cells presented the ultrastructural, genetic and biophysical characteristics of fetal ventricular myocytes, namely, the presence of sarcomeres and atrial granules around a central nucleus, the expression of a fetal cardiac gene profile and prolonged action potentials (Makino et al. 1999), as well as expression of functional adrenergic and muscarinic receptors (Jiang et al. 2002).

Several groups have provided evidence of bone-marrow-derived cardiac myocyte precursor cells. Administration of mononuclear cell preparations harvested from bone marrow has been reported to improve cardiac function in various models of myocardial injury (Tomita et al. 1999; Wang et al. 2001; Jackson et al. 2001; Orlic et al. 2001a; Toma et al. 2002; for review see Orlic et al. 2002). Toma and colleagues (2002) showed recently that transplantation of human MSC into the left ventricular wall of immunodeficient mice differentiate into cardiac myocytes without the need for myogenic differentiation prior to transplantation. Tomita et al. (1999) reported that transplantation of 4-azacytidine-treated bone marrow cells repopulated the scar and significantly improved left ventricular function in cryoinjured rat hearts. Wang et al. (2001) detected several cell types, including cardiomyocytes, endothelial cells, and fibroblasts, within and on the border of the scar 1 month after intracoronary delivery of retrovirally transduced isogenous bone marrow cells to infarcted rat hearts. This suggested that factors emanating from the injured myocardium may induce transdifferentiation of the bone marrow progenitors into the various cell types necessary for regeneration and maintenance of the infarcted myocardium. More recently, several groups reported evidence of extracardiac progenitors in necropsy specimens of hearts obtained from subjects that had undergone sex-mismatched heart (Quaini et al. 2002; Muller et al. 2002) or bone marrow transplantation (Deb et al. 2003). Quaini et al. (2002) and Muller et al. (2002) reported the presence of highly proliferative Y-chromosome-positive myocytes and vascular cells in myocardial specimens from male recipients that had received hearts from female donors. The recipient-derived cells expressed stem cell-related antigens, including c-kit, MDR1 and Sca-1 (Quaini 2002), and connected by gap junction with neighboring myocytes (Muller et al. 2002), indicating the ability of these precursor cells to develop into functional cardiomyocytes. In addition to the recipient-derived precursor cells, a significant number of highly proliferative host-derived primitive cells were detected in the infarcted myocardium, suggesting that these resident precursor cells may constitute a cardiac self-repair mechanism that may be potentiated by recruitment of marrow cardiogenic precursors. The regenerative capacity of this self-repair mechanism has, however, recently been questioned by two groups who have argued that the number of extracardiac progenitors that are capable of migrating to the heart is too small to induce effective long-term regeneration of the myocardium (Taylor et al. 2002; Laflamme et al. 2002).

Recently, systemic mobilization of bone marrow progenitors with cytokines has been investigated as a potential strategy for treatment of acute myocardial infarction. Treatment with stem cell factor (SCF) and granulocyte-colony stimulating factor (G-CSF) prior and immediately following infarction led to significant regeneration of infarcted myocardium and improvement in ventricular function, chamber dimensions and long-term survival in mice (Orlic et al. 2001b), suggesting that homing and transdifferentiation of bone marrow-derived cardiogenic precursors to sites of injury in the heart may constitute a natural repair mechanism. >From a clinical perspective, the effectiveness and simplicity of bone marrow mobilization protocols is attractive and may hold therapeutic potential as a noninvasive strategy for treatment of acute myocardial infarction.

Despite these promising findings, further work is required to establish the lineage of these precursors, the nature of the migratory and homing signals, the mechanism of transdifferentiation, and their role in myocardial regeneration. The identity of the putative bone marrow-derived cardiogenic precursor remains elusive. Jackson et al. (2001) demonstrated that injection of SP cells ($CD34^+$, $c-kit^+$) from bone marrow of LacZ-expressing Rosa26 transgenic mice into lethally irradiated mice prior to myocardial ischemia and reperfusion led to engraftment of donor-derived cells predominantly in the peri-infarct region, where they differentiate into cardiomyocytes and endothelial cells lining small vessels. However, the abundance of SP cells in the peri-infarct region was less than 0.02%, raising doubts that this population is the only source of bone marrow-derived mesenchymal cardiac progenitor cells. Orlic et al. (2001) reported that the injection of $c-kit^+/lin^-$ marrow cells from GFP-transgenic mice into the infarct border of syngeneic females with myocardial infarction formed new capillaries and cardiomyocytes and regenerated 68% of the infarcted myocardium. Up to 54% of the cells in the regenerated myocardium expressed GFP and stained positive for the Y chromosome, indicating that the regenerated tissue originated from the donor cells. The difficulty in expanding these cells in culture, however, may limit their therapeutic application. The mechanism mediating recruitment, homing and transdifferentiation of these progenitors to the injured myocardium is not known. SCF is rapidly induced in response to myocardial injury (Frangogiannis et al. 1998) and stromal-derived growth factor (SDF)-1 was shown to stimulate homing of angioblasts to ischemic myocardium (Hattori et al. 2001), suggesting that these cytokines may play a role in the migration and proliferation of bone marrow cardiac precursors in the infarcted heart. The molecular signals responsible for differentiation of the presumptive cardiogenic precursors have not been identified, and several recent studies have raised doubts about the plasticity of bone marrow-derived stem cells. In separate studies, Terada et al. (2002) and Ying et al. (2002) showed that mouse bone marrow or brain cells, respectively can fuse spontaneously with embryonic stem cells grown in co-culture and adopt the phenotype of the recipient cells. The authors suggest that this may be one mechanism by which the transplanted cells assume the phenotype of the surrounding host tissue. However, the relevance of this phenomenon in vivo, remains to be established. The optimal time for transplantation and survival of the grafted cells need to be defined.

In spite of these outstanding issues, two groups have recently treated patients that had suffered acute myocardial infarct with autologous bone marrow cells. Strauer et al. (2001) reported that intracoronary delivery of unfractionated autologous mononuclear bone marrow cells 6 days after infarction led to a reduction in infarct size and improvement in ventricular function and chamber geometry 10 weeks after transplantation. In a recent small-scale phase I clinical trial, Stamm and colleagues (2003) injected autologous AC133$^+$ bone marrow cells into the infarct border during CABG in six patients that had suffered earlier acute transmural myocardial infarction. The authors reported improved perfusion of the infarcted area and significant enhancement of global left ventricular function 3–9 months after surgery. The transplantation protocol appears to be safe and did not cause adverse cardiac effects (Galinanes et al. 2002). These findings should, however, be considered preliminary. Further characterization of the biology of these cells and clarification of the outstanding issues is necessary. Multicenter controlled trials will be needed in order to define the optimal time and method of delivery, the subpopulation and number of bone marrow cells required to achieve a sustained therapeutic benefit, and the survival of the transplanted cells. The question of whether transplantation should be performed soon after infarction or after the inflammatory process has resolved remains unsettled. Bone marrow-derived cells are very sensitive to hypoxia and inflammation, and a large number of the transplanted cells die soon after implantation (Toma et al. 2002). Strategies for improved cell survival, particularly around the time of transplantation, when the cells are most vulnerable, may need to be devised. Genetic engineering of the cells prior to grafting with vectors expressing survival genes and cytoprotective genes may help reduce peritransplantation cell death and improve the long-term survival of the graft. Zhang et al. (2001) has recently provided support for the feasibility of this approach by showing that the survival of grafted neonatal cardiac myocytes is greatly improved by adenoviral transduction of the cells with the survival gene Akt prior to transplantation.

Finally, the morphological, histological and functional complexity of the myocardium should not be overlooked when designing cell-based protocols for cardiac grafting. The myocardium consists of a variety of cell types, including cardiomyocytes, fibroblasts, vascular smooth muscle cells and endothelial cells embedded in a complex extracellular matrix that provides scaffolding for the three-dimensional alignment of the various components required for proper structural and mechanical function. This level of complexity raises a cautionary point against designing overly simplistic grafting protocols. It may be that the optimal grafting procedure for cardiac repair may require more than one cell type, for example, cardiomyocytes, fibroblasts and endothelial cells, to produce a graft that is able to recapitulate normal cardiac function.

5
Perspectives and Future Directions

The last decade has brought clarification of the molecular mechanisms underlying many of the most common cardiovascular diseases. This has led to the development of an array of gene and cell-based strategies with potential therapeutic value for treatment of these diseases. Some of these strategies have already made the transition from the preclinical phase into clinical trial and are now being considered for use in human patients, while several others are currently undergoing safety and feasibility evaluation in early phase trials. Notwithstanding these significant advances, we recognize the need for further developments in several aspects of cardiovascular gene therapy. Progress in vector and delivery technologies have not kept up with the identification of novel therapeutic targets, which continues to occur at a swift pace. All vectors currently in use for transfer of genetic material lack some of the desired features of the ideal vector. Emphasis needs to be put in the development of vectors that are safe and amenable to endogenous regulation and with the capability of conferring tissue specificity of transgene expression. Such a degree of spatial and temporal control over transgene expression will enhance the safety of human gene therapy protocols and potentially overcome many of the ethical issues that can arise as a result of nonspecific transgene expression such as germ cell line transmission. Much of this development can be carried out using current vector platforms. Rigorous systematic evaluation of the safety and efficacy of delivery strategies and improvement of delivery devices are also essential prerequisites for human gene therapy protocols.

The optimal genetic therapy for complex diseases such as coronary artery disease and myocardial infarction may require a combination of cell transplantation and pro-angiogenic gene therapy for long-term sustenance of the regenerated myocardium. Due to regulatory hurdles, such potentially synergistic combinatorial approaches have seldom been considered in the design of cardiovascular gene therapy strategies. Instead, the strategies have traditionally been developed around a single therapeutic target. We see future advances in gene and cell therapies linked to genomic research. Genomic profiling and screening is being employed for molecular phenotyping of patients and will permit the detection of disease-causing polymorphisms and the design of individualized therapies. The convergence of gene transfer technology and genomic technology will facilitate the elucidation of novel genes and may help uncover new roles for previously known genes, thereby leading to the discovery of novel therapeutic targets.

Acknowledgements. Dr. Melo is a New Investigator of the Heart and Stroke Foundation of Canada and is supported by grants from the Canadian Institutes of Health Research, Canadian Foundation of Innovation and the Health and Services Utilization and Research Commission of Saskatchewan. Dr. Dzau is supported by grants from the National Institutes of Health and is the recipient of a MERIT award from NIH. Dr. Pachori is the recipient of a postdoctoral fellowship from the American Heart Association

References

Agata J, Chao L, Chao J (2002) Kallikrein gene delivery improves cardiac reserve and attenuates remodeling after myocardial infarction. Hypertension 40:653–659

Agrawal RS, Muangman S, Melo LG et al (2001) Recombinant adeno-associated virus mediated antioxidant enzyme delivery as preventive gene therapy against ischemia-reperfusion injury of the rat myocardium. Mol Ther 3:A837

Akhtar S, Hughes MD, Khan A et al (2000). The delivery of antisense therapeutics. Adv Drug Del Rev 44:3-21

Akhter SA, Skaer CA, Kypson AP et al (1997) Restoration of beta-adrenergic signaling in failing cardiac ventricular myocytes via adenoviral-mediated gene transfer. Proc Natl Acad Sci USA 94:12100-12105

Alexander MY, Webster KA, McDonald PH et al (1999) Gene transfer and models of gene therapy for the myocardium. Clin Exp Pharmacol Physiol 26:661-668

Antos CL, McKinsey TA, Frey N et al (2002) Activated glycogen synthase kinase 3-β suppresses cardiac hypertrophy in vivo. Proc Natl Acad Sci USA 99:907-912

Aoki M, Morishita R, Taniyama Y et al (2000) Therapeutic angiogenesis induced by hepatocyte growth factor: potential gene therapy for ischemic diseases. J. Atheroscler Thromb 7:71-76

Applebaum-Bowden D, Kobayashi J, Kashyap VS et al (1996) Hepatic lipase gene therapy in hepatic lipase-deficient mice: Adenovirus-mediated replacement of a lipolytic enzyme to the vascular endothelium. J Clin Invest 97:799-805

Armstrong PW, Moe GW (1994). Medical advances in the treatment of congestive heart failure. Circulation 88:2941-2952

Asahara T, Murohara T, Sullivan A et al (1997) Isolation of putatitve progenitor endothelial cells for angiogenesis. Science 275:964-967

Asahara T, Masuda H, Takahashi T et (1999) Bone marrow origin of endothelial progenitor cells responsible for postnatal vasculogenesis in physiological and pathological neovascularization Circ Res 85:221-228

Asakura M, Kitakaze M, Taskashima S et al (2002) Cardiac hypertrophy is inhibited by antagonism of ADAM12 processing of HB-EGF: Metalloproteinase inhibitors as a new therapy. Nature Med 8:35-40

Baker Dl, Hashimoto K, Grupp IL et al (1998) Targeted overexpression of the sarcoplasmic reticulum Ca^{2+}ATPase increases cardiac contractility in transgenic mouse hearts. Circ Res 83:1205-1214

Balser JR (2002) Inherited sodium channelopathies: models for acquired arrhytmias? Am J Physiol 282:H1175-H1180

Bashir R, Vale PR, Isner JM et al (2002) Angiogenic gene therapy: pre-clinical studies and phase I clinical data. Kideny Int 61 (Suppl 1):110-114

Beeri R, Guerrero JL, Supple G et al (2002) New efficient catheter-based system for myocardial gene delivery. Circulation 106:1756-1759

Beltrami AP, Urbanek K, Kajstura J et al (2001) Evidence that human cardiac myocytes divide after myocardial infarction. New Engl J Med 344:175-1757

Bennett MR, O'Sullivan MO (2001) Mechanisms of angioplasty and stent restenosis: implications for design of rational therapy. Pharmacol Ther 91:149-166

Bers DM (2002) Cardiac excitation-contraction coupling. Nature 415:198-205

Boekstegers P, Degenfeld C von, Giehrl W et al (2000) Myocardial gene transfer by selective pressure-regulated retroinfusion of coronary veins. Gen Ther 7:232-240

Braun-Dullaeus RD, Mann MJ, Dzau VJ (1998) Cell cycle progression. New therapeutic target for vascular proliferative disease. Circulation 98:82-89

Brauner R, Nonoyama M, Laks H et al (1997) Intracoronary adenovirus-mediated transfer of immunosuppressive cytokine genes prolongs allograft survival. J Thorac Cardiovasc Surg 114:923-933

Braunwald E, Kloner RA (1985) Myocardial reperfusion: a double edged sword. J Clin Invest 76:1713-1719

Brocheriou V, Hagege AA, Oubenaissa A et al (2000) Cardiac functional improvement by a human Bcl-2 transgene in a mouse model of ischemia/reperfusion injury. J Gene Med 2:326-333

Bruneau BG (2002) Transcriptional regulation of vertebrate cardiac morphogenesis 90:509-519

Carden DL, Granger DN (2000) Pathophysiology of ischemia-reperfusion injury. Am J Pathol 190:255-266

Carmeliet P (2000a) Mechanisms of angiogenesis and arteriogenesis. Nature Medicine 6:389-395

Carmeliet P (2000b) VEGF gene therapy: stimulating angiogenesis or angioma-genesis. Nature Medicine 6:1102-1103

Cartier R, Reszka R (2002) Utilization of synthetic peptides containing nuclear localization signals for non viral gene transfer systems. Gene Ther 9:157-167

Chang MW, Barr E, Lu MM et al (1995a) Adenovirus-mediated overexpression of the cyclin/cyclin dependent kinase inhibitor, p21 inhibits vascular smooth muscle proliferation and neointima formation in the rat carotid artery model of balloon angioplasty. J Clin Invest 96:2260-2268

Chang MW, Barr E, Seltzer J et al (1995b) Cytostatic gene therapy for vascular proliferative disorders with a constitutively active form of the retinoblastoma gene product. Science 267:518-522

Channon KM, Qian HS, George SE (2000) Nitric oxide synthase in atherosclerosis and vascular injury. Insights from experimental gene therapy. Arterioscler Throm Vasc Biol 20:1873-1881

Chatterjee S, Stewart AS, Bish LT et al (2002) Viral gene transfer of the antiapoptotic factor Bcl-2 protects against chronic ischemic heart failure. Circulation 106 (Suppl): I212-I217

Chen D, Krasinski K, Sylvester A et al (1997) Downregulation of cyclin-dependent kinase 2 activity and cyclin A promoter activity in vascular smooth muscle cells by p27 (KIP1), an inhibitor of neointima formation in the rat carotid artery. J Clin Invest 99:2334-2341

Chen EP, Bittner HB, Davis RD et al (1998) Physiological effects of extracellular superoxide dismutase transgene overexpression on myocardial function after ischemia and reperfusion injury. J Thorac Cardiovasc Surg 115:450-458

Chen SJ, Rader DJ, Tazelaar J et al (2000) Prolonged correction of hyperlipidemia with familial hypercholesterolemia using an adeno-associated viral vector expressing very low density lipoprotein receptor. Mol Ther 2:256-261

Chen H, Mohuczy D, Li D et al (2001) Protection against ischemia/reperfusion injury and myocardial dysfunction by antisense-oligodeoxyynucleotide directed at angiotensin-converting enzyme mRNA. Gene Ther 8:804-810

Chien KR (2000) Genomic circuits and the integrative biology of cardiac disease. Nature 407:227-232

Chirmule N, Propert K, Magosin S et al (1999) Immune response to adenovirus and adenoassociated virus in humans. Gene Ther 6:1574-1583

Chowdhury JR, Grossman M, Gupta S et al(1991) Long-term improvement of hypercholesterolemia after ex vivo gene therapy in LDLR-deficient rabbits. Science 254:1802-1805

Chu Y, Iida S, Lund DD et al (2003) Gene transfer of extracellular superoxide dismutase reduces arterial pressure in spontaneously hypertensive rats: Role of heparin binding domain. Circ Res 92:461-468

Coffin RS, Howard MK, Cummings DV et al (1996). Gene delivery to the heart in vivo and to cardiac myocytes and vascular smooth muscle cells in vitro using herpes virus vectors. Gene Ther 3:560-566

Colucci WS (1997) Molecular and cellular mechanisms of myocardial failure. Am J Cardiol 80 (11A): 15L-25L

Daly G, Chernajovski Y (2000) Recent developments in retroviralk-mediated gene transduction. Mol. Ther 2:423-434

Datwyler DA, Eppenberger HM, Koller D et al (1999) Efficient gene delivery into adult cardiomyocytes by recombinant Sindis virus. J Mol Med 77:859–864

D'Agostino RB, Russel MW, Huse DM et al (2000) Primary and subsequent coronary risk appraisal: new results from the Framingham study. Am Heart J 139:272–281

Deb A, Wang S, Skelding KA et al (2003) Bone marrow-derived cardiomyocytes are present in adult human heart. A study of gender-mismatched bone marrow transplantation patients. Circulation 107:1247–1249

Del Monte F, Harding SE, Schmidt U et al (1999) Restoration of contractile function in isolated cardiomyocytes from failing human hearts by gene transfer of SERCA2a. Circulation 100:2308–2311

Del Monte F, Williams E, Lebeche D et al (2001) Improvement in survival and cardiac metabolism after gene transfer of sarcoplasmic reticulum Ca^{2+}-ATPase in a rat model of heart failure. Circulation 104:1424–1429

Del Monte F, Harding SE, Dec W et al (2002) Targeting phospholamban by gene transfer in human heart failure. Circulation 105:904–907

Dicheck DA, Anderson J, Kelly AB et al (1996) Enhanced antithrombotic effects of endothelial cells expressing recombinant plasminogen activators transduced with retroviral vectors. Circulation 93:301–309

Dimmeler S, Aicher A, Vasa M et al (2001) HMG-CoA reductase inhibitors (statins) increase endothelial progenitor cells via the PI3-kinase/Akt pathway. J Clin Invest 108:391–397

Dobrzynski E, Wang C, Chao J et al (2000) Adrenomedullin gene delivery attenuates hypertension, cardiac remodeling and renal injury in deoxycorticosterone acetate salt hypertensive rats. Hypertension 36:995–1001

Donahue JK, Heldman AW, Fraser H et al (2000) Focal modification of electrical conduction in the heart by viral gene transfer. Nature Med 6:1395–1398

Dzau VJ, Mann MJ, Morishita R et al (1996) Fusigenic viral liposome for gene therapy in cardiovascular diseases. Proc Natl Acad Sci U S A 93:11421–11425

Ehsan A, Mann MJ, Dell'Acqua G et al (2002) Endorhelial healing in vein grafts. Proliferative burst is unimpaired by genetic therapy of neointimal disease. Circulation 105:1686–1692.

Ennis IL, Li RA, Murphy AM et al (2002) Dual gene therapy with SERCA1 and Kir2.1 abbreviates excitation without suppressing contractility. J Clin Invest 109:393–400

Eizema K, Fechner H, Bexstarosti K et al (2000) Adenovirus-based phospholamban antisense expression as a novel approach to improve cardiac contractile function. Circulation 101:2193–2199

Feeley BT, Poston RS, Park AK et al (2000) Optimization of ex vivo pressure mediated delivery of antisense oligodeoxynucleotides to ICAM-1 reduces reperfusion injury in rat cardiac allografts. Transplantation 69:1067–74

Feldman LJ, Isner JM (1995) Gene therapy for vulnerable plaque. J Am Coll Cardiol 26:826–833

Frangogiannis NG, Perrard JL, Mendoza LH et al (1998) Stem cell factor induction is associated with mast cell accumulation after canine myocardial ischemia and reperfusion. Circulation 98:687–698

Fromes Y, Salmon A, Wang X et al (1999) Gene delivery to the myocardium by intrapericardial injection. Gene Ther 6:683–688

Fuchs S, Baffour R, Zhou YF et al (2001) Transendocardial delivery of autologous bone marrow enhances collateral perfusion and regional function in pigs with chronic experimental myocardial ischemia. J Am Coll Cardiol 37:1726–1732

Fukui T, Yoshiyama M, Hanatani A et al (2001) Expression of p22-phox and gp91-phox, essential components of the NADPH oxidase, increases after myocardial infarction Biochem Biophys Res Commun 281:1200–1206

Funk M, Krumholz, HM (1996) Epidemiology and economic impact of advanced heart failure. J. Cardiovasc. Nurs. 10:1-10

Galinanes M, Loubani M, Davies J et al (2002) Safety and efficacy of transplantation of autologous bone marrow into scarred myocardium for the enhancement of cardiac function in man. Circulation 106 (Suppl II):II-463

Giordano FJ, Ping P, McKirnan MD et al (1996) Intracoronary gene transfer of fibroblast growth factor-5 increases blood flow and contractile function in an ischemic region of the heart. Nat Med 2:534–539

Givertz MM, Colucci WS (1998) New targets for heart failure therapy: endothelin, inflammatory cytokines, and oxidative stress. Lancet 352 (Suppl 1): S134-S138

Golino P, Cirillo P, Calabro P et al (2001) Expression of exogenous tissue factor pathway inhibitor in vivo suppresses thrombus formation in injured rabbit carotid arteries. J Am Coll Cardiol 38:569–576

Griffiths I, Binley K, Iqball S et al (2000) The macrophage—a novel system to deliver gene therapy to pathological hypoxia. Gene Ther 7:255–262

Grines CL, Watkins MW, Helmer G et al (2002) Angiogenic gene therapy (AGENT) trial in patients with stable angina pectoris. Circulation 105:1291–1297

Grossman M, Rader DJ, Muller DW et al (1995) A pilot study of ex vivo gene therapy for homozygous familial hypercholesterolaemia. Nat Med 1:1148–1154

Gunn J, Holt CM, Francis SE et al (1997) The effect of oligonucleotides to c-myb on vascular smooth muscle cell proliferation and neointima formation after porcine coronary angioplasty. Circ Res 80:520–531

Hajjar RJ, Kang, Gwathmey JK et al (1997) Physiological effects of adenoviral gene transfer of sarcoplasmic reticulum ATPase in isolated rat myocytes. Circulation 95:423–429

Hajjar RJ, del Monte F, Matsui T et al (2000) Prospects for gene therapy for heart failure. Circulation 86:616–621

Hakuno D, Fukuda K, Makino S et al (2002) Bone marrow-derived regenerated cardiomyocytes (CMG cells) express functional adrenergic and muscarinic receptors. Circulation 105:380–386

Hall WD (1999) Risk reduction associated with lowering systolic blood pressure: review of clinical trial data. Am. Heart J. 138:225–230

Hammond HK, McKirman MD (2001) Angiogenic gene therapy for heart disese: a review of animal studies and clinical trials. Cardiovasc Res 49–561–567

Hannon GJ (2002) RNA interference. Nature 418:244–251.

Harris JD, Schepelmann S, Athanasopoulos T et al (2002) Inhibition of atherosclerosis in apolipoprotein-E-deficient mice following muscle transduction with adeno-associated virus vectors encoding human apoliprotein-E. Gene Ther 9:21–29

Harrison RL, Byrne BJ, Tung L (1998) Electroporation-mediated gene transfer in cardiac tissue. FEBS Lett 435:1–5

Hartigan-O'Connor D, Amalfitano A, Chamberlain JS (1999) Improved production of gutted adenovirus in cells expressing adenovirus preterminal protein and DNA polymerase. J Virol 73:7835–7841

Hattori K, Heissig B, Tashiro K et al (2001) Plasma elevation of stromal cell-derived factor-1 induces mobilization of mature and immature hematopoietic progenitor and stem cells. Blood 97:3354–3360

Herttuala S-Y, Martin JF (2000) Cardiovascular gene therapy. Lancet 355:213–222

Holly TA, Drincic A, Byun Y et al (1999) Caspase inhibition reduces myocyte cell death induced by myocardial ischemia and reperfusion in vivo. J Mol Cell Cardiol 31:1709–1715

Hoppe UC, Marban E, Johns DC (2001) Distinct gene-specific mechanisms of arrhytmia revealed by cardiac gene transfer of two long QT disease genes, HERG and KCNE1. Proc Natl Acad Sci USA 98:5335–5340

Hu, W-S, Pathak VK (2000) Design of retroviral vectors and helper cells for gene therapy. Pharmacol Rev 52:493–511

Ikeda Y, Ross J (2000) Models of dilated cardiomyopathy in the mouse and the hamster. Curr Opin Cardiol 15:197–201

Ikeda Y, Gu Y, Iwanaga Y et al (2002) Restoration of deficient membrane proteins in the cardiomyophatic hamster by in vivo cardiac gene transfer. Circulation 105:502–508

Ikenaga S, Hamano K, Nishida M et al (2001) Autologous bone marrow implantation induced angiogenesis and improved deteriorated exdercise capacity in a rat ischemic hindlimb model. J Surg Res 96:277–283

Inoue S, Egashira K, Ni W et al (2002) Anti-monocyte chemoattractant protein-1 gene therapy limits progression and destabilization of established atherosclerosis in apolipoprotein E-knockout mice. Circulation 106:2700–2706

Isner JM (2002) Myocardial gene therapy. Nature 415:234–239

Iwaguro H, Yamaguchi J, Kalka C et al (2002) Endothelial progenitor cell vascular endothelial growth factor gene transfer for vascular regeneration. Circulation 105:732–738

Jackson K, Majka SM, Wang H et al (2001) Regeneration of ischemic cardiac muscle and vascular endothelium by adult stem cells. J Clin Invest 107:1395–1402

Jiang Y, Jahagirdar BN, Reinhardt RL et al (2002) Pluripotency of mesenchymal stem cells derived from adult marrow. Nature 418:41–49

Juan SH, Lee TS, Tseng KW et al (2001) Adenovirus-mediated heme oxygenase-1 gene transfer inhibits the development of atherosclerosis in apolipoprotein E-deficient mice. Circulation 104:1519–1525

Kalka C, Masuda H, Takahashi T et al (2000) Transplantation of ex vivo expanded endothelial progenitor cells foe therapeutic neovascularization. Proc Natl Acad Sci USA 97:3422–3427

Kannel WB, Belanger AJ (1991) Epidemiology of heart failure. Am Heart J 12:951–957

Kaplitt MG, Xiao X, Samulski RJ et al (2000) Long term gene transfer in porcine myocardium after coronary infusion of and adeno-associated virus vector. Ann Thorac Surg 62:1669–1676

Katovich MJ, Gelband CH, Reaves PW et al (1999) Reversal of hypertension by angiotensin II type I receptor antisense gene therapy in the adult SHR rat. Am J Physiol 277: H1260-H1264

Kawada T, Nakazawa M, Nakauchi S et al (2002) Rescue of hereditary form of dilated cardiomyopathy by rAAV-mediated somatic gene therapy: Amelioration of morphological findings, sarcolemmal permeability, cardiac performance and the prognosis of TO-2 hamsters. Proc Natl Acad Sci USA 99:901–906

Kawamoto A, Gwon H-C, Iwaguro H et al (2001) Therapeutic potential of ex vivo expanded endothelial progenitor cells for myocardial ischemia. Circulation 103:634–637

Kereiakes DJ (1998) Preferential benefit of platelet glycoprotein IIb/IIIa receptor blockade: specific considerations by device and disease state. Am J Cardiol 81 (7A): 49E-54E

Kibbe MR, Billiar TR, Tzeng E (2000) Gene therapy for restenosis. Circ Res 86:829–833

Kim AY, Wallinsky PL, Kolodgie FD (2002) Early loss of thrombomodulin expresson impairs vein graft thromboresistance: implications for vein graft resistance. Circ Res 90:205–212

Kimura B, Mohuczy D, Tang X et al (2001) Attenuation of hypertension and heart hypertrophy by adeno-associated virus delivering angiotensin antisense. Hypertension 37:376–380

Kingma JG Jr, Plante S, Bogaty P (2000) Platelet GIIb/IIIa receptor blockade reduces infarct size in a canine model of ischemia-reperfusion. J Am Coll Cardiol 36:2317–2324

Knopp RH (1999) Drug treatment of lipid disorders. New Engl J Med 341:498–411

Kocher AA, Schuster MD, Szabolcs MJ et al (2001) Neovascularization of ischemic myocardium by human bone marrow-derived angioblasts prevents cardiomyocyte apoptosis, reduces remodeling and improves cardiac function. Nature Med 4:430–436

Krasnykh VN, Douglas JT, van Beusechem VW (2000) Gene targeting of adenoviral vectors. Mol Ther 1:391–405

Krause DS (2002) Plasticity of marrow-derived cells. Gene Ther 9:754–758

Kullo IJ, Simari Rd, Schwartz RS (1999) Vascular gene transfer. From bench to bedside. Arterioscler Thromb Vasc Biol 19:196–207

Labhasetwar V, Bonadio J Goldstein S et al (1998) A DNA controlled-release coating for gene transfer: transfection in skeletal and cardiac muscle. J Pharm Sci 87:1347–1350

Laflamme MA, Myerson D, Saffitz JE et al (2002) Evidence for cardiomyocyte repopulation by extracardiac progenitors in transplanted human hearts. Circ Res 90:634–640

Laukkanen J, Lehtolainen P, Gough PJ et al (2000) Adenovirus-mediated gene transfer of a secreted form of human macrophage scavenger receptor inhibits modified low-density lipoprotein degradation and foam cell formation in macrophages. Circulation 101:1091–1096

Laukkanen MO, Kivela A, Rissane T et al (2002) Adenovirus-mediated extracellular superoxide dismutase gene therapy reduces neointima formation in balloon-denuded rabbit aorta. Circulation 106:1999–2003

Lee RJ, Springer ML, Blanco-Bose WE et al (2000) VEGF gene delivery to myocardium. Deleterious effect of upregulated expression. Circulation 102:898–901

Lee Y, Lee WH, Lee SC et al (1999) CD40L activation in circulating platelets in patients with acute coronary syndrome Cardiology 92:11–16

Lee LY, Patel SR, Hackett NR et al (2000) Focal angiogen therapy using intramyocardial delivery of an adenovirus vector coding for vascular endothelial growth factor 121. Ann Thorac Surg 69:14–24

Leri A, Fiordaliso F, Setoguchi M et al (2000) Inhibition of p53 function prevents renon angiotensin system activation and stretch-mediated myocyte apoptosis. Am J Pathol 157:843–857

Li F, Wang X, Capasso JM et al (1996) Rapid transition of cardiac myocytes from hyperplasia to hypertrophy during postnatal development. J Mol Cell Cardiol 28:1737–1746

Li Q, Li B, Wang X et al (1997) Overexpression of insulin-like growth factor-1 in mice protects from myocyte death after infarction, attenuating ventricular dilation, wall stress, and cardiac hypertrophy. J Clin Invest 100:1991–1999

Li Q, Bolli R, Qiu Y et al (2001) Gene therapy with extracellular superoxide dismutase protects conscious rabbits against myocardial infarction. Circulation 103:1893–1898

Li RK, Mickle DA, Weisel RD et al (1997) Natural history of fetal rat cardiomyocytes transplanted into adult rat myocardial scar tissue. Circulation 96 (Suppl II): II-179-II186

Li S, Huang L (2000) Nonviral gene therapy: promises and challenges. Gene Ther 7:31–34

Lin KF, Chao J, Chao L (1995) Human atrial natriuretic peptide gene delivery reduces blood pressure in hypertensive rats. Hypertension 26:847–853

Lin KF, Chao L, Chao J (1997) Prolonged reduction of high blood pressure with human nitric oxide synthase delivery. Hypertension 30:307–3113

Lin KF, Chao J, Chao L (1998) Atrial natriuretic peptide gene delivery attenuates hypertension, cardiac hypertrophy and renal injury in salt-sensitive rats. Hum Gene Ther 9:1429–1438

Loscalzo J (2001) Nitric oxide insufficiency, platelet activation, and arterial thrombosis. Circ Res 88:756–762

Losordo DW, Vale PR, Symes JF et al (1998) Gene therapy for myocardial angiogenesis. Initial clinical results with direct myocardial injection of phVEGF$_{165}$ as sole therapy for myocardial ischemia Circulation 98:2800–2804

Losordo DW, Vale PR, Isner JM (1999) Gene therapy for myocardial angiogenesis. Am Heart J 138:S132-S141

Maass A, Leinwand LA (2000) Animal models of hypertrophic cardiomyopathy. Curr Opin Cardiol 15:189–196

Mack CA, Patel SA, Schwarz EA et al (1998) Biological bypass with the use of adenovirus-mediated transfer of the complementary deoxyribonucleic acid for vascular endothe-

lial growth factor 121 improves myocardial perfusion and function in the ischemic porcine heart. J Thorac Cardiovasc Surg 115:168–177
Magnani M, Rossi L, Fraternale A et al (2002) Erythrocyte-mediated delivery of drugs, peptides and modified oligonucleotides. Gene Ther 9:749–751
Mah C, Byrne BJ, Flotte TR(2002) Virus-based gene delivery systems. Clin Pharmacokinet 41:901–911
Makino S, Fukuda K, Miyoshi S et al (1999) Cardiomyocytes can be generated from marrow stromal cells in vitro. J Clin Invest 103:697–705
Makino N, Sugano M, Ohtsuka S et al (1999) Chronic antisense therapy for angiotensinogen on cardiac hypertrophy in spontaneously hypertensive rats. Cardiovasc Res 44:43–548
Mann DL (1999) Mechanisms and models in heart failure. A combinatorial approach. Circulation 100:999–108
Mann MJ, Gibbons GH, Tsao PS et al (1997) Cell cycle inhibition preserves endothelial function in genetically-engineered rabbit vein grafts. J Clin Invest 99:1295–1301
Mann MJ, Gibbons GH, Hutchinson H et al (1999) Pressure-mediated oligonucleotide transfection of rat and human cardiovascular tissues. Proc Natl Acad Sci USA 96:6411–6416
Mann MJ, Whittemore AD, Donaldson MC et al (1999) Ex-vivo gene therapy of human vascular bypass grafts with E2F decoy: The PREVENT single-centre, randomized, controlled trial. Lancet 354:1493–1498
Marban E (2002) Cardiac channelopathies. Nature 415:213–218
Martens JR, Reaves PY, Lu D et al (1998) Prevention of renovascular and cardiovascular pathophysiological changes in hypertension by angiotensin II type I receptor antisense gene therapy. Proc Natl Acad Sci USA 95:2664–2669
Matsui T, Li L, Del Monte F et al (1999) Adenoviral gene transfer of activated phosphatidylinositol 3'-kinase and Akt inhibits apoptosis of hypoxic cardiomyocytes in vitro. Circulation 100:2373–2379
Maurice JP, Hata JA, Shah AS et al (1999) Enhancement of cardiac function after adenoviral-mediated in vivo intracoronary β_2-adrenergic receptor gene delivery. J Clin Invest 104:21–29
McCarthy M (2001) Molecular decoy may keep bypass grafts open. Lancet 358:1703
McMurray JC, Pfeffer MA (2002) New therapeutic options in congestive heart failure. Circulation 105:2099–2106
Melo LG, Agrawal R, Zhang L et al (2002) Gene Therapy strategy for long term myocardial protection using adeno-associated virus-mediated delivery of heme oxygenase gene. Circulation 105:602–607
Mehta JL, Li DY (1999) Inflammation in ischemic heart disease: Response to tissue injury or a pathogenic villain? Cardiovasc Res 43:291–299
Miao W, Luo Z, Kitsis RN et al (2000) Intracoronary, adenovirus-mediated Akt gene transfer in heart limits infarct size following ischemia-reperfusion injury in vivo. J Mol Cell Cardiol 32:2397–2402
Michelle DE, Metzger JM (2000) Contractile dysfunction in hypertrophic cardiomyopathy: Elucidating primary defects of mutant contractile proteins by gene transfer. Trends Cardiovasc Med 10:177–182
Milano CA, Allen LF, Rockman HA et al (1994) Enhanced myocardial function in transgenic mice overexpressing the beta 2-adrenergic receptor. Science 264:582–586
Min JY, Yang Y, Converso KL et al (2002) Transplantation of embryonic stem cells improves cardiac function in postinfarcted rats. J Appl Physiol 92:288–296
Miyamoto MI, del Monte F, Schmidt U et al (2000) Adenoviral gene transfer of SERCA2a improves left-ventricular function in aortic-banded rats in transition to heart failure. Proc Natl Acad Sci USA 97:793–798
Monahan PE, Samulski RJ (2000) Adeno-associated virus vectors for gene therapy: more pros than cons? Mol Med Today 6:433–440

Morishita R, Gibbons GH, Ellison KE et al (1993) Single intraluminal delivery of antisense cdc2 kinase and proliferating cell nuclear antigen oligonucleotides results in chronic inhibition of neointimal hyperplasia. Proc Natl Acad Sci USA 90:8474–8478

Morishita R, Gibbons GH, Ellison KE et al (1994) Intimal hyperplasia after vascular injury is inhibited by antisense cdk2 kinase oligonucleotides. J Clin Invest 93:1458–1464

Morishita R, Gibbons GH, Ellison KE et al (1995) A gene therapy strategy using a transcription factor decoy of the E2F binding site inhibits smooth muscle proliferation in vivo. Proc Natl Acad Sci 92:5855–5859

Morishita R, Sugimoto T, Aoki M et al (1997). In vivo transfection of cis element "decoy" against nuclear factor factor κB binding sites prevents myocardial infarction. Nat Med 3:894–899

Morishita R, Higaki J, Tomita N et al (1998) Application of transcription factor "decoy" "strategy" strategy as a means of gene thereapy and study of gene expression in cardiovascular disease. Circ Res 82:1023–1028

Morishita R, Gibbons GH, Tomita N et al (2000) Antisense oligodeoxynucleotide inhibition of vascular angiotensin converting enzyme expression attenuates neointimal formation. Arterioscler Thromb Vasc Biol 20:915–922

Morse D, Choi AMK (2002) Heme oxygenase-1. The "emerging molecule" has arrived. Am J Resp Cell Mol Biol 27:8–16

Muller P, Pfeiffer P, Koglin J et al (2002) Cardiomyocytes of non cardiac origin in myocardial biopsies of human transplanted hearts. Circulation 106:31–35

Murohara T, Ikeda H, Duan J et al (2000) Transplanted chord blood-derived endothelial precursor cells augment postnatal neovascularization. J Clin Invest 105:1527–1536

Nathan C, Xie QW (1994) Nitric oxide synthases: roles, tools and controls. Cell 78:915–918

Napoli C, Ignarro LJ (2001) Nitric oxide and atherosclerosis. Nitric Oxide 5:88–97

Nicklin SA, Buening H, Dishart KL et al (2001). Efficient and selective AAV-2 mediated gene transfer directed to human vascular endothelial cells. Mol Ther 4:174–181

Nishida T, Ueno H, Atsuchi N et al (1999). Adenovirus-mediated local expression of human tissue factor pathway inhibitor eliminates shear-stress induced recurrent thrombosis in the injured carotid artery of the rabbit. Circ Res 84:1446–1452

Nozato T, Ito H, Watanabe M et al (2001) Overexpression of Cdk inhibitor p16 by adenovirus vector inhibits cardiac hypertrophy in vitro and in vivo: a novel strategy for the gene therapy of cardiac hypertrophy. J Mol Cell Cardiol 33:1493–1504

Numaguchi Y, Naruse K, Harada M et al (1999) Prostacyclin synthase gene transfer accelerates reendothelialization and inhibits neointimal formation in rat carotid arteries after balloon injury. Arterioscler Throm Vasc Biol 19:727–733

Nuss, HB, Marban E, Johns DC (1999) Overexpression of a human potassium channel suppresses cardiac hyperexcitability in rabbit ventricular myocytes. J Clin Invest 103:889–896

Oka K, Pastore L, Kim IH et al (2001) Long-term stable correction of low density lipoprotein receptor-deficient mice with a helper dependent adenoviral vector expressing the very low density lipoprotein receptor. Circulation 103:1274–1281

Okudo S, Wildner O, Shah MR et al (2001) Gene transfer of heat shock protein 70 reduces infarct size in vivo after ischemia/reperfusion in the rabbit heart. Circulation 103:877–881

Orlic D, Kajstura J, Chimenti S et al (2001a) Bone marrow cells regenerate infarcted myocardium. Nature 410:710–705

Orlic D, Kajstura J, Chimenti S et al (2001b) Mobilized bone marrow cells repair the infarcted heart, improving function and survival. Proc Natl Acad Sci USA 98:10344–10349

Orlic D, Hill JM, Arai AE (2002) Stem cells for myocardial regeneration. Circ Res 91:1092–1102

Pachori AS, Numan MT, Ferrario CM et al (2002) Blood pressure-independent attenuation of cardiac hypertrophy by AT(1)R-AS gene therapy. Hypertension 20:969–975

Park JL, Lucchesi BR (1999) Mechanisms of myocardial reperfusion injury. Ann Thorac Surg 68:1905–1912

Pfeffer JM, Pfeffer MA, Fletcher PJ et al (1991) Progressive ventricular remodelling in rat myocardial infarction. Am J Physiol 260:H14106–H1414

Peterson JT, Li H, Dillon L et al (2000). Evolution of metlloprotease and tissue inhibitor expression during heart failure progression in the infarcted heart. Cardiovasc Res 46:307–315

Poston RS, Tran KP, Mann MJ et al (1998) Prevention of ischemically-induced neointimal hyperplasia using ex vivo antisense oligodeoxynucleotides. J Heart Lung Transplant 17:349–1355

Poston RS, Mann MJ, Hoyt EG et al (1999) Antisense oligodeoxynucleotides prevent acute cardiac allograft rejection via a novel, non-toxic, highly efficient transfection method. Transplantation 68:825–832

Prentice H, Bishopric N, Hicks MN et al (1997) Regulated expression of a foreign gene targeted to the ischemic myocardium. Cardiovasc Res 35:567–574

Qian H, Neplioueva V, Shetty GA et al (1999) Nitric oxide synthase gene therapy rapidly reduces molecule expression and inflammatory cell infiltration in carotid artery of cholesterol-fed rabbits. Circulation 99:2979–2982

Quaini F, Urbanek K, Beltrami AP et al (2002) Chimerism of the transplanted heart. New Engl J Med 346:5–15

Rader DJ, Tietge UJ (1999) Gene therapy for dyslipidemia: clinical prospects. Curr Atheroscler Rep 1:58–69

Rebar EJ, Huang Y, Hickey R et al (2002) cInduction of angiogenesis in a mouse model using engineered transcription factors. Nature Medicine 8:1427–1432

Reinlib L, Field L (2000) Cell transplantation as future therapy for cardiovascular disease? Circulation 101:e192-e197

Rekhter MD, Simari RD, Work CW et al (1998) Gene transfer into normal and atherosclerotic human blood vessels. Circ Res 82:1243–1252

Rentrop KP (2000). Thrombi in acute coronary syndromes. Revised and revisited. Circulation 101:1619–1626

Rinaldi M, Catapano AL, Parrella P et al (2000) Treatment of severe hypercholesterolemia in apolipoprotein E-deficient mice by intramuscular injection of plasmid DNA. Gene Ther 7:1795–1801

Robbins PD, Ghivizzani SC (1998) Viral vectors for gene therapy. Pharmacol. Ther. 80:35–47

Rockman HA, Koch WJ, Lefkowitz RJ (2002) Seven-transmembrane spanning receptors and heart function. Nature 415:206–212

Rosengart TK, Lee LY, Patel SR et al (1999) Angiogenesis gene therapy: phase I assessment of direct intramyocardial administration of an adenovirus vector expressing $VEGF_{121}$ cDNA to individuals with clinically significant severe coronary artery disease. Circulation 100:468–474

Roth DM, Gao MH, Lai C et al (1999) Cardiac-directed adenylyl cyclase expression improves heart function in murine cardiomyopathy. Circulation 99:3099–3102

Roth DM, Bayat H, Drumm JD et al (2002) Adenylyl cyclase increases survival in cardiomyopathy. Circulation 105:1989–1994

Sabaaway HE, Zhang F, Nguyen X et al (2001) Human heme oxygenase-1 gene transfer lowers blood pressure and promotes growth in spontaneously hypertensive rats. Hypertension 38:210–215

Sakoda T, Kasahara N, Hamamori Y et al (1999) A high titer lentiviral production system mediates transduction of differentiated cells including beating cardiac myocytes. J Mol Cell Cardiol 31:2037–2047

Sayeed-Shah U, Mann MJ, Martin J (1998) Complete reversal of ischemic wall motion abnormalities by combined use of gene therapy with transmyocardial laser revascularization. J Thorac Cardiovasc Surg 116:763–769

Schlesinger S (2001) Alphavirus vectors: development and potential therapeutic applications. Expert Opin Biol Ther 1:177–191

Schmidt U, Hajjar RJ, Helm PA et al (1997) Contribution of abnormal sarcoplasmic reticulum ATPase activity to systolic and diastolic function in human heart failure. J Mol Cell Cardiol 30:1929–1937

Schmidt U, del Monte F, Miyamoto MI et al (2000) Restoration of diastolic function in senescent rat hearts through adenoviral gene transfer of sarcoplasmic reticulum Ca^{2+}-ATPase. Circulation 101:790–796

Schratzberger P (2001) Ultrasound enhances therapeutic gene expression in ischemic pig myocardium. J Am Coll Cardiol 37:266A

Schwartz RS (1998) Pathophysiology of restenosis: Interaction of thrombosis, hyperplasia, and/or remodeling. Am J Cardiol 81(7A):14E-17E

Shah AS, Lilly RE, Kypson AP et al (2000) Intracoronary adenovirus-mediated delivery and overexpression of the beta(2)-adrenergic receptor in the heart: prospects for molecular ventricular assistance. Circulation 101:408–414

Shah AS, White DC, Emani S et al (2001) In vivo ventricular gene delivery of a beta-adrenergic receptor kinase inhibitor to the failing heart reverses cardiac expression. Circulation 103:1311–1316

Shears II LL, Kawaharada N, Tzeng E et al (1997) Inducible nitric oxide synthase suppresses the development of allograft atherosclerosis. J Clin Invest 100:2035–2042

Shi Y, Fard A, Galeo A et al (1994) Transcatheter delivery of c-myc antisense oligomers reduces neointimal formation in a porcine model of coronary artery balloon injury. Circulation 90:944–951 (1998) Evidence for circulating bone marrow derived endothelial cells. Blood 92:362–367

Shibata T; Giaccia AJ; Brown JM (2000) Development of a hypoxia-responsive vector for tumour-specific gene therapy. Gene Ther 7:493–498

Shintani S, Murohara T, Ikeda H et al (2001) Augmentation of postnatal neovascularization with autologous bone marrow transplantation. Circulation 103:897–903

Shohet RV, Chen S, Zhou Y-T et al (2000) Echocardiographic destruction of albumin microobubbles directs gene delivery to the myocardium. Circulation 101:2554–2556

Schuster MD, Kocher A, John R et al (2001) Stromal derived growth factor (SDF)-1 augments myocardial neovascularization and cardiomycyte regeneration induced by human bone marrow angioblasts. Circulation 106 (Suppl II):II-65

Shyu KG, Wang MT, Wang BW et al (2002) Intrmyocardial injection of naked DNA encoding HIF-1α/VP16 hybrid to enhnance angiogenesis in an acute myocardial infarction model in the rat. Caridovasc Res 54:576–583

Simons M, Edeelman ER, DeKeyser JL et al (1992). Antisense c-myb oligonucleotides inhibit intimal arterial smooth muscle accumulation in vivo. Nature 359:67–70 3S-9S.

Singal PK, Khaper N, Palace V et al (1998) The role of oxidative stress in the genesis of heart disease. Cardiovasc Res 40:436–442

Song YK, Liu F, Chu S et al (1997) Characterization of cationic liposome-mediated gene transfer in vivo by intravenous administration. Human Gene Ther 8:1585–1594

Soonpaa MH, Field L (1998) Survey of studies examining mammalian cardiomyocyte DNA synthesis. Circ Res 83:15–26

Spinale FG (2002) Matrix metelloproteinases. Regulation and dysregulation in the failing heart Circ Res 90:520–530

Sposito AC, Chapman MJ (2002) Statin therapy in acute coronary syndromes: mechanistic insight into clinical benefit. Arterioscl Throm Vasc Biol 22:1524–1534

Srivastava D, Olson EN (2000) A genetic blueprint for cardiac development Nature 407:221–226

Stamm C, Westphal B, Kleine H-D et al (2003) Autologous bone marrow stem cell transplantation for myocardial regeneration. Lancet 361:45–46

St. John Sutton MG, Sharpe N (2000) left ventricular remodeling after myocardial infarction. Pathophysiology and therapy. Circulation 101:2981–2988

Steg PG, Tahlil O'Aubailly N, Caillaud JM et al (1997) Reduction of restenosis after angioplasty in an atheromatous rabbit model by suicide gene therapy. Circulation 96:408–411

Stein EA (2002) Identification and treatment of individuals at high risk of coronary artery disease. Am J Med 112(8A):3S-9S

Stepkowski SM (2000) Development of antisense oligodeoxynucleotides for transplantation. Curr Opin Mol Ther 2:304–317

Strauer BE, Brehm M, Zeus T et al (2001) Myocardial regeneration after intracoronary transplantation of human autologous stem cells following acute myocardial infarction. Dtsch Med Wochenschr 126:932–938

Strauer BE, Brehm M, Zeus T et al (2002) Repair of infarcted myocardium by autologous intracoronary mononuclear bone marrow cell transplantation in humans. Circulation 106:1913–1918

Su H, Arakawa-Hoyt J, Kan YW (2002) Adeno-associated viral vector-mediated hypoxia response element-regulated gene expression in mouse ischemic heart model. Proc Natl Acad Sci USA 99:9480–9485

Suzuki K, Sawa Y, Kaneda Y (1997). In vivo gene transfer of heat shock protein 70 enhances myocardial tolerance to ischemia-reperfusion injury in rat. J Clin Invest 99:1645–1650

Svensson EC, Marshall DJ, Woodard K et al (1999) Efficient and stable transduction of cardiomyocytes after intramyocardial injection or intracoronary perfusion with recombinant adeno-associated virus vectors. Circulation 99:201–205

Swynghedauw B (1999) Molecular mechanisms of myocardial remodeling. Physiological Reviews 79:215–262

Sylven C, Sarkar N, Insulander P et al (2002) Catheter-based transendocardial myocardial gene transfer. J Interv Cardiol 15:7–13

Symes JF, Losordo DW, Vale PR et al (1999) Gene therapy with vascular endothelial growth factor for inoperable coronary artery disease. Ann. Thorac Surg 68:830–837

Tabata H, Silver M, Isner JM (1997) Arterial gene transfer of acidic fibroblast growth factor for therapeutic angiogenesis in vivo: critical role of secretion signal in use of naked DNA. Cardiovasc Res 25:470–479

Taigen T, Windt LJ, Lim HW et al (2000) Targeted inhibition of calcineurin prevents agonist-induced cardiomyocyte hypertrophy. Proc Natl Acad Sci USA 97:1196–1201

Tamamori M, Ito H, Hiroe M et al (1998) Essential roles for G1 cyclin-dependent kinase activity in development of cardiomyocyte hypertrophy. Am J Physiol 275:H2036-H2040

Tang X, Mohuczy D, Zhang CY et al (1999) Intravenous angiotensinogen antisense in AAV-based vector decreases hypertension. Am J Physiol 277: H2392-H2399

Tangirala RK, Tsukamoto K, Chun SH et al (1999) Regression of atherosclerosis induced by liver-directed gene transfer of apolipoprotein A-1 in mice. Circulation 100:1816–1822

Taniyama Y, Morishita R, Aoki M et al (2002) Angiogenesis and antifibrotic action by hepatocyte growth factor in cardiomyopathy. Hypertension 40:47–53

Taylor DA, Atkins BZ, Hungspreugs P et al (1998) Regenerating functional myocardium: improves performance after skeletal myoblast transplantation. Nat Med 4:929–933

Taylor DA, Hruban R, Rodriguez R et al (2002) Cardiac chimerism as a mechanism for self-repair. Does it happen and if so to what degree. Circulation 106:2–4

Terada N, Hamazaki T, Oka M et al (2002) Bone marrow cells adopt the phenotype of other cells by spontaneous cell fusion. Nature 416:542–544

Tio RA, Tkebuchava T, Scheurermann TH et al (1999) Intramyocardial gene therapy with naked DNA encoding vascular endothelial growth factor improves collateral blood flow to ischemic myocardium. Human Gene Ther 10:2953–2960

Toma C, Pittenger MF, Cahill KS et al (2002) Human mesenchymal stem cells differentiate to a cardiomyocyte phenotype in the adult murine heart. Circulation 105:93–98

Tomaiyasu K, Oda Y, Nomura M et al (2000) Direct intra-cardiomuscular transfer of β_2-adrenergic receptor gene augments cardiac output in cardiomyophatic hamsters Gene Ther 7:2087–2093

Tomita S, Li RK, Weisel RD et al (1999) Autologous transplantation of bone marrow cells improves damaged heart function. Circulation 100 (Suppl): II-247-II256

Towbin JA, Bowles NE (2002) The failing heart. Nature 415:227–233

Trono D (2000) Lentiviral vectors: turning a deadly foe into a therapeutic agent. Gene Ther 7:20–23

Tulis DA, Durante W, Liu X et al (2001) Adenovirus-mediated heme oxygenase-1 gene delivery inhbits injury-induced vascular neointima formation. Circulation 104:2710–2715

Tzeng E, Shears LL, Robbins PD et al (1996) Vascular gene transfer of the human inducible nitric oxide synthase: characterization of activity and effects of myointimal hyperplasia. Mol Med 2:211–215

Ueda H, Sawa Y, Matsumoto K et al (1999) Gene transfection of hepatocyte growth factor attenuates reperfusion injury in the heart. Ann Thorac Surg 67:1726–1731

Ueno H, Li JJ, Masuda S, Qi Z et al (1997) Adenovirus-mediated expression of the secreted form of basic fibroblast groeth factor (FGF-2) induces cellular proliferation and angiogenesis in vivo. Arterioscler Thromb Vasc Biol 17:2453–2460

Vale PR, Losordo DW, Milliken CE et al (2001) Randomized, single-blind, placebo-controlled pilot study of catheter-based myocardial gene transfer for therapeutic angiogenesis using left ventricular electromechanical mapping in patients with chronic myocardial ischemia. Circulation 103:2138–2143

Van Belle E, Maillard L, Tio FO et al (1997) Accelerated endothelialization by local delivery of recombinant human vascular endothelial growth factor reduces in-stent intimal formation. Biochem Biophys Res Commun 235:311–316

Van der Heide RS (2002) Increased expression of HSP27 protects canine myocytes from simulated ischemia-reperfusion injury Am J Physiol 282:H935-H941

Vasa M, Fichtscherer S, Aicher A et al (2001) Number and migratory activity of circulating endothelial progenitor cells inversely correlate with risk factors for coronary artery disease. Circ Res 89:E1-E7

Vasa M, Fichtlschrerer S, Adler K et al (2001) Increase in circulating endothelial progenitor cells by statin therapy in patients with stable coronary artery disease. Circulation 103:2885–2890

Vincent KA, Shyu K-G, Luo Y et al (2000) Angiogenesis is induced in a rabbit model of hindlimb ischemia by naked DNA encoding an HIF-1α/VP16 hybrid transcription factor. Circulation 102:2255–2261

Von der Leyen HE, Gibbons GH, Morishita R et al (1995) Gene therapy inhibiting neointimal vascular lesion: in vivo transfer of endothelial cell nitric oxide synthase gene. Proc Natl Acad Sci USA 92:1137–1141

Von der Leyen HE, Dzau VJ (2001) Therapeutic potential of nitric oxide synthase gene manipulation. Circulation 103:2760–2765

Walter DH, Rittig K, Bahlmann FH et al (2002) Statin therapy accelerates reendothelialization: a novel effect involving mobilization and incroporation of bone marrow-derived endothelial progenitor cells. Circulation 105:3017–3024

Wang H, Katovich MJ, Gelband CH et al (1999) Sustained inhibition of angiotensin I converting enzyme (ACE) expression and long-term antihypertensive action by virally mediated delivery of ACE antisense cDNA. Circ Res 85:614–622

Wang J-S, Shum-Tim D, Chedrawy E et al (2001) The coronary delivery of marrow stromal cells for myocardial regeneration: Pathophysiological and therapeutic implications. J Thorac Cardiovasc Surg 122:699–705

Ware JH, Simons M (1997) Angiogenesis in ischemic heart disease. Nature Medicine 3:158–164

Waugh JM, Kattash M, Li J et al (1999a) Gene therapy to promote thromboresistance: Local overexpression of tissue plasminogen activator to prevent arterial thrombosis in an vivo rabbit model. Proc Natl Acad Sci USA 96:1065–1070

Waugh JM, Yuksel E, Li J et al (1999b) Local overexpression of thrombomodulin for in vivo prevention of arterial thrombosis in a rabbit model. Circ Res 84:84–92

Williams RS, Benjamin IJ (2000) Protective responses of the ischemic myocardium. J Clin Invest 106:813–818

Wilson PWF, D'Agostino RB, Levy D et al (1998) Prediction of coronary heart disease using risk factor categories. Circulation 97:1837–1847

Woo YZ, Zhang JC, Vijayasarathy C et al (1998). Recombinant adenovirus-mediated cardiac gene transfer of superoxide dismutase and catalase attenuates postischemic contractile dysfunction. Circulation 98 (Suppl):II255–II260

Wright MJ, Wightman, LML, Lilley C et al (2001) In vivo myocardial gene transfer: Optimization, evaluation and direct comparison of gene transfer vectors. Bas Res Cardiol 96:227–236

Yamamoto N, Kohmoto T, Roethy W et al (2000) Histological evidence that fibroblast growth factor enhances the angiogenic effects of transmyocardial laser revascularization. Basic Res Cardiol 95:55–63

Yan Z, Zhang Y, Duan D et al (2000) Trans-splicing vectors expand the utility of adeno-associated virus for gene therapy. Proc Natl Acad Sci USA 97:6716–6721

Yang Z, Bove CM, French BA et al (2002) Angiotensin II type 2 receptor overexpression preserves left ventricular function after myocardial infarction. Circulation 106:106–111

Yang Z, Cerniway RJ, Byford AM (2002) Cardiac overexpression of A1-adenosine receptor protects intact mice against myocardial infarction. Am J Physiol 282:H949–H955

Yasue H, Kugiyama K (1997) Coronary spasm: clinical features and pathogenesis. Intern Med 36:760–765

Yasue H, Kugiyama K (1997) Coronary spasm: clinical features and pathogenesis. Intern Med 36:760–765

Yellon DM, Baxter GF (2000) Reperfusion injury revisited. Is there a role for growth factor signalling in limiting lethal reperfusion injury? Trends Cardiovasc Med 9:245–249

Ying Q-L, Nichols J, Evans EP et al (2002) Changing potency by spontaneous fusion. Nature 416:545–547

Yonemitsu Y, Kaneda Y, Tanaka S et al (1998) Transfer of wild-type p53 gene effectively inhibits vascular smooth muscle proliferation in vitro and in vivo. Circ Res 82:147–156

Yoshida T, Watanabe M, Engelman DT et al (1996) Transgenic mice overespressing glutathione peroxidase are resistant to myocardial reperfusion injury. J Mol Cell Cardiol 28:1759–1767

Yoshida H, Zhang JJ, Chao L et al (2000) Kallikrein gene delivery attenuates myocardial infarction and apoptosis after myocardial ischemia and reperfusion. Hypertension 35:25–31

Zhang M, Methot D, Poppa V et al (2001) Cardiac myocyte grafting for cardiac repair: Graft cell death and anti-death strategies. J Mol Cell Cardiol 33:907–921

Zhang YC, Kimura B, Shen L et al (2000) New β-blocker: Prolonged reduction in high blood pressure with β1 antisense oligodeoxynucleotides. Hypertension 35:219–224

Zhao J, Pettigrew GJ, Thomas J et al (2002) Lentiviral vectors for delivery of genes into neonatal and adult ventricular cardiac myocytes in vitro and in vivo. Basic Res Cardiol 97:348–358

Zhu HL, Stewart AS, Taylor MD (2000) Blocking free radical production via adenoviral gene transfer decreases cardiac ischemia-reperfusion injury. Mol Ther 2:470–475

Zoldhelyi P, McNatt J, Xu XM et al (1996) Prevention of arterial thrombosis by adenovirus-mediated transfer of cyclooxygenase gene. Circulation 93:10–17

Zoldhelyi P, McNatt J, Shelat HS et al (2000) Thromboresistance of balloon-injured porcine carotid arteries after local gene transfer of human tissue factor pathway inhibitor. Circulation101:289–295

Zsigmond E, Kobayashi K, Tzung KW et al (1997) Adenovirus-mediated gene transfer of human lipoprotein lipase ameliorates the hyperlipidemias associated with apolipoprotein E and LDL receptor deficiencies in mice. Hum Gene Ther 8:1921–1933

Subject Index

ARCP 260, 262
Activated protein C resistance (APCR) 312
Angiotensin Converting Enzyme (ACE) 159, 203, 207, 211, 216, 267
Adeno-associated virus (AAV) 360, 367
Adenosine monophosphate deaminase 1, 214
Adenovirus 367, 376
Adhesion molecules 323-334
Adhesion cascade 324
Adiponectin (ACRP 30) 260, 262, 266
Adducin 161
Akt kinase 288
Aldosterone synthase 210
Angiotensinogen 160
Angiotensin receptors 212, 217
Antisense oligonucleotide 361, 371, 375-377, 384
Apparent mineralocorticoid excess 154
Apolipoprotein A 1 123
Apolipoprotein A 1V 123
Apolipoprotein B 115, 124
Apolipoprotein E 125
Atherosclerosis 80-96, 259, 267-269, 293-295, 323, 325, 334, 341, 346, 349, 371, 386

Bartter's syndrome 153
Beta-receptors 208, 212
Beta-blockers 215

Calpain 254
Candidate genes 31, 150, 204, 252-253
Cardiac hypertrophy 167, 178-194, 224, 383
Cardiomyopathy 178, 205, 384
CD31 331
CD36 260, 262, 265
Channelopathies 224-236, 384
Cholesterol-7α-hydroxylase 134
Cholesterol ester transfer protein 130
Common disease common variant (CDCV) hypothesis 28
Congenic strains 164
Connexin 319
Consomic strains 164
Coronary artery disease 301, 350, 375-380 (see also myocardial infarction, and atherosclerosis)
Cytochrome P450 (CYP450) 9, 40, 45-49, 59-65, 118

Diabetes mellitus (see Insulin resistance)
Drug transporters 55, 68 (see also P-glycoprotein)

Endothelin receptor 209, 213
eNOS 263, 293, 296-303, 386

Factor V 311, 313
Factor VII 315

Factor VIII   314
Factor XII   316
Factor XIIIA   314
Fibrates   116
Fibrinogen   315
Fibroblast growth factor (FGF)   377-379
Gene therapy   360-390
Genome wide association studies   32
Gitelman's syndrome   153
Glucose transporter 4 (GLUT4)   255
Glucocorticoid-mediated hypertension   154

Haplotype mapping   27, 29
HERG   230, 385 (see also long QT syndrome)
Haemoxygenase 1 (HO1)   370, 373
Heart failure   204-218, 224, 381-383, 386
Hyperlipidaemia   84, 108-137, 373
Hypertension   85, 90, 151-170, 258, 262, 297-299, 301, 371
Hypertrophic cardiomyopathy   179-189

ICAM-1   329
Inflammation   323-334, 371
iNOS   263, 290, 303, 385
Insulin resistance   245-269
Insulin receptor substrate (IRS-1)   247-251, 254-257
Integrins   332
Ion channels   (see channalopathies and potassium channels)
Ischaemic heart disease   (see coronary artery disease, myocardial infarction and atherosclerosis)

Jervell and Lange-Neilson   227

Knockout mice   166, 289-295, 247-251

Lamin A   254, 265
Lentivirus   267
Leukocyte adhesion deficiency   329, 332
Liddle's syndrome   151

Linkage analysis   27, 252
Lipoprotein a   133
Lipoprotein lipase   114, 251
Long QT syndrome   224-237, 384

Matrix metalloproteinases (MMPs)   341-354, 369
Metabolic syndrome   (see Insulin Resistance)
Myocardial gene transfer   362
Myocardial infarction   297-299, 315-317, 326, 330, 332, 333, 349, 351, 360, 367, 371
Myocardial ischaemia (see coronary artery disease, ischaemic heart disease, myocardial infarction)

Nitric oxide   282-304, 263
Nitric oxide synthase   162, 283-304, 263 (see also eNOS, iNOS and nNOS)
nNOS   263, 291, 294, 303, 386

Obesity   244 (see also Insulin Resistance)

P-glycoprotein   55, 68
P-selectin   326
P-selectin glycoprotein ligand-1   328
PAI-1   316
Paraoxinase-1 (PON1)   269
PC1   261
PECAM-1   331
Peroxisome proliferator activated receptors (α and γ)   114, 117, 134, 155, 258, 259, 264, 266
Pharmacogenetics   2-4, 6-22, 68-70, 80, 109
Pharmacogenomics   2, 4-6, 109 (see also pharmacogenetics)
Pharmacokinetics   7-9, 41-45, 118
P13 kinase   247, 257
Platelet activating factor acetylhydrolase   210
Platelet membrane glycoprotein - 1b-IX-V   317
Platelet membrane glycoprotein - IIb/IIIa   317

Platelet membrane glycoprotein - Ia/IIa   319
Population structure   25, 31
Potassium channels   229, 330, 385
Prothrombin   314

QT interval   (see long QT syndrome)

Racial labelling   31
Resistin   261
Retrovirus   365

Sample size   112
SCN5A   230, 235
Selectins   325 (see E-, L- and P- selectins)
Spontaneously hypertensive rat (SHR)   262
Sterol responsive element binding protein (SREBP)   120, 251, 252, 264
Stratification   30
Stem cell therapy   360, 365, 377, 380, 386-389

Stroke   167, 291-293, 328, 331, 333
Stromelysin-1   135, 319, 343, 350
Super oxide dismutase   370, 373

Thrifty genotype   246
Thrombomodulin   374
Thrombosis   312-318 (see myocardial infarction and venous thrombosis)
Tissue inhibitor of metalloproteinases (TIMP)   344, 349
Torsades des pointes   224
Transforming growth factor $\beta 1$ 203, 210
Transplantation
  Bone marrow   331
  Cardiac   330, 371
  Renal   330
Tumor necrosis factor   203, 211, 284

VCAM-1   330
Vectors   365
VEGF   288, 377
Venous thrombosis   312-315, 320

Printing: Saladruck Berlin
Binding Lüderitz&Bauer, Berlin